This volume looks in depth at the carcinogenic properties of N-nitroso compounds. These compounds occur widely in smoked foods and in meat and fish cured with nitrites; they are also formed in some industrial processes and in addition are found in cosmetics, tobacco and tobacco smoke. By focussing on the chemical and biological properties, the volume attempts to explain how these compounds exert their carcinogenic potential and, furthermore, goes on to explain how apparently harmless precursors such as nitrates and amines are transformed to produce carcinogens.

Cambridge Monographs on Cancer Research

Chemistry and biology of *N*-nitroso compounds

Cambridge Monographs on Cancer Research

Scientific Editors
M. M. Coombs, Imperial Cancer Research Fund Laboratories, London
J. Ashby, Imperial Chemical Industries, Macclesfield, Cheshire
M. Hicks, Science Director, United Biscuits UK Ltd, Maidenhead

Executive Editor
H. Baxter, formerly at the Laboratory of the Government Chemist, London

Books in this Series
Martin R. Osborne and Neil T. Crosby
Benzopyrenes
Maurice M. Coombs and Tarlochan S. Bhatt
Cyclopenta[a]phenanthrenes
M. S. Newman, B. Tierney and S. Veeraraghavan
The chemistry and biology of benz[a]anthracenes
Jürgen Jacob *Sulfur analogues of polycyclic aromatic hydrocarbons (thiaarenes)*
Ronald G. Harvey *Polycyclic aromatic hydrocarbons: chemistry and carcinogenicity*

Forthcoming volumes
John Higginson, Calum S. Muir and Nubia Muñoz
Human cancer: epidemiology environmental causes
W. F. Karcher *Dibenzanthracenes and environmental carcinogenesis*

Chemistry and biology of N-nitroso compounds

WILLIAM LIJINSKY

*Director, Laboratory of Chemical and Physical Carcinogenesis,
Frederick Cancer Research Facility, Frederick, Maryland, U.S.A.*

CAMBRIDGE UNIVERSITY PRESS
*Cambridge
New York   Port Chester
Melbourne   Sydney*

CAMBRIDGE UNIVERSITY PRESS
Cambridge, New York, Melbourne, Madrid, Cape Town,
Singapore, São Paulo, Delhi, Tokyo, Mexico City

Cambridge University Press
The Edinburgh Building, Cambridge CB2 8RU, UK

Published in the United States of America by Cambridge University Press, New York

www.cambridge.org
Information on this title: www.cambridge.org/9780521116978

© Cambridge University Press 1992

This publication is in copyright. Subject to statutory exception
and to the provisions of relevant collective licensing agreements,
no reproduction of any part may take place without the written
permission of Cambridge University Press.

First published 1992
First paperback edition 2011

A catalogue record for this publication is available from the British Library

Library of Congress Cataloguing in Publication data
Lijinsky, W. (William), 1928–
Chemistry and biology of N-nitroso compounds/William Lijinsky.
 p. cm. – (Cambridge monographs on cancer research)
Includes bibliographical references and index.
ISBN 0-521-34629-0
1. Nitroso compounds – Carcinogenicity. I. Title. II. Series.
RC268.7.N58L55 1992
616.99'4071 – dc20   91-22142 CIP

ISBN 978-0-521-34629-0 Hardback
ISBN 978-0-521-11697-8 Paperback

Cambridge University Press has no responsibility for the persistence or
accuracy of URLs for external or third-party internet websites referred to in
this publication, and does not guarantee that any content on such websites is,
or will remain, accurate or appropriate.

Contents

	Glossary	page x
	Foreword	xiii
	Acknowledgements	xv
1	**Introduction**	1
1.1	Historical developments	1
1.2	Sources of N-nitroso compounds	5
1.3	Estimation of carcinogenic risks	9
1.4	Modulation of nitrosamine formation	12
1.5	Chemistry and biological activity of N-nitroso compounds	13
2	**Occurrence, formation and detection of N-nitroso compounds**	19
2.1	Occurrence	19
2.2	Formation	33
2.3	Methods of detection and analysis	50
3	**Chemical properties of N-nitroso compounds**	55
3.1	Introduction	55
3.2	Solubility and partition	57
3.3	Spectroscopic properties	60
3.4	Mass spectrometry	64
3.5	Stability	68
3.6	Reducing nitrosamine exposure: destruction and degradation of compounds	76
3.7	Chemical reactivity	77
3.8	Uses of nitrosamines in syntheses	83
3.9	Activation and reactions of nitrosamines	85
3.10	Interactions of N-nitroso compounds in biological systems	93
3.11	Conclusions	95

4	**Metabolism and cellular interactions of *N*-nitroso compounds**	98
4.1	Introduction – history	98
4.2	Pharmacokinetics and organ-specific carcinogenesis	103
4.3	Nitrosamine-metabolising enzymes	108
4.4	Metabolism of cyclic nitrosamines	113
4.5	Acyclic nitrosamines	120
4.6	Nitrosamines containing oxygen substituents	130
4.7	Which enzymic activation leads to carcinogenesis?	146
4.8	Alkylation of DNA and other macromolecules	151
5	**Toxicity of *N*-nitroso compounds**	173
5.1	Introduction	173
5.2	Toxicity of nitrosodimethylamine	175
5.3	Toxicity of *N*-nitroso compounds in several species	177
5.4	Toxicity of nitrosamines with substituted alkyl groups	185
5.5	Toxicity by different routes of administration	189
5.6	Subchronic toxicity	190
5.7	Carcinogenic and toxic potency	191
5.8	Structure–activity relations in toxicity and mechanisms	192
5.9	Toxicity of cyclic nitrosamines and carcinogenicity	195
5.10	Hamsters: toxicity of nitrosamines with oxygenated alkyl groups	196
5.11	Toxicity of alkylnitrosoureas and alkylnitrosocarbamates	197
5.12	Toxicity of haloalkyl nitrosoureas	198
5.13	Conclusions	201
6	**Mutagenesis and cell transformation by *N*-nitroso compounds**	202
6.1	Introduction	202
6.2	Activation systems for nitrosamines	205
6.3	Structure of *N*-nitroso compounds related to mutagenesis	219
6.4	Mutagenesis by directly acting *N*-nitroso compounds	224
6.5	Mutagenic nitrosamines requiring metabolic activation	227
6.6	Mutagenic nitrosamines which are not carcinogens	238
6.7	Prophage induction activity by *N*-nitroso compounds	240
6.8	Diffusibility of products of metabolic activation – a limitation on detectable mutagenic activity	241
6.9	Transformation and mutagenesis in cultured cells	242
6.10	Conclusions	249

7	**Structure–activity relations in carcinogenesis by *N*-nitroso compounds**	251
7.1	Early developments	251
7.2	Alkylation of DNA in relation to carcinogenesis	255
7.3	Alkylnitrosamides (alkylnitroso-ureas, -carbamates and -guanidines)	264
7.4	Importance of α-oxidation in nitrosamine carcinogenesis	270
7.5	Organ specificity of carcinogenic *N*-nitroso compounds	320
7.6	Dose–response studies	386
7.7	Transplacental carcinogenesis by *N*-nitroso compounds	393
7.8	Response of different species to *N*-nitroso compounds	398
8	**Conclusions – The importance of *N*-nitroso compounds as environmental carcinogens and as experimental models for investigating cancer**	404
8.1	Environmental carcinogenesis	404
8.2	Models for carcinogenesis	407
	References	414
	Index	453

Glossary

HPLC	= High Pressure Liquid Chromatography
NMR	= Nuclear Magnetic Resonance
ORD	= Optical Rotatory Dispersion
LD_{50}	= Median Lethal Dose
I.P.	= Intraperitoneal
I.R.	= Intrarectal
I.U.	= Intraurethral
I.V.	= Intravenous
S.C.	= Subcutaneous
TEA	= Thermal Energy Analyser
TLC	= Thin Layer Chromatography
MFO	= Mixed Function Oxidase
ppm	= Parts per million
DNA	= Deoxyribonucleic Acid
MeG	= Methylguanine
EtG	= Ethylguanine
HEG	= Hydroxyethylguanine
MNU	= Methylnitrosourea
ENU	= Ethylnitrosourea
HENU	= Hydroxyethylnitrosourea
MNNG	= N-Methyl-N-nitroso-N'-nitroguanidine
NDMA	= Nitrosodimethylamine
NDEA	= Nitrosodiethylamine
MNEA	= Methylnitrosoethylamine
NDBA	= Nitrosodi-n-butylamine
DMNM	= 2,6-Dimethylnitrosomorpholine
NDELA	= Nitrosodiethanolamine
MNBA	= Methylnitroso-n-butylamine

Glossary

MNHEA	= Methylnitrosohydroxyethylamine
MNHPA	= Methylnitroso-2-hydroxypropylamine
MNOPA	= Methylnitroso-2-oxopropylamine
NBHPA	= Nitrosobis-(2-hydroxypropyl)amine
NBOPA	= Nitrosobis-(2-oxopropyl)amine
NHPOPA	= Nitroso-2-hydroxypropyl-2-oxopropylamine
NTA	= Nitrilotriacetic Acid
NNK	= 4-(Methylnitrosamino)-1-(3-pyridyl)-1-butanone
NNAl	= 4-(Methylnitrosamino)-1-(3-pyridyl)-1-butanol
NO-PYRR	= Nitrosopyrrolidine
NO-PIP	= Nitrosopiperidine
BCNU	= Nitrosobis-(chloroethyl)urea
CCNU	= Chloroethylnitrosocyclohexylurea
HECNU	= Chloroethylnitrosohydroxyethylurea
HENCU	= Hydroxyethylnitrosochloroethylurea

To Peter Magee
and to the memory of Dietrich Schmähl

Foreword

N-Nitroso compounds occupy a small and obscure corner of organic chemistry and would probably have remained there if it were not for the discovery that they are carcinogenic. They occur in the environment, sometimes at concentrations which might create apprehension even if they were not so biologically active. They are unique among carcinogenic agents in being active in all species and they have an unparalleled spectrum of target cells and organs in which they can induce cancer. Among the *N*-nitroso compounds are many of the most potent carcinogens we know, active in inducing tumours in animals at concentrations close to those to which humans may be exposed. There are even *N*-nitroso compounds that are synthesised by plants, but most are formed incidentally by nitrosation of amines.

Their versatility as carcinogens has fascinated many investigators, including this author, and they have provided useful models for investigating mechanisms of carcinogenesis. While we do not understand the mechanism by which any carcinogen causes cancer, *N*-nitroso compounds are providing clues more readily than do other carcinogens.

Because of the enormous volume of publications about *N*-nitroso compounds (more than 2000 about nitrosodimethylamine alone), I have had to be selective. Wherever possible, I have tried to refer to original publications of information, rather than to reviews. Most references are to primary data, but some particularly splendid reviews, including those of Challis and Challis (1982) and Saavedra (1987a,b) are referred to. Other reviews, including that of Fridman *et al.* (1970), Druckrey *et al.* (1967), Preussmann and Stewart (1984), and Scanlan (1974) contain large compilations of data, which add to what is presented in this volume.

The role of nitrate and nitrite, as they occur in the environment and in food, or are formed in the human body, is important for human health, particularly in the context of human exposure to N-nitroso compounds formed *in vivo*. However, it is not dealt with in depth here because it has been previously (Lijinsky, 1986*b*; National Academy of Sciences, 1981; World Health Organization, 1977; Lijinsky, 1976*b*) and is related to only one aspect of N-nitroso compounds.

A very large number of investigators have contributed to our knowledge of the biological effects of N-nitroso compounds, especially to carcinogenesis. For convenience – and because of familiarity – much of the specific information presented is from my own studies in animals. The chemicals used were of high purity and the treatment of the animals, and the animals themselves, were very uniform, in most cases bred in-house and maintained behind a barrier. In almost all of the chronic treatments, the animals died of or with tumours, induced or spontaneous; dissection and histopathology were carried out by the same personnel using uniform criteria. This greatly facilitates comparison of the effects of one treatment with another, in regard to potency, for example. The inclusion of these extensive data is not intended to disparage or diminish the frequently similar findings of other investigators.

The book is intended as a survey, rather than an encyclopedia, and some topics have been given more attention than others. My goal has been to attempt to answer some questions related to carcinogenesis and mutagenesis by N-nitroso compounds, but this has inevitably raised other questions. The answer to those should, in turn, bring us closer to an understanding of the two biological effects which are seen as increasingly important to the protection of human health.

Acknowledgements

I am deeply grateful to the many colleagues and associates whose work is the foundation of this book. Many of them, in whatever capacity, are co-authors of the papers referred to. The carcinogenesis data are the reference point of most of the chemical and biochemical studies, and for the excellence of the histopathology I am indebted to a number of pathologists, especially Drs M. Greenblatt, H. Garcia, M. Goodall, H. W. Taylor, M. D. Reuber and R. M. Kovatch, whom I thank for their patience with me. I thank Drs J. E. Saavedra and R. K. Elespuru for their critical evaluation of parts of the book, and several stalwart ladies for typing the manuscript.

The research was supported by the National Cancer Institute of the United States Public Health Service under a variety of contracts: NO1-CO-23909 with Litton Bionetics, Inc., and with Bionetics Research, Inc. and NO1-CO-74101 with Advanced BioScience Laboratories, Inc.

1

Introduction

1.1 Historical developments

Nitrosamines – or more specifically N-nitroso compounds – were first reported by Geuther in 1863, when organic chemistry was a new science still in its descriptive stage. They remained for almost a century a group of compounds of no great importance, used occasionally as solvents or synthetic intermediates in preparing hydrazines or diazoalkanes, the latter being useful alkylating agents (von Pechmann, 1895). The generic structures of the important types of N-nitroso compound are shown in Fig. 1.1. Since early reports of their toxic effects in humans were largely ignored (Freund, 1937), N-nitroso compounds remained in relative obscurity. However, the later report of the liver toxicity in humans and animals of nitrosodimethylamine by Barnes and Magee (1954) received wide attention. There have been two reports more recently of deliberate poisonings of people with nitrosodimethylamine, one in Germany, and the other in Nebraska, U.S.A.

The report that nitrosodimethylamine induced liver cancer when fed to rats (Magee and Barnes, 1956) prompted an expanding study of the series of compounds we call nitrosamines and when, later, it was found that they had unusual alkylating properties, interest in them quickened. Biologists were mainly interested in the use of these simple, water-soluble compounds as models for producing a wide variety of cancers, whereas chemists were more interested in the relation between chemical structure and biological activity.

The studies of Magee and his colleagues, Hultin and Farber, showing that nitrosodimethylamine was converted enzymically into a methylating agent (Magee and Hultin, 1962; Magee and Farber, 1962) sparked an interest in the mechanism of carcinogenesis through alkylation of proteins and nucleic acids by N-nitroso compounds, an extension of the work of

Brookes and Lawley (1960) on mustard gas. A small number of investigators including Druckrey and his co-workers in Freiburg, and Magee and his collaborators (including this author) took up this theme, which became a broad international effort ten years later. During the past decade or so a considerable proportion of model studies of carcinogenesis, including those using molecular biology techniques, involve the actions of N-nitroso compounds in biological systems.

Druckrey and his group, including Preussmann and Schmähl, Magee and his associates, and this author separately pursued the association between the induction of a great variety of tumours in several species, but especially in rats, and the potential of the N-nitroso compounds to form an alkylating agent. These investigators began testing a series of N-nitroso compounds administered chronically to experimental animals, and also the acute and chronic effects of single doses of these compounds, which are still in progress in a small number of laboratories. Most of our admittedly incomplete understanding of mechanisms of carcinogenesis by N-nitroso compounds and of their toxic effects is based on these studies which have spanned three decades, and the results of which are described and discussed in Chapters 5 and 7. The bulk of the chronic studies have been reviewed and summarised (Druckrey *et al.*, 1967; Preussmann and Stewart, 1984; Lijinsky, 1987). Some examples of the types of tumours induced by nitrosodiethylamine in a number of species are shown in Table 1.1.

Unlike another large group of chemical carcinogens, the polynuclear hydrocarbons, N-nitroso compounds were not suspected of being important contributors to human cancer risks until many years had

Fig. 1.1. Main types of N-nitroso compound.

Dialkylnitrosamine: R_1R_2N-NO

Cyclic Nitrosamine: $(CH_2)_n N-NO$

Aromatic Nitrosamine: Ph–N(R)–NO

Alkylnitrosourea: $R_1(ON)N-C(=O)-N(H)R_2$

Alkylnitrosocarbamate: $R_1(ON)N-C(=O)-OR_2$

Alkylnitrosoguanidine: $R_1(ON)N-C(=NH)-N(H)R_2$

Alkylnitrosamide: $R_1(ON)N-C(=O)-R_2$

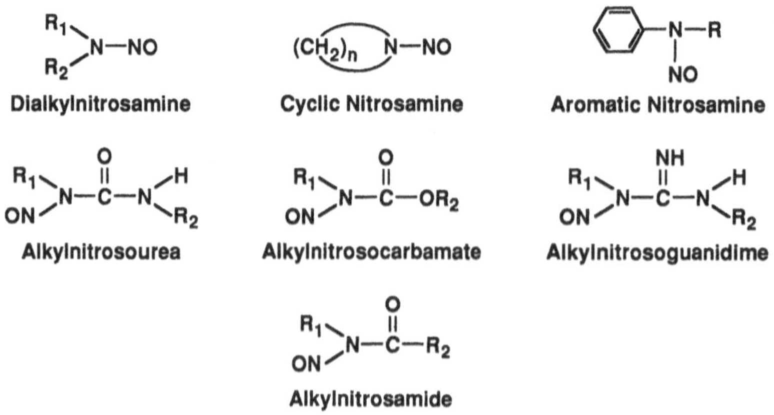

Table 1.1. *Tumours induced by nitrosodiethylamine in various species*

Species	Tumours
Mouse	Liver, Esophagus
Rat	Liver, Esophagus
Syrian Hamster	Lung, Liver
Chinese Hamster	Esophagus, Lung
European Hamster	Respiratory tract
Gerbil	Liver, Lung, Nasal mucosa
Guinea pig	Liver
Mastomys	Liver
Hedgehog	Liver, Lung
Rabbit	Liver
Cat	Liver
Dog	Liver
Pig	Liver, Reticuloendothelial system
Monkey (4 species)	Liver
Chicken	Liver, Kidney
Grass Parakeet	Liver
Frog	Liver, Blood system
Snake (python)	Liver, Kidney
Fish (4 species)	Liver
Mollusc	Liver
Bushbaby	Liver

passed. In the early 1970s there was a growing understanding that the ease of formation of N-nitroso compounds from precursors (e.g. amines and nitrites) common in our environment, including formation in the stomach, gave them a relevance heretofore unsuspected (Lijinsky and Epstein, 1970). The most common N-nitroso compounds to which humans can be exposed, and their sources (Table 1.2) are discussed in Chapter 2.

The occurrence of toxic nitrosamines in food was first indicated by the chance observation that a batch of fish meal caused sickness and death of sheep, mainly from liver damage (Koppang, 1964; Sakshaug et al., 1965). Ender et al. (1964) reported that the fish meal, which had been treated with sodium nitrite to preserve it, contained considerable quantities of NDMA. Thus, as with aflatoxins some years earlier, investigation of unexpected toxicity in domestic animals led to findings of large general significance. At that time no-one questioned how the amines in partially decomposed fish would be nitrosated in an alkaline milieu, which opposed the conventional wisdom, and was later illuminated by the studies of Keefer and Roller (1973). Previously there was little interest in N-nitroso compounds as other than model carcinogens with characteristic organ and species specificities, of no importance for human health. That they were

Table 1.2. *Human exposure to carcinogenic N-nitroso compounds*

Compound	Principal sources
Known exposures	
Nitrosodiethanolamine	Cosmetics, workplace, tobacco
Nitrosodimethylamine	Food, workplace, tobacco, environment
Nitrosonornicotine	Tobacco
NNK	Tobacco
Nitrosomorpholine	Workplace, tobacco
Nitrosopyrrolidine	Food, tobacco
Nitrosopiperidine	Food, workplace
Nitrosodiethylamine	Workplace, food
Nitrosobis-(2-hydroxypropyl)amine	Workplace
Nitrosomethyldodecylamine	Cosmetics
Nitrosomethyltetradecylamine	Cosmetics
Nitrosodi-n-butylamine	Workplace
Nitrosodiphenylamine	Workplace
1-Methylacetonylnitroso-3-methylbutylamine	Food (China)
1-Methylacetonylnitroso-2-methylpropylamine	Food (China)
Nitrosoanabasine	Tobacco
NNAl	Tobacco
Nitrosomethylethylamine	Tobacco smoke
Possible exposures	
Nitrosodi-n-propylamine	(Herbicide)
Nitrosomethylphenylamine	(Factories?)
Nitroso-N-methylcarbamates	Pesticides
Nitrosohexamethyleneimine	Drugs

so toxic, and their target so often the liver, which is rarely a site of neoplasia in Western industrialised societies (all that matter!), supported their curiosity status; to some extent, this is still true, and investigators who use them for studies of mechanisms of carcinogenesis seldom consider the broader implications (or lack of them) in their findings.

The importance of the finding of nitrosamines in nitrite-treated fish meal prompted consideration of formation of nitrosamines in all foods that were treated with nitrites or nitrates, or contained them naturally. Nitrates have been used to cure meat, and give it the pleasant pink colour, since time immemorial (probably the Romans). After discovery in the early years of this century that it was the nitrite formed by bacterial reduction of nitrate that was reactive (and the nitrate was unreactive), nitrite itself was used to cure meat. Legal limits were set by the Food and Drug Administration of the U.S.A. in 1926, although 20 years earlier its administrator, Dr Harvey Wiley, had branded sodium nitrite a poison (which it is). A surge of interest, beginning about 1970, led to large-scale analytical studies of the content of nitrosamines in cured meats and fish, and levels in the range of one to 100 parts per billion were found.

1.2 Sources of *N*-nitroso compounds

There are few *N*-nitroso compounds occurring in nature, for example the unusual methylnitrosourea derivative streptozotocin, which illustrates that biological systems can combine the precursors of these compounds. It follows that other examples probably exist, since nitrosatable amines and nitrosating agents are not uncommon.

It became clear that many nitrosamines, including a number of suspected carcinogens, could be formed in the environment by reaction of secondary amines with nitrosating agents. The common method of preparation of nitrosamines is through reaction of a secondary amine with nitrous acid (or a nitrite salt in acid solution) (Ridd, 1961). Because of the gastric secretion of hydrochloric acid, the stomach of humans provides a milieu in which nitrosamines could form from ingested nitrite and amines (Lijinsky and Epstein, 1970). Amines are widespread, and commonly ingested as food constituents, food additives, residues of agricultural chemicals and medicines, a list which is not exhaustive. Nitrites are formed by reduction of the almost ubiquitous nitrates (present in all plants and excreted by humans in urine). Nitrites are common in the diet – being used to process foods, such as cured meats and fish – and in the body, being present universally in saliva at low concentrations (Tannenbaum *et al.*, 1974; Spiegelhalder *et al.*, 1976).

There is also a more direct source of exposure to *N*-nitroso compounds. Nitrosamines are found in various foods, being formed from amines in the food and nitrosating agents which come into contact with them. For example, several nitrosamines are present in cured meats such as bacon and sausages (Sen *et al.*, 1973), including NDMA and nitrosopyrrolidine, the latter being formed when the meat is cooked. NDMA is – or was – present in beer and whisky (Spiegelhalder *et al.*, 1979), originating from malt in which the tertiary amine alkaloids gramine and hordenine interacted with nitrogen oxides in the gases used to heat the malt (Mangino and Scanlan, 1984).

Gas-phase nitrosation is probably important in leading to formation of nitrosamines in tobacco smoke. The smoke, like all products of combustion of nitrogenous organic matter, contains nitrogen oxides, and tobacco itself contains many volatile amines, which are also formed by thermal degradation of tobacco components. In addition, tobacco is cured with the aid of nitrates, which become reduced to nitrites, and react with nitrosatable amines, such as nicotine, to form nitrosamines. The curing process lasts a long time, so that these slow reactions can proceed well, resulting in the presence of quite high concentrations of nitrosonornicotine, 4-(methylnitrosamino)-1-(3-pyridyl)-1-butanone (NNK), and other nitrosamines in the tobacco (Hoffmann and Hecht, 1985). These potent carcinogens are, therefore, present in considerable concentrations (five to 80 ppm) in such products as snuff and chewing tobacco, as well as tobacco smoke (Hoffmann et al., 1984a). Local oral exposure to these nitrosamines is almost surely causally related to the oral cancer common in tobacco chewers and snuff 'dippers'. Intragastric exposure also occurs through dissolution in the saliva. Tobacco use may constitute the most important exposure of humans to nitrosamines; the carcinogenic nitrosamines in tobacco and tobacco smoke are probably the major cause of tobacco-related cancers.

Other important sources of exposure to N-nitroso compounds are particular factory environments in which nitrosamines are used, or can be formed as byproducts, for example, during manufacture and processing of rubber and leather. A survey by a U.S. Government Agency some years ago showed that nitrosamines that are potent carcinogens, NDMA, NDEA and nitrosomorpholine, are commonly present in factories making or fabricating rubber and leather (Fajen et al., 1979). In the rubber industry for example, nitrosodiphenylamine – itself a fairly weak carcinogen (Cardy et al., 1979), but a powerful nitrosating agent (Challis and Osborne, 1973) – is frequently used as a retarder of vulcanisation. Many readily nitrosated secondary and tertiary amines are used in rubber manufacture, including derivatives of morpholine, dimethylamine and diethylamine (IARC, 1982a), and these can react with nitrosodiphenylamine or nitrogen oxides in burning fuel to form the corresponding nitrosamines.

Other sources of nitrosamines relate to the use of secondary amines, including dimethylamine, diethylamine and diethanolamine, to form salts with acidic industrial products, such as 2,4-dichlorophenoxyacetic acid and other herbicides. Secondary amines, especially dimethylamine, are commonly used as bases for neutralising industrial acids, and interaction of the amines (or their salts) with nitrosating agents, such as nitrogen oxides in the fumes of burning fuel, produces nitrosamines. There have been studies of the reaction of nitrogen oxides with amines in the gas phase (Challis and Kyrtopoulos, 1979; Neurath et al., 1976) in industry, and also in tobacco smoke. It is of interest that manufacture of some N-nitroso compounds involved passing nitrogen oxides (principally nitrogen dioxide) through an amine; for example, nitrosiminodiacetic acid was made from iminodiacetic acid and nitrogen oxides (Dubsky and Spritzmann, 1916).

Because *N*-alkylcarbamate esters are inhibitors of the enzyme cholinesterase in insects, many compounds of this structure, particularly *N*-methylcarbamates, are in wide use as insecticides. Examples include some widely used insecticides, such as carbaryl, baygon, carbofuran, aldicarb and methomyl, most of which have low toxicity to mammals and other vertebrates. The *N*-methylcarbamate insecticides are manufactured in very large quantities; it was through manufacture of methyl isocyanate which is used in the synthesis of these carbamates that the Bhopal disaster in India occurred. These compounds are converted by nitrosation with nitrites or nitrogen oxides to *N*-nitroso products of quite extraordinary carcinogenic and mutagenic potency, e.g. nitrosocarbaryl (Elespuru *et al.*, 1974). As bacterial mutagens, for example, they outclass any other *N*-nitroso compounds; nitrosomethomyl, as described elsewhere, gives rise to more than a million revertants per micromol in the Ames *Salmonella* assay, compared with a hundred per micromol of the otherwise potent biological agent ethylnitrosourea. The nitrosomethylcarbamates are more similar to one another in potency as carcinogens than they are as mutagens. Most of the tests in animals have been carried out by oral administration and, at least in rats, the range of their carcinogenic action is limited to the forestomach, in which they induce carcinomas, following administration of only a few milligrams. Human exposure to them at low levels might be considerable.

Nitrosodiethanolamine (NDELA) has recently become important because of the introduction of synthetic cutting oils in machine shops to replace mineral oils, which were formerly considered hazardous because they contained carcinogenic polynuclear hydrocarbons, albeit at low concentrations. Modern synthetic cutting oils contain some mineral oil emulsified with water by means of triethanolamine, and sodium nitrite which is a corrosion inhibitor. Commercial triethanolamine contains substantial quantities of diethanolamine, which reacts with nitrite to form NDELA. Although nitrosation of secondary amines takes place most readily in acid solution, the reaction proceeds slowly in alkaline conditions (which pertain in the cutting oil emulsion) and is powerfully catalysed by carbonyl compounds, as reported almost 20 years ago by Keefer and Roller (1973). Tertiary amines, such as triethanolamine, react with nitrite in weakly acid solution, but the reaction is slow (Lijinsky *et al.*, 1972*d*) and it is clear that the NDELA in cutting oils arises mainly from diethanolamine.

Sources containing NDELA represent the highest concentrations of nitrosamine to which humans are exposed, as high as 3% in some cutting oils, although usually lower (Fan *et al.*, 1977*b*). Commercial triethanolamine, used as an emulsifier, is also the source of NDELA found at lower concentrations (several parts per million) in some cosmetics (Fan et al., 1977a). It is probable that the nitrosamine was present in the triethanolamine used to formulate those products, probably arising from nitrite lining cans in which the amine is transported and stored. The NDELA homologue nitrosobis-(2-hydroxypropyl)amine – an inducer of pancreas tumours in hamsters (Krüger et al., 1974) – is in products containing the tertiary amine *triiso*propanolamine (Issenberg et al., 1984) and seems to arise in a similar way. Shampoos and hair-care products have small concentrations (a few ppm) of such nitrosamines as methylnitrosotetradecylamine and methylnitrosododecylamine (Hecht et al., 1982a) formed similarly from the secondary amines.

Although the nitrosation of tertiary amines was not well known (Hein, 1963), about 20 years ago the possibility that NTA (nitrilotriacetic acid) would become a major environmental contaminant through its use as a detergent builder, led to serious consideration of the implications of its interaction with nitrosating agents (Lijinsky et al., 1973b). To assess the importance of this possibility, several experiments were undertaken, including the testing of the expected nitrosamine product. However, it turned out that nitrosiminodiacetic acid was neither toxic nor carcinogenic; in fact, no toxic effects in rats or mice were observed at the highest doses given, 70 g per kg body weight (Lijinsky et al., 1973b). As a corollary there were several studies of combinations of NTA with sodium nitrite fed to rats or mice, and all were negative (Lijinsky et al., 1973b; Greenblatt and Lijinsky, 1974).

As an outgrowth of the studies reported by Smith and Loeppky of reaction of tribenzylamine with nitrite in mildly acid (above pH 3) solution to form nitrosodibenzylamine and benzaldehyde (Smith and Loeppky, 1967), following similar studies with substituted hydrazines (Smith and Pars, 1959), examination of a number of important tertiary amino compounds was undertaken. Among them were trimethylamine, triethylamine, which are commonly used organic bases, and trimethylamine-N-oxide, a common constituent of fish – and the major form in which fish and other aquatic species excrete nitrogen. All of the tertiary amines studied (including 4-dimethylaminoazobenzene) gave appreciable yields of nitrosamine when incubated at pH 4 to 5 at 90 °C after several hours in phosphate buffer, acetic acid or in boric acid solution; yields of 1 % or more of nitrosamine were obtained at 37 °C after 16 hours (Lijinsky et al., 1972d).

Among the most important tertiary amines in relation to humans were many drugs which also reacted readily to give high yields of nitrosamine at 90 °C (Lijinsky et al., 1972a), including the widely used aminopyrine, oxytetracycline and disulphiram (although the last was only slightly soluble in water). At 37 °C and lower concentrations of tertiary amine and nitrite (0.01 M), the yields of nitrosamine were lower, but still significant (Lijinsky, 1974), considering their potent carcinogenicity. Thiram (tetramethylthiuram disulphide) – a homologue of disulphiram and equally insoluble – reacts with nitrous acid to form NDMA (Elespuru and Lijinsky, 1973); both are commonly used agricultural chemicals, as well as having heavy use in the rubber industry.

The widespread use of many tertiary amines in agriculture, mainly as herbicides and insecticides, makes them of great interest as potential sources of nitrosamines, since they are often persistent, and become contaminants of foods; there are tolerances set by Departments of Agriculture in various countries for many of these compounds. Reactions of many compounds, including dialkylthiocarbamates, dialkyldisulphides and dialkylarylureas, with nitrous acid (nitrite in weak acid medium) have been studied and considerable yields of the corresponding nitrosamine have been usual (Elespuru and Lijinsky, 1973). In these reactions of tertiary amines it is possible for more than one N-nitroso compound to be produced by reaction with nitrous acid, but usually only the volatile nitrosamine has been measured. Nitrosation of the antihistaminic methapyrilene produced a number of nitrosamines, in addition to the expected NDMA (Mergens et al., 1979). If such reactions were to occur in the environment or in the human stomach, the unknown products as well as the obvious volatile nitrosamine could contribute to carcinogenic risk. However, the potency of these other products of nitrosation of tertiary amines and the target organs of their action are usually not known, and might be difficult to discover, since many of them could be expected to be unstable. Ignorance of the carcinogenic risk presented by such N-nitroso products does not justify our assuming that they would lack carcinogenic activity, or even that they would be weak carcinogens.

1.3   Estimation of carcinogenic risks

The many drugs and agricultural products that are secondary amino compounds pose a potential carcinogenic risk through formation of nitrosamines by reaction with nitrosating agents. However, it cannot automatically be assumed that a secondary amine is a greater risk in this regard than a tertiary amine, even though nitrosation of the latter is usually slow. For example, the two commonly used drugs phenmetrazine (a derivative of morpholine) and methylphenidate (a derivative of piperidine) were likely targets of suspicion because they are easily nitrosated and their parent nitrosamines are among the most potent of carcinogens. A simple study of the reaction of phenmetrazine with nitrite in the stomach of rats and of guinea pigs, showed that the reaction was rapid and yields as great as 40% of theoretical of nitrosophenmetrazine were obtained (Greenblatt et al., 1972). However, the importance of this finding to human health – apart from general information about nitrosation of amines *in vivo* – was diminished when nitrosophenmetrazine was found not to be carcinogenic (Lijinsky and Taylor, 1976c). Nitrosomethylphenidate also failed to give rise to an increased incidence of any tumour in rats to which it was given chronically by gavage (Lijinsky and Taylor, 1976c).

The presence of nitrite in food can contribute to exposure of humans to carcinogenic N-nitroso compounds. Interactions within the treated food can lead to formation of N-nitroso compounds. One of the first reactions of nitrite with food to which it is added is with hemoglobin and myoglobin, forming nitrosohemoglobin and nitrosomyoglobin. This rapidly leads to oxidation of the iron in the haem from the biologically active ferrous state (2^+) to the inactive ferric state (3^+). The fate of the NO during this process is not clearly understood, but it might well react with amines to form nitrosamines (or other N-nitroso compounds). Among them are nitrosoproline and nitroso-hydroxyproline from the amino acids; nitrosation of the imino groups in peptides is also possible, as suggested by Challis and others (1984).

A most important source of exposure of humans to N-nitroso compounds is through endogenous formation from amines and nitrite in the stomach. This could lead to exposure to many more nitrosamines than are found preformed (Lijinsky, 1986b). Nitrite is used to manufacture cured meats and fish. It has been known for a considerable time that nitrite is present in saliva (it has also been reported in blood of humans). Therefore, it is probable that people are exposed to N-nitroso compounds through interaction of nitrite with ingested secondary and tertiary amines in the stomach. The easy formation of the non-carcinogenic nitrosoproline, and its measurement in urine, has been applied to estimation of nitrosating capacity in humans (Ohshima and Bartsch, 1981). The method has been widely used, for example, to monitor people in areas of China with a high incidence of esophageal cancer.

This straightforward chemical reaction takes place at a rate depending on the concentration of nitrite and of the amine, the nature of the amine, and the pH of the stomach contents. Nitrite in saliva probably arises by bacterial reduction of nitrate present in blood and lymph. The nitrate is ingested in vegetables, or is formed endogenously as described by

Tannenbaum et al. (1978). Nitrite in saliva seems to vary from a few ppm to 10 ppm, although it can be higher (Tannenbaum et al., 1974). Two factors make the nitrite in food an equal or more important source of nitrosamines than nitrite in saliva. Firstly, the slow dripping of saliva into the stomach makes the nitrite concentration of the contents quite low, and, secondly, the heterogeneous nature of food particles containing nitrite can make local concentrations high in some cases, and low in others. These factors, because the rate of nitrosation is proportional to the square of the nitrite concentration (Ridd, 1961), make the yield of nitrosamine much greater from high concentrations of nitrite than from low concentrations; hence nitrite in food is the more important.

Tertiary amines, because of their structure and their relatively low rate of reaction with nitrous acid, even at optimum pH (Jones et al., 1974), are more difficult to deal with in estimating the carcinogenic risk posed by their nitrosation than are secondary amino compounds. Although procedures have been discussed (Coulston and Dunne, 1980) for estimating the risks in human exposure to nitrosatable drugs which are ingested, estimates can be in error by large margins.

The rate of nitrosation of tertiary amines, which is very dependent on pH – optimum 3.8–4.2 (Mirvish, 1975) – is very much slower than that of secondary amines. However, it can be assumed that many tertiary amines can be an important source of nitrosamines through endogenous nitrosation, because considerable nitrosamine formation could occur during an hour or more in the stomach, and in the absence of competing reactions which would draw down the concentration of nitrite.

The strongly basic secondary amines, such as dimethylamine, react slowly with nitrite in acid solution, unless catalysts are present; tertiary amines react even more slowly. Experiments in which diets containing amines and nitrite were taken by humans, have led to detection of, for example, nitrosodimethylamine in the blood after a few hours (Fine et al., 1977). An earlier observation involving an experiment in humans was reported by Sander and Seif (1969), in which formation of nitrosodiphenylamine resulted from administration of diphenylamine and nitrite, the nitrosamine being detected in the urine at low concentrations. There was also some evidence that nitrosamines could form in people with infected bladders by reaction of nitrite (from bacterial reduction of urinary nitrate) with amines in urine (Hawksworth and Hill, 1974), followed by easy absorption of the nitrosamine (Lijinsky, 1990d).

1.4 Modulation of nitrosamine formation

There has been much investigation into the possibility of interfering with the formation of *N*-nitroso compounds, endogenously and in commercial materials, both by accelerators and inhibitors of nitrosation. Fortuitously, it was discovered that ascorbic acid used as a preservative in a medicinal sample had inhibited the formation of nitrosodimethylamine from oxytetracycline and nitrite (Mirvish *et al.*, 1972). This was due to the competition of ascorbic acid with the amine for reaction with nitrite; the reaction of nitrous acid with ascorbic acid had been known for some time (von Dahn *et al.*, 1960), but had not previously been considered in this context. This finding suggested that ascorbic acid, and possibly other naturally occurring substances, might retard or prevent *in vivo* nitrosation by blocking reaction of amines with nitrite. Among other inhibitors considered have been tocopherols, tannins and glutathione. In addition, phenols have been reported as inhibitors of nitrosation, although the evidence is conflicting because the investigator who reported this also reported elsewhere that certain phenols accelerate nitrosation (Challis, 1973; Challis and Bartlett, 1975); this dilemma has not been resolved.

In addition to inhibitors of nitrosation, there are accelerators of the process, many of them nucleophilic anions, such as thiocyanate. Chloride and bromide also accelerated reaction of tertiary amines with nitrite (Lijinsky *et al.*, 1972*d*). An important finding by Keefer and Roller (1973) was that the reaction of secondary amines with nitrous acid (but not of tertiary amines) was accelerated by certain carbonyl compounds, especially by acetaldehyde and chloral, and took place in neutral or alkaline solution. This reaction might be of considerable importance in the formation of nitrosamines, since carbonyl compounds are common and are quite likely to be present in conditions under which nitrites react with amines.

Transnitrosation is the transfer of the nitroso group from a nitroso compound to an amine, and is another mechanism by which carcinogenic *N*-nitroso compounds may be formed. Common among the transnitrosating compounds are several that are weakly or non-carcinogenic. It has long been known that aromatic nitrosamines, such as nitrosodiphenylamine, are somewhat unstable and readily donate NO to a secondary amine, such as morpholine, forming nitrosomorpholine (Challis and Osborne, 1973). Several aliphatic nitrosamines are good transnitrosating agents, and several secondary amines are good recipients, especially at acid pH and their nitrosation is accelerated in the presence of thiocyanate as catalyst (Singer *et al.*, 1978).

The weak carcinogens 1-methyl-4-nitrosopiperazine and nitrosophenylbenzylamine, and the non-carcinogens nitrosoproline and nitrosohydroxyproline, all transnitrosated morpholine (as the hydrochloride) when fed to rats. Concurrent feeding during most of a rat's lifetime gave rise to sufficient nitrosomorpholine to induce a significant incidence of liver tumours (Lijinsky and Reuber, 1982b). This presents the possibility that humans could be exposed to N-nitroso compounds formed by transnitrosation through ingestion of nitrosating compounds, such as nitrosoproline and nitrosohydroxyproline, which are present in cured foods.

1.5 Chemistry and biological activity of N-nitroso compounds

Since there are many possibilities of human exposure to N-nitroso compounds – and N-nitroso compounds might well be the most important exposure of humans to any type of carcinogen, or the largest contributor to increased carcinogenic risk – a study of their biological effects has become of increasing interest and importance. Their chemical simplicity and ease of formation, together with the variety of structures they comprise (which is limited only by the ingenuity of chemical synthesisers) and the small number of metabolic pathways in which they are involved, make them the most accessible carcinogens with which to investigate mechanisms of carcinogenic action. The largest and smallest molecules having carcinogenic activity are illustrated in Fig. 1.2.

Although the N-nitroso group is not without reactivity, there are few reactions it undergoes. One is cleavage to the parent amine and NO which can interact with a suitable receptor. This is not likely to be related to the biological activity of N-nitroso compounds, since there is only a relatively small difference in rate of loss of NO between similar compounds, which would hardly explain the huge difference in biological activity (mutagenesis, carcinogenesis, toxicity, etc.) shown by the various members of the family, ranging from very large effects to almost zero. The nitroso group can also be reduced to amino, giving rise to a di-substituted hydrazine, and it can be oxidised to a nitro group, giving rise to a nitramine. While nitramines resemble in carcinogenic activity the corresponding nitrosamines, but are much weaker (Goodall and Kennedy, 1976), they are much less toxic and have quite different mutagenic properties. It seems likely that, in living systems, nitramines are reduced to nitrosamines, which then have all of the effect attributable to the nitrosamine.

The unsymmetrical dialkylhydrazines formed by reduction of N-nitroso compounds (a reduction which can be effected chemically with a variety of agents, or electrochemically) are relatively weak carcinogens (Toth, 1973), and less toxic than the *N*-nitroso compounds. Several dialkylhydrazines are produced commercially; one large-scale use of dimethylhydrazine is as a rocket fuel. It is probable that, under normal conditions, dialkylhydrazines will be oxidised, at least in part, to nitrosamines. Alkylhydrazides will be formed, of course, by reduction of alkylnitrosamides, but the former have attracted little interest.

Alkylnitrosamides are not stable, and quite unstable at alkaline pH, giving rise to diazoalkanes. It is improbable that their biological activities are mediated through formation of diazoalkanes, although it is possible. Alkylnitrosamides decompose readily on heating, and the solids amongst them (and most are solids at room temperature) melt with decomposition. Many alkylnitrosocarbamates, however, can be distilled *in vacuo* without decomposition. Methylnitrosonitroguanidine decomposes violently on heating. There have been several reports of violent decomposition of methylnitrosourea stored in bottles at room temperature. These compounds should always be stored in small quantities, and at temperatures at least as low as refrigerator temperature, preferably at $-20\,°C$. The residue after such explosions is likely to be harmless, unlike the nitroso compounds, which are usually potent carcinogens.

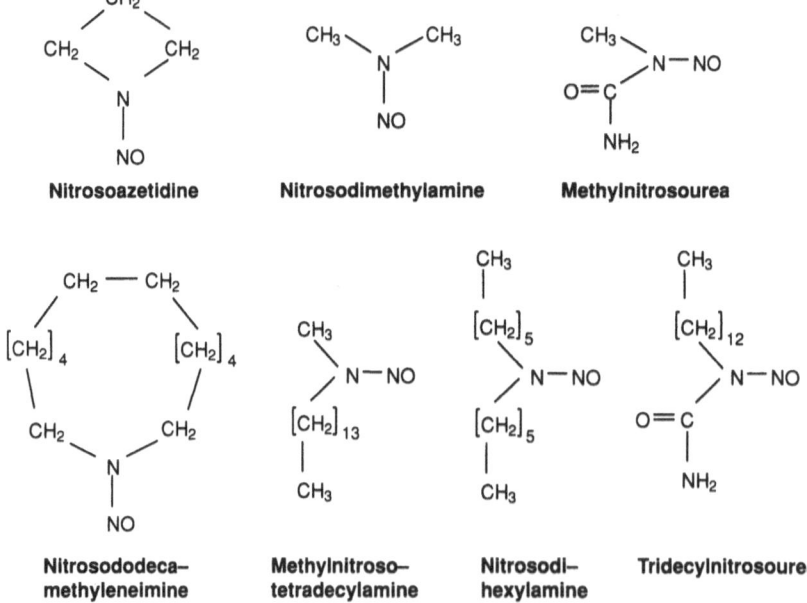

Fig. 1.2. Largest and smallest carcinogenic *N*-nitroso compounds.

Nitrosamines are stable at room temperature, and many of them can be distilled *in vacuo* with little decomposition. Above 200 °C, however, most nitrosamines decompose, and NO is liberated. This is the basis of the Thermal Energy Analyzer used for detecting nitrosamines, which it does by measurement of decay of excited NO formed thermally from the nitrosamine, emitting infrared energy of a discrete frequency. Some nitrosamines, especially those derived from weakly basic amines such as diphenylamine, lose NO readily at room temperature; they are good nitrosating agents for this reason. On the other hand, *N*-nitroso derivatives of basic secondary amines are very stable, and have been stored intact in neutral aqueous solution in the dark over several years. They are also very stable in solution in most organic solvents and in aqueous alkali, but they slowly hydrolyse in acid.

When exposed to visible light, solutions of nitrosamines are converted to the amine and nitrite. It is probable that other forms of radiation would have similar effects – but few studies have been published. One well tried method of measuring *N*-nitroso compounds is by photolysis and measurement of the nitrite released. Another procedure uses the so-called Griess reagent and involves hydrolysing the nitrosamine with hydrobromic acid and estimating the nitrite formed by diazotization and coupling with an aromatic compound.

None of these properties of *N*-nitroso compounds is a good guide to possible mechanisms by which they produce their biological effects. This is because the chemical reactions are general, while the biological actions are particular. Notably, the alkylating properties of a number of *N*-nitroso compounds which are methylating agents are no guide to their relative activities as bacterial mutagens, which are quantitatively very different (Elespuru, 1976). Their carcinogenic activities are even more disparate, both quantitatively (as assessed by potency) and qualitatively, as shown by large variations between them in the type of tumour they induce, even in the same species and strain of animals. An enigmatic finding is that the cyclic nitrosamines – a large group of *N*-nitroso compounds studied – are not alkylating agents in the usual sense. Since many of them are mutagenic to bacteria when activated by microsomal enzymes from liver, it is assumed that the enzymic activation forms reactive intermediates which interact with, and mutate, the bacterial DNA. However, the difficulty in identifying reactive products of cyclic nitrosamines, or to identify – even detect – adducts with DNA *in vivo*, except in the case of nitrosopyrrolidine, makes this an unproven hypothesis. It is possible that mutagenesis by cyclic nitrosamines comes about through an entirely novel and unknown process, perhaps an indirect one. This possibility is fortified by the finding that cyclic nitrosamines, such as nitrosoazetidine, nitrosopyrrolidine, nitrosohexamethyleneimine and nitrosomorpholine are often considerably more potent mutagens than their analogous acyclic counterparts methylnitrosoethylamine, nitrosodiethylamine, nitrosodi-n-propylamine and nitrosodiethanolamine. Methylnitrosoethylamine and nitrosodiethanolamine are, in fact, not mutagenic at all in the usual bacterial systems.

There is, in addition, a large list of carcinogenic N-nitroso compounds, mostly nitrosamines, which are not mutagenic for reasons not easy to discern. Several cyclic nitrosamines containing sulphur are carcinogenic but not mutagenic. For example, nitrosothiomorpholine, equally carcinogenic with nitrosomorpholine, although it induces tumours of the esophagus rather than the liver in rats, is not mutagenic, even with microsomal activation. Nitrosothialdine and nitrosodithiazine are also non-mutagenic, although the former is carcinogenic and the latter not (Lijinsky *et al.*, 1988*a*).

The latter two compounds illustrate another facet of the nitrosamine problem, the difficulty of predicting carcinogenic activity from the chemical structure. Nitrosodithiazine, as a featureless cyclic nitrosamine, would be expected to be carcinogenic, while nitrosothialdine, bearing methyl substituents in the positions α to the N-nitroso function would be expected to be non-carcinogenic or weakly so, as a comparison of the non-carcinogenicity of 2,5-dimethylnitrosopyrrolidine with the potent carcinogenicity of nitrosopyrrolidine suggests (Lijinsky and Taylor, 1976*d*). In fact, the reverse is the case; nitrosodithiazine is without carcinogenic activity and nitrosothialdine is a fairly potent carcinogen, although weaker than nitrosodiethylamine, of which it can be considered a partially activated derivative (Lijinsky *et al.*, 1988*a*). The difficulty of attempting to forecast carcinogenicity from chemical structure is illustrated by nitrosothiazolidine, predicted by Rosenkranz and Klopman (1987) to be a potent carcinogen based on calculations from computerised structure–activity data. Contemporary animal tests showed it not to be a carcinogen (Lijinsky *et al.*, 1988*a*).

Examination of the relationship between chemical structure and carcinogenic activity suggests that there are receptors in cells, for which N-nitroso compounds are the substitute for the normal substrate. Binding of the N-nitroso compound to the receptor might block normal activity, sending an abnormal signal to the nucleus and initiating cell division at an inappropriate time. This could result in permanent adverse consequences to the host, if not necessarily for the transformed cell. The receptors involved might be enzymes (perhaps concerned with the maintenance of the integrity of the cell surface) or cell surface receptors involved in signal transduction, as are believed to be important by Weinstein (1987) and others. Whether or not changes in the immunological properties of cells are involved is not clear.

Whatever the precise mechanism of carcinogenesis by N-nitroso compounds is, there seems to be a strong relationship between chemical structure and carcinogenic activity, although of a type that we cannot at present elucidate. Certain chemical structural relationships to particular types of cancer in one or another species will be described in Chapter 7. The alkylnitrosourea structure is invariably related to induction of hemangiosarcomas of the spleen in Syrian hamsters, a tumour not induced by other classes of N-nitroso compounds. Neoplasms of the pancreas ducts in Syrian hamsters are almost always related to nitrosamines (but not alkylnitrosoureas) containing the 2-oxopropyl structure, or convertible into it. It is probably not without relevance that a 2-oxopropylnitrosamine is a form of activated acetone, which may have a significant relationship to important biochemical intermediates.

In rats, nitrosodimethylamine is almost the only methylnitrosamine which does not induce tumours of the esophagus. This suggests that there is something special about NDMA whereby it lacks affinity for an esophageal receptor with which the other methylnitrosamines interact, including methylnitroso-n-propylamine and most of the higher homologues. This receptor also appears to respond to many ethylnitrosamines, including nitrosodiethylamine and ethylnitrosomethylamine (i.e. methylnitrosoethylamine). These compounds differ in potency as esophageal carcinogens, and there is evidence that activation to an alkylating agent might be involved, since Von Hofe *et al.* (1987) have shown that methylation of rat esophageal DNA by a series of methylnitrosoalkylamines administered orally corresponds roughly with their potency as esophageal carcinogens. However, alkylation of DNA is probably not the most important event, because many cyclic nitrosamines, including numerous derivatives of nitrosopiperidine and dinitrosopiperazine, as well as nitrosomorpholines, are very potent esophageal carcinogens in rats, without the ability to alkylate DNA detectably (Lijinsky *et al.*, 1973c).

Although we are far from an understanding of the mechanisms by which N-nitroso compounds cause cancer, there is hope in some of our observations. It might be necessary to go back to first principles, and to examine the pharmacokinetics of compounds of similar structure, but different carcinogenic activity, to approach some solutions. As J. M. Barnes suggested many years ago, this might be the greatest importance of N-nitroso compounds, as models with which to study the mechanisms of organ- and species-specific carcinogenesis. Even if this is true, the human exposure to *N*-nitroso compounds, and the consequent contribution to human cancer risk because of their great potency, cannot be ignored. The reader is left to decide which of these facets of *N*-nitroso compounds is the more interesting.

2

Occurrence, formation and detection of N-nitroso compounds

2.1 Occurrence

Until comparatively recently N-nitroso compounds were not thought to occur in Nature and there was little speculation about the matter before the discovery that they had interesting toxicological properties. In fact, although nitrosamines were known laboratory chemicals, it was the knowledge that they belonged to the group of chemicals that are alkylating agents which made them interesting. Before this time, there were occasional studies of their chemical properties and many of them were prepared in the last century when whole groups of analogues and homologues of certain chemical structures were prepared simply to find out their basic properties. This must be the explanation for the preparation by Marckwald and Droste-Huelshorff in 1898 of nitrosoazetidine, which cannot conceivably be a very useful compound – and is costly to prepare. But it was the simplest cyclic nitrosamine, since the nitroso derivative of ethyleneimine is not stable, although its existence has been reported at low temperatures.

The preparation of methylnitrosourea (von Brüning, 1888) and of methylnitrosourethane (von Pechmann, 1895) were related to their usefulness as precursors of the valuable methylating agent diazomethane. This was also a purpose for which N-methyl-N-nitroso-N-nitroguanidine (MNNG) was prepared (McKay, 1948), although its instability also suggested its use as an explosive. These rather unstable N-nitroso compounds have been replaced for generation of diazomethane by more stable precursors, such as methylnitrosotoluenesulphonamide. Other alkylnitrosamides are used, of course, to prepare other diazoalkanes, and many of these are in common use.

Any circumstance in which a nitrosating agent and a nitrosatable amino compound come into contact is a potential source of N-nitroso compounds. This explains the occurrence of N-nitroso compounds (usually nitrosamines) in the atmosphere in or surrounding certain factories, as a result of industrial processes, in many foodstuffs, and in our own bodies. The chemistry of formation of N-nitroso compounds favours high rates of reaction in acidic conditions; metal ions and nucleophilic anions are catalysts of nitrosation (Douglass et al., 1978). Membrane-simulating micelles also accelerate formation of nitrosamines (Okun and Archer, 1977). In many conditions times is not of the essence, as in formation of nitrosamines in synthetic cutting oils (Fan et al., 1977b), when the reactions can occur over periods of weeks or months. Even in exposure of amines in industrial settings to nitrogen oxides in flue gases – a common condition – many hours can elapse during which formation of nitrosamines can occur. An old report describes the formation of nitrosiminodiacetic acid, and other nitrosamino acids, by passing nitrogen oxides over, or into, the amino acid (Dubsky and Spritzmann, 1916). Inhibitors of nitrosation by nitrous acid are usually ineffective in inhibiting gas phase nitrosation.

2.1.1 In food and beverages

Nitrosamines in foods are mainly derived from secondary and tertiary amines by reaction with a nitrosating agent. Secondary amines are not uncommon in Nature, and they were long prepared by distilling vegetable matter with alkali. Dimethylamine for a long time was manufactured by distilling with alkali sugar cane residues ('bagasse') after extraction of molasses; undoubtedly part of the dimethylamine arose from destruction of choline and betaine. Derivatives of piperidine and pyrrolidine, especially as alkaloids, are quite common in plants, and it can be expected that their N-nitroso derivatives would be formed by nitrosation of those products. Certainly the presence of nitrosopiperidine in pickling spice can be attributed to the nitrosation of piperine and other piperidine derivatives by the nitrite added to the spice mixture (Lijinsky et al., 1972b).

Nitrosamines – particularly nitrosodimethylamine – were responsible for the mysterious toxicity of fish meal to domestic fowl and to other domestic animals, first reported by Koppang (1964). The animals died of severe liver damage and it was discovered that the fish meal had been treated with sodium nitrite as a preservative (much as nitrite is used in cured meats), and then stored (Ender et al., 1964). Because the fish meal was likely to be an alkaline medium, rather than an acid one in which nitrosation would be expected, more knowledge might have led the investigators away from thinking that formation of nitrosamines could be involved. It is now common knowledge that nitrosation can occur, albeit slowly, in alkaline as well as in acid conditions, particularly in the presence of carbonyl compounds as catalysts (Keefer and Roller, 1973). Some decomposition by bacterial action of the tertiary amines, such as trimethylamine-N-oxide common in fish probably produced dimethylamine, which was nitrosated in alkaline conditions. Nitrosation of the tertiary amines might also contribute to formation of NDMA. Since that time there have been many reports of deaths of domesticated and farm animals, such as mink (Koppang and Helgebostad, 1987), after they had been fed nitrite-treated fish meal containing nitrosamines which caused fatal liver damage in the animals.

Nitrosamines are present at low concentrations in beer and in some distilled spirits, those made with malt, such as various kinds of whisky. Nitrosodimethylamine was first discovered in beer more than ten years ago, by Preussmann, Eisenbrand and their colleagues (Spiegelhalder et al., 1979), during a systematic examination of foodstuffs for the presence of N-nitroso compounds. There was considerable variation between beers in their content of nitrosodimethylamine, which was found at particularly high concentrations in so-called 'Rauch-bier', which is a dark malty brew. There were, incidentally, some beers with very little nitrosamine, a fact prominently proclaimed in their advertising. By some standards the levels of nitrosamine in the beers, and to a lesser extent in whisky, would be considered low, of the order of 0.1 ppm at the highest. However, because of the huge amounts of beer drunk, several liters a day per person in Germany for example, the exposure of these people to nitrosodimethylamine (and possibly other nitrosamines) could be several hundred μg a day. This consumption could continue for a lifetime, making it a major contributor to the cancer risk of those people. Nitrosodimethylamine and nitrosomorpholine given to rats at concentrations of the order of 0.1 ppm in drinking water induced a significant incidence of tumours (of the liver) within the two-year lifespan of the rats (Peto et al., 1984; Lijinsky et al., 1988b); nitrosamines in beer must be considered an important source of carcinogens to humans. The nitrosamines in whisky and other liquors are present at a similar concentration to that in beer (Goff and Fine, 1979; Walker et al., 1979), but is of less significance to humans, because whisky is drunk usually in much smaller quantities than beer.

The solution to the problem of nitrosamines in beer is something of a detective story. It was first traced to the malt, in many batches of which considerable concentrations of nitrosodimethylamine were found (Preussmann and Eisenbrand, 1984; Spiegelhalder et al., 1980). The malt containing nitrosamines that was used to make beer conveyed the nitrosamine to the beer. Beers made with batches of malt having low or negligible concentrations of nitrosamine had little nitrosamine in them.

The problem was solved by Preussmann and his colleagues in Germany, and Scanlan and his colleagues in the United States, who used the fact that only certain batches of malt contained appreciable concentrations of nitrosamine. In preparing this malt the heating was by direct exposure to gases from burning fuel, and these gases contained nitrogen oxides. Samples of malt heated in closed containers, preventing contact of the nitrogen oxides in gases from burning fuel with the malt, contained virtually no nitrosamine. Scanlan and his colleagues focussed further on precursors of nitrosodimethylamine, and postulated that the alkaloids hordenine and gramine present in barley were the source of the nitrosamine (Mangino and Scanlan, 1984). Scanlan showed that, indeed, gramine and hordenine, which are tertiary amines, reacted readily in solution with nitrous acid to form nitrosodimethylamine in quite high yield (see Fig. 2.1). The finale of the story is that the manufacture of malt has been changed in most countries so that formation of nitrosamine is minimised, and as a consequence levels of nitrosodimethylamine in beer now are of the order of five parts per billion (0.005 ppm = 5 μg/L) or less. While this concentration cannot be considered completely without risk, it is certainly safer than levels common ten years ago.

Nitrosamines in food are a less tractable problem, because the major source of them is the nitrite (or nitrate) added to food as a colourant and flavouring. In some circumstances, such as when meat is vacuum-packed without being pasteurised or thoroughly cleansed, sodium nitrite is used as a preservative. The origin of the use of nitrate (saltpetre) to preserve meat and fish, an ancient practice dating back at least to the Romans, is probably the observation that meat so treated did not spoil. The pink colour of meat treated with nitrate is due to the nitrosomyoglobin formed from the nitrite produced by bacterial reduction of nitrate. All of this has been known for many years, and nitrite has been permitted as an additive to meat at a level of 200 ppm since 1926, even though it was considered by some, including Harvey Wiley, the early Commissioner of Food and Drugs of the United States, an adulterant of food and therefore unacceptable as a food additive. The 200 ppm permissible level of sodium

Fig. 2.1. Structures of hordenine and gramine.

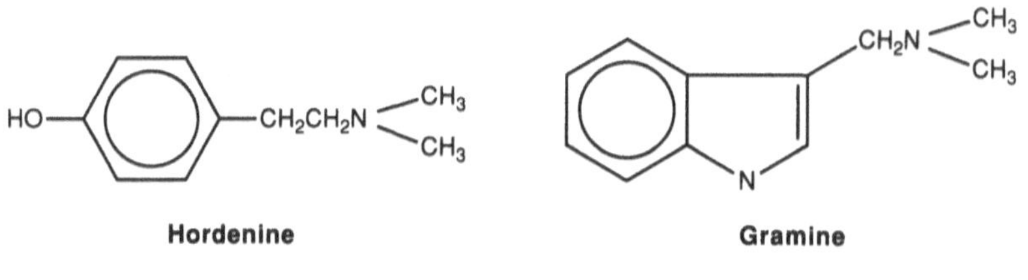

Hordenine

Gramine

nitrite in cured foods was apparently not based on scientific evidence of the efficacy and safety of this concentration of nitrite, but simply on the customary usage of quantities of nitrite which resulted in approximately 200 ppm as a residue in the meat or fish. Nitrate was also used as the additive in certain products, such as Lebanon bologna. In this instance, as in ancient usage, nitrate was reduced to nitrite by bacterial action and the nitrite then interacted with the meat pigments.

Nitrite is a preservative partially owing to its ability to prevent outgrowth of spores of *Clostridium botulinum*, and there has been much work during the past 15 years or so to establish the necessity of using nitrite to treat meat which is not pasteurised, or frozen, and might be kept in anaerobic conditions favouring outgrowth of Clostridia spores. It appears that the minimum concentration of nitrite for this purpose is 150 ppm, but the concentration needed to impart flavour and colour to the meat is much smaller. After much urging it was agreed by the United States Department of Agriculture and by the U.S. Food and Drug Administration, that the heretofore permitted level of 200 ppm of nitrite might present a cancer risk to humans – largely through formation of N-nitroso compounds in the stomach of people ingesting nitrosatable amines together with the nitrite-cured meat or fish. As a consequence, lower concentrations of nitrite were mandated in some foods, especially bacon, and there was a stronger requirement that the meat be processed under clean conditions (in which bacterial contamination would be minimal), that it be pasteurised whenever possible, and that it be refrigerated, but not vacuum packed unless that was essential for other reasons.

It is interesting that many countries with high meat consumption (such as Germany and Scandinavian countries), have greatly reduced, for health reasons related to exposure to nitrosamines, the permissible level of nitrite in meat – even banning it altogether in some products in Norway. This is perhaps excessive in view of the impossibility of avoiding completely exposure to nitrite from other sources. There is normally a low concentration of nitrite in saliva, arising from reduction of nitrate by bacteria in the mouth.

It is unlikely that all of the products formed in food by reaction of nitrite with amines are carcinogenic, but surely some of them are and others might by easily converted into carcinogens. For example, nitrosoproline (Fig. 2.2) can be thermally decarboxylated to form nitrosopyrrolidine, a powerful carcinogen. This is one of the sources of the nitrosopyrrolidine found in fried bacon (Sen *et al.*, 1973; Janzowski *et al.*, 1978) and in other cooked products containing nitrite. The elegant chemical studies of Loeppky *et al.* (1984) have demonstrated other ways by which nitrite and amines, in the presence of some other common components of biological mixtures, are converted to nitrosamines.

There have been numerous studies of the nitrosamine content of foods of various kinds, including, of course, cured meats and fish. There have also been reports of nitrosamines in cheese (Gough et al., 1977), in vegetable oils (Hedler and Marquardt, 1974), in dried milk (Libbey et al., 1980), and as products of the 'Amadori' reaction (Röper et al., 1984). Many of the earlier reports are suspicious, if not completely spurious, based as they were on analytical methods which were not specific for nitrosamines, but on the similarity of retention time of some chromatographic component of an extract with a particular nitrosamine. For example much of a substance identified as nitrosodimethylamine in a Zambian 'beer' was later discovered to be furfural, which had the same chromatographic properties (McGlashan et al., 1968). The likelihood of some of the more exotic nitrosamines 'identified' in this way being in the foods was quite remote. Greater sophistication of analytical techniques during the past 10 years or so has led to the demand that acceptance of identification of a nitrosamine in food be based on mass spectrometric confirmation of the structure of the component having the appropriate retention time. In cases of a suspected nitrosamine found at low concentrations and without confirmation of the chemical structure by mass spectrometry (Rainey et al., 1978), the presence of that nitrosamine is considered moot.

In Table 2.1 is a summary of the types of food in which nitrosamines have been found and the highest concentrations reported; most have been confirmed by mass spectrometry. The concentrations found have usually been low, five to 100 parts per billion, the latter high values being relatively uncommon. In many cases it was quite probable that other nitrosamines were also present, but they were not confirmed by mass spectrometric identification. A major gap in our knowledge is that non-volatile nitrosamines were not identified or measured, because no satisfactory methods for separation and identification of non-volatile

Fig. 2.2. Formation of nitrosopyrrolidine from nitrosoproline.

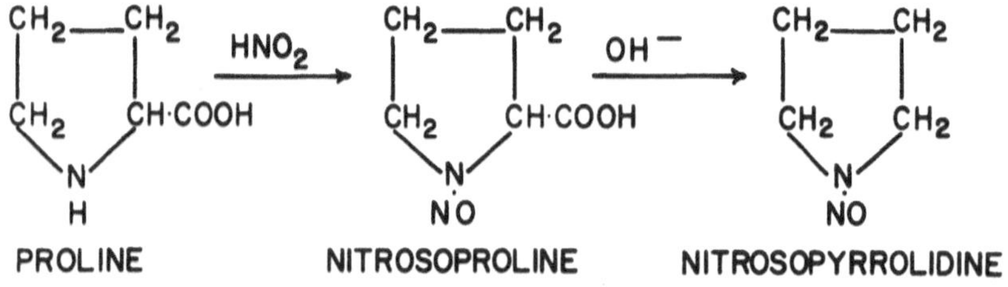

Table 2.1. *Nitrosamines in foods and beverages worldwide*

Type of food (source)	Nitrosamine identified	Concentration (μg/kg)	
Bacon, fried (Canada)	NDMA	17	
	NO-PYRR	22	(a)
	NDEA	6	
Bacon, fried (U.S.A.)	NDMA	5	
	NO-PYRR	84	(b)
Meat and sausage (Germany)	NDMA	12	
	NO-PYRR	45	(c)
Meat, smoked (Canada)	NDMA	2	
	NO-PYRR	10	(d)
	NO-PIP	59	
Meat, salted (U.S.S.R.)	NDMA	54	
	NO-PYRR	12	(e)
	NO-PIP	7	
Mutton, in oil (Tunisia)	NDMA	23	
	NO-PYRR	3	(f)
Fish, dried (Greenland)	NDMA	38	(f)
Fish, salted (Japan)	NDMA	26	(g)
Fish, dried salted (China)	NDMA	133	(f)
Squid, dried broiled (Japan)	NDMA	313	
	NO-PYRR	10	(g)
Pickled cabbage (China)	NDMA	6	
	NO-PYRR	96	(f)
Stewing base (Tunisia)	NDMA	12	
	NO-PYRR	6	(f)
	NO-PIP	45	
Coffee, instant (Canada)	NDMA	0.4	
	NO-PYRR	1.7	(d)
Milk, powdered (Canada)	NDMA	0.7	(a)
(U.S.A.)	NDMA	4.5	(b)
Cheese (Canada)	NDMA	0.7	
(Germany)	NDMA	5	(c)
(Denmark)	NDMA	1	
Beer (Canada)	NDMA	3	(d)
(U.S.A.)	NDMA	14	(h)
(Germany)	NDMA	68	(c)
Whisky	NDMA	1	(h)

(a) Sen *et al.*, 1980; (b) Hotchkiss *et al.*, 1980; (c) Spiegelhalder *et al.*, 1980; (d) Sen *et al.*, 1982; (e) Aidjanov and Sharmanov, 1982; (f) Poirier *et al.*, 1987; (g) Kawabata *et al.*, 1980; (h) Fazio and Havery, 1982.

nitrosamines at low concentrations have been developed, at least for routine use. A highly specific detector for nitrosamines (not entirely without artefactual interferences, however) is the Thermal Energy Analyzer (Fine and Rufeh, 1974), developed about 15 years ago, and in quite wide use now. It is most easily – and usually – used as a detector coupled to a gas-liquid chromatograph, and has very high sensitivity, of the order of picograms of nitrosamine in a sample. Coupling of this very sensitive detector to a liquid chromatograph, usually a high pressure system, has been achieved, but is more difficult and complicated. This is required for separation and identification of non-volatile N-nitroso compounds, preferably with the possibility of identifying the compound by mass spectrometry. In special circumstances success has been considerable, but the procedure is far from routine.

The alternative for non-volatile N-nitroso compounds is to convert them into volatile derivatives, such as trimethylsilyl esters or ethers, or fluorinated derivatives in the case of alkylnitrosoureas. Some of these procedures are quite involved, in the case of nitrosoureas for example, so that the precision of the measurements of concentration is questionable. It is apposite to mention that several non-volatile N-nitroso compounds that might be expected to form or to be present in foods or other environmental materials are potent carcinogens, certainly comparable with most of the volatile nitrosamines encountered. Furthermore, we know little of the concentrations of these non-volatile nitrosamines which might occur, and they could be considerably higher than those of volatile nitrosamines, which are usually found at low concentrations. Examples are hydroxylated nitrosamines and alkylnitrosamides, including alkylnitrosoureas. Although N-alkylamides are not uncommon, there are few reports of the occurrence of their N-nitroso derivatives, and some of them are not reliable.

2.1.2 Occupational exposures

The highest concentrations of N-nitroso compounds are found in industrial settings. Some chemical processes favour formation of N-nitroso compounds, although none were so designed deliberately. The problems arise because easily nitrosatable amines are able to encounter nitrosating agents. In the rubber industry many amines of this type are commonly used (IARC, 1982a; Lijinsky, 1990a), for example as accelerators of vulcanisation. These include derivatives of morpholine, 2,6-dimethylmorpholine, dimethylamine, diethylamine, di-n-butylamine, methylphenylamine, pyrrolidine and piperidine. All of these can react with nitrosating agents to form nitrosamines which are potent carcinogens, having a variety of organs in rats and hamsters as their targets. The

Table 2.2. *Human exposures to N-nitroso compounds*

Compound	Source of exposure	Reported highest exposure*/day
Nitrosodimethylamine	Leather tanning	470
	Rubber/tyre manufacture	1300
	Rocket fuel	360
	Pesticide manufacture	400
	Fish processing	8
	Cigarette smoker (2 packs)	1
	Bar	2
Nitrosodiethylamine	Rubber/tyre manufacture	50
	Cigarette smoker	0.5
Nitrosodi-*n*-propylamine	Herbicide formulation	150 ppm
Nitrosodiethanolamine	Cutting fluids	up to 3%
	Cosmetics	4 ppm
Nitrosomorpholine	Rubber/tyre manufacture	1200
	Leather tanning	20
Nitrosopyrrolidine	Rubber/tyre manufacture	2
	Cigarette smoker	1
	Bacon frying	5
Nitrosonornicotine	Cigarette smoker	12
	Snuff	200
Methylnitrosamino-3-pyridylbutanone (NNK)	Cigarette smoker	6
	Snuff	50

*Values are μg except where otherwise stated.

nitrosating agents with which these amines come into contact include nitrogen oxides in combustion gases and, in some cases, nitrosating agents used in processing of the materials. The nitrosating agent is sometimes an inorganic nitrite but can often be a nitrosamine, such as nitrosodiphenylamine, used as a retarder of vulcanisation. Nitrosodiphenylamine, in addition to being itself a bladder carcinogen in rats, is a powerful nitrosating agent. By transnitrosation this nitrosamine interacts with a secondary or tertiary amino compound to form an aliphatic nitrosamine (Challis and Osborne, 1973), in many cases a more potent carcinogen than nitrosodiphenylamine itself. Many nitrosamines and nitrosamides are able to act as transnitrosating agents in this way, especially when catalysed by nucleophilic anions, such as halide or thiocyanate. Such transnitrosation reactions are partially responsible for the nitrosamines found in rubber, which include the potent carcinogens nitrosomorpholine, nitrosodimethylamine and nitrosodi-*n*-butylamine. Typical concentrations of several nitrosamines associated with industrial processes are shown in Table 2.2.

It is notable that many nitrosamines which are excellent transnitrosating agents, such as nitrosoproline, nitrosohydroxyproline, nitrosodiphenylamine, 1-nitrosopiperazine and nitroso-N-methylpiperazine, are relatively weak carcinogens (Singer et al., 1978). However, when they transnitrosate a weakly basic amine they can form a potently carcinogenic nitrosamine, such as the ones found in factories in which rubber or leather is processed (Fajen et al., 1979).

The non-carcinogenic nitrosoproline and nitrosohydroxyproline, present in cured meats as a result of the action of nitrite on the amino acids, might act as stable nitrosating agents which can be precursors of the nitrosamines found in those products. They might also have a function as nitrosating agents in vivo in the stomach of people ingesting them. As will be discussed later, in vivo formation of nitrosamines is an important source of human exposure to these carcinogens.

The customary presence of nitrogen oxides in combustion gases of factories explains the common observation of nitrosodimethylamine in the atmosphere of many areas in which dimethylamine or its salts are used. This nitrosamine has also been observed in the atmosphere near and in factories (Fine et al., 1976). There is little doubt that other nitrosamines are also present in these circumstances, but analysis has usually focussed on nitrosodimethylamine. Dimethylamine and other amines are often employed to neutralise commercially important organic acids, such as dichloro- or trichloro-phenoxyacetic acid (Silvex) used as herbicides. Such materials often contain nitrosamines, due to exposure to nitrogen oxides. Another contributor is the sodium nitrite used (as a corrosion inhibitor) to line cans in which these organic amine salts are stored, and which also leads to nitrosation of the amine. There are many pesticides which may contain substantial concentrations of nitrosamines (Zweig et al., 1980). The nitrosodiethanolamine in tobacco derives from the diethanolamine salt of maleic hydrazide, which is sprayed on tobacco plants as an antisuckering agent, remains in the leaf, and is converted to nitrosodiethanolamine by reaction with nitrite formed during the fermentation of the tobacco.

Diethanolamine is a very widely used organic amine, inexpensive, a good solvent completely miscible with water and with many organic solvents, and non-volatile, in many ways an ideal alkali. Unfortunately, this has led to perhaps the most severe exposure of humans to carcinogenic nitrosamines, that to nitrosodiethanolamine in cutting oils. Modern synthetic cutting oils contain mineral oils and water, with triethanolamine as an emulsifier and sodium nitrite as a corrosion inhibitor. Triethanolamine and the diethanolamine it contains (Fig. 2.3), react with the nitrite over time to form nitrosodiethanolamine (even though the solution can be quite basic). Concentrations of nitrosodiethanolamine as high as

3% have been found in some samples (Fan et al., 1977b). This concentration of nitrosodiethanolamine represented a substantial carcinogenic risk to those exposed. Nitrosodiethanolamine is readily absorbed through the skin (Lijinsky et al., 1981a), so the risk exists from splashing on the skin, as well as from inhalation of the non-volatile nitrosodiethanolamine through aerosols. The number of people exposed to nitrosodiethanolamine in cutting oils is large, perhaps several millions in the U.S.A. alone, making this probably the greatest single source of exposure of the human population to nitrosamines. Nitrosodiethanolamine, although a weaker carcinogen than nitrosodimethylamine, is quite broadly acting, inducing in rats tumours of the esophagus, nasal mucosa, lung and kidney, as well as liver (Lijinsky and Reuber, 1984b); in hamsters it induced only tumours of the nasal mucosa (Pour and Wallcave, 1981; Lijinsky et al., 1988e).

2.1.3 Domestic sources

Another source of exposure to nitrosodiethanolamine, again mainly by absorption through the skin, is due to the use of triethanolamine as an emulsifier in creams and cosmetics of various kinds. In these cases the nitrosodiethanolamine is most likely formed from diethanolamine during processing or during storage in contact with inorganic nitrite, or with nitrosating agents such as the bacteriocide Bronopol (Ong and Rutherford, 1980). Concentrations of nitrosodiethanolamine in cosmetics (Fan et al., 1977a) are very much smaller (a few ppm) than those in cutting oils, but are nevertheless important because of the large amounts of these cosmetics used on the skin, and the vast numbers of people using them.

The complexity of reactions of hydroxyalkylamines with nitrite, particularly in the presence of other substances, suggests the possibility of the presence of other nitrosamines in crude mixtures containing these

Fig. 2.3. Formation of nitrosodiethanolamine from ethanolamines and nitrite.

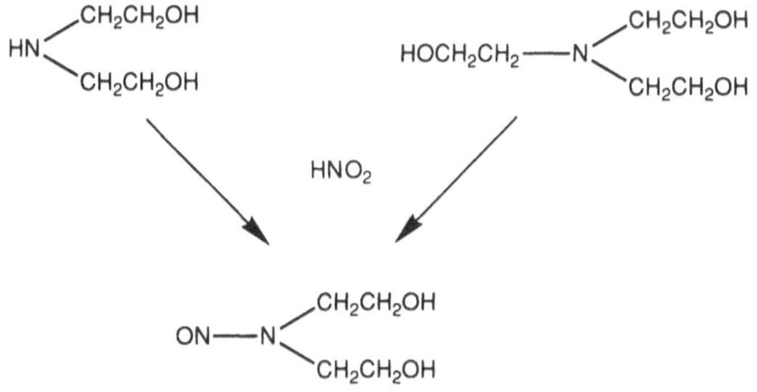

compounds. For example, in the presence of aldehydes a variety of cyclic oxygen-containing nitrosamines can be formed, including nitrosooxazolidines (Eiter et al., 1972) and nitrosooxazines. Some of these compounds have been tested for carcinogenic (Lijinsky and Reuber, 1982c; Lijinsky et al., 1984a) and mutagenic (Andrews and Lijinsky, 1980) activity, and have proved to be very potent. Analysis of commercial materials for these products has not been thorough, although 5-methylnitrosooxazolidine has been reported in cutting oil (Stephany et al., 1978). Our knowledge of human exposure to these less-common nitrosamines is incomplete, if not scanty. Like nitrosomorpholine, they are likely to be easily absorbed through the skin (Lijinsky et al., 1981a; Bronough et al., 1981).

Exposure to incidentally formed nitrosamines in toiletries is not limited to products derived from hydroxyalkylamines. Several long-chain aliphatic amines are used as detergents, including dimethyldodecylamine and dimethyltetradecylamine. These commercial chemicals contain secondary amines as impurities, and they are often exposed to nitrites (as well as to nitrogen oxides) during manufacture or storage. Both secondary and tertiary amines lead to formation of such compounds as methylnitrosododecylamine and methylnitrosotetradecylamine which are found in shampoos (Hecht et al., 1982a), and also probably occur in other products. They might form readily in alkaline solution because of the presence of aldehydes or ketones which act as catalysts. The methylnitrosoalkylamines mentioned are not among the most potent carcinogenic nitrosamines, but are reasonably active and induce tumours of the urinary bladder in both rats and hamsters (Lijinsky et al., 1981c; Althoff and Lijinsky, 1977). They, too, can be expected to be readily absorbed through the skin.

While no comprehensive assessment of human exposure to these preformed nitrosamines can be made, nor of the carcinogenic risk they pose, the exceptional potency of these systemic carcinogens mandates that they cannot be dismissed as insignificant, even when the concentrations at which they occur are small. A list of the nitrosamines and the products or locations in which they have been found is given in Table 2.2. Concentrations range from less than 1 μg per kg to milligrams per kg, and from less than 0.1 μg/m^3 to 25 μg/m^3 in some factories. These concentrations must be set against the findings that some nitrosamines have induced significant incidences of tumours in animals at concentrations of 0.1 ppm or lower, in deciding their importance as contributors to human cancer risk. It is not within the purview of this book, nor of the experience of the author, to make calculations or estimates of risk to

2.1.4 Tobacco

humans that are realistic. It is possible that none of us has sufficient knowledge to make estimates that are not likely to be wildly in error, perhaps by several orders of magnitude.

Because smoking is one of the most common human habits, with well known adverse health consequences, the identification of toxic and carcinogenic compounds in tobacco is a most important achievement. Polynuclear compounds were a focus of concern for many years, but the association of nitrosamines with tobacco use, including snuff and chewing tobacco, has provided fresh insight into the great importance of tobacco as a factor in human cancer causation. It was suspected that amines in tobacco and tobacco smoke could be swallowed and become converted to carcinogenic nitrosamines by nitrite in the stomach (Lijinsky and Epstein, 1970; Taylor and Lijinsky, 1975a), thereby contributing to induction of lung cancer. Nitrosodimethylamine was identified in tobacco smoke by Rhoades and Johnson (1972), but the opening of the problem began with the incisive and ground-breaking studies of Hecht, Hoffmann and their colleagues at the American Health Foundation.

They found that a number of nitrosamines were constituents of tobacco smoke, including nitrosonornicotine (Boyland et al., 1964a) and some that were previously unknown compounds, such as 4-(methylnitrosamino)-1-(-3-pyridyl)-1-butanone, which was abbreviated to NNK. This is one of the most potent carcinogens among nitrosamines, comparable with nitrosodimethylamine, and inducing a variety of tumours in rats, mice and hamsters, including tumours of liver, lung and possibly kidney, as well as in the nasal mucosa. It was shown that NNK was formed by nitrosative dealkylation (followed by ring opening) of nicotine, which is a major alkaloid of tobacco and a tertiary amine. Another product of nitrosative dealkylation of nicotine is nitrosonornicotine (Fig. 2.4), and this is a carcinogen also, but considerably less potent than NNK and induces mainly tumours of the nasal mucosa and esophagus in rats (Hoffmann and Hecht, 1985).

The most surprising finding of these investigators, and arguably the most important, was that NNK and nitrosonornicotine are also present at high concentrations in cured tobacco, although not in the growing plant. There is considerable evidence that the nitrate used in curing tobacco is the source of the nitrosating agent followed by bacterial reduction of the nitrate during fermentation of the tobacco and probably complemented by nitrogen oxides in flue gases used to dry the leaf. This finding has ominous implications for users of smokeless tobacco (chewing or snuff)

Table 2.3. *Nitrosamines in tobacco and tobacco smoke (highest concentrations reported)*

N-Nitroso-	In tobacco (snuff) (ng/g)	In cigarette smoke (ng/cigarette)
Dimethylamine	215	20
Methylethylamine	—	3
Diethylamine	—	8
Pyrrolidine	360	110
Morpholine	690	—
Diethanolamine	3,300	36
Nornicotine	30,000	950
Methylamino-3-pyridyl-butanone (NNK)	8,300	770
Methylamino-3-pyridyl-butanol (NNA1)	140	—
Anabasine	1,900	150
Anatabine	40,000	990

since they do not evade the known carcinogenic risks of tobacco smoking, but merely convert one risk to another. Some typical amounts of the tobacco-associated nitrosamines available in tobacco and in cigarette smoke are shown in Table 2.3. Although the idea is contested, it is difficult to believe that the nitrosamines in tobacco are not a major contributor to

Fig. 2.4. Nitrosation of nicotine and nornicotine.

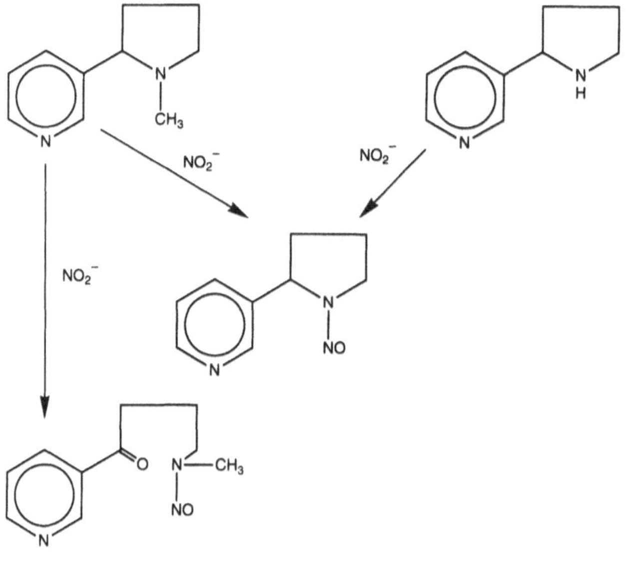

the known risk of users of smokeless tobacco developing cancer in organs of the oral cavity (Winn, 1984). While it is not reliable to compare risks in the area of carcinogenesis, because there are so many unknowns, it seems likely that the nitrosamine-related cancer risk of snuff-dippers or tobacco chewers might be larger than any other, comparable with exposure of workers in rubber factories.

There are many other nitrosamines in tobacco and in tobacco smoke, but their contribution to an increased risk of cancer in those exposed is considerably smaller than that posed by NNK and NNN. This is because the concentrations are lower and, except for nitrosodimethylamine, nitrosopyrrolidine and methylnitrosoethylamine, they are considerably less potent carcinogens in experimental animals (see Chapter 7). In addition to the compounds just named, they include NNA1 – the secondary alcohol corresponding to NNK – nitrosoanabasine and nitrosoanatabine, of which the latter two have long been known and have been tested for carcinogenic activity (Boyland *et al.*, 1964*b*).

2.2 Formation

Formation of *N*-nitroso compounds by reaction of a nitrosating agent with a secondary amine, as presented by Ridd (1961), is fairly straightforward. The pH optimum in aqueous solution is 3.4 with nitrite and the reaction is second-order in nitrite.

By a strange artefact, it was popularly supposed that tertiary amines do not react with nitrous acid or nitrosating agents, a myth which has entered many textbooks. However, it was recognised a long time ago that tertiary amines can be nitrosated to form nitrosamines (Solanina, 1906), and the truth was more recently uncovered (Hein, 1963; Smith and Loeppky, 1967); this reaction is of great importance in assessing the risk of exposure of humans to nitrosamines. The truth is that the reaction of tertiary amines with nitrous acid is normally very slow, even at elevated temperatures, as normally studied using strongly acid media. However, at pH 3 to 4 the reaction is favoured, and the rate then falls off somewhat at higher pH, but is measurable at pH near neutrality. In aqueous solution nitrosation of tertiary amines is usually much slower than of secondary amines (Jones *et al.*, 1974). Other nitrosating agents, such as nitrogen oxides, have long been known to interact with tertiary amines to form nitrosamines. For example, nitrosiminodiacetic acid was manufactured by passing nitrogen oxides into nitrilotriacetic acid as long ago as 1916. At about that time, the formation was reported of a number of products which, however, were not identified, by reaction of nicotine with

Table 2.4. *Types of nitrosatable amines important to humans*

Amine type	Examples	Use
Amine salts	Dimethylamine dichlorophenoxyacetate	Herbicide
	Maleic hydrazide, Diethanolamine	Anti-suckering agent
Alkaloids	Hordenine, gramine,	In barley
	nicotine	In tobacco
Alkanolamines	Triethanolamine	Emulsifiers
	Tri-*iso*propanolamine	Acid neutralisers
Alkylthiocarbamates	Thiram, Disulfiram, Zineb	Pesticides, animal repellants, rubber chemicals
N-Morpholino and piperidino compounds	Morfax, Santocure	Rubber manufacture
Methyl-*N*-alkylamines	Laurylamine	Cosmetics
N,*N*-Dialkylaminoethyl compounds	Chlorpheniramine Diphenhydramine Chlorpromazine	Antihistamines and other drugs
Pyrazolones	Aminopyrine Dipyrone	Analgesics
Amino acids	Proline, hydroxyproline	In cured meats

nitrosating agents. It is only comparatively recently that the products of nitrosation of nicotine have been identified (reviewed in Hoffman and Hecht, 1985).

2.2.1 Formation of N-*nitroso compounds* in vivo

In addition to the exposure of humans to preformed *N*-nitroso compounds, there is another source of exposure to them, and that is formation in the body through what is called endogenous nitrosation. The principal nitrosating agent in this context is the nitrite ion, usually derived from sodium or potassium nitrite in food, but also present in the saliva and probably other body fluids; nitrite was measured in blood almost a century ago. The nitrite in saliva arises from reduction of the nitrate always present in blood and lymph by bacteria in the mouth; bacterial activity in the infected bladder also is responsible for reduction of nitrate in urine (also always present) to nitrite, and might lead to formation of *N*-nitroso compounds in the bladder (Hawksworth and Hill, 1974). There is ready absorption of *N*-nitroso compounds from the urine through the bladder wall, and many experiments in which solutions of *N*-nitroso compounds have been administered intravesically have led to induction of tumours in a number of organs, such as kidney, lung, liver and esophagus (Lijinsky and Kovatch, 1989a; Hashimoto et al., 1974).

The interaction of a number of secondary and tertiary amines with nitrite in the stomach or in the bladder can lead to formation of N-nitroso compounds, the effects of which will vary with the chemical structure of the N-nitroso compound and the amount of it produced. The number of amines that could contribute to this exposure is large, including a great variety of drugs and medicines, as well as food constituents and food additives, which include a number of agricultural chemicals. A list of some of the pertinent types of compound is given in Table 2.4.

The commonly used secondary amine, piperazine, widely used for treatment of worm infestation of the gastrointestinal tract of both humans and domestic animals, presents a special and interesting case. Piperazine is easily nitrosated, and this reaction has been explored in chemical systems and in animals, including humans (Bellander and Osterdahl, 1983). The first product, formed at low concentrations of nitrite, is 1-nitrosopiperazine. Whether or not this is a carcinogenic nitrosamine is difficult to establish. As a bacterial mutagen, 1-nitrosopiperazine is inactive (Andrews and Lijinsky, 1984; Elespuru, 1976), whether tested as the free base or as a salt, with or without metabolic activation, although this cannot necessarily be equated with a lack of carcinogen activity, for reasons discussed elsewhere (Rao et al., 1984). More recently, a fortuitous finding during a test of 1-nitrosopiperazine, which gave a strange result, indicated that this aliphatic nitrosamine was a nitrosating agent (Love et al., 1977). Further investigation showed that pure 1-nitrosopiperazine underwent an intermolecular transnitrosation with formation of 1,4-dinitrosopiperazine and piperazine; this process also took place in solution.

An early test of the carcinogenic activity of 1-nitrosopiperazine showed it to induce a high incidence of tumours in rats when administered in large doses in drinking water (Garcia et al., 1970). However, examination of the stored compound (free base) after several years showed that a considerable portion of it had been converted to 1,4-dinitrosopiperazine (Love et al., 1977). This led to an investigation of transnitrosation by nitrosopiperazine and other nitrosamines, which has produced an increased understanding of the possible role of the process of transnitrosation by aliphatic nitrosamines in forming a powerful carcinogenic N-nitroso compound from a weak or inactive one.

While it is not worthwhile to speculate how large the relative risks are of exposure to N-nitroso compounds formed from amines in Table 2.4, because we are lacking in knowledge of the amounts formed in normal body conditions, some tentative estimates can be made based on observations in chemical systems. Even with the recently devised methods of measuring nitrosation capacity which have been widely applied in people by administering proline and measuring the amount of nitrosoproline excreted in urine (Ohshima and Bartsch, 1981), estimates of the carcinogenic risk presented by *in vivo* nitrosation are very imprecise. In particular, there is a background level of nitrosoproline excretion which, unlike the nitrosoproline arising from ingested nitrate, is not decreased by administration of large amounts of ascorbic acid. It must be assumed that the background nitrosoproline comes from a different source than nitrosation in the stomach, but its presence makes the use of urinary nitrosoproline as a measure of *in vivo* nitrosation capacity quite imprecise. It has been suggested by Tsuda *et al.* (1988) that nitrosothioproline (Fig. 2.5) might be a better marker, since it is easily measured and the rate of nitrosation of thioproline is perhaps two orders of magnitude greater than that of proline, leading to higher yields of the nitrosamino acid and more precise measurement.

Another complication in the assessment of risk in exposure of humans to *N*-nitroso compounds formed in the stomach is the presence of inhibitors of the reaction of amines with nitrite. Mirvish *et al.* (1972) reported that ascorbic acid blocked the reaction of oxytetracycline with nitrous acid. This inhibition was ascribed to the competitive reaction of ascorbic acid with nitrous acid, a reaction previously described by von Dahn *et al.* (1960). Subsequently, Mirvish and others have described the similar, but less effective, inhibition of formation of *N*-nitroso compounds by a variety of common substances, including tocopherols, tannins, glutathione and phenols; the latter have variously been reported to inhibit (Challis, 1973) and to accelerate (Challis and Bartlett, 1975) reaction of amines with nitrous acid. Ascorbate and erythorbate have long been used by meat processors in making bacon to increase the effectiveness of nitrite in the cure and to reduce the amount of nitrosamines formed (Fiddler, 1973).

In animal studies the effect of adding ascorbic acid to mixtures of

Fig. 2.5. Nitrosoproline, nitrosohydroxyproline, and nitrosothioproline.

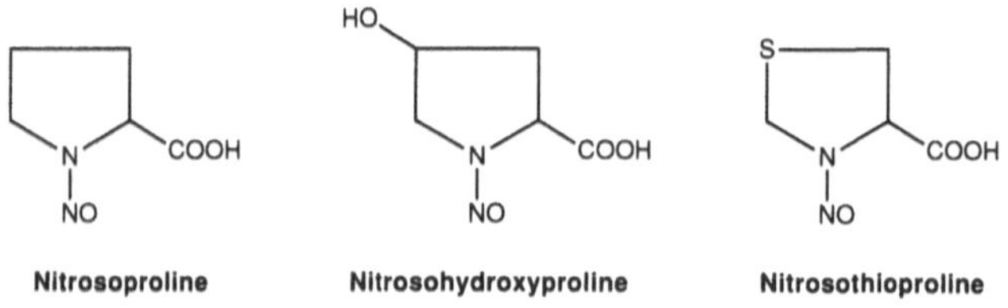

Nitrosoproline Nitrosohydroxyproline Nitrosothioproline

Formation 37

amines and nitrite fed has been to reduce the hepatotoxicity as a consequence of the nitrosamine formed *in vivo* (Kamm *et al.*, 1973). There are similar implications for the effects of inhibitors on nitrosation of amines at concentrations that give rise to tumours. However, it appears that quite large doses of inhibitor would be needed to eliminate formation of nitrosamines and reduce the carcinogenic risk to zero, but the concept is helpful in reducing exposure to carcinogenic *N*-nitroso compounds.

The significance for humans of nitrosation of secondary and of tertiary amino compounds (Fig. 2.6) can best be considered separately.

2.2.1.1 Secondary amines

The first evidence that *N*-nitroso compounds could be formed by nitrosation of amines in the stomach was obtained in humans by Sander and Seif (1969), who administered *N,N*-diphenylamine and detected nitrosodiphenylamine in the urine. This was not at the time considered a dangerous experiment, since nitrosodiphenylamine was thought to be a non-carcinogenic nitrosamine, unfortunately based on an unproved hypothesis and inappropriate data. But also in 1969, Sander and Bürkle reported an experiment in which two moderately basic amines, *N*-methylbenzylamine and morpholine were each fed to rats together with sodium nitrite, which resulted in high incidences of tumours of the esophagus and liver, respectively. These are the tumour types which are induced by feeding methylnitrosobenzylamine (Fig. 2.7) and nitroso-

Fig. 2.6. Nitrosation of tertiary amines.

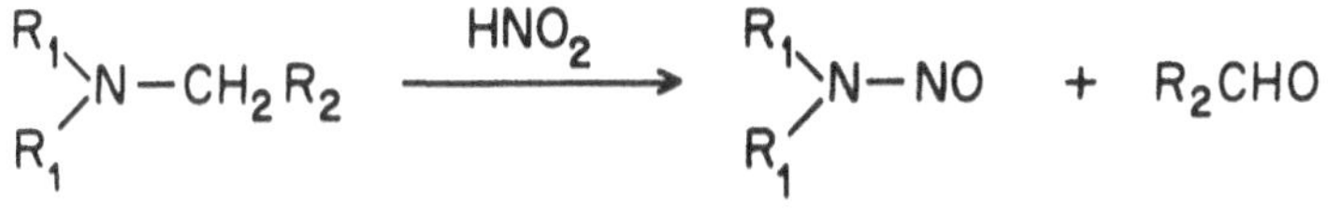

$$R_1\!\!>\!\!N-CH_2R_2 \xrightarrow{HNO_2} R_1\!\!>\!\!N-NO + R_2CHO$$

Fig. 2.7. Induction of tumours in rats by feeding N-methylbenzylamine with nitrite.

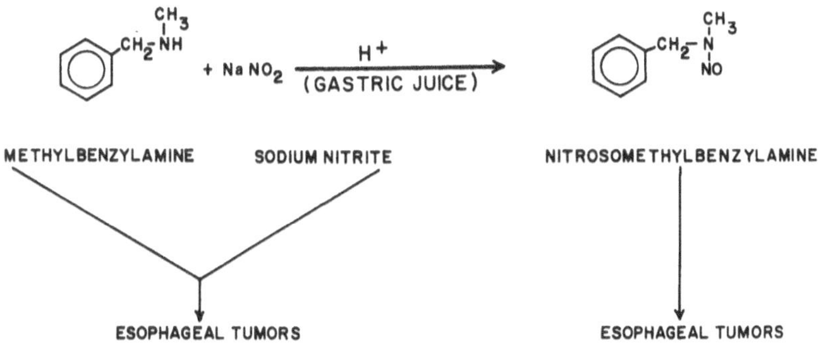

morpholine to rats, and the results are strong evidence that nitrosation of the two amines had occurred in the rat stomach; such nitrosation had already been demonstrated in gastric juice. These positive results superseded a previous similar but negative study by Druckrey et al. (1963), in which feeding of diethylamine together with nitrite to rats had failed to induce tumours; this is now known to be due to the slow reaction rate of the strongly basic diethylamine with nitrous acid. The implication of these and other results for the exposure of humans to endogenously formed N-nitroso compounds and the consequent increased risk of cancer was discussed by Lijinsky and Epstein (1970). Further experiments, by Sander, Lijinsky, Mirvish and others, fortified and extended the early findings, although the results were not always positive, and in some cases rather indirect. For example, the studies of Greenblatt et al. (1971) in mice reported an increase in lung adenomas following administration of a number of secondary amines together with sodium nitrite by mouth, the amine being mixed with food and the nitrite dissolved in drinking water. Lung adenomas are spontaneous in these mice, but are accelerated and increased in number following treatment with a variety of carcinogens, as shown in a general abbreviated test proposed by Shimkin et al. (1966) for carcinogenicity. Sander and Schweinsberg (1972) described tests of this type with a number of other secondary amines, and some gave rise to tumours that might be expected with the corresponding N-nitroso compound, but several, including N-methylguanidine and trimethylamine, failed to induce tumours under these conditions.

In the laboratories of this author a study of *in vivo* nitrosation of the widely used drug phenmetrazine demonstrated extensive conversion to the N-nitroso derivative (Fig. 2.8) when administered with nitrite to rabbits and rats (Greenblatt et al., 1972), although fortunately nitrosophenmetrazine was subsequently found to be non-carcinogenic (Lijinsky and Taylor, 1976c). A number of secondary amines fed to rats together with sodium nitrite gave rise to significant incidences of tumours; these

Fig. 2.8. Nitrosation of phenmetrazine.

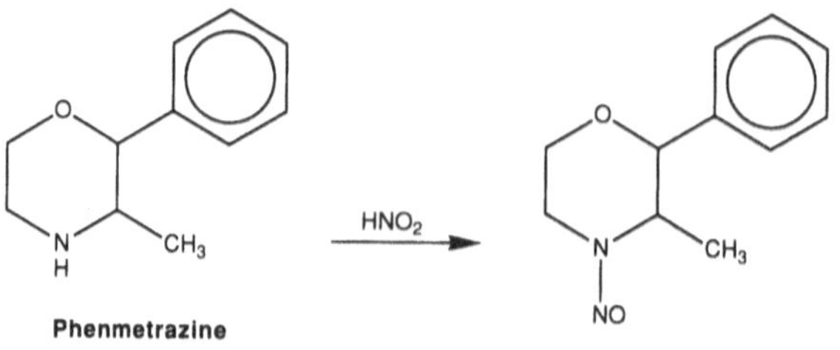

included heptamethyleneimine (Fig. 2.9), which induced a high incidence of lung carcinomas under these conditions (Taylor and Lijinsky, 1975a). This result suggested that part of the increased cancer risk of smokers might be due to swallowing amines in the tobacco smoke; these compounds could be nitrosated in the stomach by nitrite present in the gastric juice (from food or saliva). Earlier feeding studies of several amines and amino acids together with nitrite had failed to increase tumour induction in mice or rats; amines included pyrrolidine, piperidine, proline, hydroxyproline and nitrilotriacetic acid (Lijinsky et al., 1973b, 1974; Greenblatt and Lijinsky, 1972a, 1974; Garcia and Lijinsky, 1973). The secondary amino compounds tested by feeding together with sodium nitrite are listed in Table 2.5, with the positive or negative results.

Another secondary amino compound which was readily nitrosated was the anti-ulcer drug cimetidine, whose N-nitroso derivative although quite mutagenic, was not carcinogenic in rats or mice (Lijinsky and Reuber, 1984c). This was one of a large number of secondary amine drugs which would be expected to give rise to an N-nitroso derivative by reaction with nitrite in the stomach, but whose N-nitroso derivative is non-carcinogenic for steric or other reasons; many of these have been listed (Lijinsky, 1971). For example, the widely used drug methylphenidate (ritalin) yields a non-carcinogenic nitrosamine (Lijinsky and Taylor, 1976c).

Fig. 2.9. Induction of tumours in rats by feeding heptamethyleneimine with nitrite.

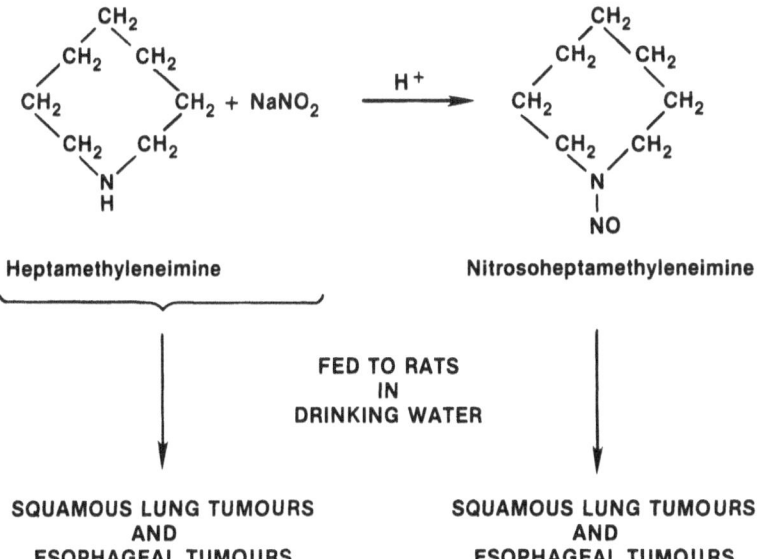

Table 2.5. *Feeding tests of secondary amines with nitrite*

Amine	Species	Tumours
Allantoin	Rat	Forestomach ±
Arginine	Rat	—
Carbaryl	Rat	—
Chlordiazepoxide	Rat	±
Diethylamine	Rat	—
Dimethylamine	Mouse	—
N,N'-Dimethylurea	Rat	Nervous system
Ethyleneurea	Rat	Nervous system, kidney
Ethylurea	Hamster	Spleen hemangiosarcoma
	Rat	Nervous system
Glycylglycine	Rat	—
Heptamethyleneimine	Rat	Lung, esophagus
Hydrochlorothiazide	Rat	—
N-Methylacetamide	Rat	—
N-Methylbenzylamine	Rat	Esophagus
N-Methylguanidine	Rat	—
N-Methylphenylamine	Mouse	Lung adenomas
Methylurea	Rat	Nervous system
Morpholine	Rat	Liver
Piperazine	Rat	—
	Mouse	Lung adenomas
Piperidine	Rat	—
	Mouse	Lung adenomas
Tolbutamide	Rat	Forestomach, liver

The mechanism of reaction of secondary amines with nitrous acid was definitively reviewed by Ridd (1961). In this reaction the unprotonated secondary amine reacts with nitrous anhydride (nitrogen trioxide) at an optimum pH near 3.4, near the pK_a for nitrous acid; the reaction is first-order in amine and second-order in nitrous acid, and the rate of formation of the nitrosamine is much more dependent on the concentration of nitrite-nitrous acid than on the concentration of amine. N-Alkylamides are usually neutral compounds and their reaction with nitrous acid is dependent on pH because the nitrosating agent is the nitrous acidium ion (Fig. 2.10), the concentration of which increases at lower pH. This reaction is, therefore, very dependent on pH and is first-order in both alkylamide and in nitrite-nitrous acid concentration. In formation of N-nitroso compounds in the stomach, therefore, the yield of product depends greatly on the pH of the stomach contents, the concentration of

Fig. 2.10. Decomposition of nitrous acid.

$$NO_2^- + H_2ONO^+ \rightleftharpoons 2HNO_2 \rightleftharpoons N_2O_3 + H_2O$$

nitrite and the concentrations of amine or alkylamide. Modelling this system becomes very complicated when the mixture of nitrosatable amino compounds consists of many components, and when the contents of the stomach are heterogeneous, as during digestion of a meal; in these conditions there might be areas of high concentration and areas of low concentration of reactants, the former being much more important because of the second-order parameter of nitrite concentration when amines are nitrosated.

2.2.1.2 Tertiary amines

The rediscovery that tertiary amines, as well as secondary amines, reacted with nitrite in acid solution (albeit in weakly acidic conditions), led to the examination of chemical formation of nitrosamines from a number of widely used drugs, food additives, etc., which are tertiary amines.

Nitrosation of tertiary amines shows an important difference in consequences from nitrosation of secondary amines. That is that tertiary amines usually form at least one nitrosamine that is a carcinogenic compound, in many cases a known potently carcinogenic nitrosamine. Many of these in wide use as drugs have been classified according to the common nitrosamine formed by reaction with nitrosating agents. For example the widely used analgesic aminopyrine (pyramidone) reacts more readily than most secondary amines to form nitrosodimethylamine (Lijinsky *et al.*, 1972*a*). Most of these simple nitrosamines formed are potent carcinogens which, even though formed slowly and in low yield, might present a considerable carcinogenic risk to people exposed frequently. An extensive list of tertiary amines that can be predicted to form carcinogenic nitrosamines by reaction with nitrosating agents was culled from the *Merck Index*, 9th Edition, and was part of the hearing record of a Committee of the Senate of the United States in 1976. Most of the compounds listed are not in wide use, and not of importance, but some that are commonly used as drugs and medicines, agricultural chemicals and food additives, as well as some food components, are discussed in a recent review of the subject (Lijinsky, 1990*b*).

Tertiary amines vary broadly in their reactivity with nitrosating agents, and it was widely believed that they did not react at all with nitrous acid (or nitrite in acid solution), a 'fact' that appeared in many textbooks of organic chemistry. Hein (1963) pointed out that there were several reports of nitrosation of tertiary amines which had been overlooked. Smith and Loeppky (1967) nailed down quite conclusively that tertiary amines – in their study tribenzylamine – did react with nitrous acid to form nitrosamines, but only above pH 3, not in strongly acid solution. Their paper caught the attention of many chemists studying nitrosamines as a new class of carcinogens, and gave impetus to a large series of studies of nitrosation of tertiary amines. One of the amines was the anti-suckering agent called 'Penar' used on plants, particularly on tobacco plants through which it could be inhaled as part of tobacco smoke. Its structure is N,N-dimethyldodecylamine, and it underwent ready nitrosation at elevated temperature in aqueous solution (Fig. 2.11) to form methylnitrosododecylamine (a bladder carcinogen in rats and hamsters) and nitrosodimethylamine, the two possible nitrosamine products; we prepared methylnitrosododecylamine in this way for testing in animals.

The mechanism of reaction of tertiary amines with nitrous acid is not well understood, unlike that of secondary amines. A study of the reaction of a number of tertiary amines with nitrite in mildly acid solution was published in 1972 (Lijinsky et al., 1972d), and a comparison was shown of the variation with tertiary amine structure of reactivity leading to formation of nitrosamines. Among the amines shown to react with nitrous acid was the common emulsifier triethanolamine, and it formed the carcinogen nitrosodiethanolamine which, as mentioned previously, is a common constituent of synthetic cutting oils (Zingmark and Rappe, 1976; Fan et al., 1977b), to which there is heavy human exposure. The yield of nitrosamine in these reactions was pH-dependent and there was a profound influence of nucleophilic anions, such as halide and thiocyanate; these are also catalytic in nitrosation of secondary amines by nitrous acid.

A number of naturally occurring alkaloids have been shown to undergo nitrosative dealkylation to form nitrosamines, by reaction with nitrous acid and other nitrosating agents. For example nicotine in tobacco has given rise to nitrosonornicotine and to the potent acyclic carcinogenic nitrosamine NNK, which is the parent of a number of other nitrosamines also of importance in tobacco carcinogenesis. The formation of these nitrosamines takes place during the curing of tobacco leaf, so that they are

Fig. 2.11. Nitrosative dealkylation of dimethyldodecylamine (Penar).

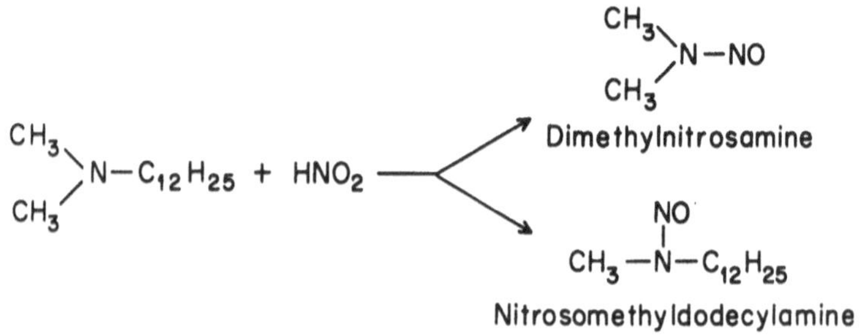

present in chewing or smokeless tobacco (Hoffmann et al., 1984a). They are also present in tobacco smoke, where they probably arise through reaction of nicotine with nitrogen oxides. The tobacco-related nitrosamines are surely the most important human carcinogens so far identified. Similarly a potent carcinogen, methylnitrosopropionitrile, has been reported as a product of nitrosation of the betel-nut alkaloid arecoline (Fig. 2.12), and found in the saliva of chewers of betel nuts (Prokopczyk et al., 1987).

The nitrosodimethylamine found in beer and whisky also arises through nitrosation by gaseous nitrogen oxides, this time in the flue gases used to heat the malt which is a prime ingredient of these alcoholic drinks. Mangino and Scanlan (1984) reported that the tertiary amine alkaloids hordenine and gramine in barley are the amines which, through nitrosation, form NDMA. Hordenine in particular reacts readily with nitrosating agents due to its enamine structure, which facilitates reaction.

Fig. 2.12. Nitrosation of arecoline.

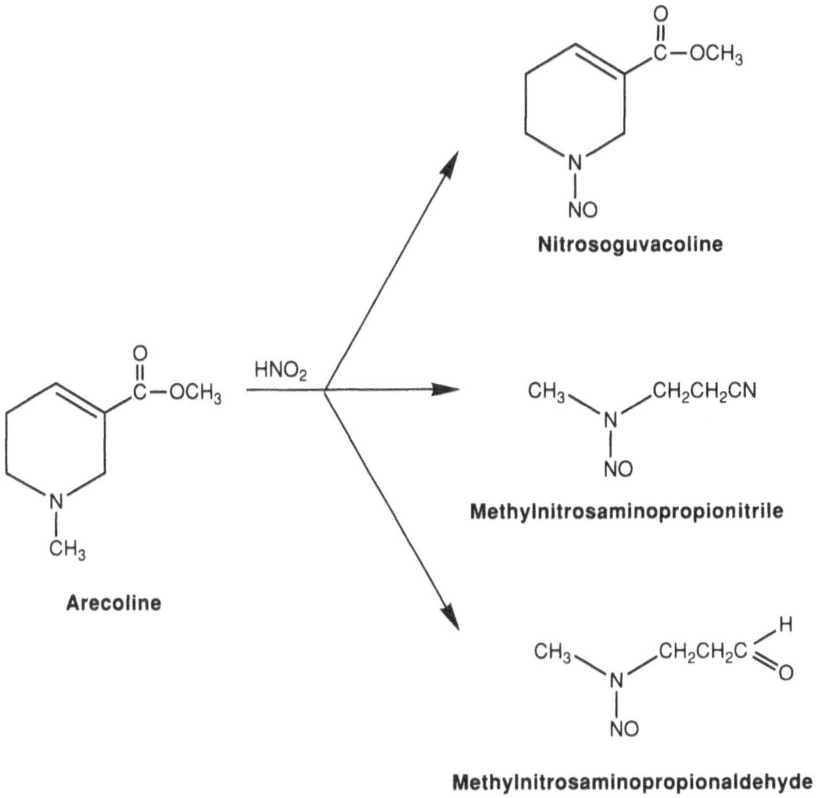

Probably the most important tertiary amines which could pose a carcinogenic risk to humans through formation of nitrosamines are many very widely used drugs. For example, the analgesic aminopyrine (Pyramidone) is a methylated enamine and reacts very readily with nitrous acid to form nitrosodimethylamine (Fig. 2.13); this reaction is as rapid as nitrosation of any secondary amine, and many samples of aminopyrine have been found to contain nitrosodimethylamine, presumably due to reaction with nitrogen oxides in the air. Administration of a single dose of aminopyrine with sodium nitrite orally to rats caused death from liver damage due to the NDMA formed in the stomach (Lijinsky and Greenblatt, 1972). Another tertiary amine drug of unusual structure which reacts exceptionally readily with nitrous acid is oxytetracycline (and presumably other tetracyclines of closely similar structure); a quantitative yield of nitrosodimethylamine was obtained in one reaction of oxytetracycline with sodium nitrite at pH 4 (Lijinsky et al., 1972b). Several other tertiary amine drugs of more conventional structure reacted considerably less readily with nitrous acid and, in similar conditions, gave much smaller yields of nitrosamines than aminopyrine or oxytetracycline, which would nevertheless be sufficient to increase the cancer risk of someone exposed to them.

Among the drugs examined were cyclizine, disulphiram, chlorpromazine, lucanthone, methapyrilene, chlorpheniramine, diphenhydramine

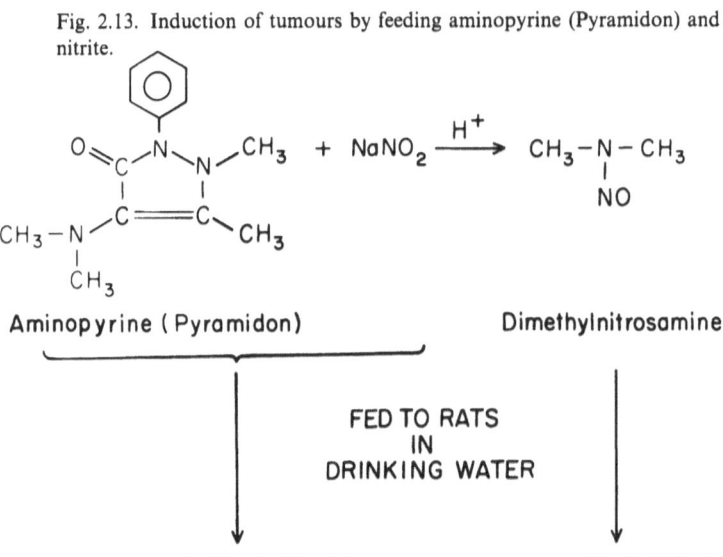

Fig. 2.13. Induction of tumours by feeding aminopyrine (Pyramidon) and nitrite.

Formation

and tolazamide (Fig. 2.14). In chemical reactions all of these gave rise to the expected nitrosodialkylamine (Lijinsky, 1974).

When these drugs and other amines were reacted with sodium nitrite in weakly acid solution and the reaction products were tested for mutagenic activity in the *Salmonella* system (McCann et al., 1975), there was frequently more mutagenic activity than would be accounted for by formation of the simple nitrosamine expected. This points to the

Fig. 2.14. Some nitrosatable tertiary amine drugs.

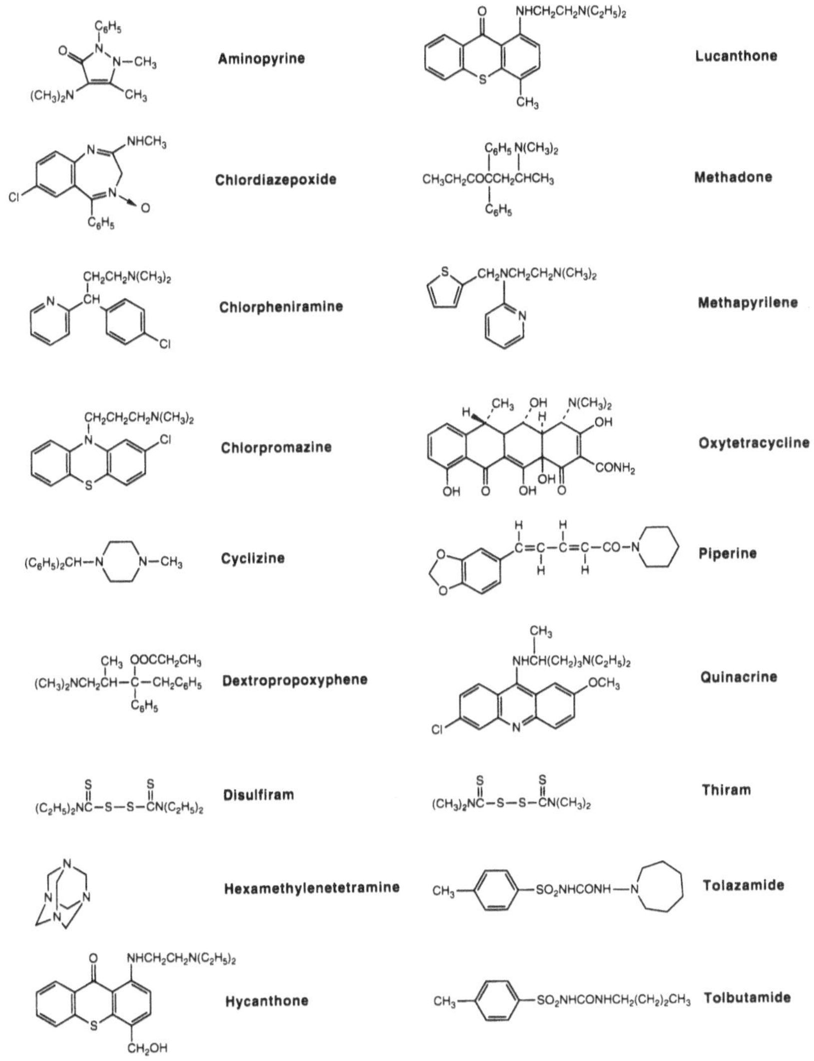

- Aminopyrine
- Lucanthone
- Chlordiazepoxide
- Methadone
- Chlorpheniramine
- Methapyrilene
- Chlorpromazine
- Oxytetracycline
- Cyclizine
- Piperine
- Dextropropoxyphene
- Quinacrine
- Disulfiram
- Thiram
- Hexamethylenetetramine
- Tolazamide
- Hycanthone
- Tolbutamide

probability that other nitrosamine products formed from these complex tertiary amines were contributing to the mutagenic activity, and presumably might also contribute to the carcinogenic activity of the products of these reactions. One difficulty in interpreting the mutagenicity results which is inherent in the 'black box' nature of bacterial mutagenesis systems is the usual failure to identify the mutagenic compounds in the complex reaction mixture. In addition some of the drugs themselves, such as lucanthone and quinacrine (Andrews *et al.*, 1980, 1984), are mutagenic, and some, including lucanthone and methapyrilene, proved to be carcinogenic. The reaction of methapyrilene with nitrite was studied in considerable detail (Mergens *et al.*, 1979). Because of the relatively large quantities of some of these drugs taken by people, often every day or several times a week for many years, it must be assumed that their reaction with nitrite in the stomach could lead to formation of sufficient carcinogenic nitrosamine to increase the risk of cancer.

Another large group of tertiary amines that are commercially important and entail in their use considerable exposure of humans, with possible important consequences because of nitrosamine formation, are agricultural chemicals. For example thiram (tetramethylthiuram disulphide) (Fig. 2.15) and its homologue disulphiram form the corresponding nitrosamines (NDMA and NDEA, respectively) in considerable yield when reacted with nitrite in mildly acid solution, although both have very

Fig. 2.15. Some nitrosatable agricultural chemicals.

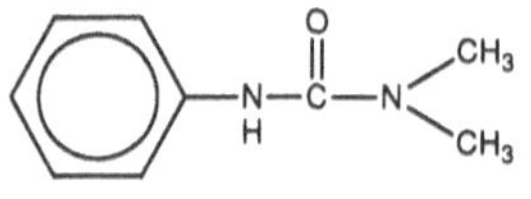

Fenuron

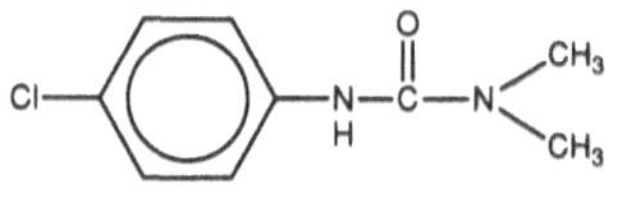

Monuron

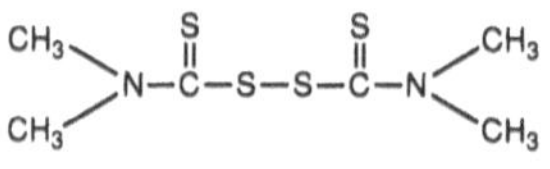

Thiram

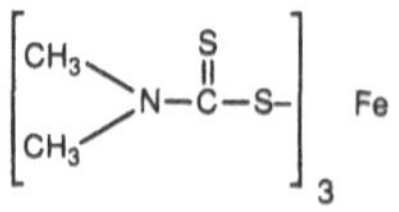

Ferbam

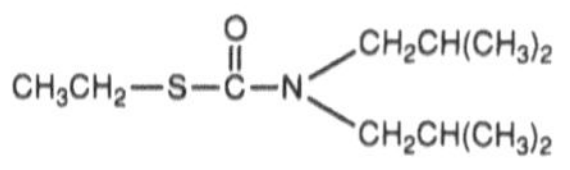

Butylate

low solubility in water (Elespuru and Lijinsky, 1973). Other agricultural chemicals in this category are several dialkylarylureas, such as fenuron, monuron and others; those examined (Elespuru and Lijinsky, 1973) gave rise to the expected nitrosodialkylamine in considerable yield, suggesting that people might be at risk even if exposed to small quantities of residues of them on crops.

Finally there are food components that are tertiary amines, such as piperine in pepper and trimethylamine-N-oxide in fish, that also react with nitrous acid to form, respectively, nitrosopiperidine and nitrosodimethylamine (Lijinsky *et al.*, 1972*b*; Lijinsky and Singer, 1974). There are other tertiary amines, known and unknown, in food, which can participate in the same reaction.

Many experiments have been designed during the past 15 years or so to gain some insight into the size of the problem posed by nitrosation of tertiary amines in the stomach to form carcinogenic nitrosamines. In addition to the chemical studies, which can suggest part of the answer, experiments have been undertaken in which an amine, together with sodium nitrite, has been fed to experimental animals, usually rats, for a prolonged period, usually most of the animals' lifetime. The compounds studied in this way, most of which have been mentioned already, are listed in Table 2.6. The outcome of the experiments is stated as tumours induced by the treatment, or no tumours induced; the nitrosamine formed is also named if it has been identified.

It can be seen that many of the tertiary amines did not form sufficient nitrosamine by reaction with nitrite in the stomach to give rise to increased incidence of tumours within the lifetime of the rat. However, since most of these compounds were shown in chemical systems to form carcinogenic nitrosamines, it cannot be assumed that their ingestion by humans poses no risk of exposure to these carcinogenic nitrosamines. This is always a difficult gap to bridge, between tumour induction in experimental animals and increased cancer risk in humans, although exposure of humans can occur over a lifespan of up to 70 years. The carcinogenic effects of early exposures can last almost a lifetime, and humans are exposed to more than one source of carcinogens, which can act additively or perhaps synergistically.

There is less question about the interpretation of those treatments which do result in an increased incidence of tumours related to the treatment, of which there are also many listed. For example, aminopyrine, chlorpheniramine and oxytetracycline produce sufficient liver tumours in rats when fed together with nitrite that the combination can be considered a serious risk. Similarly with disulphiram and thiram which give rise

Table 2.6. *Feeding tests of tertiary amines with nitrite*

Amine	Species	Tumours
Aminopyrine	Rat	—
Chlorpheniramine	Rat	Liver
Chlorpromazine	Rat	—
Cyclizine	Rat	—
Dimethyldodecylamine	Rat	Bladder ±
Dimethyldodecylamine-N-oxide	Rat	Liver
Dimethylphenylurea	Rat	—
Diphenhydramine	Rat	—
Disulphiram	Rat	Esophagus
Hexamethylenetetramine	Rat	—
Lucanthone	Rat	Liver with or without nitrite
Methapyrilene	Rat	Liver with or without nitrite
Monuron	Rat	Liver ±
Oxytetracycline	Rat	Liver
Piperine	Rat	Forestomach ±
Thiram	Rat	Nasal cavity
Tolazamide	Rat	—
Trimethylamine-N-oxide	Rat	—

through *in vivo* nitrosation to tumours of the esophagus (Lijinsky and Reuber, 1980*d*) and nasal cavity in rats (Lijinsky, 1984*a*). The results with some other tertiary amines are statistically significant, but more marginal, including diphenhydramine, dimethyldodecylamine-N-oxide, and piperine. It is interesting that in some of these experiments the female control rats given sodium nitrite alone had an elevated incidence of liver tumours (Lijinsky *et al.*, 1983*a*), while usually it was male rats given combinations of tertiary amines and sodium nitrite that gave the greatest increase in incidence of tumours, especially when liver tumours were the consequence (Lijinsky, 1984*b*). With methapyrilene and lucanthone, which were carcinogenic and induced liver tumours in rats by themselves, the additional exposure to nitrosamines formed by *in vivo* nitrosation did not measurably increase the carcinogenic effect when nitrite was fed along with the drug; the additional effect of the nitrosamine might have been revealed, however, in different conditions.

The evidence is convincing, therefore, that human exposure to nitrosatable secondary or tertiary amines by ingestion when nitrite is present in the stomach (from food or in saliva) can increase cancer risks through formation of N-nitroso compounds. The risk, however, is difficult to quantify, although attempts to assess 'nitrosation capacity' by feeding known amounts of proline and measuring the nitrosoproline excreted in the urine, which circumvents measurement of nitrite concentration, have been discussed above.

While the mechanism of formation of nitrosamines from secondary amines and nitrosating agents is fairly well understood, this is certainly not the case with tertiary amines. In spite of the attempts of many investigators to elucidate the mechanism of nitrosation of tertiary amines, beginning with Smith and Loeppky (1967), little is known beyond the composition of the mixture of products, which include nitrosamines, carbonyl compounds and nitric oxide. It has not been possible to devise a plausible stoichiometric equation for the reaction (Lijinsky and Singer, 1974; Singer, 1980). Although the reaction has been characterised as nitrosative dealkylation (Lijinsky et al., 1972d), and the formation of an iminium ion as an intermediate has been plausibly postulated (Fig. 2.16), it is by no means certain that this is sound. The intermediacy of formation of a secondary amine by dealkylation is similarly postulated, but with inadequate evidence. This is certainly not the mechanism by which nitrosodimethylamine is formed from aminopyrine, since that reaction is much more rapid than nitrosation of dimethylamine under the same conditions. There is no doubt that nitrosation of tertiary amines is a multi-step process, and possibly several reactions or pathways occur concurrently, which makes elucidation of the kinetics of the reaction a daunting task. When this is compounded by the uncertainties in dosimetry in exposure of humans or animals, the precise estimation of risks of human exposure to N-nitroso compounds formed *in vivo* is an impossible task; however, the risk is not zero.

Fig. 2.16. Nitrosative dealkylation of tertiary amines.

$$R_3N + N_2O_3 \rightleftharpoons R_3\overset{+}{N}-NO + NO_2^-$$

$$R_3\overset{+}{N}-NO \longrightarrow \dashrightarrow R_2\overset{+}{N}=CH-R'$$

$$R_2\overset{+}{N}=CH-R' \xrightarrow[H_3O^+]{N_2O_3} R_2NNO + R'CHO$$
$$\uparrow$$
$$R_2NH + R'CHO$$

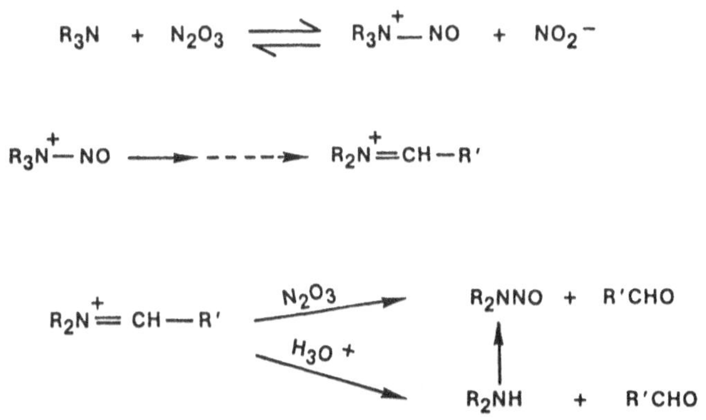

2.2.1.3 Sites of nitrosamine formation

Although the focus has been on formation of N-nitroso compounds in the stomach with its acid medium, reactions of amines with nitrite and other nitrosating agents occur elsewhere. For example, nitrite formed from urinary nitrate by bacterial reduction can react with excreted amines in the bladder (Hawksworth and Hill, 1974), although not necessarily through bacterial nitrosation; the N-nitroso compounds so formed are readily absorbed through the bladder wall and can act systemically to induce tumours in distant organs, as has been shown with a number of intravesically administered nitrosamines (Hashimoto *et al.*, 1974; Thomas *et al.*, 1988; Lijinsky and Kovatch, 1989*a*). The excretion of nitrosopiperazines has been investigated in the urine of humans who have taken the anti-helminthic piperazine (Bellander and Osterdahl, 1983).

Gas-phase nitrosation is probably responsible for a good deal of atmospheric contamination with nitrosamines, especially nitrosodimethylamine, and has been extensively studied by Challis and his colleagues (Challis and Kyrtopoulos, 1979). The formation in solution of oxygen-containing cyclic nitrosamines from hydroxyamines, aldehydes and nitrite has been extensively studied by Eiter *et al.* (1972), as has the catalytic role of carbonyl compounds, especially aldehydes, in reaction of secondary amines with nitrite, even in alkaline conditions (Keefer and Roller, 1973). These broaden our understanding of the many sources of human exposure to N-nitroso compounds; but we still do not know enough to make anything but educated guesses about the magnitude of their contribution to our risk of cancer.

2.3 Methods of detection and analysis

Progress in understanding the importance of human exposure to N-nitroso compounds, particularly nitrosamines, has depended on identifying them in the environment and on the availability of methods for measuring them. Most methods are closely related to their chemical properties. Twenty years ago a committee of the W.H.O. International Union Against Cancer (UICC) meeting in Kingston, Jamaica issued a report *The Quantification of Environmental Carcinogens*, edited by Philippe Shubik, in which it was stated that the minimum detectable concentration of nitrosamines was approximately 1 ppm. The analytical methods were cumbersome and involved polarography as used by the Carshalton M.R.C. group in their original studies of nitrosamine toxicology (Heath and Jarvis, 1955), or photochemical cleavage of the N-NO bond followed by colourimetric measurement of nitrite, or by gas chromatography following removal of interfering substances. None of these methods was sensitive, nor specific for *N*-nitroso compounds, but all have been used successfully in particular situations, such as investigation of metabolism of a nitrosamine, often in conjunction with radiolabelling of the nitrosamine.

Many of these methods were used for a long time, with variations designed to improve reliability and specificity. Thin layer chromatography (TLC) followed by decomposition of the *N*-nitroso compounds in the fractions with hydrobromic acid was used extensively by Preussmann and his group in Germany. Gas chromatography combined with a Hall, Coulson or other nitrogen-specific detector has been used. The perils of using a non-specific detector were revealed to the author who was estimating the yield of nitrosodimethylamine from reaction of tertiary amines with nitrous acid in acetic acid solution. When in some reactions the yield was twice theoretical the procedure was evaluated carefully and it transpired that traces of acetic acid in the neutralised extract were responsible, because in the columns used acetic acid and nitrosodimethylamine had the same retention time.

There are numerous reports in the literature of discoveries of nitrosamines at unexpectedly high concentrations, which seem likely to have been due to inappropriate or insufficiently selective analytical procedures. For example, nitrosamines were reported in vegetable oils (Hedler and Marquardt, 1974), in products of the Maillard reaction of amino acids and carbohydrates (Devik, 1967; Heyns and Koch, 1971; Heyns and Röper, 1974), in the human vaginal vault (Harington *et al.*, 1973) and in alcoholic drinks of certain regions of South Africa (McGlashan *et al.*, 1968). The latter claim was an artefact due to the coincidence of chromatographic retention time of the naturally occurring furfural and nitrosodimethylamine. It was soon realised that all of the chromatographic analyses for nitrosamines were subject to artefacts and that confirmation of the identity of the compound by mass spectrometry was needed, although the rarity of this expensive equipment made it a difficult requirement to satisfy.

The realisation that one of the principal sources of exposure to *N*-nitroso compounds was food that had been cured by treatment with sodium nitrite (Lijinsky and Epstein, 1970), such as bacon, ham and sausages, led to considerable interest in the analysis of such foods for nitrosamines, and methods were soon developed for determining them by Fazio, Sen, Issenberg, Tannenbaum, and others. The methods were based on selective extraction of nitrosamines with solvents such as chloroform and methylene chloride, concentration of the extracts without loss of the often-volatile nitrosamines and chromatographic separation, preferably coupled with mass spectrometric identification. The establishment of general patterns of behaviour of nitrosamines in the mass spectrometer was important in this development (Rainey et al., 1978). It is unfortunate that most alkylnitrosamides, including alkylnitrosoureas and alkylnitrosocarbamates, are too unstable to give molecular ions in the mass spectrometer. Various methods of derivatising these N-nitroso compounds, which are often also applicable to nitrosamines – especially non-volatile ones – have been developed and applied. Other chromatographic methods to deal with alkylnitrosamides have been developed by Singer et al. (1977), Iwaoka and Tannenbaum (1976) and others; they involved cleavage of the N-NO bond. However, alkylnitrosamides are not commonly found in the environment, so application of these methods has been mainly in a research setting, rather than for monitoring.

High pressure liquid chromatography (HPLC) has become the dominant method for resolving mixtures of N-nitroso compounds, because it can deal with both volatile and non-volatile compounds. Separation is so good on suitable matrices that rotamers of asymmetric nitrosamines are easily seen (Farrelly et al., 1988), and have sometimes been regarded erroneously as two different compounds. The UV absorption bands of N-nitroso compounds due to the nitroso group provide easy detection of fractions eluting from HPLC columns. Coupling of liquid chromatographs to a mass spectrometer has proved quite difficult, so that confirmation of identity of an N-nitroso compound separated by this means is not as easy as that of compounds separated by gas chromatography, but it can be done.

By far the most useful method for detecting and measuring an N-nitroso compound has been the Thermal Energy Analyzer (TEA), developed by Fine and Rufeh (1974) of the Thermo-Electron Corporation. This instrument, which is a detector, uses the transient excited form of NO produced by pyrolysis of N-nitroso compounds, which decays with emission of discrete infrared radiation, to detect and measure the N-nitroso compound introduced into its reactor. As pointed out by Walters (1983) the method is not foolproof, and there are many compounds other than N-nitroso compounds that release excited NO under those conditions, including nitrite and nitrate esters, nitrolic acids, C-nitroso compounds, etc., many of which could be present in the same sources as contain the N-nitroso compounds. Therefore, good extraction methods and mass spectrometric confirmation are important in analysis. An enormous advantage of the TEA is its great sensitivity, down to a level of nanograms or picograms of NO-forming compound. Artefacts are not uncommon and must be avoided, for example the use of too high an injection temperature into the instrument, which ignores the sensitivity of nitrosamines to decomposition near 200 °C; artefactual reporting of impurities in nitrosamine standards was common in the early years of use of the TEA.

Now it was possible to undertake surveys of food, air in factories and elsewhere in the environment, because analysis was rapid and the limits of detection were much lower than previously, permitting easier sampling. The applications were particularly appealing for air and water sampling, since relatively large volumes could be extracted and the concentrates then analysed by chromatography, coupled with detection using the TEA. At that time gas chromatography was used, limiting the identification of *N*-nitroso compounds to those that are volatile. It is still difficult to carry out comparable analyses of non-volatile *N*-nitroso compounds, because of the problems in coupling a liquid chromatograph (such as HPLC) to the TEA.

One of the important applications of the TEA to practical problems, taking advantage of its sensitivity and specificity, was analysis of blood for nitrosamines. Thus, Fine *et al.* (1977) were able to estimate the amount of nitrosodimethylamine formed in a human who had eaten a meal of bacon, lettuce and tomato, with a bottle of beer to follow. Unfortunately, the controls were not adequate so that the estimate is suspect, and more or less nitrosamine than that measured could have been formed in the stomach. Similarly, analysis of a small sample of blood or other body fluids facilitated the measurement of the time course of absorption of nitrosomorpholine and nitrosodiethanolamine through the skin, and from the digestive tract (Lijinsky *et al.*, 1981*a*). This avoided the use of the radiolabelled compounds, which would have been much more expensive. Further applications of these analytical techniques have been limited, but Keefer and his associates have used them in their studies of nitrosamine pharmacokinetics, which are such an important parameter of organ-specific carcinogenesis (Mico *et al.*, 1985).

One impediment in the use of the TEA response alone as a method of identification of *N*-nitroso compounds, is that the instrument is not entirely specific to them. However, many of the TEA-responsive compounds that are not *N*-nitroso compounds are quite esoteric and unlikely to be encountered (Walters, 1983). In some cases mass spectrometry might not help, since the other positive compounds are isomeric with *N*-nitroso compounds and have similar mass spectra. Nevertheless, the TEA coupled with chromatography provides the best means of identification and estimation of *N*-nitroso compounds, particularly at low concentrations, that is available at present. It is possible that modifications will improve the specificity of those systems for detecting N-nitroso compounds, without sacrifice of sensitivity. Unfortunately, more specific methods, such as NMR spectrometry, require much larger samples of compound than are likely to be found in the environment.

As has been described elsewhere, TEA-chromatography has been used on a large scale for surveying factories and food and water sources for the presence of N-nitroso compounds, and much of our information about their distribution has come from such surveys. Examination of household products and tobacco products has also benefited greatly from this sensitive detection system.

3

Chemical properties of *N*-nitroso compounds

3.1 Introduction

In the burst of zeal in organic chemistry at the end of the nineteenth century every derivative of every new compound was prepared and catalogued, or so it seems. Yet the nitroso derivatives of secondary amino compounds were of so little interest that only a handful of them are listed alongside the analogous secondary amine in *Heilbron's Dictionary of Organic Compounds*, Revised Edition of 1953. Nitrosodiethylamine (first reported by Geuther in 1863) has an entry of its own, as has nitrosodimethylamine (first reported by Van Romburgh in 1886). The remaining few include nitrosopiperidine, nitrosopyrrolidine, nitrosomethylaniline and dinitrosopiperazine, but not several other compounds known then, such as nitrosoazetidine, reported by Marckwald and Droste-Huelshorff in 1898. The lack of interest was as likely because most nitrosamines were oils at room temperature, and of no use for characterising the parent, as because they were not known to have important biological properties – or chemical properties. Nitroso-*N*-methylurethane was reported in 1894 (von Pechmann) and nitroso-*N*-methylurea in 1888 (von Brüning). Neither of the corresponding *N*-ethyl derivatives was considered worth mentioning. In accordance with the long line of discoveries of toxic or pharmacological effects of groups of chemicals through the observation of the effects of accidental or incidental exposure of humans or animals, interest in *N*-nitroso compounds awaited Barnes and Magee's (1954) report of the hepatotoxicity of nitrosodimethylamine in humans.

There was little interest in the chemistry of *N*-nitroso compounds for almost a century following the first description of a compound of this structure. They lay in a backwater of chemistry, only surfacing when one was found to have interesting properties, for example the finding that methylnitrosourethane was converted to diazomethane, a very useful intermediate, in alkaline solution (von Pechmann, 1894), and later that a similar reaction took place with methylnitrosourea (Werner, 1919). Similarly, the potent mutagenic properties of methylnitrosonitroguanidine (MNNG), when tested in bacteria, led to an interest in its properties as a methylating agent; it is also converted to diazomethane by alkali (McKay, 1948). Both compounds, because of their properties, had a profound influence on thinking about carcinogenesis by N-nitroso compounds.

There are very few naturally occurring N-nitroso compounds and they include streptozotocin (from *Streptomyces*) (Herr *et al.*, 1967).

Nitrosamines were occasionally used as solvents, but general knowledge of them by chemists was restricted to their formation by reaction of secondary amines with nitrous acid. For generations, nitrosamines were the yellow oil that formed in mixtures of primary, secondary and tertiary amines, demonstrating the presence of secondary amines, and into which they could be reconverted by acid hydrolysis. Nitrosodimethylamine has been used to prepare 1,1-dimethylhydrazine, a rocket fuel. It was the use of nitrosodimethylamine as a solvent which led to the earliest report of its toxicity to humans, by Freund in 1937, and was also the source of exposure of workers to the nitrosamine reported by Barnes and Magee (1954). This episode began the interest in nitrosamines which continues to the present.

As is frequently the case, interest in the biological properties of N-nitroso compounds has led to an increase in interest in other aspects of these compounds, including their chemical properties and characteristics, as well as in mechanisms by which they are formed and can be synthesised. An impressive amount of information about these compounds has been assembled, most of which is not very helpful towards possible explanations of their carcinogenic and mutagenic activities. This will be discussed in some detail in later chapters.

The most notable property of methylnitrosourea is its conversion to diazomethane in the presence of alkali, the other products being bicarbonate ion and ammonia. Nitroso-N-methylurethane is similarly unstable in alkali, again producing diazomethane together with bicarbonate ion. MNNG also decomposes in alkali with formation of diazomethane, together with nitrourea (McKay and Wright, 1947). Each of these alkylnitrosamides is the simplest and smallest of a large homologous series containing different alkyl groups, but all of which decompose in alkali with generation of a diazoalkane or diazoalkane precursor. Alkylnitrosamides form metal alkanediazotates which are precursors of azoxyalkanes (Moss *et al.*, 1972). Decomposition of cyclic nitrosoureas or nitrosocarbamates is more complicated, but essentially similar to the acyclic compounds. It is fairly certain now that formation of a diazoalkane is not the mechanism by which these compounds exert their biological effects, toxic, mutagenic or carcinogenic.

3.2 Solubility and partition

Only the N-nitroso compounds of lowest molecular weight are soluble in water to any appreciable extent. Most of those containing three carbon atoms or less are freely soluble in water, including nitrosodimethylamine, methylnitrosoethylamine, methylnitrosourea and methylnitrosourethane. Beginning with nitrosodiethylamine, solubility in water decreases, and nitrosodi-n-propylamine has low solubility. The alkylnitrosamides with alkyl groups larger than ethyl (C-2) have solubilities in water that rapidly decrease as the molecular size increases. Almost all nitrosamines having unsubstituted alkyl groups have low solubility in water, for example that of methylnitroso-n-octylamine is less than 200 mg per L; nitrosamines larger than this are essentially insoluble in water and experiments with them are not possible in aqueous solution. Exceptions are nitrosamines with alkyl groups containing carbonyl or hydroxyl functions, for example, nitrosodiethanolamine, nitroso-bis-(2-hydroxypropyl)amine and nitrosobis-(2-oxopropyl)amine, all of which are very soluble in water. The bladder carcinogen butylnitroso-4-hydroxybutylamine is soluble in water, in contrast with the low solubility of nitrosodi-n-butylamine. Nitrosamines containing carboxyl functions are very soluble in water, whether cyclic or acyclic. Excepting nitrosoazetidine, nitrosopyrrolidine, and the oxygen-containing cyclic nitrosamines, nitrosomorpholine and nitrosooxazolidine, cyclic nitrosamines have low solubility in water, unless they contain hydroxyl or carboxyl groups.

Most nitrosamines are very soluble in the common organic solvents, such as alcohols, ketones, esters and halogenated hydrocarbons. There are a few exceptions, for example alkylnitrosoureas are largely insoluble in methylene chloride or chloroform, whereas alkylnitrosocarbamates are very soluble in the chlorinated solvents. This difference facilitates the separation of hydroxyalkylnitrosoureas from nitrosooxazolidones which are always present as impurities formed during the preparation of the former (Lijinsky and Reuber, 1983b).

In addition to their solubility, the partition characteristics of N-nitroso compounds between two solvents are likely to be important arbiters of their toxic and carcinogenic properties. Many nitrosamines, whether very soluble in water or not, have favourable partition coefficients between chloroform or methylene chloride and water. As an example, the preferred solvent for extracting nitrosodimethylamine from the aqueous media in which it is often found is methylene chloride, into which its partition coefficient is greater than 1 to 1 (it is less than 1:1 in other solvents, including chloroform). Most other nitrosamines have favourable partition coefficients into most organic solvents versus water, and a study of a selected number of them attempted to relate 'liposolubility' to carcinogenic effectiveness in certain organs, for example the esophagus of rats (Singer *et al.*, 1977); partition into lipid solvents did not, however, explain why the same nitrosamines failed to induce tumours of the esophagus in hamsters. The same failing pertains to the studies of Mirvish *et al.* (1976) of the air/water partition of a number of nitrosamines which induce tumours of the esophagus in rats and for which there were correlations for some compounds.

It is probable that partition into lipids (as, for example, across the plasma membrane of cells) plays a role in facilitating the toxic or carcinogenic activities of N-nitroso compounds (and other carcinogens) – thereby modulating their potency – but the nature of this effect is not known. Water-soluble hydroxylated cyclic nitrosamines partition quite favourably into chlorinated solvents, such as chloroform, but nitrosamino acids do not; the latter are well extracted from aqueous solutions with ether or ethyl acetate. Some acyclic nitrosamines with hydroxylated alkyl groups are not readily extracted by chlorinated solvents, but are readily extracted by ethyl acetate, with the singular exception of nitrosodiethanolamine, which has an unfavourable partition coefficient in ethyl acetate versus water. Since nitrosodiethanolamine and nitrosobis-(2-hydroxypropyl)amine induce tumours, depending on the dose and dose rate, in a similar variety of organs of rats and hamsters to those induced by non-hydroxylated analogues, the partition properties of the nitrosamines do not seem to determine the target organs of these carcinogens.

This is further borne out in studies of alkylnitrosoureas, the simplest of which, methylnitrosourea, induces tumours of the nervous system, uterus and mammary gland in rats, depending on the conditions of administration, and of the spleen in hamsters, but does not have a favourable partition coefficient into organic solvents from water. Ethylnitrosourea and homologues with larger alkyl groups have increasingly favourable partition into organic solvents, as do dialkyl- and trialkyl-nitrosoureas, yet all seem to induce tumours with similar effectiveness in several organs, the selection of which seems to depend largely on the identity of the alkyl group neighbouring the N-nitroso function (see Chapter 7). Methylnitrosourea, dimethylnitrosourea, methylnitrosoethylurea and trimethylnitrosourea (Fig. 3.1) show greatly increasing solvent–water partition as the molecular weight increases, yet all have the same relatively limited range of target organs for their carcinogenic effect in rats. Hydroxyethylnitrosourea is something of an anomaly. It is very water-soluble and partitions into ethyl acetate, but not at all into chloroform, yet it is one of the most broadly acting of all carcinogens, inducing tumours in about a dozen organs of rats, as many as six types of tumour in a single animal (Lijinsky and Reuber, 1983b); in hamsters, like all alkylnitrosoureas (Lijinsky and Kovatch, 1989c) it induced almost exclusively hemangiosarcomas of the spleen and forestomach tumours.

These results point to chemical structure, as distinct from physical properties, as being the important determinants of the selection of particular organs and cell types as the targets for induction of tumours. If such physical properties as solubility and partition coefficient have any role it must be minor and limited perhaps to influences on potency. It even seems likely that chemical stability does not have a very important influence on carcinogenicity of N-nitroso compounds, since among alkylnitrosoureas having the same alkyl group neighbouring the N-nitroso function, there is little difference in carcinogenic effect between the monoalkylnitrosourea with a half-life at pH 7 of about one hour, the alkylnitrosourea with a half-life of one or two days (Lijinsky et al., 1987a) or the trialkylnitrosourea with a half-life (Lijinsky and Taylor, 1975d) of several weeks or months. This points again to the structure of the alkyl group adjacent to the nitroso group as the important determinant of the carcinogenic effect, the remainder of the molecule being important only insofar as it is a nitrosamide or a nitrosamine. The interaction of that

Fig. 3.1. Structures of methyl- and ethyl-nitrosoureas.

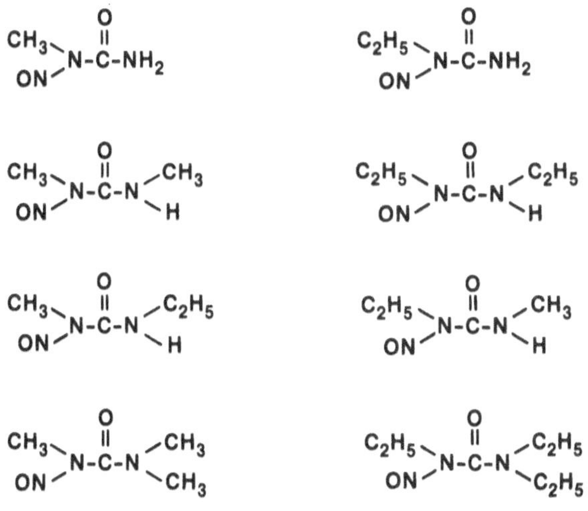

chemical structure with a particular type of cell in a given species converts or transforms it to the neoplastic state, while cells in other organs (often including cells of the same type as are affected in the 'target' organ, for example endothelial cells) are either unaffected, or affected only in ways which do not result in induction of tumours.

The alkylnitrosocarbamates have quite different carcinogenic properties from the alkylnitrosoureas, being almost limited to the upper gastro-intestinal tract, mainly the forestomach, in rats, hamsters or mice. As mutagens and alkylating agents, alkylnitrosocarbamates are equally or more effective than the corresponding alkylnitrosoureas, so these properties alone do not account for the differences in carcinogenic effectiveness between the two types of directly acting agent, which have such close structural similarities. Alkylnitrosourethanes are more stable than the corresponding alkylnitrosoureas and the alkylnitrosoguanidines are even more stable in aqueous media, but the carbamates are more lipophilic. The latter property is not likely to explain the lack of tumourigenic activity of alkylnitrosocarbamates in the nervous system, in contrast with the common induction of tumours of the nervous system in rats by alkylnitrosoureas. The stabilities of a number of alkylnitroso-ureas, -carbamates, and -nitroguanidines at neutral and slightly alkaline pH are shown in Table 3.4 (p. 71).

3.3 Spectroscopic properties

All N-nitroso compounds show absorption in the ultraviolet region, a more intense band in the 230–240 nm region showing $\pi \to \pi^*$ transition (log ε approximately 4) and a less-intense band usually in the 330–350 nm region for a $n \to \pi^*$ transition (log ε approximately 2), due to the N=O bond, varying somewhat with the nature of the alkyl or aryl substituents. The longer-wavelength band is usually a singlet in water or other hydroxylic solvents, but in non-hydroxylic solvents such as methylene chloride or hydrocarbons there is commonly a triplet. Among alkylnitrosamides, such as alkylnitrosoureas or alkylnitrosocarbamates, the longer-wavelength band is usually a triplet in the 370–415 nm region, even in hydroxylic solvents. The absorbance of alkylnitrosocarbamates is at longer wavelengths than that of alkylnitrosoureas, and the former are usually pink in colour. There is a bathochromic shift of 6 to 10 nm in non-polar solvents of nitrosamines and alkylnitrosamides, accompanying the appearance of a triplet in the case of the former. The same absorbance properties are found in solutions of nitrite salts or nitrite esters.

The UV absorption spectrum of nitrosamines having an aromatic ring attached to the amino nitrogen reflects the presence of the aromatic

Spectroscopic properties 61

Table 3.1. *UV absorbance of some N-nitroso compounds*

	Solvent	λ_{max} (nm)	log ε
Nitrosamines			
N-nitroso-			
dimethylamine	H$_2$O	230, 332	3.86, 1.98
diethylamine	H$_2$O	230, 340	3.87, 1.93
diallylamine	H$_2$O	233, 344	3.88, 1.90
dicyclohexylamine	75% EtOH	236, 357	3.88, 1.91
diphenylamine	75% EtOH	208, 295, 360	4.26, 2.77, 1.90
dibenzylamine	75% EtOH	203, 237, 353	4.22, 3.89, 1.92
methylethylamine	H$_2$O	230, 335	3.85, 1.90
sarcosine	H$_2$O	234, 337	3.77, 1.89
methylvinylamine	H$_2$O	200, 270, 376	3.65, 3.83, 2.25
methylcyclohexylamine	50% EtOH	230, 340	3.89, 2.00
methylphenylamine	50% EtOH	207, 271	4.04, 2.89
ethyl-*iso*propylamine	H$_2$O	230, 341	3.85, 1.83
butylamylamine	50% EtOH	236, 347	3.86, 1.95
diethanolamine	H$_2$O	234, 345	3.80, 1.93
pyrrolidine	H$_2$O	230, 333	3.91, 2.03
piperidine	H$_2$O	235, 337	3.98, 1.92
morpholine	H$_2$O	237, 346	3.90, 1.93
Nitrosamides			
N-nitroso-			
methylurethane	H$_2$O	237, 400	3.77, 2.11
ethylurethane	H$_2$O	240, 405	3.84, 2.10
methylformamide	H$_2$O	403	1.72
methylurea	H$_2$O	231, 392	3.77, 1.96
ethylurea	H$_2$O	233, 395	3.74, 1.87
dimethylurea	50% EtOH	255, 385	3.85, 1.98
trimethylurea	H$_2$O	238, 375	3.85, 1.94
diethylurea	H$_2$O	234, 391	3.81, 1.91
n-butylurea	H$_2$O	238, 394	3.74, 1.90
methylnitroguanidine	H$_2$O	275, 398	4.26, 2.19
ethylnitroguanidine	H$_2$O	277, 404	4.30, 2.16

structure, and the absorbance is strongly increased. Other substituents have only a marginal effect on the absorption wavelength or intensity, except in the case of vinyl nitrosamines (see Table 3.1), but a comparison of ethyl with methyl compounds shows almost invariably that the wavelength of the absorption band of ethyl compounds is higher, and the intensity lower, than of the analogous methyl compound.

Substituents in the α-methylene of nitrosamines often caused a bathochromic shift in absorption wavelength and a concomitant decrease in absorptivity. This was evident in nitrosodi-*iso*propylamine compared with nitrosodi-*n*-propylamine, with nitroso-2,5-dimethylpyrrolidine com-

Table 3.2. *UV/visible absorption maxima (nm) of* N-*nitroso compounds in various solvents*

Solvent	Compound			
	Nitroso-diethylamine	Nitroso-morpholine	Ethylnitroso-urea	Ethylnitroso-urethane
Water	337	342	395	(417), 402, (388)
EtOH	349	354	416, 398, 383	422, 404, 388
EtOAc	(371), 360	369, 358	413, 395, 380	422, 402, 388
CH_2Cl_2	356	358	413, 396, (380)	420, 402, (387)
Iso-octane	377, 365, (354)	375, 363	416, 398, 382	424, 404, 388

Numbers in parentheses represent shoulders.

pared with nitrosopyrrolidine, nitroso-2,6-dimethylpiperidine compared with nitrosopiperidine and dinitroso-2,3,5,6-tetramethylpiperazine compared with dinitrosopiperazine, each α-substituted nitrosamine showing a shift of 7 to 10 nm towards longer wavelengths. The effect was particularly marked when both α-methylenes were completely substituted, as in nitroso-2,2,6,6-tetramethylpiperidine, which showed a bathochromic shift of almost 20 nm compared with nitrosopiperidine, and a drop in molar absorptivity from 87 to 58; a similar effect was seen when the absorption spectra of nitroso-4-piperidone and nitroso-4-piperidinol were compared with those of their 2,2,6,6-tetramethyl derivatives. The effects of sulphur substituents in the ring of cyclic nitrosamines were surprising in that the absorptivity of nitrosodithiazine and its trimethyl derivative, nitrosothialdine, was very low compared with nitrosopiperidine, and the absorbance maxima were at considerably higher wavelengths (Lijinsky *et al.*, 1988*a*).

The effects of solvents on the absorption spectra of *N*-nitroso compounds were diverse (Table 3.2). In water there was commonly a single maximum between 333 and 345 nm for nitrosamines (there were some exceptions as noted above), and for alkylnitrosamides between 390 and 405 nm; alkylnitrosocarbamates displayed shoulders near 420 and 390 nm. In ethanol nitrosamines showed a single maximum 12 nm higher than in water, and in methylene chloride there was a single maximum approximately 5 nm higher still. In ethyl acetate, nitrosamines showed a doublet near 370 and 360 nm, and in the non-polar isooctane there was again a doublet, but at approximately five nm higher wavelength, the lower wavelength absorbance being the more intense. Both alkylnitrosoureas and alkylnitrosocarbamates, on the other hand, displayed triplets in all solvents but water; the middle band was the most intense and was near 398 nm for alkylnitrosoureas and near 402 nm for alkylnitrosocarbamates. The upper absorbance band was typically 18 to 20 nm higher, and the lower band 15 to 18 nm lower than the middle maximum. There was little bathochromic shift in less-polar solvents in the case of the alkylnitrosamides, although the upper band for ethylnitrosourethane was 424 nm in *iso*-octane. The influence of the neighbouring carbonyl function is strong enough to overcome the solvent effect, as it moves the absorbance towards the visible range and increases the intensity in the case of the alkylnitrosocarbamates.

The two absorbance regions of alkylnitrosoamides can be used to monitor the stability of these compounds, as has been demonstrated by Leung and Archer (1984) for 2-oxopropylnitrosourea. Disappearance of the longer-wavelength band, at which there is little interference from products of the decomposition, has been used as a measure of decomposition rate, and hence stability, of these compounds.

Few, if any, *N*-nitroso compounds are fluorescent, and there appear to be no reports describing this characteristic of the compounds.

The physical properties and spectroscopic properties of a number of *N*-nitroso compounds are given by Druckrey *et al.*, 1967 (65 compounds) and by Rainey *et al.*, 1978 (131 compounds). The spectroscopic properties, including infrared and NMR data, of a number of *N*-nitroso compounds synthesised during the last ten years or so for biological testing have been published along with the description of their preparation.

In the infrared region *N*-nitroso compounds show a characteristic but relatively weak band at 1445–1490 cm^{-1}, attributed to N=O stretching, considerably different from the analogous band at 1605–1620 cm^{-1} in *C*-nitroso compounds and in nitrite esters. Spectroscopic evidence indicated a partial 1,3-dipolar ion structure in nitrosamines, intermediate between the structures below (Fig. 3.2).

The nuclear magnetic resonance (NMR) spectra of *N*-nitroso compounds (Karabatsos and Taller, 1964) are straightforward, and in asymmetric compounds usually reveal the presence of two rotamers, the ratios of which are often near 1:1, but sometimes vary considerably from that. At low temperatures the rotamers are frozen and can be separated,

Fig. 3.2. Dipolar ion structure of nitrosamines.

taking considerable time to come to equilibrium. This has biological significance, as will be discussed in Chapter 4. As interest has developed recently in this aspect, it has become apparent that only one rotamer (usually E) is metabolised readily, and possibly the carcinogenic potency of a nitrosamine is modulated by the relative proportions of the two isomers in the mixture given to the animals, being possibly greater when the proportion of readily metabolised isomer is greater. Unfortunately, neither rotamer is sufficiently stable at body temperature to allow a meaningful comparison of their carcinogenic potencies, but studies of the relative mutagenic effectiveness of the two rotamers of some nitrosamines might be helpful. Effects of stereochemistry on the stability of α-anions have also been studied by NMR (Lyle et al., 1976; Fraser et al., 1975). NMR spectroscopy has been an invaluable tool in the conformational analysis, as well as configurational assignments of heterocyclic nitrosamines (Fraser et al., 1975; Chow and Colon, 1968).

A new sector rule for the N-nitroso chromophore was developed by Polonski and Prajer, 1976. The nitrosamino acids and their derivatives are a special case and there has been considerable interest in ORD (Djerassi et al., 1961; Gaffield et al., 1972) and their NMR spectra (Lijinsky et al., 1970b). There has been a more recent study by Gaffield (Gaffield et al., 1981). They exist as rotamers and as stereoisomers, leading to quite complex spectra. On the other hand, the stereoisomers are usually easy to separate, and those derived from the naturally occurring stereochemically pure amino acids (proline, hydroxyproline, pipecolic acid) are easy to evaluate. It is of some interest that the nitrosamino acids exist in solution almost entirely as a single rotamer which, curiously, is opposite in nitrosoazetidine-carboxylic acid to that in nitrosoproline.

3.4 Mass spectrometry

The interest in studying possible human exposure to strongly carcinogenic N-nitroso compounds, principally nitrosamines, has provoked a concern to develop methods for detecting and measuring them in trace amounts, as might occur in food, air or other environmental materials. Chromatographic methods for separating N-nitroso compounds isolated from complex mixtures have been combined with the ultrasensitive Thermal Energy Analyzer to accomplish this. However, the TEA responds to any compound releasing NO when heated, and identification of a compound chromatographically relies on comparison of retention time with that of a known standard, neither of which observation permits positive identification of a particular compound to be made; several erroneous reports of nitrosamines have appeared. We have

Table 3.3. *EI mass spectra of nitrosamines; relative intensity of fragment ions*

N-Nitroso-	Molecular weight	M^+	M^+-17	M^+-30	M^+-31
dimethylamine	74	100	—	24	58
methylethylamine	88	100	2	4	20
methyl-n-propylamine	102	100	31	15	23
methyl-n-butylamine	116	100	56	3	3
methyl-n-amylamine	130	27	100	9	4
methyl-n-hexylamine	144	19	100	17	0
methyl-n-heptylamine	158	14	100	22	0
methyl-n-octylamine	172	8	100	19	3
methyl-n-nonylamine	186	3	100	25	5
methyl-n-decylamine	200	2	100	32	0
methyl-n-undecylamine	214	2	100	36	0
methyl-n-dodecylamine	228	1	100	29	1
methyl-n-tetradecylamine	256	0.4	100	29	0.7
methyl-n-octadecylamine	312	4	100	21	0.3
diethylamine	102	100	31	2	23
di-n-propylamine	130	100	12	0	0
di-isopropylamine	130	100	0	0	0
di-n-butylamine	158	100	25	3	0
di-sec-butylamine	158	100	8	0	0
di-isobutylamine	158	100	32	2	4
di-n-octylamine	270	5	100	27	0
di-n-decylamine	326	2.2	100	3	1
di-n-dodecylamine	382	1.6	100	1	3
diallylamine	126	22	100	6	4
dicyclohexylamine	210	100	19	13	13
dibenzylamine	226	100	0	5	0
azetidine	86	100	24	33	67
pyrroline	98	100	14	50	64
pyrrolidine	100	100	0	17	54
Δ^3-dehydropiperidine	112	100	1	25	6
piperidine	114	100	3	30	29
hexamethyleneimine	128	100	13	15	7
heptamethyleneimine	142	22	6	100	4
octamethyleneimine	156	9	2	100	2
dodecamethyleneimine	212	55	90	100	9
morpholine	116	100	0	67	11
thiomorpholine	132	100	2	37	11
piperazine	115	2	0	100	18
di-nitrosopiperazine	144	100	0	50	0

come to rely on a very sensitive and specific method of trace analysis, mass spectrometry.

It was necessary to examine the mass spectra of a number of N-nitroso compounds to establish some criteria by which the mass spectrum of an unknown N-nitroso compound could be used to determine its chemical structure. It can also be done, of course, using NMR spectroscopy, but this requires much larger quantities than are normally available from extraction of a complex environmental mixture. At Oak Ridge National Laboratory (ORNL) we undertook to examine systematically the mass spectra of many N-nitroso compounds available to us. Some were also prepared labelled with deuterium which facilitated interpretation of the mass spectrum in several cases. In two ORNL technical reports (TM-4359, Lijinsky *et al.*, 1973a, and TM-5500, Rainey *et al.*, 1976) the electron impact (EI) mass spectra (70 eV) of more than 130 compounds are shown. An evaluation of these mass spectra relating to the chemical structures of the N-nitroso compounds is in Rainey *et al.* (1978). The patterns of fragmentation of those compounds followed a quite regular system, so that subsequent mass spectra of newly synthesised N-nitroso compounds could be easily rationalised. Mass spectra have usually been reported along with other spectroscopic criteria for more recently synthesised compounds.

There have been clear differences between the mass spectra of nitrosamines and of nitrosamides, and between cyclic and acyclic nitrosamines, although the latter differences were not in all cases large. For example, most alkylnitroso-ureas, -carbamates, -guanidines, etc. give a small or insignificant molecular ion by electron impact, although they often show a substantial molecular ion by chemical ionisation. Cyclic nitrosamines usually lose NO, to give a prominent M-30 ion, whereas acyclic nitrosamines lose HNO to give an M-31 ion (the proton coming from the α-methylene, as shown by labelling with deuterium). Nitrosamines usually give a prominent molecular ion, but sometimes it is small or even non-existent, particularly with nitrosamines of high molecular weight. A common fragmentation of acyclic nitrosamines is loss of OH, giving an M-17 ion (Table 3.3); in the case of several nitrosamines of high molecular weight M-17 is the base peak of the spectrum (e.g. methylnitrosododecylamine, nitrosodidecylamine). This ion, M-17, is small or non-existent in cyclic nitrosamines, except the large, flexible nitrosododecamethyleneimine. The hydrogen of the OH does not come from the α-methylene, because nitrosodi-(n-butyl-1-d_2)-amine has the same M-17 ion as the undeuterated compound. It is likely that loss of OH occurs with formation of a four-membered ring (Fig. 3.3) involving the two nitrogen atoms and the α- and β-carbons of the alkyl chain (Rainey et al., 1978). Cleavage then follows the pattern established for aliphatic amines, principally α-cleavage, leading to $CH_2 = N = CH_2$ (m/z 42) or $CH_3 = N = CH_2$ (m/z 43). Branched chains in nitrosamines undergo successive methyl α-cleavage ending, for example, in $CH_3\text{-}N = C$ (m/z 41) from methylnitroso-*tert*-butylamine. The relative intensities of these ions in the EI mass spectra of a number of nitrosamines are compared in Table 3.3. Some typical mass spectra are shown in Fig. 3.4.

Substitution in aliphatic nitrosamines led to various types of fragmentation, depending on the nature of the substituent. Hydroxylated nitrosamines did not give significant molecular ions, and the fragmentations were complex, differing from one to another. There was sometimes

Fig. 3.3. Intermediate ions in mass spectral fragmentation of nitrosamines.

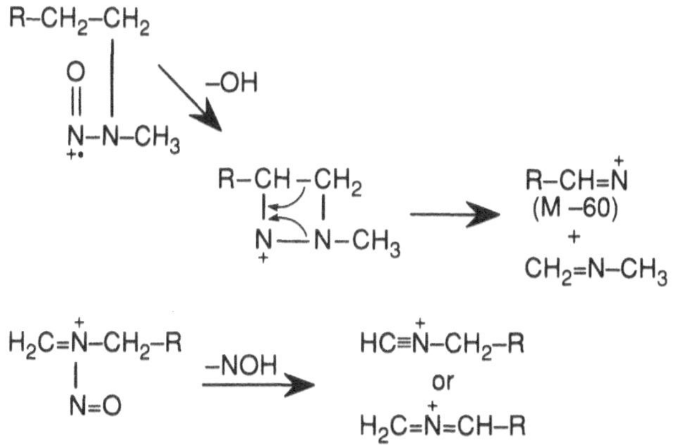

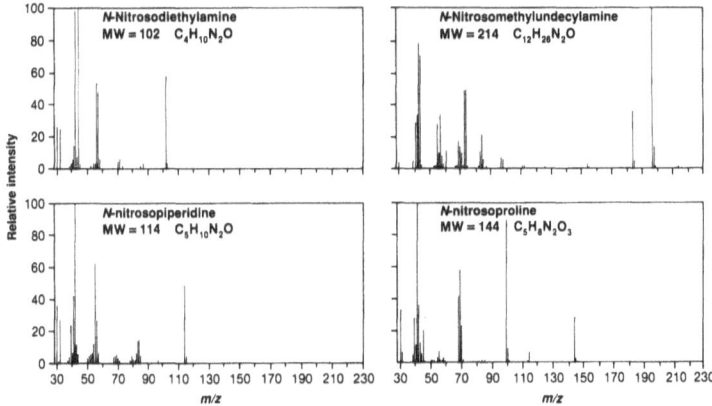

Fig. 3.4. Mass spectra of nitrosodiethylamine, methylnitrosoundecylamine, nitrosopiperidine and nitrosoproline.

an M-1 fragment, as there was from hydroxylated cyclic nitrosamines, which, however, gave prominent molecular ions. Carboxyl substituents (nitrosamino acids) usually produced a small molecular ion, but a very prominent M-45 ion, consistent with loss of COOH; this was especially true of cyclic nitrosamino acids such as nitrosoproline and nitrosopipecolic acid; when deuterium was substituted in the ring, COOH was still lost.

Loss of halogen from nitrosamines containing two halogen atoms was common in both cyclic and acyclic nitrosamines; the ion arising from loss of one halogen was more prominent than the molecular ion. There was no ion resulting from loss of the second halogen, nor from monohalogenated nitrosamines. Some substituted acyclic nitrosamines had unhelpful mass spectra, for example nitrosobis-(2-oxopropyl)-amine, which showed no molecular ion, a very small M-30 ion, and the largest significant fragment was at m/z 86; there was a large m/z 86 ion in the mass spectrum of the corresponding alcohol, nitrosobis-(2-hydroxypropyl)-amine.

Most of the alkylnitrosamides are unstable (usually melting with decomposition) and would not be expected to show molecular ions by electron impact. However, even trialkylnitrosoureas, which can be distilled under reduced pressure, do not show molecular ions. In trimethylnitrosourea and triethylnitrosourea, the largest ions are at m/z 72 and m/z 100, respectively, corresponding to loss of R-(methyl or ethyl)-N-NO.

In summary, whereas electron impact mass spectrometry is usually diagnostic of nitrosamines, it is not so for alkylnitrosamides. Chemical ionisation mass spectrometry, on the other hand, can provide molecular ions of both nitrosamines and nitrosamides, but the lack of fragmentation in those spectra reduces their usefulness for determining the structure of an unknown N-nitroso compound. Nevertheless, mass spectrometry is a most useful tool for confirmation of identification of an N-nitroso compound that has been tentatively identified by its chromatographic retention time, and which gives a positive response in a nitrogen-specific detector or in the Thermal Energy Analyzer.

3.5 Stability

Nitrosamines are mostly stable compounds, remaining unchanged in the pure state or in neutral aqueous solution at room temperature and in the dark for years; they are slowly decomposed by light, especially in solution. Some nitrosamines containing functional groups in the alkyl chains, such as halogen atoms or carbonyl groups, and some unsaturated compounds, are not very stable (although more so than alkylnitrosamides). As pure compounds, they decompose slowly to a variety of products at ambient temperatures, and more rapidly in aqueous solution. Such nitrosamines are more stable at low temperatures and in organic solvents, particularly non-polar ones.

Alkylnitroso-ureas, -carbamates and -guanidines are more stable either in solution or as pure substances. Methylnitrosonitroguanidine was, in fact, once considered useful as a detonator for explosives, a property promoted by its high content of nitrogen. Methylnitrosoguanidine itself, without the stabilising effect of the nitro group, is very unstable. We obtained evidence for its existence in solution from the mutagenic activity of a solution containing N-methylguanidine and sodium nitrite which had been acidified, then neutralised. The mutagenic potency of the compound in solution – assuming all of the methylguanidine had been converted into the N-nitroso derivative – was very similar to that of an equimolar solution of MNNG. All attempts to isolate methylnitrosoguanidine, the presence of which was shown by the absorption spectrum of the solution, failed whether by extraction with a variety of solvents or by evaporation of the solution *in vacuo*; the oily residue in the latter case no longer contained a nitroso compound.

The well known drug cimetidine also contains an N-methylguanidine structure and its N-nitroso derivative (Fig. 3.5) which is stabilised by a cyano group, is also a strong mutagen, but is not carcinogenic. Another common N-alkylguanidine is the amino acid arginine, but its N-nitroso derivative has not been isolated and apparently is unstable. It is notable that in feeding studies designed to demonstrate the formation *in vivo* of carcinogenic N-nitroso compounds, neither methylguanidine (as sulphate) nor arginine, fed together with sodium nitrite to rats for most of their lifespan, gave rise to any significant incidence of tumours (Lijinsky and Taylor, 1977c). The evidence that methylnitrosoguanidine is formed under such acid conditions as exist in an animal's stomach (and which with other amines, such as ethylurea or morpholine, result in induction of large numbers of tumours), suggests that methylnitrosoguanidine is too unstable to be biologically effective in these conditions. Nitrosoarginine, on the other hand, might well be non-carcinogenic because of its structure, particularly the presence of the carboxyl group. Most basic or acidic nitrosamines (such as nitrosamino acids) are weakly toxic and weakly carcinogenic or non-carcinogenic, probably because ionised molecules are less able to enter cells than neutral ones.

Alkylnitrosocarbamates (Fig. 3.6) are more stable than alkylnitroso-ureas (Table 3.4), although they also decompose at high pH to form diazoalkanes and at low pH to nitrite and the secondary amino compound. Several of them melt at moderate temperatures without decomposition, whereas alkylnitrosoureas melt with decomposition evidenced by the evolution of bubbles of gas (nitrogen and probably carbon dioxide). The simplest members of the class, methyl- and ethyl-nitrosourethane are pink liquids, which can be distilled without decomposition under reduced pressure. A number of nitroso-N-methylcarbamate esters have been investigated, because their parent esters are widely used insecticides, an activity they have because they are strong inhibitors of cholinesterase; the N-nitroso derivatives are not cholinesterase inhibitors and are less toxic to animals than are the N-methylcarbamate esters.

The N-nitroso derivatives of the insecticides are solids, often strongly

Fig. 3.5. Methylnitrosoguanidine, methylnitroso-N'-nitroguanidine, nitrosocimetidine and nitrosoarginine.

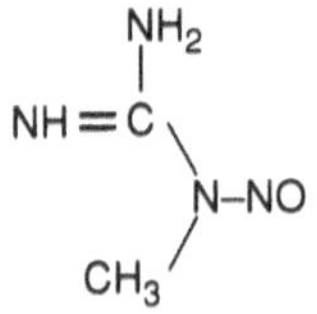

METHYLNITROSOGUANIDINE

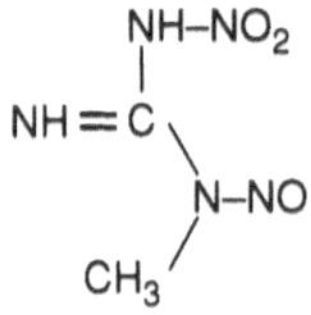

METHYLNITROSO-N'-NITROGUANIDINE

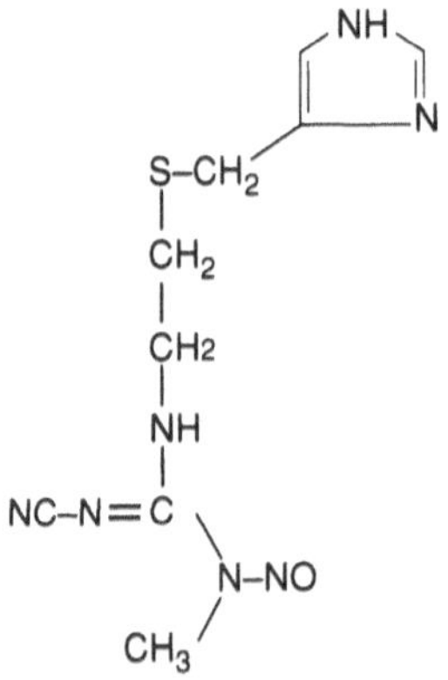

NITROSOCIMETIDINE

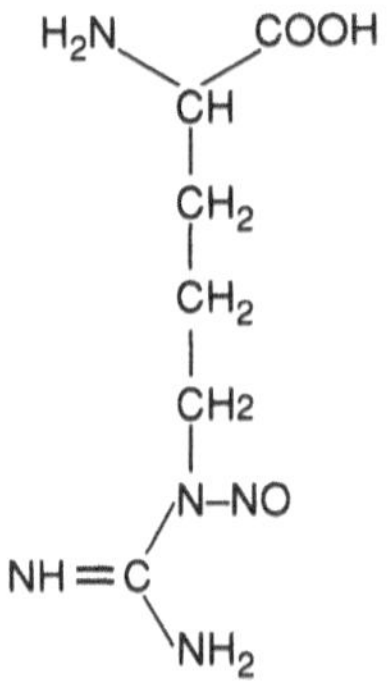

NITROSOARGININE

Fig. 3.6. Alkylnitrosourea and alkylnitrosocarbamate ester.

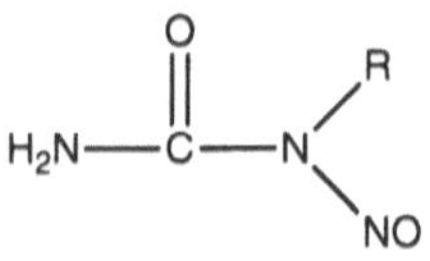

Alkylnitrosourea

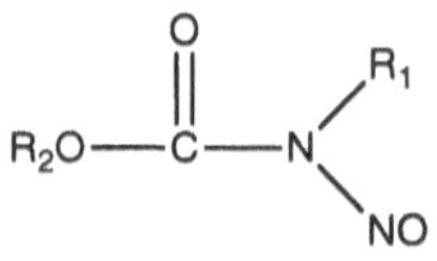

Alkylnitrosocarbamate–Ester

Table 3.4. *Stability of alkylnitrosamides: half-life at pH 7 and pH 8 (22 °C)*

N-nitroso-	$t_{\frac{1}{2}}$ (h)	
	pH 7	pH 8
methylurea	2.3	
ethylurea	2.4	
n-propylurea	2.3	
isopropylurea	0.1	
allylurea	1.9	
n-butylurea	2.7	
sec-butylurea	0.1	
isobutylurea	1.7	
n-amylurea	3.3	
n-hexylurea	2	
n-undecylurea	1	
n-tridecylurea	0.5	
cyclohexylurea	< 0.1	
benzylurea	0.3	
2-phenylethylurea	1.1	
2-hydroxyethylurea	1.1	
2-methoxyethylurea	1	
2-fluoroethylurea	0.7	
2-hydroxypropylurea	0.3	
2-oxopropylurea	2.5	
3-hydroxypropylurea	1.5	
methyl-N'-nitroguanidine		2.5
ethyl-N'-nitroguanidine		7
methylurethane	70	2.5
ethylurethane	100	4
oxazolidone	3	
5-methyloxazolidone	1	
tetrahydro-oxazone	6	
1-ethyl-3-hydroxyethylurea	40	2.4
1-ethyl-3-(2-oxopropyl)urea	40	9
1-hydroxyethyl-3-ethylurea	60	2.5
1-hydroxyethyl-3-chloroethylurea	45	1.4
1-chloroethyl-3-hydroxyethylurea	11	1.7
1-chloroethyl-3-(2-hydroxypropyl)urea	20	1.8
bis-(chloroethyl)urea	13	9
1-(2-hydroxypropyl)-3-chloroethylurea	38	1.9
1-(2-oxopropyl)-3-chloroethylurea	235	58
1,3-dimethylurea	300	43
1,3-diethylurea	87	18
1-methyl-3-ethylurea	230	38
1-ethyl-3-methylurea	84	18
1-(dimethylaminoethyl)-3-ethylurea	15	5

coloured, and they have absorption maxima in the visible range (420 nm). They include the nitroso derivatives of carbaryl, baygon, carbofuran, landrin, Bux-10, methomyl and aldicarb (Fig. 3.7), all of them very potent bacterial mutagens and cell-transforming agents. Some cyclic nitrosoalkylcarbamates have also been examined and they are similar in properties, biological as well as chemical, to their acyclic analogues. Nitroso-1,3-oxazolidone, nitroso-5-methyl-1,3-oxazolidone and nitrosotetrahydro-1,3-oxazone are pink solids, which are potent bacterial mutagens. The first two have been tested for carcinogenicity in rats and hamsters and like all alkylnitrosocarbamates, induce tumours of the forestomach and virtually no other tumours when administered orally.

Fig. 3.7. Nitroso derivatives of methylcarbamate insecticides.

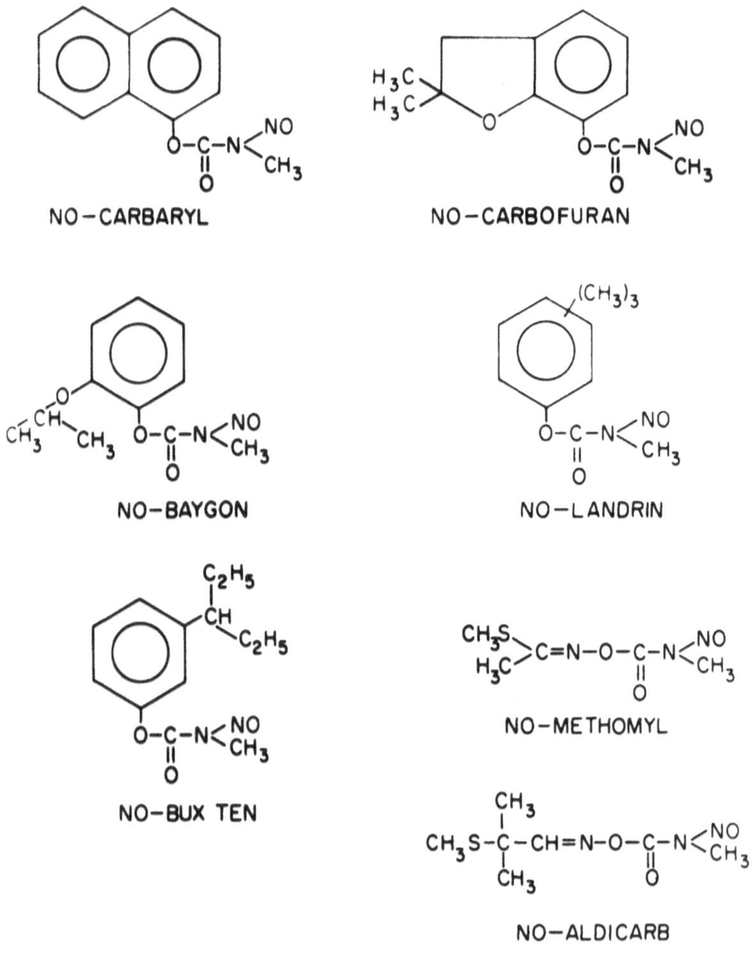

The alkylnitrosocarbamates have not attracted much attention, for reasons not easily understood, and less of their chemistry is known than of alkylnitrosoureas.

The alkylnitrosoureas have been extensively studied because of their interesting and varied organ-specific carcinogenic activity and also because several dialkylnitrosoureas have been used as cancer-therapeutic agents. For use as drugs, of course, those alkylnitrosoureas have undergone extensive examination of their chemistry and pharmacology, so we know more about them. Their mechanisms of action as therapeutic agents are not completely understood, although they are powerful alkylating agents and carbamoylating agents. Whether their alkylation of DNA in tumour cells is the reason for the death of cells in tumours susceptible to such compounds as BCNU, CCNU and HECNU is not clear. These compounds, which have in common the structure of a 1-chloroethyl-1-nitroso-3-alkylurea, will be discussed in relation to their biological properties elsewhere. Almost all, if not all, alkylnitrosoureas are solids, usually almost colourless or pale yellow. They have absorption maxima in the region of 385 to 400 nm, barely in the visible range.

Alkylnitrosoureas and dialkylnitrosoureas are both directly acting alkylating agents, although the latter are much less reactive than the former. Trialkylnitrosoureas, on the other hand, are not directly acting alkylating agents; they are liquids usually and very much more stable than the other two classes (Lijinsky and Taylor, 1975d).

The variety of alkylnitrosoureas investigated is large and, while they have properties in common because of the nitrosourea structure, their properties are also affected by the nature of the alkyl group. The smaller members, for example, are water-soluble (methyl- and ethyl-), as are nitrosoureas with hydroxylated alkyl groups. As the size of the alkyl group increases the solubility in water decreases, but many of the smaller ones are sufficiently soluble in water for them to be tested for carcinogenicity by administration to animals in drinking water; this has been especially useful for dialkylnitrosoureas, but alkylnitrosoureas are usually so unstable at neutral pH that it is not possible to quantify the dose. Nevertheless, when the objective is only the induction of tumours, administration of methylnitrosourea in drinking water, for example, has been used to induce high incidences of tumours of the glandular stomach in rats (Hirota *et al.*, 1987); no tumours of other organs, such as the nervous system, were induced under these conditions, as has been the case when other modes of administration have been employed (Lijinsky and Kovatch, 1989b). Although the lower molecular weight alkylnitrosoureas are not highly soluble in lipids, such as natural oils and fats, it seems that they readily enter cells of many organs, including the nervous system (which nitrosamines are said not to do). Like nitrosamines, alkylnitrosoureas are readily soluble in most organic solvents, and are very stable in most of them, although undergoing slow decomposition in polar solvents, particularly when they contain traces of water.

Alkylnitrosoureas are reasonably stable in the pure state at moderate temperatures, but they are best stored below 0 °C; at -20 °C they remain unchanged for years. On the other hand, there have been several reports of explosions of bottles of commercially supplied methylnitrosourea stored at room temperature (Werner, 1919; Sparrow, 1973). The amounts in the bottles were usually relatively large, 100 g, and few other alkylnitrosoureas are transported or kept in such large amounts. This practice is distinctly unwise with alkylnitrosoureas, although there are no reports of mishaps with those dialkylnitrosoureas used medicinally, which are made in relatively large quantities. Alkylnitrosoureas are not volatile and heating of even small quantities of them can be dangerous. An explosion occurred during preparation of 1-methyl-1-nitroso-3-hydroxyethylurea, probably due to cyclisation; such instability of a dialkylnitrosourea is very unusual (Saavedra, 1990).

It is probable that in animals trialkylnitrosoureas are oxidatively dealkylated to the corresponding dialkylnitrosourea (Khoda et al., 1986). Therefore, discussion of the special properties of trialkylnitrosoureas is not valuable, except that they are stable in alkaline solution (Lijinsky and Taylor, 1975d) and less stable in acidic solution, which is reminiscent of dialkylnitrosamines. That they can be distilled under reduced pressure without decomposition makes them unusual among alkylnitrosoureas. The N-nitroso derivatives of aryldialkylureas, which include a number of widely used herbicides (monuron, fenuron, etc.) are unstable, like phenylnitrosourea, which latter has decomposed spontaneously at room temperature. The biological properties of these arylnitrosoureas have been little studied, therefore.

Monoalkylnitrosoureas are quite unstable in aqueous solution, but the range of instability is large. For example, methylnitrosourea and ethylnitrosourea have half-lives at pH 7 of one to two hours, and of only a few minutes at pH 8 or higher, so that their stability is difficult to measure above pH 8. Isopropylnitrosourea and cyclohexylnitrosourea have half-lives of six minutes or less at pH 7. The other alkylnitrosoureas examined (Lijinsky et al., 1987a) have stabilities lying in between (Table 3.4). However, even so large a difference in decomposition rate as 50- or 60-fold is not reflected in large differences in carcinogenic potency (see Chapter 7).

Radiolabelled alkylnitrosoureas become rapidly and fairly uniformly distributed around the animal body following oral administration, for example, suggesting that they might be stabilised, perhaps by conjugation with some biomolecule, and then transported round the body. It seems unlikely that the physiological half-life of MNU and ENU claimed by Druckrey et al. (1967) of five to seven minutes can be due entirely to simple solvolysis, and it is inconsistent with the specifically targeted tumour induction by alkylnitrosoureas in organs of a particular species.

Apart from formation of an alkylating species, usually considered to be an alkyldiazonium ion, little is known of the chemical properties of alkylnitrosoureas that is helpful in understanding their biological activity. They can form carbamoyl entities. In a study of the decomposition in aqueous solution at pH 7 of hydroxyethylnitrosourea, 2-hydroxypropyl-nitrosourea and the corresponding nitrosooxazolidones (cyclic carbamates), Singer (1984) showed that acetaldehyde and acetone were formed from the respective 2- and 3-carbon nitroso compounds, which also produced the glycols. The cyclic carbonates were formed in substantial yield from the two nitrosooxazolidones, but not from the hydroxyalkyl-nitrosoureas. These reactions are not possible for the non-hydroxylated alkylnitroso compounds.

As shown by Singer et al. (1977), Singer (1980), and Singer and Cole (1981) alkylnitrosoureas can lose their nitroso group and they are, therefore, nitrosating agents. In the case of dialkylnitrosoureas in solution, particularly in the presence of nucleophilic anions such as thiocyanate, the nitroso group can migrate from one amino nitrogen to the other. Among trialkylnitrosoureas the loss of the nitroso group is accompanied by dismutation of the molecule and formation of several products including dialkylcarbamate, tetraalkylurea, dialkylglycine (Fig. 3.8) and in certain conditions, the dialkylnitrosamine (Singer, 1982). This reaction is not

Fig. 3.8. Dismutation of trialkylnitrosoureas.

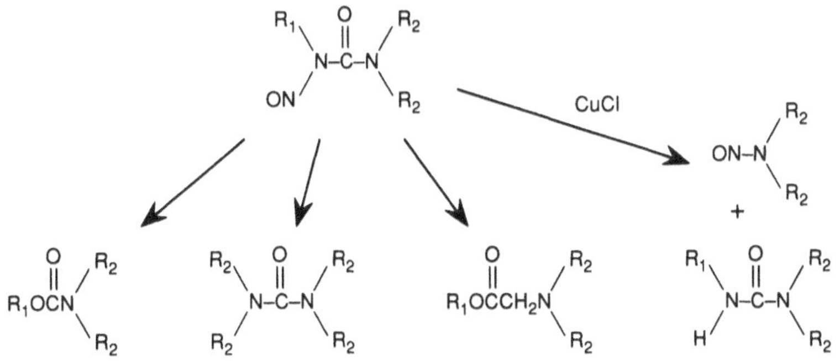

76 *Chemical properties*

likely to be biologically relevant, but is an index that trialkylnitrosoureas are not entirely unreactive. Their potent carcinogenicity is perplexing, since they act biologically like alkylnitrosoureas, but biochemically like nitrosamines.

3.6 Reducing nitrosamine exposure: destruction and degradation of compounds

The exposure of humans to N-nitroso compounds is very common and would not be important in other than occupational settings, were it not that they are such potent carcinogens. A no-effect or 'safe' level cannot be assumed for any of the nitrosamines to which people are exposed, based on the results of their carcinogenic effects in animal tests. It is certainly possible to reduce exposure to the tobacco-associated nitrosamines, by ceasing to smoke and by avoiding close association with smokers or with areas in which smoking takes place; however, use of smokeless tobacco does not eliminate nitrosamine exposure because it also contains carcinogenic nitrosamines (Hoffman and Hecht, 1985). Similarly, exposure to nitrosamines in the workplace can be reduced or avoided by changing jobs or by actions of management and workers to reduce the amount of nitrosamine in the factory or workshop; this might also involve changes in processes or materials, for example replacing nitrite-containing cutting fluids with those that do not contain nitrites. The same might apply to reducing air emissions of nitrosatable secondary and tertiary amines, for example in pesticide formulations. Some of these changes have been alluded to earlier. It is unlikely to be feasible to reduce the amount of nitrogen oxides, which can lead to formation of nitrosamines, that is present in air due to burning of fuels for heat or in engine emissions, so reduction of nitrosatable amines is the only alternative.

Other sources of exposure to nitroso compounds are food and waste materials, possibly including sewage. Food presents a particular difficulty, since the N-nitroso compounds usually arise from nitrite that has been added to the food during processing. The solution to this is to use less nitrite (or its precursor nitrate). This has long been advocated for cured meats, especially bacon, and has been gradually accepted by the meat processing industry, together with the use of ascorbic or erythorbic acid. At present the concentrations of nitrosamines in bacon are considerably lower than they were 20 years ago, when the problem was recognised.

In the removal of N-nitroso compounds from waste materials several procedures have been recommended. For example, Gangolli *et al.* (1974) used aluminium foil to remove small amounts of nitrosamines, which were reduced to hydrazines or amines; hydrazines are not themselves free of carcinogenic risk, although they are much weaker carcinogens than are nitrosamines. Other means of destruction of N-nitroso compounds rely on the susceptibility of these compounds to photolysis, especially in acid solution (Chow, 1973; Polo and Chow, 1976). These methods have been reviewed by Emmett et al. (1980), and focus especially on the use of nitrite scavengers to remove the nitrite formed by photolysis, thereby hastening the photolytic cleavage. The scavengers include vitamin C, urea, hydrazoic acid and guanidine. Depending on the complexity of the N-nitroso compound there are products of photolysis in addition to nitrite and the amine. The toxicity of such products is assumed to be less than that of the N-nitroso compounds, but might not be lower in all cases.

3.7 Chemical reactivity

The chemical reactions of N-nitroso compounds are not a particularly vital area of chemistry. If it were not for their biological activity, it would probably be mostly ignored. However, the close association between chemical structures of N-nitroso compounds and their induction of particular types of cancer demonstrates that some chemical or physical characteristic is intimately related to induction of cancer, and not merely possession of the nitroso group (since few C-nitroso compounds are carcinogenic at all, and those are very weak carcinogens – Goodall and Lijinsky, 1976), or even the N-nitroso group. What, then, are some of the chemical properties of N-nitroso compounds that might lend themselves to induction of this unusual biological effect, of such great interest and importance to humans? Structural characteristics, such as molecular size and shape are also important and will be considered later. An excellent review by Challis and Challis (1982) contains much that is known about chemical reactions of N-nitroso compounds.

The N-nitroso function has similarities to the carbonyl group, withdrawing electrons so as to make protons on neighbouring carbon atoms more acidic. This in turn leads to a tendency for the compound to form a carbanion (Keefer and Fodor, 1970), producing intermediates very useful in synthesis (Saavedra, 1983) and facilitating replacement of hydrogen atoms at the α-position to the N-nitroso function with deuterium or tritium. This isotopic replacement has been used for preparation of nitrosamines labelled with both isotopes of hydrogen (Singer and Lijinsky, 1979) for studies of metabolism and mechanisms of carcinogenesis of some nitrosamines (Lijinsky and Reuber, 1980a, b; Lijinsky et al., 1980f; Underwood and Lijinsky, 1982). It is likely that the biological actions of nitrosamines are related to reactions at the α-carbon atoms, beginning with formation of a carbanion. It is almost certain, for example, that the toxic effects of nitrosamines are due to formation of reactive products other than the alkyldiazonium ions popularly believed to be the most important intermediates in induction of tumours, and the alternative products are likely to be those formed by oxidation or substitution at the α-carbons. Alkylnitrosoureas give rise to alkyldiazonium ions directly, but are much less toxic than the analogous nitrosamines. The toxicity of N-nitroso compounds is discussed in Chapter 5.

The presence of the N-nitroso function is essential for the carcinogenic effects of these compounds, but it seemingly must be in conjunction with an adjacent methylene group on either side. Few nitrosamines which have one or more substituents in the α-methylene groups are carcinogenic. One exception is nitrosodi-*iso*propylamine, which is a very weak carcinogen, however (see Chapter 7) and another is nitrosodiphenylamine, which is a special case. Nitroso-2,6-dimethylpiperidine is not carcinogenic, and nitroso-2-methylpiperidine is markedly less potent than the unsubstituted nitrosamine (Lijinsky and Taylor, 1975a). Methylnitroso-*tert*-butylamine is non-carcinogenic, and not very toxic. The lack of carcinogenic activity of 3-methyl-2-phenylnitrosomorpholine (Lijinsky and Taylor, 1976c) is more difficult to explain, since it has an unsubstituted α-methylene, although it might have much to do with the presence of the bulky phenyl substituent.

It is easy to assume that the presence of an α-substituent in a nitrosamine prevents access of the α-methylene to the active site of the activating enzyme – and that is probably part of the explanation of inactivity of substituted nitrosamines – but there might be other effects on the chemistry of the nitrosamine that are not readily understood. Hindered alkylnitrosoureas are also relatively inactive biologically, for example phenylnitrosourea, cyclohexylnitrosourea and even benzyl-nitrosourea; it has not been possible to prepare *tert*-butylnitrosourea; *iso*propylnitrosourea and *sec*-butylnitrosourea are unstable and very weakly carcinogenic, if active at all on mouse skin (Lijinsky and Winter, 1981).

The implication is clear that the presence of a methine next to the N-nitroso function is not sufficient to make a nitrosamine biologically active, although theoretically it can be activated by removal of the hydrogen. Except in the case of nitrosodi-*iso*propylamine there is no acyclic nitrosamine with substituents in both α-methylenes that is biologically active. Methylnitrosoaniline and methylnitrosocyclohexylamine contain one unsubstituted alkyl group and one blocked, yet they are both potent carcinogens in rats, inducing esophageal tumours, but very weak carcinogens in hamsters, in which the esophagus is unresponsive. The reason for the reactivity of these two nitrosamines is not clear, but seems to be related to their particular chemical structure, since the structurally similar methylnitroso-*tert*-butylamine is completely inactive. This and other findings make it difficult to generalise about mechanisms of carcinogenesis by nitrosamines. Michejda has speculated that oxidation of the methyl group of methylnitrosoaniline could give rise to a phenyldiazonium ion which might couple with an amine (such as guanine) to form a triazene (Koepke et al., 1991).

The nucleophilic nature of the carbon atoms α- to the N-nitroso function makes nitrosamines amenable to substitution at those positions. Metalation, followed by alkylation (Fig. 3.9) has been described by Seebach and Enders (1975), who built on the earlier findings of Keefer and Fodor (1970). Another synthesis of consequence was developed by Saavedra (1978), who demonstrated the ease of conversion of α-nitrosamino acids to the α-acetoxynitrosamines by treatment with lead acetate.

Nitrosamines with functional groups can, of course, undergo the usual reactions of those substituents, quite apart from the properties bestowed by the N-nitroso function. Thus, halogen substituents can be replaced, alcohols oxidised to ketones or aldehydes, and ketones reduced to secondary alcohols. All of these might play some role in organ-specific carcinogenesis by these compounds. It is difficult, however, to relate this type of reaction to the patterns of tumours produced in rats and hamsters, for example. Although the chemical reactions occurring are mirrored in the metabolism that takes place in the animals – which is not usually very different between species – there are large differences in carcinogenic effects between one species and another. To take a simple example, nitrosodiethylamine is considerably more potent than nitrosodimethylamine in rats (and induces a different pattern of tumours); in hamsters nitrosodimethylamine is much more potent than nitroso-

Fig. 3.9. Alkylation of nitrosamines; a = acceptor, d = donor of electrons.

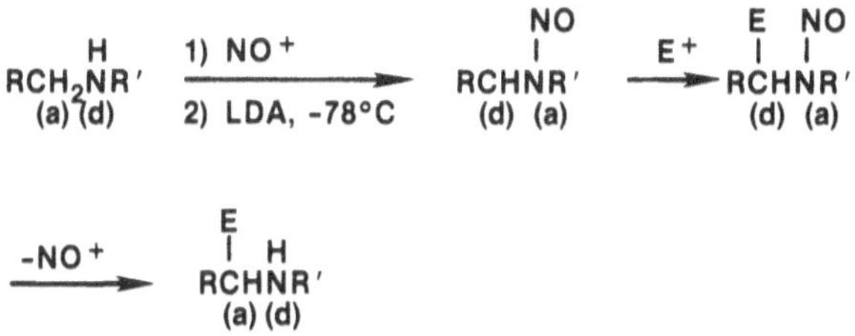

$$\underset{(a)(d)}{\underset{H}{RCH_2NR'}} \xrightarrow{\text{1) NO}^+ \atop \text{2) LDA, -78°C}} \underset{(d)(a)}{\underset{NO}{RCHNR'}} \xrightarrow{E^+} \underset{(d)(a)}{\underset{E\ NO}{RCHNR'}}$$

$$\xrightarrow{-NO^+} \underset{(a)(d)}{\underset{E\ H}{RCHNR'}}$$

diethylamine. However, the set of reactions leading to formation of the respective alkylating agents, methylating and ethylating, appear to be similar, since the extents of methylation and ethylation, respectively, of nucleic acids are similar in the two species (Lijinsky, 1988b).

Studies of nitrosamines containing hydroxylated alkyl groups have illuminated possible reactions of nitrosamines related to their biological actions, especially experiments of Michejda and his co-workers. For example, tosylated derivatives of methylnitrosohydroxypropylamines can cyclise to form a five-membered ring system, which is mutagenic (Michejda and Koepke, 1978; Koepke et al., 1979). Activation of methylnitrosohydroxyethylamine leads to formation of a cyclic oxadiazolium ion (Fig. 3.10), which is mutagenic and might be a reactive intermediate in carcinogenesis by this and similar compounds (Michejda et al., 1979).

Regarding oxidation of nitrosodiethanolamine or nitrosobis-(2-hydroxypropyl)amine and reduction of nitrosobis-(2-oxopropyl)amine, these reactions occur readily with the usual reagents, and derivatives are prepared chemically in this way. However, the several oxidation products of nitrosodiethanolamine that have been tested in animals (Fig. 3.11), including the hemiacetal (nitroso-2-hydroxymorpholine), the hydroxy-acid, and the diacid (nitrosiminodiacetic acid), are all non-carcinogenic, even though the hemiacetal is a directly acting mutagen (Hecht, 1984). The dialdehyde has not been isolated, but its bis-diethyl acetal was neither carcinogenic nor mutagenic (Lijinsky and Taylor, 1978c). It is, therefore, unlikely that formation of any of these products is involved in carcinogenesis by nitrosodiethanolamine. The products of oxidation of NBHPA, nitrosohydroxypropyl-oxopropylamine and NBOPA, are both more potent carcinogens than the parent alcohol, although they induce a different pattern of tumours, and therefore, are not probable intermediates

Fig. 3.10. Formation of oxadiazolium ion from hydroxyalkylnitrosamine.

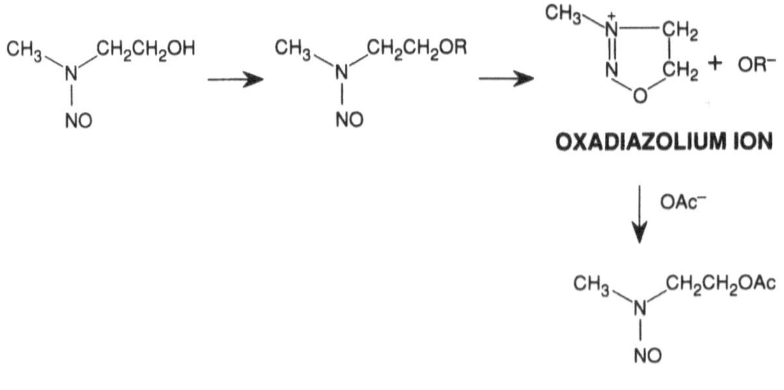

Chemical reactivity 81

in the induction of all the tumours induced by NBHPA. Oxidation of the alcohol to the two ketones is readily accomplished by oxidation with chromic oxide, as described by Saavedra *et al.* (1988). Esterification of the alcohols also can be accomplished, and there is the possibility that sulphate esters or glucuronides act as transport forms of the compounds, as suggested by Kokkinakis *et al.* (1985). However, there is now some doubt that sulphation is an important process of activation of the compounds.

It might be expected that *N*-nitroso compounds which are ketones and aldehydes would condense readily with reactive groups in cellular components, but this possibility has not been much explored, although hydrazones, phenylhydrazones and oximes have been prepared (Farrelly *et al.*, 1984). It is also probable that the acidity of the methylene group between the nitroso group and the carbonyl group in such compounds as methylnitroso-2-oxopropylamine makes it especially reactive; exchange of deuterium for hydrogen in this position is rapid. It is possible that this structure is related to the propensity of 2-oxopropylnitrosamines to induce tumours of the pancreas in hamsters, although nothing of the mechanism is understood. Another possibility for oxopropylnitrosamines is a Baeyer-Villiger oxidation to an ester which, on hydrolysis gives rise to a highly reactive α-hydroxynitrosamine. The product from methylnitroso-

Fig. 3.11. Common oxidation product of nitrosodiethanolamine and nitrosomorpholine.

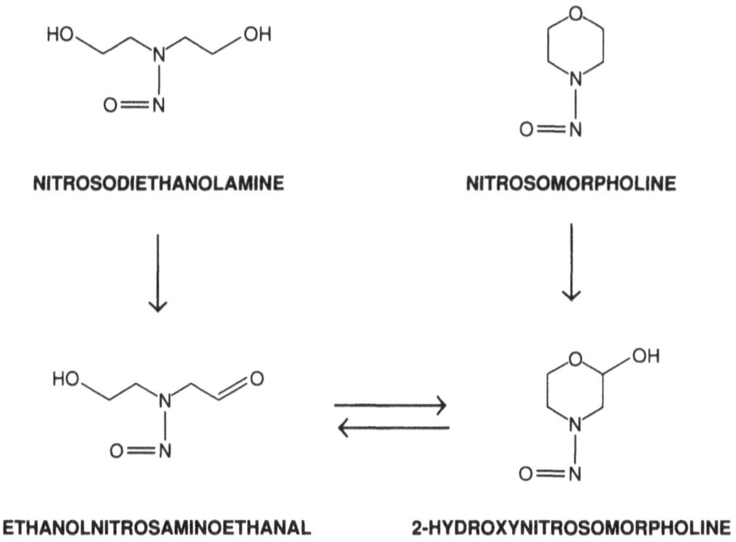

NITROSODIETHANOLAMINE

NITROSOMORPHOLINE

ETHANOLNITROSAMINOETHANAL

2-HYDROXYNITROSOMORPHOLINE

82   *Chemical properties*

2-oxopropylamine, for example, would be hydroxymethylnitrosomethylamine, the putative unstable intermediate in formation of a methylating agent from nitrosodimethylamine *in vivo*. Evidence supporting a Baeyer-Villiger oxidation of MNOPA (Fig. 3.12) leading to methylation of cellular nucleic acids has been obtained using deuterium-labelled MNOPA (Liberato *et al.*, 1989). It is possible, of course, that the activated nitrosamine represented by the Baeyer-Villiger oxidation product itself reacts with important cellular compounds, precipitating those events leading to neoplasia, as a complement to, or in place of, methylation.

Fig. 3.12. Formation of methylating agent from methylnitroso-2-oxopropylamine through α-oxidation or Baeyer-Villiger oxidation.

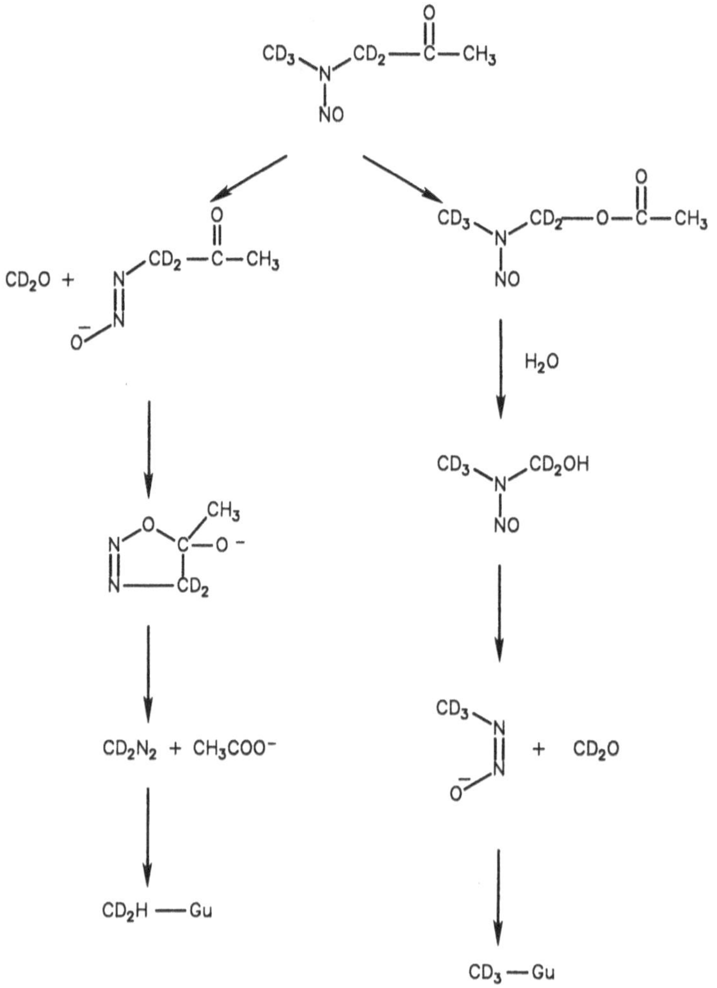

In addition to α-oxidation of nitrosodialkylamines, oxidation to nitramines is possible. Nitramines are well known and easily prepared from N-nitroso compounds by the action of peroxides (Emmons, 1954), but they are less biologically active than the analogous N-nitroso compounds as toxins, mutagens or carcinogens (Goodall and Kennedy, 1976), and so are quite unlikely to be intermediates in those biological activities. Nitramines are more stable than nitrosamines, and it seems that whatever carcinogenic activity they have is due to their partial reduction *in vivo* to the nitrosamines.

Reduction of N-nitroso compounds is easy, with $LiAlH_4$ or electrochemically (Hanna and Schueler, 1952; Iversen, 1971), and offers the most convenient way of preparing N,N-dialkylhydrazines; with strong reagents reduction goes all the way to amines. The rocket fuel 1,1-dimethylhydrazine has long been prepared by reduction of nitrosodimethylamine. The hydrazines have totally different properties from the nitrosamines; they are bases, for example, and form stable salts. Biologically their activities are quite different from the nitrosamines. Hydrazines are considerably less potent toxins and carcinogens than the analogous nitrosamines, and they are not mutagenic in the usual systems. This makes it implausible that the carcinogenic activities of N-nitroso compounds are mediated in any way through reduction to the hydrazines, or through further reduction which gives rise to amines. It seems quite unlikely, therefore, that metabolic oxidation to nitramines or reduction to hydrazines plays any role in the carcinogenic or mutagenic action of N-nitroso compounds, although both types of product might be formed to some extent *in vivo*.

3.8 Uses of nitrosamines in synthesis

In parallel to the work of Seebach, J. E. Saavedra has undertaken extensive studies of the usefulness of nitrosamines in organic synthesis, particularly that of amines, and the field has been reviewed by him recently (Saavedra, 1987a, b). These studies were based on the early observation of Keefer and Fodor (1970) of the formation of α-nitrosamino carbanions from nitrosamines in base. In turn this suggested the usefulness of nitrosamines as α-secondary amino carbanionic synthons. Seebach and Enders (1975) reported that metalation of nitrosodimethylamine with lithium di-*iso*propylamide at −80 °C formed the corresponding lithionitrosamine which can undergo alkylation or acylation with electrophiles (Fig. 3.13). The alkylated or acylated product could be denitrosated to the secondary amine homologue. This reversal of polarity, or 'umpolung' of the secondary amine through conversion to the derivative, which can be used for establishment of a new C-C bond (Fig. 3.14), is one example of a general class of 'umpoled' synthons, extensively explored by Seebach and others. As well as showing the usefulness of nitrosamines for synthesis of substituted amines, it also illustrates the chemical similarity between N-nitroso compounds and other classes of organic chemicals, especially carbonyl compounds, including aldehydes, ketones, amides and esters. The resemblance hints at possible activation pathways of N-nitroso compounds that might relate to their biological activities, particularly those that differ markedly between one similar nitrosamine structure and another. In particular, the reaction of an activated N-nitroso compound directly with some biomolecule to form an adduct with possibly greatly altered function, is an alternative to the commonly accepted S_n1 formation of an alkylating agent, which then alkylates DNA (and other biomole-

Fig. 3.13. Metalation and alkylation/acylation of nitrosamines.

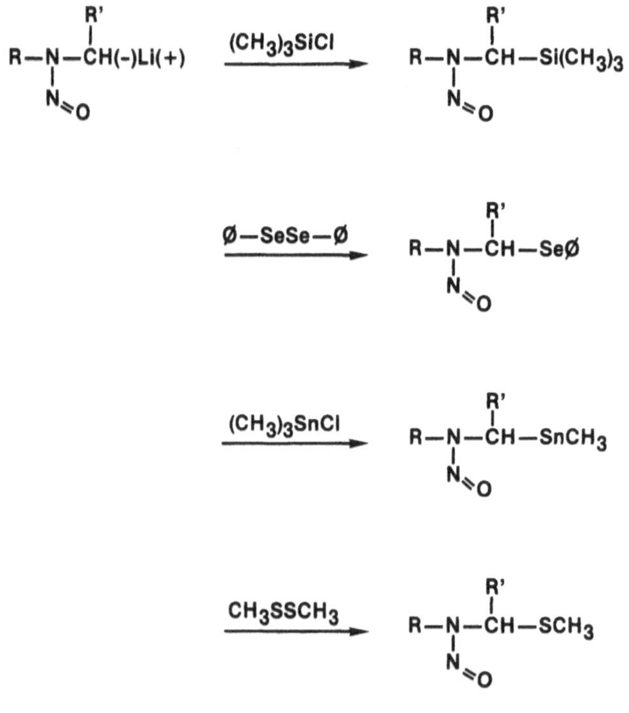

Fig. 3.14. Umpolung of amines/nitrosamines.

cules); the latter does occur in many cases, but it is not necessarily the most important reaction in toxicity and carcinogenesis by N-nitroso compounds.

An example of the usefulness of nitrosamino carbanions is the preparation of several naturally occurring alkaloids, such as pseudoconhydrine, a derivative of piperidine (Renger et al., 1977), and macostomine, a derivative of pyrrolidine (Wykypiel and Seebach, 1980). A vast number of naturally occurring compounds that have biological activity (particularly toxins) are secondary amines, for synthesis of which nitrosamino carbanions are useful. The latter also provide a route to synthesis of sydnones from nitrosamino acids formed from α-metalated nitrosamines. Conversely, heating of nitrosamino acids with lead tetraacetate gives rise to the α-acetoxy nitrosamines (Saavedra, 1979), which are reactive, mutagenic, alkylating derivatives of nitrosamines, and which are stable forms of the α-hydroxynitrosamines assumed to be formed by metabolic oxidation of nitrosamines.

Nitrosamines are useful for synthesis of more complex and useful organic compounds, particularly amines. This is not, however, a strong reason for interest in them, or in their chemistry. Instead, that interest is attributable to their unique biological activities, including the positive attributes of useful therapeutic action against some cancers. The chemistry and reactivity of N-nitroso compounds is broadly interesting because of guidance it might give to understanding the mechanisms by which N-nitroso compounds exert their mutagenicity and organ-specific carcinogenicity. Thus far, our understanding of these biological actions is meagre, even rudimentary, perhaps reflecting the minor research activity in the chemistry of N-nitroso compounds, restricted to a small number of laboratories in the world.

3.9 Activation and reactions of nitrosamines

There is no doubt that nitrosamines must be metabolised in order to become toxins, mutagens or carcinogens. Without activation they are simply good solvents, since most of the unsubstituted ones are liquids at room temperature, whether symmetrical or unsymmetrical, cyclic or acyclic. Those with appropriate substituents are reactive to various degrees and often have other properties, for example the toxicity of haloalkylnitrosamines and of α-acetoxynitrosamines to micro-organisms or to cells in culture. However, when activated by enzymes in the liver of rats or hamsters, nitrosamines become reactive and interact with DNA of test organisms so as to induce mutations, by alkylation of DNA. The nature of the alkylating agent is fairly well identified in the case of some simple nitrosamines, such as nitrosodimethylamine or nitrosodiethylamine, because methylated or ethylated purine and pyrimidine nucleotides or bases have been isolated and identified, in some cases by mass spectrometry. Furthermore, the use of test micro-organisms susceptible to mutagenesis at particular sites in their DNA has shown that several N-nitroso compounds expected to give rise to a methyldiazonium ion (NDMA, MNEA, methylnitrosourea) produce the same spectrum of mutants, and those that are expected to give rise to an ethyldiazonium ion (NDEA, MNEA and ethylnitrosourea) give rise to a different set of mutants or mutational fingerprints (Elespuru et al., 1991).

In the same vein, analysis of the DNA of rats or hamsters given these compounds also show the formation of, respectively, methylated or ethylated bases, thereby providing a link between the mutational results in bacteria and the chemical interaction in cells. What is missing, at present, is a connection between the DNA alkylation in cells and the induction of tumours in them. The matter becomes yet more complicated when our attention broadens beyond the simplest methylating and ethylating agents, to cyclic nitrosamines and those with more complex, sometimes substituted, alkyl groups. Many of these have almost identical carcinogenic actions, insofar as the same tumours are induced with a similar efficiency, to the simple alkylating agents. Yet, the cyclic nitrosamines, although they are often strong mutagens following appropriate activation, do not usually give rise in cells to alkylated bases which can be readily identified; at least, such alkylations have not been identified until now.

Cyclic and acyclic nitrosamines are chemically very similar. It is both theoretically and practically possible for α-oxidation of cyclic nitrosamines to take place with formation of α-hydroxyl derivatives, or of α-carbonyl derivatives which, of course, are nitroso-lactams. Nitrosopyrrolidone is such a compound and is more stable than the nitroso derivatives of the higher homologous lactams; nitrosopyrrolidone, however, is not stable enough to have been tested adequately for carcinogenic activity, but it is a directly acting mutagen. The nitroso-peptides (Fig. 3.15) described by Challis et al. (1984) are analogous acyclic compounds and these, too, are relatively unstable and not conveniently tested for carcinogenic activity. The reactions of the nitrosamides, particularly the cyclic nitrosolactams, are not well characterised, except for their fairly rapid decomposition in acid media. For this reason they might not be expected to survive if formed in the most likely conditions, in the mammalian stomach containing the peptides and a nitrite salt.

Oxidation at other positions in a cyclic nitrosamine occurs with formation of stable hydroxy or keto derivatives. These are in many cases known and of no particular interest. Since they are usually less potent carcinogens than the parent compound, they are unlikely to be important intermediates in the activation of cyclic nitrosamines to carcinogenic forms. It is possible that some of the hydroxy compounds become conjugated with glucuronic acid or sulphuric acid – and some such derivatives have been isolated – but it is more probable that they are simply excreted, as has been found in the case of nitrosopiperidines. The oxidation of cyclic nitrosamines remains a poorly understood matter, although it is very likely to be important in their biological activity and, in the case of a few substituted compounds, such as 2,6-dimethylnitrosomorpholine, amenable to study.

Alpha oxidation seems to be the most important reaction of nitrosamines that is common to all of them and it gives rise to unstable products from acyclic nitrosamines, but to rather more stable derivatives from cyclic nitrosamines. Little is known about the subsequent reactions, which could include esterification or further oxidation, perhaps to the corresponding carbonyl compound; such a product from NDMA, methylnitrosoformamide, is more toxic and a more potent mutagen than NDMA, but more stable then hydroxymethylnitrosomethylamine (Huisgen and Reimlinger, 1956). The α-hydroxy compound might also interact with a receptor molecule, perhaps a protein or nucleic acid. Little study has been made of the reactions of α-hydroxy nitrosamines, although in a few examples their formation has been demonstrated (Fig. 3.16) and their stabilities measured (Mochizuki et al., 1980, 1982). They can also be generated by hydrolysis of esters, such as the α-acetate of nitrosodimethylamine, which is relatively easily formed (Wiessler, 1974) from simple precursors. Analogous products, including phosphates and

Fig. 3.15. Nitrosopyrrolidone and a nitrosodipeptide.

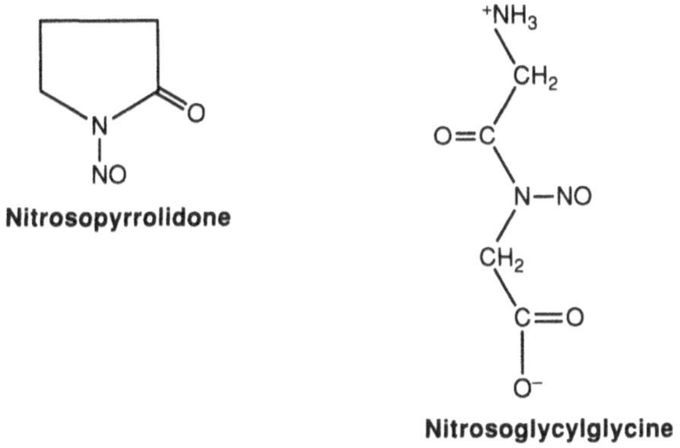

Nitrosopyrrolidone

Nitrosoglycylglycine

sulphates, might be reactive intermediates derived from nitrosamines in biological systems. This has been modelled by irradiation of nitrosamines in the presence of phosphate (Shimada et al., 1987). Some α-hydroxynitrosamines, indeed, have been prepared and tested for mutagenic activity (Mochizuki et al., 1987), although this is not necessarily an index of the mechanism by which they induce tumours, especially in the organ-specific aspects.

The α-oxidation of nitrosamines is an example of the resemblance between nitrosamines and aldehydes and ketones, as first demonstrated by Keefer and Fodor (1970). The metalation of nitrosamines has been used as a synthetic tool by Seebach and Enders (1975). Little attention, however, has been given to the possibilities that these reactions reveal for alternative mechanisms of carcinogenesis to the usually accepted conversion into simple alkylating agents. That there are interactions with biomolecules not involving alkylation is known from experiments with radiolabelled nitrosamines, by measuring the radioactivity associated with material not precipitable as biopolymers by trichloracetic acid or methanol from tissue homogenates. The resemblance of hydroxymethylnitrosomethylamine to hydroxyacetone, or of methylnitroso-2-oxopropylamine to activated acetone, suggests that activated nitrosamines might simulate some common metabolic intermediates and compete with them for active sites on enzymes or membranes.

The presence of the N-nitroso group is the common thread connecting all N-nitroso compounds, biologically active or not. Examination of structure–activity relations, using a variety of computer programs, has not yet shown mathematical predictability of carcinogenicity, in spite of such claims (Rosenkranz and Klopman, 1987). Physical properties, such as liposolubility, have not shown convincing correlations outside a relatively small group of compounds (Singer et al., 1977). It seems, then, that the influence of the N-nitroso group on the chemical reactivity of the compound has a direct and powerful relationship to organ-specific carcinogenesis, and possibly to mutagenesis. The size of the molecule has some influence, although many compounds of large molecular weight are

Fig. 3.16. Formation of α-hydroxynitrosamines.

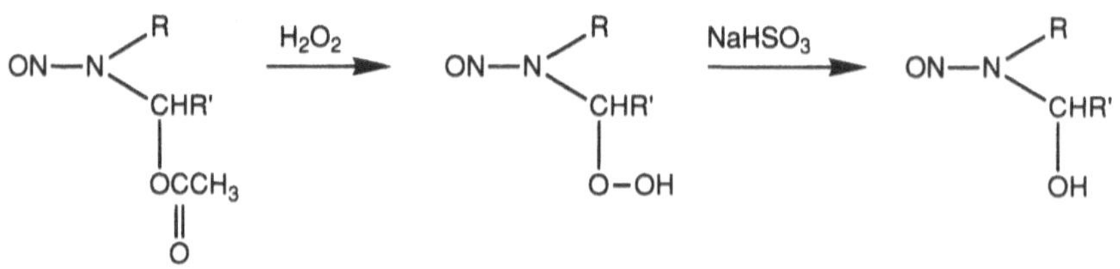

quite potent carcinogens, if they satisfy other structural requirements. It is also known, and quite logical, that the E-rotamer of an asymmetrical nitrosamine is more susceptible to oxidation than the Z-rotamer (Fig. 3.17), as evidenced by the more rapid metabolism of the E-isomer (Farrelly et al., 1988). Chemical reactivity has been not well studied in this regard, however. When the compounds are metabolised in animals, reactions are very rapid and it is quite likely that the initial biological effects will result from metabolism of the E-isomer, and that the effects of the Z-isomer will be quite different, because the rate of equilibrium is often low. Therefore, the proportion of E-isomer in the original mixture will determine the biological effect, be it toxicity or those reactions leading, cumulatively, to induction of tumours at some distant time.

It is quite conceivable that the product of oxidation of a nitrosamine is itself toxic to cells, for instance formation of an α-hydroxynitrosamine, rather than an alkyldiazonium ion resulting from further breakdown of that product. The reason for this assumption is that alkylnitrosoureas, which give rise to an alkyldiazonium ion by solvolysis, are much less toxic than nitrosamines that give rise to the same alkyldiazonium ion. It is unlikely that the toxicity of a nitrosamine to cells derives from alkylation of nuclear DNA; most toxic cell death is due to reactions with cytoplasmic constituents which disturb or distort the structure of the cytoplasm.

The interactions of nitrosamines with other molecules that could serve as models for their reactions in biological systems are not well known. They can be alkylated by powerful alkylating agents, such as triethyloxonium fluoroborate or dimethyl sulphate, to form salts. Less-reactive alkylating agents, such as methyl iodide, have been reported to form N-alkylated amino products (Schmidtpeter, 1963). This is probably via the O-alkyl salt which rearranges to the N-alkyl product.

In analogous fashion, acylation of nitrosamines probably proceeds via O-acylation. Few reactions of this kind have been reported, but the result is replacement of the NO group by acyl. Neither acylation nor alkylation of this type, although offering an alternative route of reaction of nitrosamines, seems to be a likely pathway of interaction with biological

Fig. 3.17. E and Z isomers of methylnitroso-2-oxopropylamine.

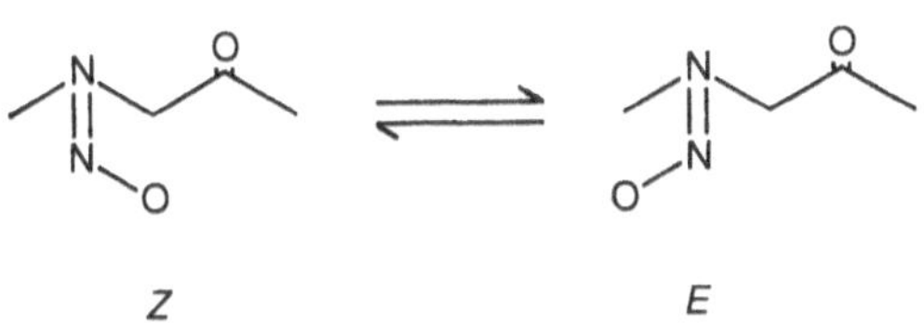

consequences in living systems. However, one must not be dogmatic and must leave open the possibility that cellular alkylating or acylating agents, whatever they might be, could react in this way with nitrosamines.

N-nitroso compounds can be denitrosated, and the structure of some of them is such that they lose NO spontaneously. Such are N-nitroso derivatives of aromatic amines, including nitrosodiphenylamine (Fig. 3.18), nitrosoatrazine and nitrososimazine (the latter two being derivatives of widely used pesticides). These N-nitroso compounds are, of course, good nitrosating agents for receptive amines, and nitrosodiphenylamine is a major source of the nitrosamines formed in rubber manufacture and processing (Lijinsky, 1990a). A number of cyclic nitrosamines, such as 1-nitrosopiperazine and nitroso-N-methylpiperazine, nitrosoproline, nitrosohydroxyproline, sterically hindered nitrosamines such as 2,2,6,6-tetramethylnitrosopiperidine and 2,6-dimethylnitrosopiperidine, lose NO readily and are good nitrosating agents, but they are weak carcinogens or are not carcinogenic. Alkylnitrosamides are also nitrosating agents, but they are usually potent carcinogens. Some of the nitrosamines that are readily denitrosated have been shown to give rise to tumours when fed to rats in combination with the readily nitrosated morpholine (Lijinsky and Reuber, 1982b). It seems unlikely, although not impossible, that the

Fig. 3.18. Nitrosodiphenylamine, N-nitrosoatrazine and N-nitrososimazine.

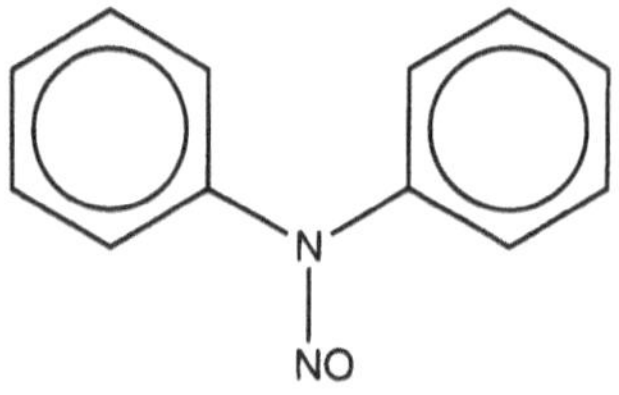

N-nitrosodiphenylanine

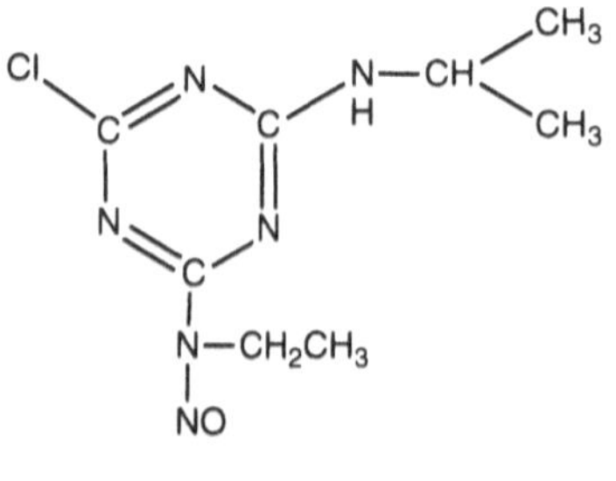

N-nitrosoatrazine

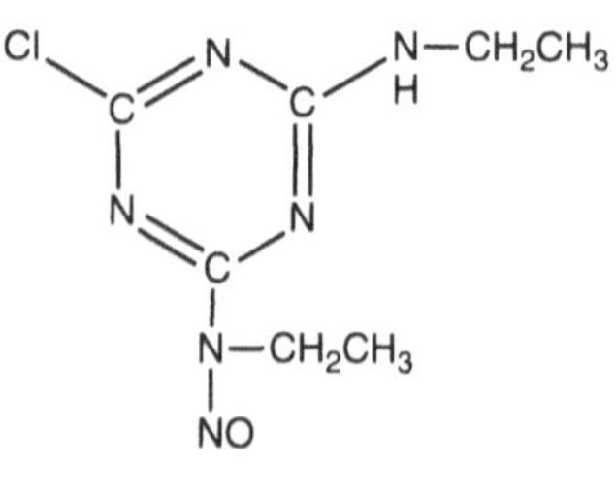

N-nitrososimazine

Activation and reactions 91

denitrosated residues of *N*-nitroso compounds play a role in tumour induction by these carcinogens. Similarly, it is not likely that nitrosation of cellular components, leading, for example, to deamination of amino acids or nucleotides, plays a role in tumour induction. There is little evidence that nitrous acid or nitrites are carcinogenic, although they are mutagenic. One study has revealed an increased incidence of liver tumours in female rats treated orally with sodium nitrite at high concentrations for most of their lifetime (Lijinsky *et al.*, 1983*a*).

One set of studies initiated by R. Loeppky has provided insight into the chemistry and reactions of β-hydroxynitrosamines, and that is their tendency to undergo a retroaldol cleavage in solution under the influence of base (Loeppky *et al.*, 1980). These studies began with model compounds (Fig. 3.19), but later included some that are important environmental carcinogens. Initial experiments in chemical systems were followed by biochemical studies *in vitro* using enzyme preparations (Loeppky and Outram, 1982). The results clearly showed that such reactions could take place in the body and might be an unexplored facet of the relation of nitrosamine carcinogenicity to chemical structure. One example is the formation, under favourable conditions, of the potent carcinogens methylnitroso-2-hydroxypropylamine and nitrosodimethylamine from the relatively weak carcinogen (but a common environmental contaminant) nitrosobis-(2-hydroxypropyl)amine. The importance of such reactions in giving rise to potent carcinogens in commercial materials containing nitrosating agents and amines is obvious. Their importance in our understanding of mechanisms of carcinogenesis by nitrosamines is less apparent, although if a β-ketonitrosamine were partially converted by retro-claisen cleavage to nitrosodimethylamine (Saavedra *et al.*, 1989), for

Fig. 3.19. Retro-Claisen reaction of methylnitroso-2-oxopropylamine.

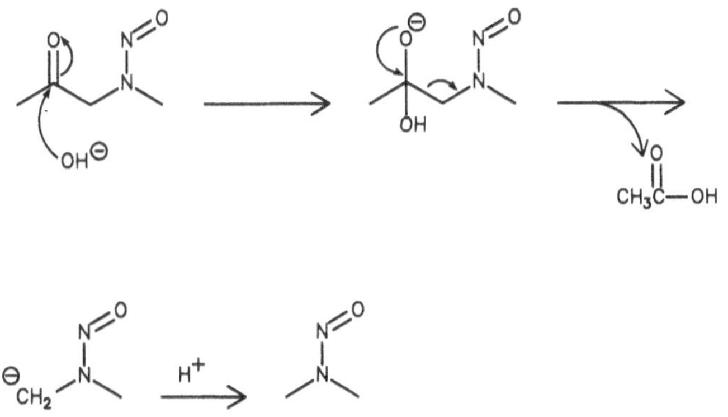

example, that might help explain the broad potency of the former in comparison with the latter. It is notable that in their studies of the reactions of β-hydroxynitrosamines in basic solution, Loeppky et al. (1980) found that the very common environmental contaminant nitrosodiethanolamine gave rise to nitrosodimethylamine, the latter being a very much more potent carcinogen than the former. Loeppky postulated that dehydration to the vinyl compound might also occur, and he reported that methylnitrosoethanolamine and methylnitrosovinylamine were found in old stored samples of nitrosodiethanolamine, and in cutting oils containing nitrosodiethanolamine. These preliminary studies are likely to be followed by findings of exposure of humans to unexpected nitrosamines from environmental sources.

Other possible chemical reactions of nitrosamines that could be related to induction of tumours have received little attention because they are not very reactive compounds. Nitrosamines with substituents that render them more reactive, such as hydroxyl, carbonyl or halogen, are not qualitatively different in carcinogenic effect from those without substituents. Such reactivity, then, even when permitting activation through formation of sulphate esters, for example, is not the germ of an explanation for their carcinogenic activity. It is more likely that their interaction with some cellular constituent, perhaps an enzyme, is the fundamental reaction leading eventually to the biological effect, and this they could have in common with the directly acting alkylnitrosamides. However, in the case of the latter, energy-requiring oxidative metabolism is unnecessary. It is possible that our focus on the detoxifying cytochrome P450 enzymes as the generators of the proximate carcinogenic metabolites of nitrosamines is misplaced, since many carcinogenic nitrosamines are not metabolised by them; other enzymes or receptors might be more important reactants.

What these might be is uncertain, but some aspects of the structure of nitrosamines cannot be ignored, for example the resemblance between the biologically potent 2-oxopropylnitrosamines and acetone, between nitrosoethanolamines and choline, and between α-hydroxymethylnitrosamines and hydroxyacetone. These analogies suggest that the activated N-nitroso compounds might compete with those biologically important molecules for the active sites of enzymes or other macromolecular entities, and thereby cause derangement of critical structures or impede normal essential biochemical pathways. Such damage might emerge as death of cells as a manifestation of the toxicity of the compound, or as part of the slow process of neoplastic transformation leading to the emergence of autonomous tumours, arising from a few cells of many affected. This process is no better known for N-nitroso compounds than it is for any other carcinogenic agent, including chemicals, viruses or radiation.

3.10 Interactions of N-nitroso compounds in biological systems

The decomposition of alkylnitrosamides in aqueous solution has been discussed, and this decomposition is rapid at alkaline pH, with formation of diazoalkanes and other products. Both alkylnitrosamides and nitrosamines decompose in acid conditions, to form mainly the amino compound and nitrous acid. It is notable, on the other hand, that as pure substances most nitrosamines are very stable, not decomposing detectably after many years in the dark at room temperature. Alkylnitrosamides are more stable as pure substances than in solution, especially if kept at lower temperatures, although decomposition in both states occurs readily at elevated temperatures.

The reactions of alkylnitrosamides are largely a consequence of their ready protonation and formation of an alkyldiazonium ion, which is a reactive electrophile; this is the same putative electrophilic alkylating agent that is reported to be formed following oxidation of nitrosamines at the methylene α to the N-nitroso function. The alkyldiazonium ion alkylates any nucleophile, including primarily water, and cellular macromolecules such as proteins and nucleic acids; these latter interactions have been studied as probable mechanisms by which the alkylnitrosamides exert their biological effects, especially mutagenicity and carcinogenicity, but also possibly certain types of toxicity. Reactions of alkylnitrosamides with nucleic acids (or nucleotides and synthetic polynucleotides) and with proteins or polypeptides results in formation of a variety of alkylated products. The comprehensive studies by Singer (1976) of the reactions of ethylnitrosourea with DNA have shown that every possible nucleophilic site in the polynucleotide is ethylated (Fig. 3.20), including extensive formation of phosphate esters.

These chemical studies are interesting and important in themselves, but their relation to genetically mediated biological effects in living organisms is even more important. Indeed, much that we know about chemical mutagenesis and carcinogenesis has been revealed through examination of the effects of alkylnitrosamides as alkylating agents, and is discussed in other chapters. The crucial problem in evaluating the mass of chemical information is deciding which of the alkylated products (or sites in macromolecules) are biologically important – assuming that any of them are – in face of a large number of possibilities. The work of Loveless (1969) was decisive in drawing a distinction between those sites on DNA that were merely susceptible to alkylation (guanine-N^7), and those that were likely when alkylated to produce mispairing when the DNA replicates, and thence lead to mutation (e.g. guanine-O^6). Since then other mutational lesions have been discussed, such as thymidine-O^4 (Swenberg et al., 1984).

The identification of alkylated bases in DNA produced by alkylnitrosamides has proceeded apace, and large numbers of papers have been published dealing with the quantitative aspects of alkylation, the extents of alkylation in several organs of various species, differences in the extent of alkylation produced by various alkylating agents, and differences in the rates of repair of the alkylated damage (i.e. enzymic removal of the alkyl group) in relation to carcinogenesis (see, e.g. Goth and Rajewsky, 1974; Pegg, 1983 a, b; Lickhachev et al., 1983 b). However, none of this has appreciably increased our understanding of organ-specific tumour induction by alkylnitrosoureas, the elegant studies of Barbacid and his group in identifying a mutated gene in the rat mammary gland produced by treatment with methylnitrosourea notwithstanding (Zarbl et al., 1985).

Much less attention has been given to the alkylation by alkylnitrosoureas of proteins and other cellular components, including membranes, which might play an important role in neoplastic transformation if damaged. In the beginning it was alkylation of cellular proteins by nitrosodimethylamine that first engaged the attention of Magee (Magee and Hultin, 1962). Because of the amount of protein in a cell, the alkylation of protein is much greater than that of nucleic acids. There have been studies of interaction of alkylnitrosamides with histones and other nuclear proteins, but little has been reported recently.

It was reported some time ago by Sugimura et al. (1968) that the guanido portion of the MNNG molecule interacts with proteins to form

Fig. 3.20. Nucleophilic sites (marked with asterisks) in DNA bases susceptible to alkylation.

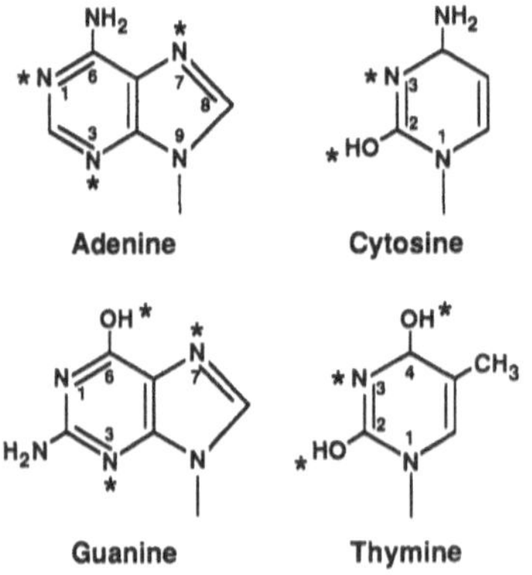

Adenine

Cytosine

Guanine

Thymine

stable adducts, in addition to the *N*-alkylation and *O*-alkylation of nucleic acids that has been given so much attention. It is also reasonable, in view of the predilection of alkylnitrosoureas with particular alkyl groups (e.g. methyl vs. ethyl) for particular organs or tissues, that alkylnitrosamides might react directly with certain molecules in cells, to form relatively stable structures. These might remain and interfere with the normal functions of these molecules (perhaps critical to the functioning of the cells), or they might undergo intramolecular rearrangements with elimination, leaving the alkyl group attached to a particular nucleophilic site. This might, in turn, have the damaging consequences that have been attributed to alkyl groups in DNA. This could be a partial explanation for the differences between the large number of alkylated sites found by Singer in DNA treated with ethylnitrosourea *in vitro* and the relatively small number of such alkylated sites found *in vivo*. The yield of alkylated products in DNA from a given amount of alkylnitrosamide *in vivo* is quite small, and most of the alkylnitrosamide is apparently consumed in other reactions (perhaps important) that have received little or no attention.

The decomposition and solvolysis of some dialkylnitrosoureas and trialkylnitrosoureas have been studied by Singer (1982), but have not provided much insight into the mechanisms by which these compounds exert their organ-specific carcinogenic effects. However, they have clarified some aspects of the chemistry of these compounds, having a bearing on the chemistry of the less-stable monoalkylnitrosoureas, which are much more difficult to study because of their instability. There is some evidence of migration of the nitroso group in a dialkylnitrosourea, particularly in acid solution, from the side with one alkyl group to the other side. Dismutation of a trialkylnitrosourea has also been shown, forming a dialkylamine or a nitrosodialkylamine and an alkyl carbonate or carbamate (Singer, 1982). For the same reasons alkylnitrosamides are frequently good nitrosating agents, a chemical property which will be discussed elsewhere.

3.11 Conclusions

Some aspects of the chemistry of nitrosamines will be described in Chapter 4 dealing with metabolism and biological activation of nitrosamines to presumptive mutagens and carcinogens, most notably alkylating intermediates. These aspects have preoccupied the attention of investigators, including chemists and biochemists, and the remainder of the chemistry of these hitherto uninteresting chemicals has been relatively unexplored. However, what is known of their chemistry might provide some leads to explaining their unique biological properties, most notably their almost universal activity as carcinogens in the animal kingdom. The influence of the nitroso group on the tendency of α-hydrogens to be replaced has been described. This is related to oxidation of nitrosamines at the α-position, which forms a product that is unstable and can decompose with formation of a carbonyl compound and an alkylating agent, claimed to be the same as that formed directly from the corresponding alkylnitrosamide.

The difference in biological activity between the two types of N-nitroso compound is considerable. The toxicity of nitrosamines is usually to the liver and lungs, occasionally to the kidneys and, in the case of some cyclic nitrosamines, to the central nervous system, although the latter have not induced tumours in the nervous system. Alkylnitrosamides, on the other hand, are not usually toxic to the liver, but they are toxic to the lungs, gastrointestinal tract and sometimes the kidneys. Some are toxic to the central nervous system, and that is often a target of their carcinogenic effects.

There are exceptions to these general observations, for example the great toxicity of methylnitrosourethane administered intravesically to rats, which caused death of three of 12 rats following a single dose of less than 2 mg; a dose larger than this given by gavage was well tolerated. Ethylnitrosourethane administered intravesically at an equimolar dose was not noticeably toxic to rats, another illustration of the large difference in effect between methyl and ethyl homologues of the same structure. Alkylnitrosoureas were less toxic than the urethanes when given intravesically. Since both types of nitrosamide are principally alkylating agents donating the same alkyl group, it is unclear why there should be such a difference in toxic effect, unless the greater partition coefficient into lipid of the nitrosourethane is a critical factor. Assuming that alkylation of some component of cells is responsible for the toxic effects of alkylnitrosourethanes, which components are the critical targets is unknown. These nitroso compounds showed little toxicity when painted on the skin of mice, and they did not induce skin tumours, whereas alkylnitrosoureas showed some skin toxicity and did induce skin tumours. However, in rats several alkylnitrosoureas gave rise to skin tumours even when given orally, whereas alkylnitrosourethanes did not. Unlike alkylnitrosourethanes, which are volatile, nitrosooxazolidones, which are non-volatile, induced skin tumours.

Reactions of N-nitroso compounds related to induction of tumours are equally difficult to specify. Here, the vast differences in carcinogenic activity between alkylnitrosamides and nitrosamines make it unlikely that the products of direct action of the former and of metabolic activation of the latter leading to carcinogenesis are the same, or even similar. Excluding formation of an alkylating agent such as an alkyldiazonium ion, there are few alternative reactions by which an alkylnitrosamide might be expected to produce neoplastic transformation. There is a possibility of direct interaction of the alkylnitrosamide with cellular macromolecules, but what they might be is beyond speculation. They can also lose their nitroso group, or undergo oxidation or reduction, neither of which seems likely to be important in their biological action, and neither of which is peculiar to alkylnitrosamides as distinct from nitrosamines. Whether oxidation of the carbon atom α- to the N-nitroso group in alkylnitrosamides can occur or does occur is not known, but it is an alternative route of activation that might explain the peculiar carcinogenic activity of alkylnitrosoureas, for example. It would hardly explain, however, the large differences between the organ-specific carcinogenic effects of alkylnitrosoureas and alkylnitrosourethanes, which can obviously not be entirely attributed to differences in partition coefficient or uptake.

The reactivity of alkylnitrosoureas with organelles within cells, including membranes, or macromolecules that are components of them is an unexplored subject, but appears to be a key process in carcinogenesis. The results of the structure–activity studies with alkylnitrosoureas show that structural characteristics determine their affinity for some molecules in cells of certain types, and suggest that these molecules might play a key role in controlling cell division. The nature of these receptors (which must have normal substrates, that are presumably displaced) is not known, nor where they are located. Neither is the process by which the alkylnitrosoureas interact, whether by simple addition or following activation of some kind.

4

Metabolism and cellular interactions of N-nitroso compounds

4.1 Introduction – history

Shortly after the discovery of the toxicity of nitrosodimethylamine, colleagues of Magee and Barnes began to study the metabolism of the compound, because it was believed that metabolites were responsible for the toxic effects (Heath, 1961); the nitrosamine itself was thought to be an inert solvent. Early experiments by Heath and Jarvis (1955) estimated the disappearance of the nitrosamine in rats by measurement of the concentration using the then novel technique of polarography, which was sensitive and specific for reducible compounds. They used ^{14}C-labelled NDMA to measure the rate of excretion of carbon dioxide, which was a measure of metabolism of the nitrosamine (Dutton and Heath, 1956). The same radiolabelling was subsequently used by Magee and his co-workers to discover that the proteins (Magee and Hultin, 1962) and nucleic acids (Magee and Farber, 1962) of rat liver, the target organ of the carcinogenic NDMA, were radiolabelled by products of NDMA metabolism. Further experiments, based on related studies by Brookes and Lawley (1960) of interactions of mustard gas with macromolecules in animals, led to identification of 7-methylguanine as an end product of NDMA *in vivo*, present in the DNA and RNA of the liver of the treated rats. The initial studies of Magee and Farber (1962) used the chromatographic identity of the radiolabelled product isolated from the nucleic acid hydrolysates with standard 7-methylguanine to identify it. Later, deuterium-labelled NDMA (NDMA-d$_6$) was used in similar experiments to show unequivocally that the alkylated base formed in nucleic acids was 7-methylguanine by determination of its accurate mass by mass spectrometry, and thence its molecular composition (Lijinsky *et al.*, 1968). Using large toxic doses of NDMA as much as 2% of the guanine was methylated.

The first experiments used RNA for the analysis, because it was isolated

Introduction – history

in larger amounts than was DNA using the relatively crude procedures then available; relatively large quantities of the alkylated bases were required for the identification, also. It was assumed that alkylation of RNA would reflect the same reaction with DNA. Later this was found to be not necessarily true, particularly after the suggestion by Loveless (1969) that alkylation of guanine in DNA at the 7-position was unlikely to produce a mutagenic change (and therefore was unlikely to lead to induction of tumours), while methylation of the phenol at the 6-position of guanine (O^6-methylguanine) would lead to mispairing and through subsequent changes to a mutation. Loveless' proposal also explained why such potent mutagens as methylmethanesulphonate were such weak carcinogens in animals, because this alkylating ester methylates ring nitrogens (quaternising them), but has little tendency to methylate phenols.

Similar mass spectrometric methods were used to show that nitrosodiethylamine-d_{10} gave rise to ethylation of guanine in RNA at the 7-position and in DNA; it was assumed that O^6-ethylation of guanine in DNA took place in the same way (Ross et al., 1971).

The studies using deuterium-labelled NDMA and NDEA served another purpose. Because a fully deuterated methyl or ethyl group was found in the guanine from the nucleic acid, it was concluded that the alkyl group was transferred intact, through perhaps an alkyldiazonium ion, rather than through formation of a diazoalkane, as long believed (Schoental, 1960; Druckrey et al., 1967). The latter would have entailed loss of a deuteron from the deuterated alkyl group, forming an alkylguanine with one mass unit less, which was not observed (Fig. 4.1). Even methylnitrosourea, which at alkaline pH forms diazomethane directly, was found to transfer a fully deuterated methyl group when methyl-d_3-nitrosourea was administered to rats and the liver nucleic acid hydrolysate was analysed by mass spectrometry (Lijinsky et al., 1972c). Similar studies showed that the methyl group of methylnitrosocyclohexylamine was transferred intact to guanosine in rat liver nucleic acids (Lijinsky et al., 1973c), although this nitrosamine did not induce liver tumours in rats under a variety of circumstances in which it has been tested; instead, it gave rise to tumours of the esophagus, even in very low doses (Lijinsky and Reuber, 1980b; Lijinsky et al., 1989a). The alkylation of DNA and mutations so produced were believed to cause neoplastic transformation leading to tumour formation, but not always (e.g. methylnitrosocyclohexylamine).

From the beginning of studies of N-nitroso compounds it was known that they were of two types, nitrosamines requiring metabolic activation and nitrosamides which were directly acting. The nitrosamides were unstable and acted as direct alkylating agents towards nucleophiles such as DNA, so that they were directly acting mutagens and carcinogens. On the other hand nitrosamines are stable compounds and biologically and chemically inert unless activated. In biological systems they can be activated by a variety of enzymes, which oxidise them to aldehydes and intermediates which are themselves alkylating agents (Fig. 4.2). Most nitrosamines can, then, be metabolically activated to mutagenic forms; some, however, are not demonstrably mutagenic, even though they are carcinogenic, for example nitrosodiethanolamine, 2-hydroxypropylnitrosoethanolamine, methylnitrosoaniline, nitrosothialdine and nitrosothiomorpholine. These nitrosamines and others have not been converted to mutagens by oxidative enzymes from either rat or hamster liver. The hydroxylated nitrosamines appear not to be metabolised by P450 enzymes (Farrelly et al., 1984) (which is also true of some other types of carcinogen, including azoxyalkanes (Lijinsky et al., 1985a); sulphur-containing cyclic nitrosamines are not activated to mutagens, although they may be metabolised, and several are carcinogenic.

Little attention has been paid to other metabolites of nitrosamines, apart from those intermediate in formation of alkylating agents. There are

Fig. 4.1. Derivation of trideuteromethyl-guanine from NDMA-d_6 and MNU-d_3.

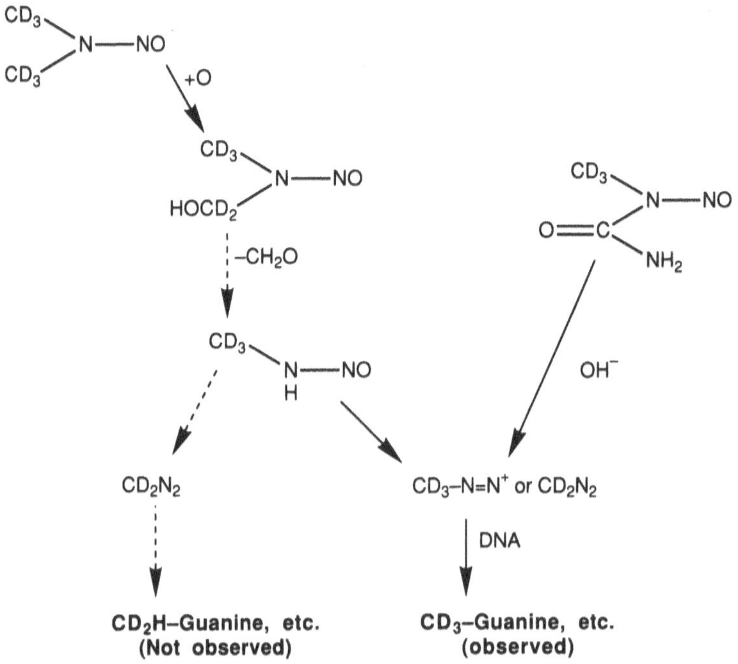

Introduction – history

certainly other metabolites, since only 60% or so at most of the dose of ^{14}C-labelled nitrosamine (or nitrosamide) is excreted as CO_2 and only a minute proportion forms the alkylated macromolecules. Only a small proportion is excreted in the urine (Table 4.1) and usually a negligible proportion in the faeces. The metabolites accounted for by the remaining 30 to 40% of the dose might well play an important role in tumour development, and explain why alkylation of DNA – even at mutagenic sites – does not inevitably cause tumours to arise in most organs.

Among the N-nitroso compounds whose urinary excretory products have been examined is nitrosodimethylamine, notably by Hemminki (1982). Several radioactive products were tentatively identified with compounds having the same retention time on ion-exchange columns, and were 7-methylguanine, methylhistidine and S-methylcysteine, the last two reflecting the considerable methylation of protein reported by Magee and

Fig. 4.2. Oxidative activation of dialkylnitrosamines.

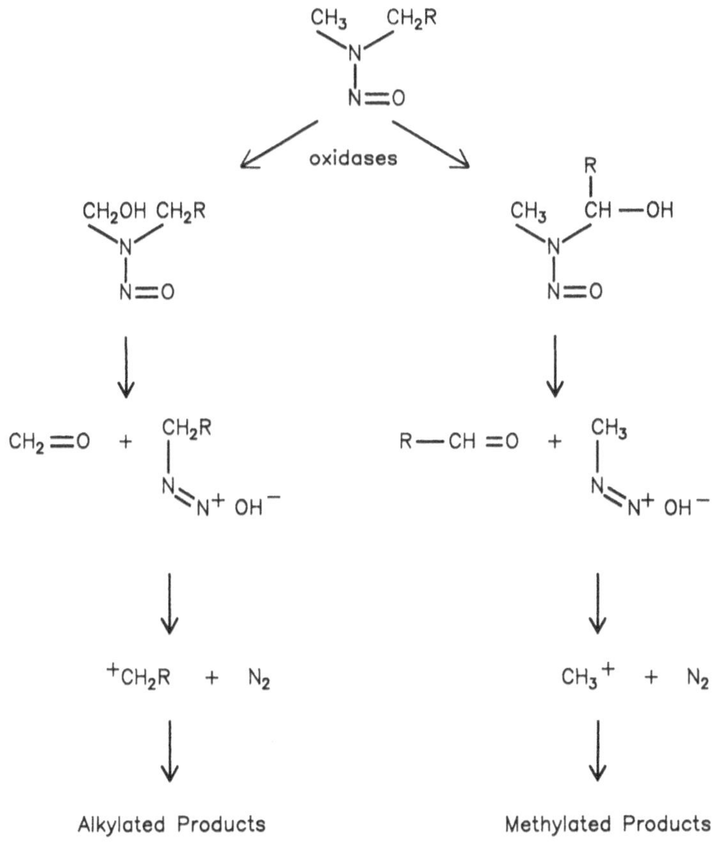

Table 4.1. *Metabolism of ^{14}C-N-nitroso compounds*

Compound	Dose	Animal	% of dose excreted in 6 h as CO_2	% of dose excreted in 6 h in urine
N-nitroso-				
dimethylamine	1.3 mg/5 μCi	♂ Rat	57	2
		♀ Rat	55	3
		♂ Hamster	55	4
methylurea	2 mg/4 μCi	♂ Rat	53	7
		♂ Hamster	30	2
^{14}C-methylethylamine	1.7 mg/5 μCi	♂ Rat	45	4
		♀ Rat	45	5
		♂ Hamster	18	3
methyl-^{14}C-ethylamine	1.7 mg/36 μCi	♂ Rat	73	2
		♂ Hamster	77	4
methyl-^{14}C-ethyl-d_2-amine	1.7 mg/46 μCi	♂ Rat	40	15
		♂ Hamster	51	27
^{14}C-methyloxopropylamine	2 mg/5 μCi	♂ Rat	42	7
	2 mg/5 μCi	♀ Rat	41	7
	2 mg/5 μCi	♀ Hamster	25	11
1-^{14}C-diethylamine	2 mg/11 μCi	♂ Rat	59	7
		♂ Hamster	36	6
1-^{14}C-ethylurea	20 mg/22 μCi	♂ Rat	42	13
	2.5 mg/12 μCi	♂ Hamster	52	12
1-^{14}C-diethanolamine	50 mg/100 μCi	♀ Rat	0.4	96
1-^{14}C-hydroxyethylurea	2.8 mg/5.3 μCi	♂ Rat	10	26
		♂ Hamster	11	31
4-^{14}C-oxazolidone	2.4 mg/4.7 μCi	♂ Rat	6	47
		♂ Hamster	8	38
1-^{14}C-hydroxyethyl-N'-ethylurea	3.6 mg/6 μCi	♂ Rat	6	26
		♂ Hamster	14	70
morpholine 3-^{14}C or 2-^{14}C	80 mg/2.4 μCi	♀ Rat	< 1	80
pyrrolidine-2,5-^{14}C	4 mg/10 μCi	♀ Rat	60	11
hexamethyleneimine-2,6-^{14}C	2 mg/10 μCi	♀ Rat	31	33
heptamethyleneimine-2,7-^{14}C	2 mg/10 μCi	♀ Rat	17	43
^{14}C-methylnitrosoethylurea	5 mg/24 μCi	♀ Rat	35	11
		♀ Hamster	30	15
1-^{14}C-diethylnitrosourea	3 mg/96 μCi	♂ Rat	42	12
		♀ Hamster	58	12

Hultin (1962). This report was of limited use because only the radioactivity dose was stated, but not the amount of NDMA administered, so that it is not known whether there was hepatotoxicity in the rats.

Many nitrosamines are oxidised by a variety of P450 enzymes of the liver and of other organs to α-hydroxy compounds. These were formerly believed to be so unstable as not to be isolable, but recent innovative work by Mochizuki (Okada *et al.*, 1980) has shown that they can be formed from hydroperoxides. The α-hydroxy-nitrosamines have half-lives in water from several seconds to several minutes, depending on their structure, quite sufficient time for them to be distributed in the body. Furthermore, these alcohols could be stabilised through conversion to esters (e.g. phosphates, sulphates or glucuronides), but there has been little study of these derivatives.

Generalisations from these studies of a few interesting nitrosamines has been risky, because many nitrosamines (including those containing alkyl groups with oxygen substituents) are not oxidised by cytochrome P450 enzymes and are not mutagenic to *Salmonella* in the rat-liver-microsome-activated 'Ames' assay (McCann *et al.*, 1975). They are obviously metabolised by other enzymes *in vivo*, possibly to metabolites of quite different structure (Guttenplan, 1989).

4.2 Pharmacokinetics and organ-specific carcinogenesis

It has become increasingly probable that pharmacokinetics play an important, probably vital, role in carcinogenesis by nitrosamines; it is less clear in the case of the directly acting alkylnitrosoureas, etc., for which the stability of the compound is likely to be a main determinant. Little attention has been paid to pharmacokinetics, even of the simplest nitrosamines, such as nitrosodimethylamine, e.g. by Keefer (Mico *et al.*, 1985) and Swann *et al.* (1983). Their results show that first-pass metabolism in the liver removes most of the nitrosamine and that deuterium-labelling of the nitrosamine has only a small effect on metabolism in the liver. On the other hand, deuterium-labelling reduces considerably the carcinogenic effect of NDMA, indicating that pathways other than α-hydroxylation might be important. First-pass metabolism in the liver is assumed to be saturable, so that large oral doses of NDMA are more likely to circulate systemically and induce kidney tumours in rats instead of, or in addition to, liver tumours.

The pharmacokinetics of methylnitrosohydroxyethylamine (MNHEA) were examined by Keefer's group (Streeter *et al.*, 1989) in relation to the sex difference in response of rats to this carcinogen. Female rats had a higher incidence of liver tumours and a greater hepatic clearance of the nitrosamine than did males; the latter, perhaps concomitantly, had a higher incidence of tumours of the nasal mucosa (Koepke *et al.*, 1988*a*). The rate of metabolism of MNHEA was much slower in both sexes than that of NDMA, which is attributed to the modulating effect of the hydroxyl group, making it resemble NDELA to some extent. This

Table 4.2. *Tumour response of rats to N-nitroso compounds administered by different routes*

Compound	Route	Dose rate (µmol/wk)	Total dose (mmol)	Median week of death	Tumours induced
Nitrosodimethylamine	Water	45	1.3	31	Liver hemangiosarcoma
	Gavage	27	0.8	59	Kidney mesenchyme
					Lung, Liver
	Inhalation	15	0.8	58	Nasal cavity
	S.C. inj.	400	0.4	53	Kidney mesenchyme
	Intravesicular	54	1.6	59	Kidney mesenchyme
					Lung
Methylnitrosoethylamine	Water	170	5	31	Liver hepatoma
					Esophagus
	Gavage	52	1.6	30	Liver hepatoma
					Lung
	Intravesicular	52	1.6	54	Liver hepatoma
					Lung
Nitrosodiethylamine	Water	44	1.0	26	Esophagus
					Liver
	Gavage	49	1.0	29	Liver
					Esophagus, Kidney
	I.V.	20	1.5	75	Liver
	Intravesicular	49	1.2	31	Liver
					Esophagus
Methylnitrosourea	Water	500	8	35	Glandular stomach
	Gavage	20	0.4	33	Forestomach
					Nervous system
	I.V.	100	1.0	69	Uterus, Mammary, Lung
	Intravesicular	10	0.2	37	Bladder

Compound	Administration				Organs
Dimethylnitrosourea	Water	41	1.2	71	Nervous system Kidney mesenchyme
	Gavage	41	1.2	40	Nervous system, Lung
	Intravesicular	41	1.2	46	Nervous system Kidney mesenchyme
Diethylnitrosourea	Water	41	1.2	32	Mammary gland
	Gavage	41	0.9	24	Mammary gland
	Intravesicular	41	0.9	29	Mammary gland, Lung
Methylnitroso-*n*-butylamine	Water	14	0.3	24	Esophagus
	Gavage	41	0.8	20	Esophagus
	Intravesicular	52	0.9	22	Esophagus Liver hemangiosarcoma
Methylnitroso-2-hydroxypropylamine	Water	85	1.9	29	Esophagus, Lung
	Gavage	85	0.4	40	Kidney, Esophagus, Lung, Thyroid
Methylnitroso-2-oxopropylamine	Intravesicular	34	1.0	83	Bladder, Nasal, Esophagus
	Water	86	1.8	24	Esophagus, Hepatoma
	Gavage	34	0.6	22	Esophagus
	Intravesicular	34	1.0	43	Bladder, Nasal
Nitrosomorpholine	Water	86	2.2	28	Liver hemangiosarcoma Esophagus
	Gavage	86	1.9	26	Liver Hemangiosarcoma Esophagus
	Intravesicular	86	2.6	35	Liver hepatoma Nasal cavity
Nitrosobis-(2-oxopropyl)-amine	Water	35	1.7	61	Liver Hepatoma Lung
	Gavage	32	1.1	59	Liver hepatoma Lung
	Intravesicular	32	0.3	87	Bladder, Lung, Kidney

resemblance probably accounts for the lesser potency of MNHEA compared with NDMA in rats. There are published studies of the distribution and excretion of several other N-nitroso compounds (e.g. NBHPA by Mori et al., 1984), but these have been less detailed or informative than those above.

The importance of route of administration, dose rate and disposition of the compound in determining the organs in which tumours appear and the timing and frequency of those tumours is well known. For example, in Table 4.2 are shown the tumour responses of rats to some simple methylating and ethylating nitrosamines administered to rats in similar doses, but by different routes of administration; the effect of dose rate is apparent by comparing treatment by gavage with treatment by the carcinogen dissolved in drinking water. For comparison, the effects of the isomeric azoxyalkanes give rise to the same alkylating agent as the nitrosamine (Fig. 4.3), yet produce different tumourigenic effects (Lijinsky et al., 1985d, 1987b).

As will be described later, when alkylation of DNA in various organs by these compounds is measured, the results are very similar for the same dose of the methylating carcinogens, as would be predicted on chemical grounds. However, as seen in Table 4.2, NDMA induced only liver tumours in rats when given in drinking water, whereas, given at the same rate by gavage, unusual mesenchymal kidney tumours and lung tumours were common and liver tumours less common, in both males and females; large single doses of NDMA to rats gave rise only to kidney tumours. The analogous, isomeric azoxymethane gave rise to kidney and colon tumours, as well as liver tumours when given in drinking water, although gavage treatment gave rise only to tumours of the colon and kidney, but not liver

Fig. 4.3. Activation of nitrosamines and azoxyalkanes.

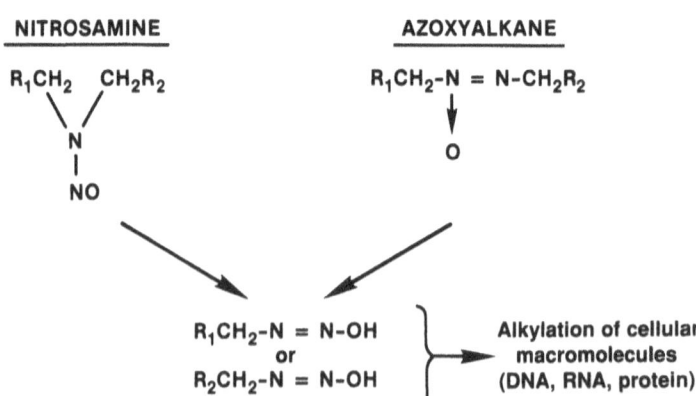

tumours or lung tumours. Clearly the metabolism of the two compounds is different in ways important for their carcinogenic action, although both give rise to similar amounts of the putative methylating intermediate, the methyldiazonium ion. Similar doses of the two compounds are converted to the ultimate metabolite, carbon dioxide, at very similar rates (approximately 60% of the dose as $^{14}CO_2$ in six hours following administration of approximately 1.5 mg of ^{14}C-labelled carcinogen by gavage). Therefore, the carcinogenically relevant metabolism of the two compounds appears to be quite different and probably related to the oxidation of azoxymethane to methylazoxymethanol and the corresponding aldehyde, which are believed to be important in the affinity of azoxymethane for the colonic mucosa (Zedeck et al., 1979). This is a pathway which NDMA cannot follow, and it has not given rise to tumours of the colon in any of the multitude of experiments in which it has been administered to animals; neither has NDMA been shown to give rise to alkylation of DNA in the colon, which has been demonstrated with azoxymethane (Fiala et al., 1977).

Methylnitrosoethylamine labelled with ^{14}C in the methyl group is converted to excreted CO_2 at essentially the same rate as NDMA or azoxymethane, and it leads to a similar pattern and extent of DNA methylation in the liver, lung and kidney of rats as does NDMA (Lijinsky, 1988b). Yet MNEA given to rats in drinking water or by gavage at very similar dose rates to NDMA has given rise to liver tumours and lung tumours (the latter by gavage only), but neither kidney tumours nor colon tumours were seen. Assuming that the ethyl group is oxidised according to the usual pattern shown by Magee, Heath and others for acyclic nitrosamines (Magee and Barnes, 1967), acetaldehyde would be formed and a methyldiazonium ion, identical to that formed from NDMA. From the rate of conversion of 1-^{14}C-ethylmethylnitrosomethylamine to CO_2 it would seem that the rate at which the methyldiazonium ion forms is also similar. However, since kidney tumours are not induced by MNEA in rats, it seems likely that the methyldiazonium ion is not the key intermediate in induction of kidney tumours, perhaps by either MNEA or NDMA.

The fact that the rate of metabolism of the methylating compounds overall is very similar suggests that the principal routes of metabolism might have much in common, but the differences in target organ specificity for their carcinogenic action suggests equally that the main metabolic routes are not related to tumour induction. Nevertheless, the 'normal' metabolism of these compounds, through which they become incorporated into the 1- or 2-carbon pool of the body, and thereby are 'lost', is important in understanding carcinogenesis because it represents a diversion of the compound from its toxic pathways. Administration of relatively large doses of ethanol, for example, slows metabolism of NDMA sufficiently for some of it to be excreted in the urine of animals given small NDMA doses (Kraft et al., 1981), which does not occur in the absence of ethanol. While measurement of CO_2 excretion does not provide detailed information about the intermediary metabolism of N-nitroso compounds, it does enable comparison to be made between one compound and another of the 'throughput' of the compounds, and in some cases demonstrates the effects of stimulators or inhibitors of metabolism (for example, the effects of disulphiram or ethanol on metabolism of NDMA). It was possible, using ethanol as an inhibitor of metabolism, to measure the formation of NDMA in humans eating a meal containing nitrosamine precursors, by measuring the nitrosamine that persisted in the blood (Fine et al., 1977).

4.3 Nitrosamine-metabolising enzymes

Many of the simpler nitrosamines are metabolised by the mixed function oxidases of the endoplasmic reticulum (microsomal enzymes) particularly in the liver, and considerable work has been done on the role of these enzymes in activation of nitrosamines to mutagenic oxidation products. Some of the enzymes involved have even been termed 'NDMA-demethylase' or 'nitrosamine-demethylase', although the enzyme or enzymes are surely not present in the liver to oxidise nitrosamines. Several investigators have been involved in studies of the microsomal oxidation of nitrosamines because of interest in the mutagenic activities of the products. They include Archer, Guttenplan, Yang, Michejda and Keefer. C. S. Yang has been particularly interested in the several isozymes of NDMA-demethylase in the liver of various species (Yang et al., 1984).

Nitrosodimethylamine demethylase is reported to be the cytochrome P450 enzyme responsible for activating NDMA, but in a particular form that is not induced by the usual inducers (phenobarbital, methylcholanthrene), or inhibited by the usual inhibitors (SKF 525A), but is induced by pyrazole (Tu et al., 1981). NDMA-demethylase seems to be a P450-dependent mono-oxygenase system, but it is not equally active towards all methylnitrosoalkylamines. The differential effects of various inducers in microsomal activation of several nitrosamines suggest that there are a number of cytochrome P450 enzymes (or isozymes) with high or low activities towards nitrosamines of different structures. This complex matter is far from resolution, but the information available suggests reasons for the failure of activation of some nitrosamines by rat liver enzymes, while they are activated by similar preparations from hamster or mouse liver. It is not clear whether various isozymes are responsible for the variations among species, or different enzymes.

It has been known for some time that there is at least one high-affinity and one low-affinity enzyme able to oxidise nitrosamines, and it is of course the high-affinity enzyme which is most important in carcinogenic activation of the nitrosamines, since relatively small doses of nitrosamines, leading to low concentrations in the blood, are most effective in inducing tumours in animals. While it is probable that in experiments in which large single doses of nitrosamines are given to animals the low-affinity enzyme comes into play, and the activity of the high-affinity enzyme might be so exceeded that the unchanged nitrosamine escapes the liver to circulate in the blood, this is not a usual occurrence. Nevertheless, it is the effect of nitrosamines administered by mouth which reach the liver first, but induce tumours in distant organs which they reach via the circulation, that are most interesting in studying mechanisms of carcinogenesis by these compounds. It is, of course, possible that primary oxidation of the nitrosamine occurs in the liver and that the product or products are converted to relatively stable transport forms, which circulate to reach the 'target' organs. This and other possibilities will be discussed later. Other enzymic reactions that nitrosamines can undergo, depending on their structure, are reduction of ketones and conjugation of those containing hydroxyl groups. Some of the latter, nitrosodiethanolamine for example, are oxidised in the β-position to the aldehyde, which can be a directly acting alkylating agent and mutagen (Fig. 4.4); however, the hemiacetal of NDELA appears to be a very weak carcinogen (Hecht et al., 1989; Lijnsky et al., 1991) although it is a metabolic intermediate (Eisenbrand et al., 1984; Hecht, 1984). This is not the only example of an intermediate that seems likely to be the 'proximate' carcinogenic product, but turns out not to be. There can also be enzymic esterification of the alcohol to a sulphate or phosphate, which can be transported or can interact with cellular constituents.

There are, apparently, multiple pathways of metabolism of such nitrosamines, some of which yield products that are carcinogenic and others that are not. It is difficult with present knowledge to determine which of the metabolic products is the important carcinogenic intermediate, since the apparently most reactive, most electrophilic and most extensively interacting with cellular macromolecules, might not be the metabolite related to induction of neoplastic transformation. Nevertheless, understanding as much as possible about the routes of metabolism of the carcinogen might provide insight into those products more likely to be involved in carcinogenesis. Testing of the various products for carcinogenic activity is a laborious and expensive process. Unstable products that might be very important cannot be isolated and tested separately.

It can in some cases be inferred that certain intermediates are likely to be involved, such as an alkyldiazonium ion, but the next steps are quite unclear, although a number of macromolecules – nucleic acids and proteins – become alkylated, with consequences that are assumed (mutations leading to abnormal expression of some genes, leading to neoplasia). This might be a major part of the story for some very simple nitroso compounds, but for most carcinogens of this class, activation is much more complex. For example, in methylalkylnitrosamines with a long carbon chain, the latter is oxidised extensively by successive β-oxidations in the manner of fatty acid chains, finally yielding nitrosamino acids or nitrosaminoketones, which have their own pattern of tumours related to their particular pathways of metabolism. This explains, at least partly, the common property of methylnitrosoalkylamines with even

Fig. 4.4. Metabolites of nitrosodiethanolamine.

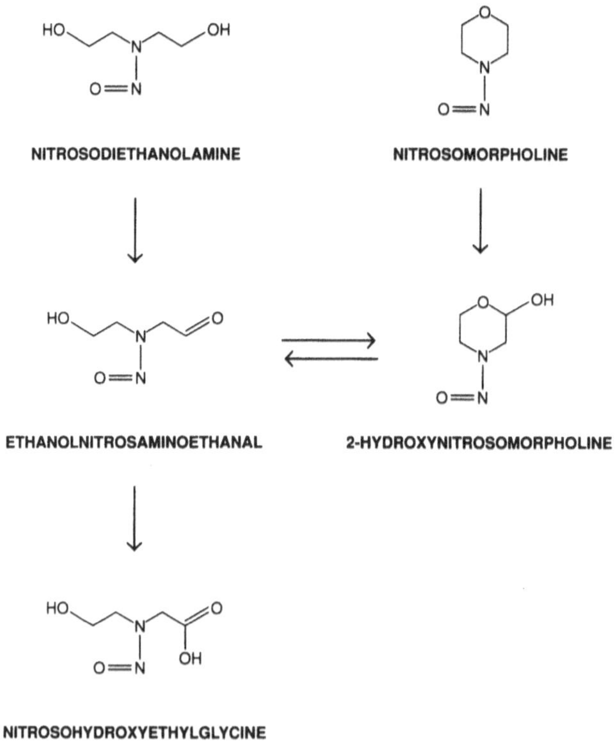

carbon chains larger than C-6 to induce tumours of the urinary bladder (Singer *et al.*, 1981). The process begins with omega oxidation to the alcohol, then to the acid, which is β-oxidised successively (cf. Knoop oxidation of fatty acids) to a C-4 compound (Okada *et al.*, 1976*b*), methylnitrosopropionic acid or methylnitroso-3-carboxy-propylamine (Fig. 4.5). The intermediate carboxylic acids have been isolated from the urine and identified, from nitrosamines as large as methylnitroso-*n*-octadecylamine. The final step is β-oxidation of the methylnitrosopropionic acid to form, following decarboxylation, methylnitroso-2-oxopropylamine, possibly in the bladder mucosa. Methylnitroso-2-oxopropylamine (MNOPA) is a potent bladder carcinogen (Thomas *et al.*, 1988) when introduced directly into the bladder in solution (although not when given orally), so that this compound can be considered a locally acting bladder carcinogen. However, it is not known what the subsequent

Fig. 4.5. Knoop oxidation of methylnitroso-*n*-octylamine.

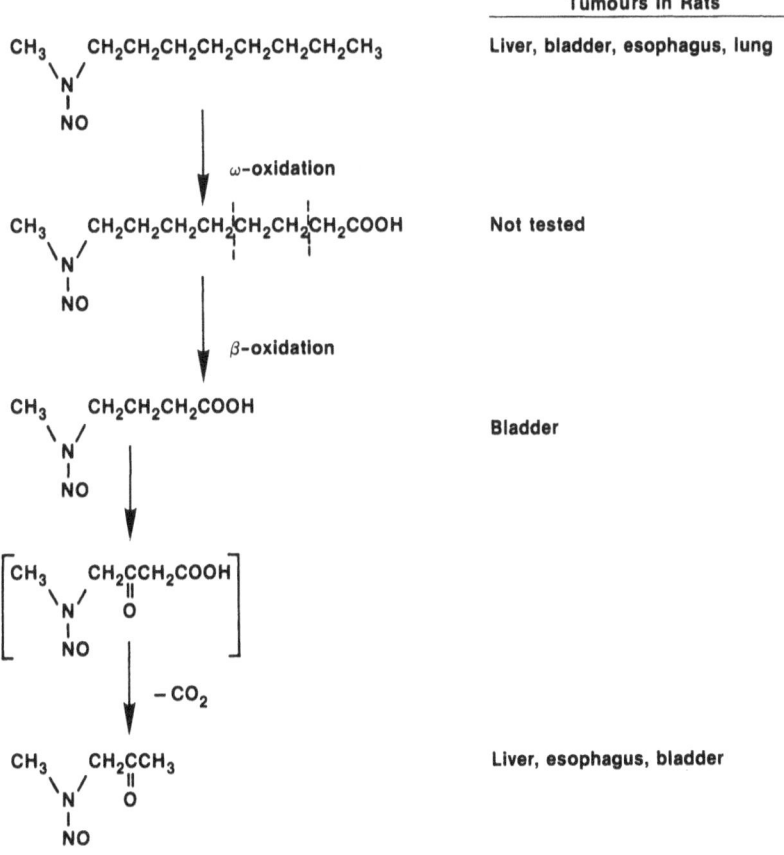

steps in carcinogenesis by MNOPA are, but it is a methylating agent in liver (Liberato et al., 1989); MNOPA gives rise to liver tumours in hamsters and in female rats. Methylnitroso-2-hydroxypropylamine (MNHPA) also gives rise to bladder tumours when given intravesically to rats (Thomas et al., 1988), and it is present together with MNOPA in the urine of rats given bladder tumour-inducing methylnitroso-n-alkylamines, but MNHPA is a much weaker carcinogen. It seems likely, therefore, that MNHPA is partially oxidised to produce MNOPA, in order to induce tumours, and not the reverse.

Nitrosodi-n-butylamine, which was of great interest to Druckrey et al. (1967) because it induced bladder tumours in rats, undergoes a similar series of reactions to methylnitroso-n-alkylamines. NDBA is oxidised to 4-hydroxybutylnitroso-n-butylamine and then to the 3-carboxypropyl derivative, which are both bladder carcinogens, active when administered intravesically (Hashimoto et al., 1974). The carboxypropyl compound is excreted in the urine and is presumably converted to the 2-oxopropyl analogue, which is the proximate bladder carcinogen.

The effects of rate of metabolism and distribution of the products (i.e. pharmacokinetics) on induction of tumours by a nitrosamine are illustrated by the results of experiments with methylnitroso-n-octylamine. Administered by gavage to rats or hamsters this nitrosamine induced many tumours of the bladder, in addition to high incidences of lung and liver tumours; many animals had all three. However, when given to rats in drinking water (a saturated solution containing 175 mg/L) not only were there virtually no bladder tumours or liver tumours, but all of the animals had tumours of the esophagus, which were not seen at all in rats treated by gavage, and most had lung tumours. These results suggest that the lower concentrations provided in drinking water altered metabolism in the liver so that less of the metabolites including nitrosamino acids and MNOPA were excreted in the urine to give rise to bladder tumours by local action (Singer et al., 1981).

It is very likely that activation of MNOPA to a proximate bladder carcinogen is brought about by enzymes in the bladder mucosa, and that these enzymes have a certain specificity for this particular nitrosamine structure. Nitrosodimethylamine, readily converted to a methylating agent in a number of organs and tissues, did not induce any bladder tumours when administered intravesically to rats, although it gave rise to tumours of the kidney, lung and thyroid, but not liver (Lijinsky and Kovatch, 1989a).

Some other nitrosamines introduced into the bladder have not induced tumours of the bladder mucosa, although several have induced tumours of other organs; for example, methylnitroso-n-butylamine has been as effective in inducing tumours of the esophagus when administered to rats intravesically as when given by gavage. Methylnitroso-n-hexylamine induced lung tumours but few bladder tumours in rats by intravesicular administration, although it induced many bladder tumours in hamsters treated orally (Lijinsky and Kovatch, 1988b), as did methylnitrosododecylamine (Althoff and Lijinsky, 1977). It is unlikely that they are activated in the bladder mucosa. The tobacco-specific nitrosamine NNK induced tumours of the nasal mucosa, lung and liver after administration of 1.5 millimol intravesically to rats. More perplexing was the failure of the directly acting methylating compound dimethylnitrosourea, given intravesically, to induce bladder tumours, although it gave rise to a high incidence of tumours of the nervous system. Methylnitrosourea injected into the bladder readily induces bladder tumours (Hicks and Wakefield, 1972). This is one of many indications that methylation of cellular constituents is probably not the main factor in tumour induction by methylating agents.

4.4 Metabolism of cyclic nitrosamines

A large proportion of the nitrosamines that have been tested have a cyclic structure, and many of them have very similar tumour-inducing properties to acyclic nitrosamines. There have, however, been few studies of the metabolism of cyclic nitrosamines that is related to their conversion to reactive metabolites responsible for the carcinogenic effects. As related elsewhere, many cyclic nitrosamines are mutagenic when suitably activated, for example with enzymes from rat or hamster liver, but with few exceptions (e.g. 2,6-dimethylnitrosomorpholine (DMNM) and nitrosopyrrolidine) it has not been possible to demonstrate the formation of covalent adducts with DNA or other macromolecules in target organs of animals in $vivo$ (Lijinsky, 1976; Farrelly and Hecker, 1984). This raises the question whether the minute or absent covalent DNA adducts of cyclic nitrosamines, or the extensive formation of covalent adducts from acyclic nitrosamines, such as nitrosodimethylamine and nitrosodiethylamine, is the more significant result in relation to induction of tumours by N-nitroso compounds.

As with acyclic nitrosamines, the initial step in metabolism of cyclic nitrosamines appears to be oxidation at the carbon α- to the N-nitroso function. Replacement of hydrogen with deuterium slowed metabolism of, for example, nitrosomorpholine, and also diminished its carcinogenic potency (and mutagenic potency) (Hecht et $al.$, 1981). The α-acetoxy derivatives of cyclic nitrosamines such as nitrosopyrrolidine were powerful bacterial mutagens, presumably following hydrolysis to the highly reactive α-hydroxy compounds. Ring opening of the α-hydroxynitrosamine is followed by loss of nitrogen from the diazonium ion readily formed (Fig. 4.6), to give finally a hemiacetal, which is highly unlikely to be the proximate carcinogenic intermediate. It is more probable that the diazonium compound is the reactive alkylating intermediate, but there is little evidence of the alkyl group which results becoming covalently bound to DNA *in vivo*. However, the α-acetoxy derivative of nitrosopyrrolidine

Fig. 4.6. Metabolism of cyclic nitrosamines.

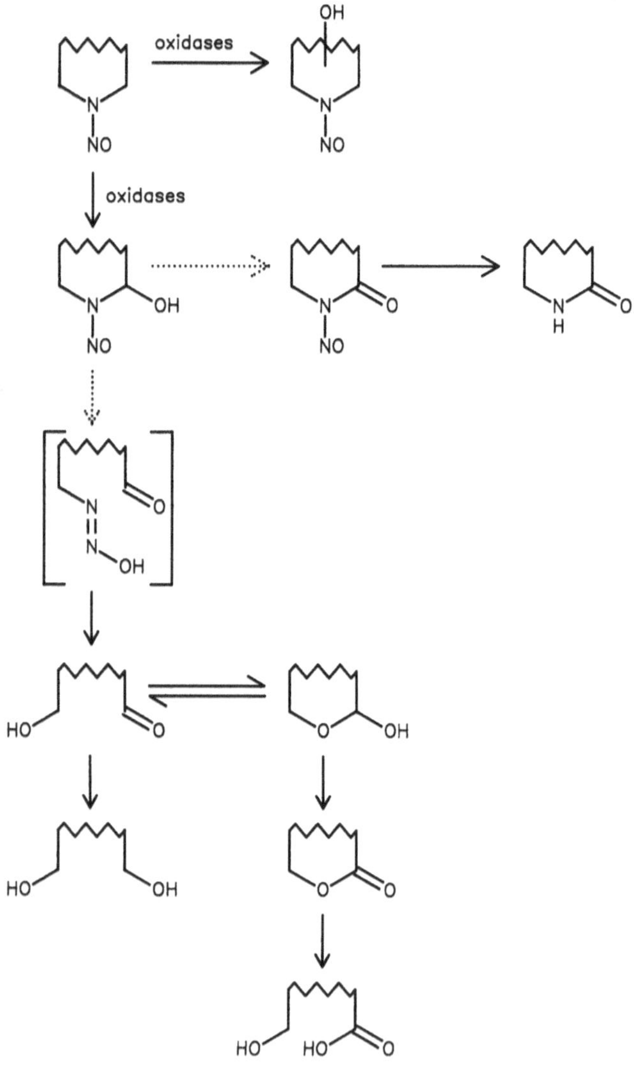

forms a cyclic adduct with deoxyguanosine *in vitro* (Chung and Hecht, 1983). In spite of considerable searching, a radiolabel at the α-carbons of a cyclic nitrosamine has usually not been detected in DNA of a target organ, such as the liver of a rat or hamster treated with the nitrosamine. An exception is the formation *in vivo* from nitrosopyrrolidine of cyclic deoxyguanosine adducts detectable at low levels by ^{32}P postlabelling (Chung *et al.*, 1989*b*). It can be concluded that, if such an adduct is formed from other cyclic nitrosamines, it is at an extremely low level, many orders of magnitude lower than those formed from NDMA or NDEA.

The metabolism of the several cyclic nitrosamines studied by Hecht and his co-workers, including nitrosopyrrolidine and the tobacco-specific nitrosamine nitrosonornicotine, all follow a similar and analogous pattern, a hemiacetal comprising the carbon chain of the nitrosamine being an end product, together with the corresponding carboxylic acid. An exception seems to be DMNM, which undergoes β-hydroxylation and consequent ring opening to a major extent, but even in that case a hydroxyacid analogous to that from nitrosopyrrolidine is a major urinary metabolite.

As with acyclic nitrosamines, *in vivo* almost any carbon atom in the ring of a cyclic nitrosamine can be oxidised. This has been much studied by Farrelly and Hecker (1984), who isolated β- and γ-hydroxyl derivatives of nitrosohexamethyleneimine. Although these derivatives have not been tested for carcinogenicity, it is unlikely based on analogy with nitrosopiperidine and its 3- and 4-hydroxy derivatives (Lijinsky and Taylor, 1975*f*) that any hydroxy derivatives of nitrosohexamethyleneimine would be more carcinogenic than the parent nitrosamine, or that the spectrum of tumours induced would be very different. Ross and Mirvish (1977) reported 1,6-hexanediol in the hydrolysate of liver RNA of rats treated with nitrosohexamethyleneimine, indicating formation of an intermediate reactive in the α- and ω-positions.

Study of the metabolism of many cyclic nitrosamines has been hampered by the difficulty of preparing suitably radiolabelled samples of the compounds. In the absence of radiolabel, only small yields of metabolites of, for example, several 4-substituted nitrosopiperidines having quite different carcinogenic activities, were obtained. The compounds studied were nitrosopiperidine and its 4-phenyl-, 4-*tert*-butyl- and 4-cyclohexyl-derivatives (Fig. 7.3a); the first and third gave rise to tumours of the esophagus in rats, the second to tumours of liver and esophagus, and the cyclohexyl compound was inactive (Lijinsky *et al.*, 1981*d*). Only small amounts of what were tentatively identified as hydroxy compounds were isolated from rat urine. There were metabolites which were unextracted and might have been conjugates in the aqueous portion, but none has been identified (Singer and MacIntosh, 1982). These experiments did not contribute significantly to understanding the metabolism of cyclic nitrosamines that might relate to their organ-specific carcinogenesis, which remains an aspect of the biological behaviour of N-nitroso compounds for which there is no hint of an adequate explanation. The suggestion of Lyle et al. (1976) that there was a correlation between induction of esophageal tumours in rats by cyclic nitrosamines and their tendency to exist in a rigid chair form with axial protons on the α-carbons, did not explain why these same compounds failed to induce tumours of the esophagus in hamsters. But it was by and large the case for a series of derivatives of nitrosopiperidine and nitrosomorpholine. In hamsters tumours of the nasal mucosa were often formed instead, and there were some cyclic nitrosamines that had the appropriate structure (the chair conformation) but did not induce esophageal tumours in rats. Considering those cyclic nitrosamines for which there was a positive correlation, it seems likely that the 'correct' configuration created an affinity for some receptor molecule, possibly an enzyme in the cells of the rat esophagus. But there was the question of the nature of the subsequent step or steps in activation, and an additional question of why such a configuration of the molecule did not also favour induction of tumours in other organs of rats, liver for example.

A comparison was made of the metabolism of nitrosopyrrolidine, nitrosohexamethyleneimine and nitrosoheptamethyleneimine following administration by gavage of small doses (2 to 4 mg), similar to those given twice weekly to induce tumours in rats. By 24 h after treatment the rate of excretion of CO_2 was levelling off (Snyder et al., 1977), and the cumulative excretion of CO_2 decreased with larger ring size from 77% of the dose of nitrosopyrrolidine to 27% of the dose of nitrosoheptamethyleneimine. Conversely, the proportion of radioactive metabolites in the urine was smallest, 11%, for nitrosopyrrolidine and greatest, 43%, for nitrosoheptamethyleneimine. At these low doses there were only traces, less than 1%, of unchanged nitrosamine in the urine, indicating that metabolism was almost complete (Singer and MacIntosh, 1984). At much higher doses, close to the LD_{50} (approximately 100 mg per animal), the proportion of the dose of the three cyclic nitrosamines excreted as CO_2 was approximately 10% or less, suggesting that the metabolic capacity of the rat, presumably mainly in the liver, was exceeded or at least strained. It has been observed that single doses of nitrosamines, however large, seldom give rise to tumours, although there are exceptions, including the offspring of pregnant females, but not usually the mothers.

The fact that nitrosopyrrolidine is metabolised with eventual formation of CO_2 as readily as its acyclic counterpart nitrosodiethylamine, implies that the cyclic configuration poses no obstacle to the metabolic enzyme systems of the body. Subtle differences in metabolism and activation must be responsible for the differences in carcinogenic potency between nitrosopyrrolidine and nitrosodiethylamine. The former is a moderate liver carcinogen in rats and induces lung tumours in hamsters, while the latter is possibly the most potent carcinogenic nitrosamine in rats, inducing tumours of the liver, esophagus and nasal mucosa; strangely, NDEA is considerably less potent in hamsters, by comparison with nitrosodimethylamine (Lijinsky et al., 1987b). It is perplexing that microsomes from rat esophagus metabolised nitrosopyrrolidine, which does not induce esophageal tumours, but did not metabolise dinitroso-2,6-dimethylpiperazine, which does (Scanlan et al., 1980).

Few metabolites of cyclic nitrosamines have been identified, and several are obvious in relation to the structure of the nitrosamine. For example, one of the metabolites in the urine of rats given nitrosomorpholine is nitrosodiethanolamine, obviously formed by opening of the ether bond (Manson et al., 1978); this reaction is probably related to the identification of 7-hydroxyethylguanine as a DNA adduct in rats treated with nitrosomorpholine (Stewart et al., 1974). Nitroso-3- and -4-piperidinol (Fig. 4.7) have been identified as metabolites of nitrosopiperidine (Rayman et al., 1975), and are themselves carcinogens of comparable potency with

Fig. 4.7. 3- and 4-Hydroxy-nitrosopiperidine, nitroso-4-piperidone and 3-hydroxy-nitrosopyrrolidine.

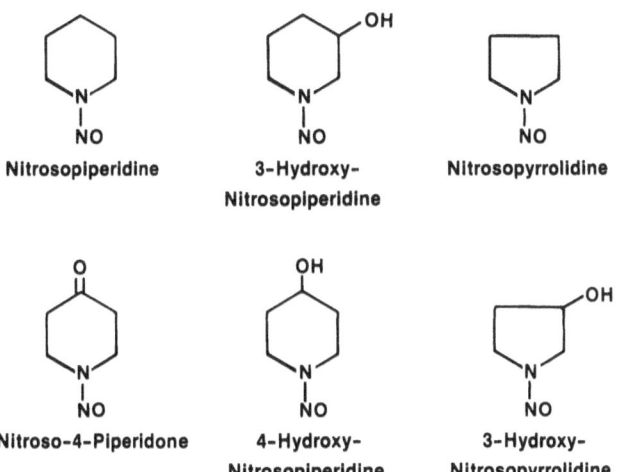

nitrosopiperidine; however, further metabolites of the nitrosohydroxypiperidines have not been identified and reported (Singer and MacIntosh, 1984). Nitrosopyrrolidine is oxidised to the 3-hydroxy derivative, which is also a carcinogen of comparable potency with its parent (Krüger and Bertram, 1973). Analogously, nitrosonornicotine is oxidised at the β-position of the pyrrolidine ring to 3'- and 4'-hydroxynitrosonornicotine, both of which are carcinogenic in the mouse lung adenoma assay (Castonguay et al., 1983). In contrast, although nitrosomorpholine is oxidised at the β-position to 2-hydroxynitrosomorpholine (Manson et al., 1978), this derivative seems to be very weakly carcinogenic; 2-hydroxynitrosomorpholine is also a metabolite of nitrosodiethanolamine, and is a directly acting mutagen to bacteria (Hecht, 1984). As mentioned, nitrosohexamethyleneimine and nitrosoheptamethyleneimine are also oxidised to β- and γ-hydroxy derivatives, which have been identified in *in vitro* systems (Farrelly and Hecker, 1984), but whether or not they retain carcinogenicity, and how potent they are relative to the parent nitrosamines, has not been determined.

The cyclic nitrosamines usually follow a similar pathway of metabolism, beginning with α-oxidation (Fig. 4.6), followed by ring opening and presumed formation of a diazonium ion or diazo-hydroxide. This could interact with cellular constituents or lose nitrogen to form a hemiacetal, which can be oxidised to the hydroxy acid and then to the dicarboxylic acid. This is the source of the major metabolite of nitrosoheptamethyleneimine found in the urine of rats, to the extent of 27% of the dose, pimelic acid. The yield of adipic acid from nitrosohexamethyleneimine is smaller (Fig. 4.8), but there was virtually no glutaric acid in the urine of rats given nitrosopiperidine (Singer and MacIntosh, 1984). Only traces of the hydroxy acids were found in the urine in the same study. As already mentioned the analogous hydroxy acids derived from nitrosomorpholine and 2,6-dimethylnitrosomorpholine are prominent in the urine of rats treated with them, while the corresponding dicarboxylic acids have not been reported.

The fact that all of the cyclic nitrosamines follow a very similar pattern of metabolism suggests that the enzymes responsible are also similar, although they have not been identified. They might be involved in the activation of these compounds to carcinogenic forms, but a close similarity is confounded by the large differences in potency and in target organ between the cyclic nitrosamines. It is further likely that the activating enzymes are present mainly in the liver, and indeed the pathways of metabolism have been established from studies in liver or liver fractions *in vitro*. It is, of course, possible that proximate carcinogens are formed in the liver and transported to the target organs, but this begs the question. It is equally reasonable to assume that the target organs themselves have susceptibility to the cyclic nitrosamines of particular structure.

Fig. 4.8. Metabolism of cyclic nitrosamines to dicarboxylic acids.

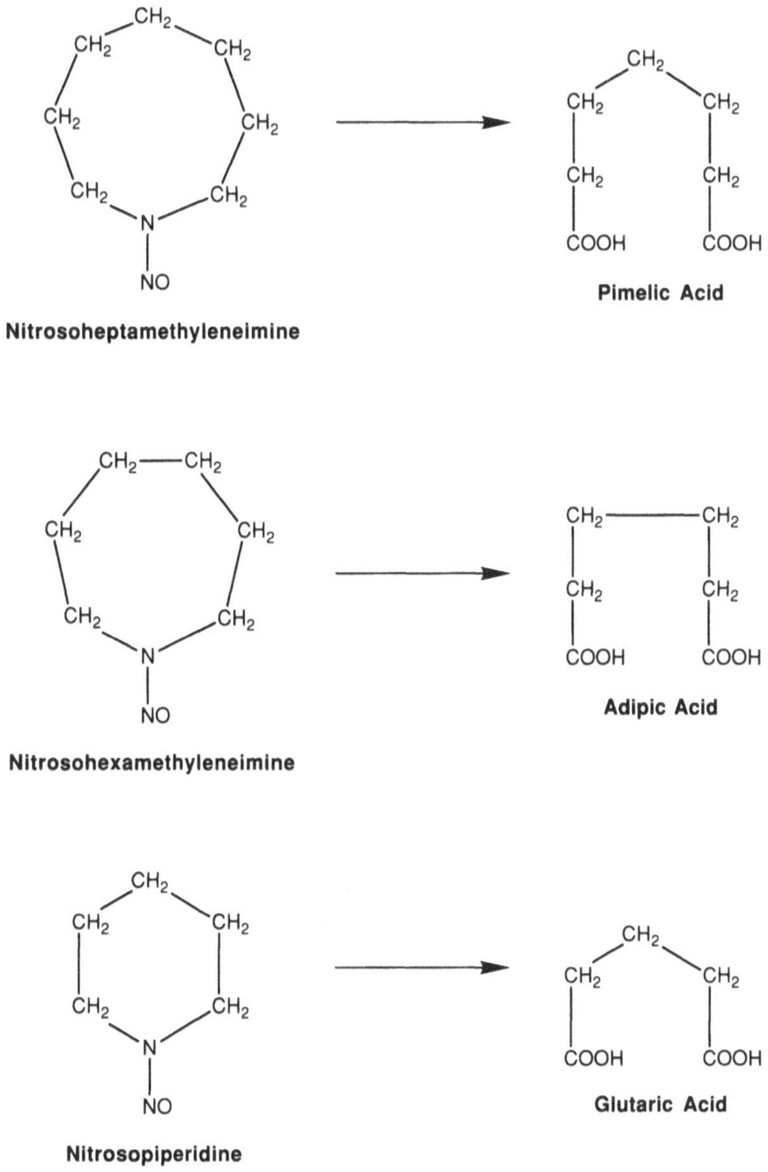

Nitrosoheptamethyleneimine → **Pimelic Acid**

Nitrosohexamethyleneimine → **Adipic Acid**

Nitrosopiperidine → **Glutaric Acid**

120   *Metabolism and interactions*

It seems that present knowledge of the metabolism of cyclic nitrosamines, and of the structure of their metabolites, does not help substantially in our understanding of mechanisms of carcinogenesis, in particular the organ and species specificity of these carcinogens. While there is a requirement for rigidity of the molecule in favouring induction of tumours of the esophagus in rats, and possibly tumours of the nasal mucosa also, this characteristic does not increase the tendency to induce tumours of the nasal mucosa in hamsters – and of course not tumours of the esophagus, which do not appear in hamsters. Also, the flexible forms of such cyclic nitrosamines as 3,5-dimethylnitrosopiperidine, 2,6-dimethyl-dinitrosopiperazine and 2,6-dimethylnitrosomorpholine (Fig. 4.9) induce the same tumours in rats as the rigid chair forms; they are simply less potent than the latter. The differences appear to be kinetic and, since knowledge of the nature of the critical enzymes involved is lacking, no firm conclusions can be drawn about the relation of the activity of the metabolising enzymes to induction of particular tumours. The lack of information about interaction with macromolecules, particularly the difficulty in detecting and identifying products of interaction with especially DNA or with proteins, or with cellular membranes, makes it difficult to place the activity of cyclic nitrosamines alongside that of acyclic nitrosamines, and to generalise from one to the other. About the latter much more is known, because the biochemical information is more accessible.

4.5   Acyclic nitrosamines
4.5.1   Overview

Study of the metabolism of acyclic nitrosamines began with the experiments of D. F. Heath on nitrosodimethylamine, shortly after the discovery of the toxic effects of the compound, in the early 1950s. These early investigations showed that following administration to animals

Fig. 4.9. From left to right, β-methylated nitrosomorpholine, dinitrosopiperazine and nitrosopiperidine.

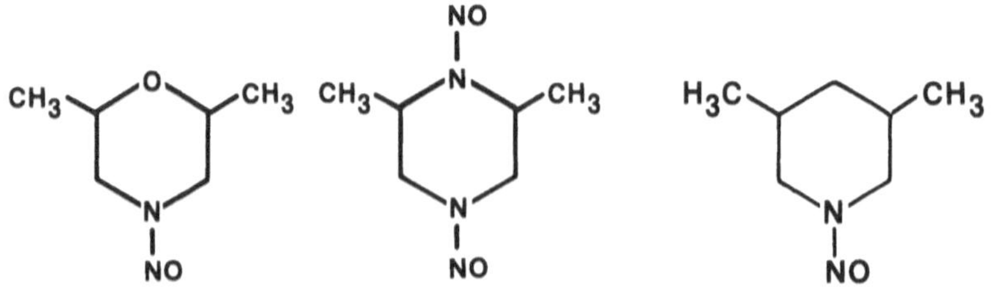

NDMA rapidly disappeared and CO_2 arising from the carbon atoms of the nitrosamine was exhaled (Heath and Dutton, 1958). In liver slices *in vitro* NDMA was also converted to CO_2 and formaldehyde was a principal product. Heat inactivation of the enzymes stopped metabolism. Other metabolites found included methylamine and nitrite, the latter identified through the use of ^{15}N-labelled nitrosamine. Although study of the metabolism of NDMA has continued for the past 35 years, and was pursued most persistently lately by L. K. Keefer, the biochemistry and chemistry of NDMA metabolism is still not completely understood, particularly in relation to the toxic actions and tumour-inducing properties of this simple molecule.

The products of metabolism of NDMA are CO_2 and a number of compounds which appear in the urine of rats or hamsters, including methylamine and dimethylamine, as well as some presently unidentified substances. Radioactive fractions chromatographically coincident with these were found in the urine of animals treated with ^{14}C-NDMA, but methylurea was not one of them. Keefer *et al.* (1987) showed that in the metabolism of NDMA by rat liver microsomes *in vitro* methylamine and nitrite were formed in equimolar amounts and to the extent of about 10% of the NDMA (0.1 mM). Formaldehyde was produced to the extent of more than 70% of the NDMA in the incubation mixture, and nitrogen was evolved. Previous studies by Michejda *et al.* (1982) using ^{15}N-labelled nitrosamine showed that approximately 70% of the theoretical yield of nitrogen was obtained from the NDMA consumed. Keefer reasoned (Mico *et al.*, 1985) that there were two pathways of metabolism of the nitrosamine (Fig. 4.10). One was by α-oxidation to methylnitrosohydroxymethylamine, which lost formaldehyde to form a methylating moiety such as a methyldiazonium ion, which in turn lost nitrogen simultaneously with methylation of a nucleophile (nucleic acids, proteins, etc.); the latter reactions with DNA are believed essential in initiating carcinogenesis. A second pathway involves demethylation (possibly again producing formaldehyde) to a metastable methylnitrosamine, which denitrosates quantitatively to form methylamine and nitrite ion (Keefer *et al.*, 1987). This would plausibly explain the analytical data.

It is assumed that enzymes able to metabolise NDMA (and other nitrosamines) exist in many organs and tissues, and are perhaps responsible for the organ-specific carcinogenesis of this compound and other nitrosamines. Much of the study of nitrosamine metabolism has, however, been carried out in the liver, both because of its large complement of detoxifying enzymes, and because many nitrosamines cause liver toxicity, presumably related to the formation of reactive toxic metabolites. As discussed above, the important toxic intermediates have not yet been identified. The enzymes commonly believed to metabolise NDMA are cytochrome P450's, the mixed function oxidases of the endoplasmic reticulum, usually studied in the artificial creation known as microsomes, or even as the crude mixture called 'S_9'. The latter are acceptable for activation of compounds to products that are mutagenic or transforming agents, but not for serious studies of the enzymology of nitrosamine activation. Some time ago, Frantz and Malling (1975) showed that the activity of microsomes to oxidise and activate NDMA declined as the S_9 fraction was purified, suggesting that co-factors, or additional enzymes were required. Prival and Mitchell (1981) exhaustively studied the microsomal metabolism of NDMA, and found that the enzyme activities (called NDMA-demethylase) varied considerably from one species to another, and could be altered by treating the animals from which the microsomes were prepared with a variety of inducers, such as phenobarbital, Aroclor, etc. These substances induce formation of different isozymes of cytochrome P450, with quite different activities. C. S. Yang (1984) and others have also extensively studied the activities of the different liver cytochrome P450 isozymes in oxidising NDMA and other dialkylnitrosamines, and have tried to relate these activities to

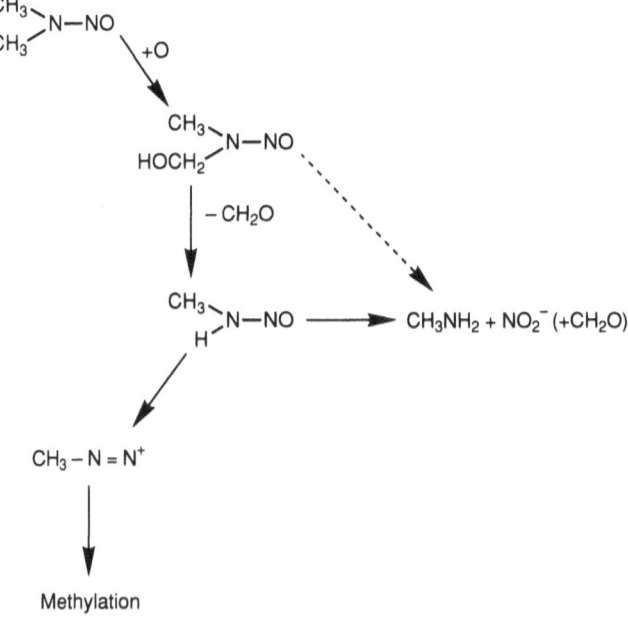

Fig. 4.10. Two pathways of metabolism of NDMA.

bacterial mutagenic effects. These studies and those of Prival have been somewhat complementary.

However, Prival has gone further and found that the activity of the NDMA-demethylase did not necessarily parallel the mutagenic activity; in some cases there was a large disparity. In particular, Prival and Mitchell (1984) found that there is a component of mouse liver microsomes that inhibits activation of NDMA to mutagens by hamster liver microsomes, which are otherwise much more active than rat liver microsomes or mouse liver microsomes. The nature of this inhibitor is not known, but it is destroyed by heating to 70 °C, and apparently is not itself a demethylase enzyme. It has been apparent for a considerable time that there are two enzymes, at least, in rodent liver that metabolise NDMA. One is a low affinity enzyme, which obviously has little physiological importance in oxidising and activating NDMA at low, carcinogenic doses. The other is a high-affinity enzyme and can be considered the most important in mediating carcinogenesis by NDMA except when it is applied in large, toxic doses. Since the low affinity enzyme is not present simply to oxidise NDMA, one might speculate what its normal function is.

The lower affinity enzyme in the hamster liver is considerably more active than that in the rat liver, which corresponds to some extent with the greater carcinogenic effectiveness of NDMA in the hamster compared with the rat; NDMA is also more toxic to the hamster liver than to the rat liver, and it is possible that the same enzymic activation is responsible for both activities. However, large single doses of NDMA in rats do not result in liver tumours, while in hamsters they do. That the large toxic doses of NDMA in rats result in liver necrosis, but kidney tumours and no liver tumours, has been attributed to spillover of the unmetabolised NDMA so that it can be activated in the kidney and induce tumours there. However, these conditions do not result in kidney tumours in hamsters; the hamster kidney seems resistant to induction of tumours by nitrosamines, although not by other carcinogens, such as steroid hormones.

The picture is complicated by the finding that considerably lower doses of NDMA than those used to induce kidney tumours in earlier days, can have the same effect under certain conditions. Thus, feeding of a low protein diet to rats, presumably reducing synthesis of the activating enzymes, increases the LD_{50} dose. The greater survival of rats given 20 mg/kg of NDMA in a single dose – there being much less liver damage – caused every rat to develop kidney tumours, but still no liver tumours were induced. This system was used by Hard and Butler (1970) to study the morphogenesis of NDMA-induced kidney tumours in rats. While kidney tumours have rarely appeared in rats following administration of

NDMA in drinking water or food, similar weekly doses given by gavage (i.e. pulsed doses) have led to appearance of kidney tumours in high incidence, without excessive liver damage, and often accompanied by liver tumours and sometimes lung tumours (Lijinsky et al., 1987b). It is certain that the pulsed doses by gavage are not completely metabolised in the first pass, so that the nitrosamine circulates to other organs, but why this should result in a major difference in distribution of tumours is not clear. When administered intravesically, again in pulsed doses which must be absorbed through the bladder wall, NDMA induced only tumours of the kidney and lung, but none in the liver or in the bladder (Lijinsky and Kovatch, 1989a). NDEA and MNEA given intravesically under the same conditions induced mainly liver tumours. Azoxymethane, the isomer of NDMA, also induced liver tumours when given to rats in drinking water, but these were accompanied by kidney tumours and colon tumours. In contrast to NDMA, gavage treatment of rats with azoxymethane gave rise to kidney tumours and colon tumours, but none in liver or lung. The kidney tumours induced by both compounds were mesenchymal, not tubular cell in origin, and it is surprising that these target cells, which are relatively sparse in the kidney, should be so susceptible to these methylating compounds; ethylating compounds, if they induce kidney tumours, do not cause mesenchymal tumours. The possibility arises that there are enzymes in rat kidney mesenchyme with particular affinity for NDMA and azoxymethane.

Accelerators of metabolism and activation of carcinogenic nitrosamines have been studied intensely in relation to their mutagenic activity *in vitro*, including enzyme inducers, such as Aroclor (polychlorinated biphenyls) and phenobarbital. Inhibitors of nitrosamine metabolism have been found, including pregnenolone-16α-carbonitrile (Somogyi et al., 1972), disulphiram (Schmähl et al., 1976), and ethanol (Swann et al., 1984). The effect of ethanol in inhibiting metabolism of nitrosodimethylamine is important, since it has allowed identification and measurement of ingested NDMA in humans, who simultaneously drank alcoholic beverages (Fine et al., 1977). Ethanol in animals has also changed the target organ of NDMA, presumably by altering its metabolism and distribution in the animals, and changing the site of tumours induced from liver to nasal mucosa (Griciute et al., 1981). Administration of ethanol does not seem to affect the metabolism of any but a few simple nitrosamines; nitrosodiethylamine, for example, is much less affected. Disulphiram, on the other hand, seems to inhibit metabolism of a number of nitrosamines, and lowers the incidence of tumours when given to animals together with NDMA, NDEA and others. It is certainly possible that components of the

Acyclic nitrosamines 125

diet can also affect metabolism and activation of nitrosamines, and this might account for some of the differences found between laboratories in the response of the same animals (species and strain) to particular nitrosamines. The effects of ethanol, pyrazole and other compounds on the activity of enzymes that metabolise NDMA, and on the distribution of the isozymes, have been the subject of considerable investigation by Tu *et al.* (1981, 1983), Garro *et al.* (1981) and others, but often without any particular relation to the carcinogenic effects of the compounds. Therefore, the relevance of much of this work, except to the activity of particular cytochrome P450's, and therefore to detoxication in general, is not clear in the context of nitrosamine carcinogenesis.

4.5.2 Metabolism by microsomes

It is unclear why metabolism of many nitrosamines is not possible with rat liver microsomal enzymes, but is with hamster liver enzymes. Many nitrosamines that are resistant to oxidation by rat liver microsomal enzymes nevertheless induce tumours in the liver of rats, indicating that the enzymes in the microsomal preparations (presumably cytochrome P450's) are not involved in activation of these nitrosamines to carcinogenic forms in the rat liver, but other enzymes are. In several cases it was shown (Farrelly *et al.*, 1984, 1987a) that no metabolism by rat liver microsomal enzymes occurs (there is no change in the concentration of the nitrosamine). The same was true for azoxyalkanes (Lijinsky *et al.*, 1985a). It is also perplexing – and is dealt with in Chapter 6 – that methylnitrosoethylamine is oxidised in either alkyl group by rat liver microsomes, by enzymes with similar characteristics to those oxidising NDMA and NDEA (Farrelly *et al.*, 1982), yet MNEA is not mutagenic to bacteria when activated by liver microsomes from either rat or hamster. The higher homologues, methylnitroso-*n*-propylamine and methylnitroso-*n*-butylamine are also readily oxidised by rat liver microsomes to both of the expected aldehydes (and, presumably, forming both respective alkyldiazonium ions) (Farrelly *et al.*, 1982), but are not mutagenic when activated by those same enzymes in rat liver microsomes; they are mutagenic, however, with hamster liver microsome activation (Lijinsky and Andrews, 1983). These disparities indicate that the metabolites responsible for the mutagenic action of these nitrosamines (and perhaps others) are not those produced by the metabolism usually measured. The cytochrome P450 enzymes certainly give rise to aldehydes and, by inference, to alkyldiazonium ions.

Farrelly and Stewart (1982) showed that there was a correspondence between the oxidation of the methyl group of the series of methylnitroso-

Table 4.3. *Apparent kinetic constants for oxidation of methylalkylnitrosamines by rat liver microsomes to formaldehyde and a carbonyl compound*

Compound	HCHO		2nd carbonyl	
	K_m (mM)	V^*_{max}	K_m (mM)	V^*_{max}
Methylnitroso-				
methylamine	2.7	1.4	—	—
ethylamine	4.2	0.46	2.7	2.2
trifluoroethylamine	10.4	1.6	—	—
n-propylamine	4.4	0.26	2.0	0.50
n-butylamine	3.4	0.64	14.2	3.1
n-pentylamine	2.6	0.82	1.2	1.7
neopentylamine	2.5	3.5	5.2	2.6
n-hexylamine	1.5	0.49	2.8	1.1
n-heptylamine	1.1	0.42	5.5	2.7
cyclohexylamine	1.4	0.53	11.1	3.9
benzylamine	1.4	1.1	11.5	1.7
phenylethylamine	2.22	0.84	2.8	1.3

* Maximum velocity, nmol/min/mg protein.

alkylamines and the length of the carbon chain of the other alkyl group (Table 4.3). The Michaelis constant (K_m) increased from NDMA to MNEA, remained constant for the n-propyl compound, then declined. The rate of formation of formaldehyde, however, was greater for NDMA than for the homologues with longer chains. The K_m for the formation of the second carbonyl compound (i.e. for the enzyme oxidising the α-methylene of the longer alkyl group) showed no particular pattern and varied considerably from one compound to the other, but the rates of formation of the larger aldehyde were within a narrow range. The same was true for the methylnitrosoalkylamines with alkyl groups that are not n-alkyl chains, such as neopentyl, benzyl, phenylethyl and trifluorethyl, as well as cyclohexyl. The aromatic analogue of the last, methylnitrosoaniline, has been extensively studied by Michejda and his group (Kroeger-Koepke and Michejda, 1979). They found that, as in the studies by Farrelly of methylnitrosoalkylamines, formaldehyde is a major product, formed by oxidation of the methyl group. However, whereas all of the compounds examined by Farrelly and others gave rise, in addition, to a carbonyl compound, this was not possible for methylnitrosoaniline. Indeed, all of the other compounds could form a methylating agent by oxidation of the α-carbon of the longer alkyl group (thereby forming the carbonyl compound), but this was not possible for methylnitrosoaniline.

Therefore, although induction of tumours in the esophagus of rats has been related by Von Hofe et al. (1987) to methylation of esophageal DNA in rats by several methylnitrosoalkylamines examined, this cannot be the mechanism by which methylnitrosoaniline induced esophageal tumours in the same rats. It is notable that methylnitrosoaniline does not induce tumours of other organs in rats, although it does in hamsters (Lijinsky and Kovatch, 1988b).

Asymmetric nitrosamines are mixtures of rotamers, with different susceptibility to enzymic oxidation, as studied extensively by J. G. Farrelly. The E-isomer of methylnitroso-n-amylamine was metabolised more readily by rat liver microsomes than the Z-isomer (He and Farrelly, 1989), although this nitrosamine does not induce liver tumours in rats; studies of metabolism of this compound by hamster liver microsomes might be more informative about mechanisms of carcinogenesis, since it does induce liver tumours in hamsters. On the other hand, methylnitroso-2-oxopropylamine does induce liver tumours in rats and Farrelly et al. (1988) found that the E- and Z-forms (Fig. 3.17, p. 89) respond differently to rat hepatocytes, the E form being oxidised readily, but the Z-form hardly at all. The relative reactivity of the two forms could be an important determinant of the effectiveness of asymmetric nitrosamines administered in pulsed doses. The two isomers might well differ in mutagenic activity, but this has not yet been studied.

4.5.3 Reactive metabolites/organ specificity

The many methylalkylnitrosamines that induce tumours of the esophagus in rats are assumed to act through formation of a methyldiazonium ion which methylates DNA in cells of the esophagus. Methylnitrosoaniline cannot act in this way, although it is quite a potent esophageal carcinogen in rats. Michejda has suggested that the phenyldiazonium ion formed by oxidation of the methyl group of methylnitrosoaniline might interact with the amino group of one of the purine or pyrimidine bases to form a transient triazene (Koepke et al., 1991), which would easily decompose, leaving no residue that could be identified as coming from the nitrosamine; however, the presence of the triazene residue could interfere with normal function of the DNA, even though temporarily. The possibility has not been proved, but is being actively investigated, as part of a continuing study of the chemistry and biology of di- and tri-alkyltriazenes. Many of these compounds are mutagenic (Sieh et al., 1980) and at least one dialkyltriazene, diethyltriazene, has induced esophageal and forestomach tumours in rats (Smith et al., 1987).

Support for an unusual mechanism of activation and carcinogenesis of methylnitrosoaniline (which is not mutagenic) comes from consideration of the non-mutagenic benzylnitrosophenylamine, which also causes esophageal tumours in rats and could also, through oxidation to benzaldehyde, give rise to a phenyldiazonium ion (Lijinsky and Reuber, 1982b). The question arises why the same tumours should be induced by such disparate effects on DNA in the esophagus (i.e. methylation and triazene formation), and also why the methylnitrosoalkylamines so active in inducing tumours in the rat esophagus do not induce tumours elsewhere in the rat, particularly in the liver, in which extensive alkylation of DNA occurs (Von Hofe et al., 1987). Neither do they induce tumours in the hamster esophagus (Lijinsky and Kovatch, 1988b). This dilemma has been investigated in some depth by Archer (Labuc and Archer, 1982; Mehta et al., 1984), who found that metabolism of methylnitrosobenzylamine is comparable in esophagus and liver of rats, and is also comparable to that in liver by NDMA. The so-called preneoplastic foci in the rat liver are also produced by both nitrosamines, although NDMA induces liver tumours and methylnitrosobenzylamine does not. The strong implications, as discussed elsewhere, are that formation of an alkylating agent and alkylation of DNA in an organ do not, except in some cases, lead to induction of tumours; where they do, it seems that other actions, primary or secondary, of the nitrosamine play the major role in determining that tumours ensue.

The metabolism of nitrosamines (such as methylnitrosobenzylamine) that induce esophageal tumours in rats has been studied in cultured esophageal tissue (Autrup and Stoner, 1982) and has confirmed the formation of aldehydes and alkylation of DNA in esophagus cells. There has been no comparison, however, with metabolism in the esophagus of other species, such as the hamster, in which these compounds do not induce tumours; such control studies could demonstrate that the observed metabolism in the rat is absent from the hamster, and therefore more likely to be related to tumour induction. This, unfortunately, is a common gap in cancer research: demonstration that what takes place in situations in which tumours result, does not take place when tumours do not result.

Similar studies with nitrosamines have been carried out in the nasal mucosa, which is a common site of induction of tumours in both rats and hamsters. This coincidence suggests that metabolism and activation related to induction of tumours might be similar in the nasal cavity of both species, which is obviously not the case in the esophagus. It has been claimed that the activity of metabolising enzymes in the rat nasal mucosa is higher than in any other organ for metabolising nitrosamines (Tjälve et al., 1985). This might explain in part the exceptional susceptibility of the rodent nasal mucosa to induction of tumours by nitrosamines. However, only nitrosamines with particular structures induce tumours in the nasal mucosa, including some cyclic nitrosamines and some acyclic nitrosamines. In addition, many nitrosamines induce nasal mucosa tumours in rats, but not in hamsters, and vice-versa. For example, nitrosomorpholine, methylnitroso-*n*-propylamine, -*n*-hexylamine, -*n*-heptylamine and -*n*-octylamine induce high incidences of tumours of the nasal mucosa in hamsters (Lijinsky and Kovatch, 1988*b*), but not in rats. Hydroxypropylnitrosooxopropylamine, dinitroso-2,6-dimethylpiperazine and nitroso-3,4,5-trimethylpiperazine have induced tumours of the nasal mucosa in rats, but not in hamsters.

Unfortunately, there have been no experiments comparing metabolism of these nitrosamines in the nasal mucosa of rats with that in hamsters, to determine those routes of metabolism that are most likely to be related to tumour induction. There are continued contradictions in the tumour responses of various species to nitrosamines of similar, but slightly different, structure (see Chapter 7).

Another perplexing fact is the failure of any directly acting *N*-nitroso compound to induce tumours of either the esophagus or nasal mucosa in either rats or hamsters, although they can give rise directly to the same reactive metabolites as are formed metabolically from analogous nitrosamines, that is, alkyldiazonium ions. The role of metabolism in the induction of tumours by nitrosamines of particular structures in esophagus of rats, and nasal mucosa of rats or hamsters, is not known. It seems probable that the metabolism that has been studied until now, resulting in formation of carbonyl compounds and alkyldiazonium ions – usually by mixed function oxidases – is not of major relevance to tumour induction in those organs. It is more likely that interaction of the nitrosamine with other enzymes for which they have affinity is a critical event. It is most unlikely that the carcinogenic action of nitrosamines in the nasal mucosa or in other organs does not involve metabolism of the compounds, because nitrosamines are almost devoid of chemical reactivity in the absence of activation. There is no clear relationship between the metabolism of so many nitrosamines that induce tumours of the esophagus and nasal mucosa and their potency as carcinogens even when, as in the case of the esophageal carcinogens, these tumours are the cause of death of the animals, and so quantifiable in their effects. There is a large discrepancy between, for example, the similar metabolism of methylnitrosocyclohexylamine and methylnitroso-*n*-heptylamine (Farrelly and Stewart, 1982) and the much more potent carcinogenic effect of the former (see Chapter 7); it is hard to envisage that there would be a large difference between the two compounds in metabolism in the liver or esophagus of the rat, when they are administered by the same route.

The reason for the failure of so many nitrosamines to induce tumours in the liver, when they are so readily metabolised by liver enzymes (*in vivo* and *in vitro*) – and often in very similar fashion to nitrosamines that do induce liver tumours – is not readily apparent. The explanations that have been offered, such as relative instability of the reactive products, or repair of alkylated lesions formed in the liver DNA by the alkylating agents produced metabolically, seem like alibis. They fail to explain the discrepancies, because the metabolites formed from the liver carcinogens and the non-carcinogens are often the same (e.g. a methyldiazonium ion), and in the same quantities; so are the alkylated lesions in the liver DNA (e.g. O^6-methylguanine).

The studies of metabolism of nitrosamines have followed conventional methods involving subcellular fractions, urinary metabolites and sometimes whole animals or cells or organs in culture. The metabolites sought are the obvious ones and those identified are those that appear in considerable concentrations, including carbon dioxide, nitrogen and various aldehydes. The possibility has been recognised that minor products of metabolism, or transient reactive metabolites, might play an important role in the carcinogenic process, but their very nature makes them difficult to find and identify. Therefore, it is necessary to consider the 'obvious' metabolites, but to view them with reservations.

4.6 Nitrosamines containing oxygen substituents

It is clear from the studies of Farrelly and others that oxygen substituents in the alkyl groups greatly affect the metabolism of these nitrosamines. For example, the metabolism of methylnitrosoethanolamine and nitrosodiethanolamine is entirely different from that of nitrosodiethylamine. The latter is oxidised in the expected way by microsomal enzymes to the α-hydroxy derivative which decomposes to produce acetaldehyde and the ethyldiazonium ion, a powerful alkylating agent. Therefore, extensive ethylation of cellular constituents occurs following *in vivo* metabolism of NDEA. Other routes of metabolism operate as indicated by the finding of 2-hydroxyethylated porphyrins in the blood of rats given NDEA (White *et al.*, 1983), showing that β-oxidation of one or both ethyl groups had occurred, followed by release of an hydroxyethylating species, possibly a diazonium ion, through α-oxidation. The importance of this reaction is lessened by the results of Von Hofe *et al.* (1986*b*), who found only very small amounts of hydroxyethylated products in rat liver DNA following administration of NDEA or methylnitrosoethylamine, indicating that β-oxidation of the ethyl group was not a favoured pathway of metabolism. To the extent this did occur, α-oxidation of the other alkyl group would give rise to a hydroxyethyldiazonium ion.

Another way in which a hydroxylated dialkylnitrosamine could be activated is through conjugation with a good leaving group, such as sulphate or phosphate, in the manner proposed by Michejda *et al.* (1979). Such a structure would easily solvolyse with cyclisation to an oxadiazolium ion (Fig. 3.10, p. 80), which is an alkylating agent. This might explain the carcinogenicity of methylnitrosoethanolamine, which is not oxidised by microsomal enzymes (and is therefore not mutagenic in the microsomally activated bacterial mutagenicity assays – Chapter 6), yet induces liver tumours in rats. Nitrosodiethanolamine might also undergo similar conjugation (Sterzel and Eisenbrand, 1986) with sulphate. It is possible that the reactive sulphate conjugate, which is mutagenic (in contrast with NDELA), might induce tumours. A similar process of conjugation was suggested by Kokkinakis *et al.* (1986) as the mechanism by which 2-hydroxypropylnitroso-2-oxopropylamine (NHPOPA) was activated to a carcinogenic form. Methylnitroso-2-oxopropylamine is not metabolised *in vitro* by microsomal enzymes from rat liver, although it is metabolised (and activated to bacterial mutagens) by hamster liver microsomes (Andrews and Lijinsky, 1984), and by rat hepatocytes (Farrelly *et al.*, 1988). Nitrosobis-(2-oxopropyl)amine (NBOPA) shows the same pattern, although it too induces tumours in rat liver. Both compounds are metabolised in the liver of rats and hamsters *in vivo*, as shown by the extensive methylation of liver nucleic acids which both compounds produce (Lijinsky, 1988b). Both compounds are readily metabolised by hepatocytes in culture, suggesting that there are activating enzymes in the cytosol (Farrelly *et al.*, 1987a, 1988). It has been suggested that NBOPA is converted into MNOPA in the course of carcinogenesis (Lawson *et al.*, 1981b), although Farrelly has found no evidence of formation of MNOPA by hepatocytes (Farrelly *et al.*, 1987a). On the other hand, there seem to be some common metabolites between the two β-keto nitrosamines, since MNOPA induces bladder tumours when applied directly to the bladder, while NBOPA induces bladder tumours when given orally to rats (Lijinsky *et al.*, 1988f), or when administered intravesically.

There is a link between the acyclic nitrosamines and cyclic nitrosamines in their metabolic activation in that NHPOPA is the acyclic form of 2-hydroxy-2,6-dimethylnitrosomorpholine, which arises from β-oxidation of 2,6-dimethylnitrosomorpholine (DMNM) (Gingell *et al.*, 1976). This cyclic nitrosamine is the only one which has induced tumours of the pancreas in Syrian hamsters, and it is probable that this is so because of its interconversion with NHPOPA and so with NBHPA and NBOPA (Fig. 4.11). The four nitrosamines are not equally carcinogenic in the hamster, NBHPA being considerably the least potent, possibly because it is the most hydrophilic and likely to be excreted from the body in the urine, which is the case (Underwood and Lijinsky, 1984b). The contrast with the effects of these four compounds in the rat is quite large, for in the rat NBOPA and NHPOPA are potent liver carcinogens, while DMNM is a potent esophageal carcinogen; NBHPA also induced tumours in the esophagus, as well as in lung, thyroid, liver and nasal mucosa, but is much weaker, because much of the dose is excreted unchanged in the urine. This hydrophilicity is shared by nitrosodiethanolamine, which also is largely excreted unchanged, and which therefore appears to be much less potent than it is.

The effects of substitution of deuterium for hydrogen in the positions α- to the N-nitroso function in DMNM, which reduced carcinogenic activity (potency) in the rat, but increased potency in the hamster, indicated that the rate-limiting step in carcinogenesis in rats was α-oxidation, and in the hamster β-oxidation. Similarly, the opposite relative potencies of the *cis* and *trans* isomers of dimethylnitrosomorpholine (Fig. 4.12) in the rat and in the hamster, the *trans* being more potent in the rat (Lijinsky and Reuber, 1980c) and the *cis* more potent in the hamster (Rao et al., 1981) and also in the guinea pig (Lijinsky and Reuber, 1981c), pointed to the same conclusion. Yet, although there were some differences in the pattern of urinary metabolites of DMNM between rats, hamsters and guinea pigs (Underwood and Lijinsky, 1982), the difference in pattern of metabolites between the *cis* and *trans* isomers was similar (Fig. 4.13) in all three species (Underwood and Lijinsky, 1984a). There are some reservations in interpreting the results of examination of metabolites appearing in the urine, because they represent end products, but nevertheless, they indicate pathways of metabolism that led to their formation. They provide complementary information to that obtained by examination of metabolism in isolated organs or subcellular fractions, which is also incomplete.

Most of the urinary metabolites of DMNM have been identified, at least tentatively, and include the expected NBHPA and NHPOPA, as described by Gingell et al. (1976). Other metabolites are *iso*propanolamine, bis-(2-hydroxypropyl)amine, bis-(2-oxopropyl)amine, 2,6-dimethylmorpholine-3-one and 2'-hydroxy-1'-methyl-2-ethoxypropionic acid (Fig. 4.14). The last is analogous to the 2-hydroxyethoxyacetic acid found in the urine of rats given nitrosomorpholine (Hecht et al., 1981), and is a product of α-oxidation of DMNM. Although it is prominent in

Nitrosamines with oxygen substituents 133

the urine of rats, hamsters and guinea pigs given the *trans* isomer of DMNM, and very little is present in the urine of those animals given the *cis* isomer of DMNM, the greatest excretion of it was by the rat (Underwood and Lijinsky, 1982, 1984a). This confirmed α-oxidation as the dominant pathway of metabolism of DMNM in the rat, which is shown also by the effects of deuterium substitution on its carcinogenicity. The formation of the hydroxy acid from the *cis* isomer of DMNM in the three species is very small, supporting the results of the deuterium labelling experiments which indicate β-oxidation of DMNM as the major pathway of metabolism of the *cis* isomer. Accordingly, more NBHPA and

Fig. 4.11. Metabolism of 2,6-dimethylnitrosomorpholine.

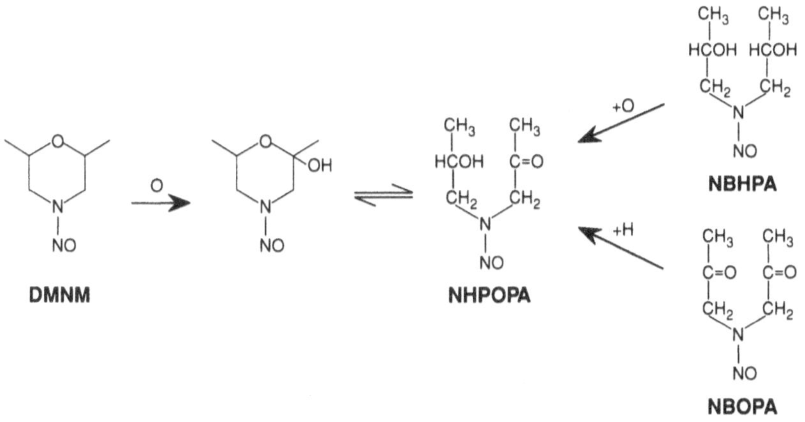

Fig. 4.12. *Cis* and *trans* dimethylnitrosomorpholine.

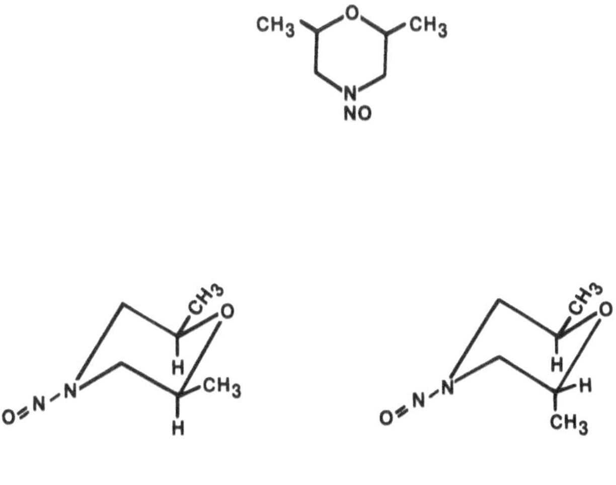

NHPOPA is excreted into the urine following treatment with the *cis* isomer of DMNM, compared with very small amounts from the *trans* isomer. The two pathways of metabolism are shown in Fig. 4.15.

Although the lactam is also a product of oxidation of DMNM at the α-position followed by denitrosation, there is little difference in the amount of that product excreted by animals given either the *cis* or *trans* isomer. It appears that considerable denitrosation occurs of the acyclic nitrosamines formed by β-oxidation of DMNM, since bis-(2-hydroxypropyl)amine and bis-(2-oxopropyl)amine are present, the latter in especially large amounts in the urine of hamsters given the *cis* isomer of DMNM, whereas it is essentially absent from the urine of rats. The formation of this amine

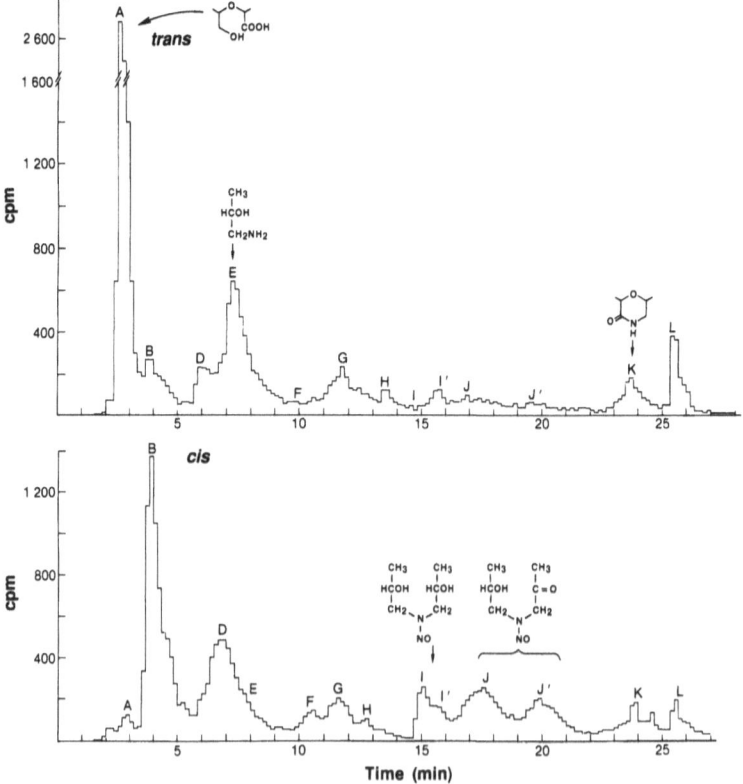

Fig. 4.13. Urinary metabolites of *cis* and *trans* dimethylnitrosomorpholine HPLC chromatograms of rat *trans* and *cis* urinary metabolites: approx. 40,000 cpm on a Dupont Zorbax-ODS column, eluted 0–5 min with 0.01 M sodium acetate buffer (pH 5.6); 5–6 min with 0–10% methanol in acetate buffer; 6–7 min with 10% methanol in acetate buffer; 17–20 min with 10–60% methanol in acetate buffer; 20–27 min with 60% methanol in acetate buffer; at 1.5 ml/min.

Fig. 4.14. Urinary metabolites of dimethylnitrosomorpholine in rats, hamsters and guinea pigs. – As in Fig. 4.13.

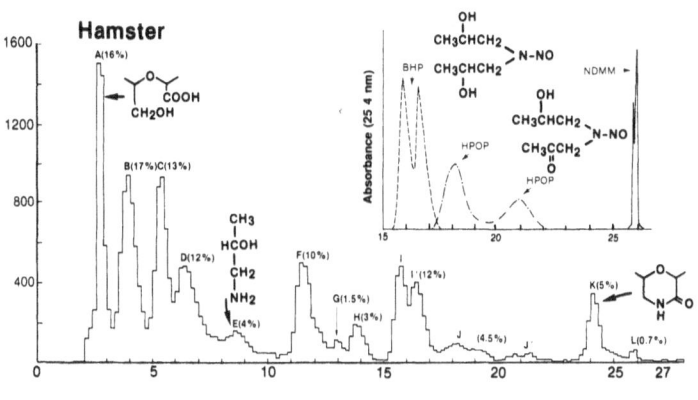

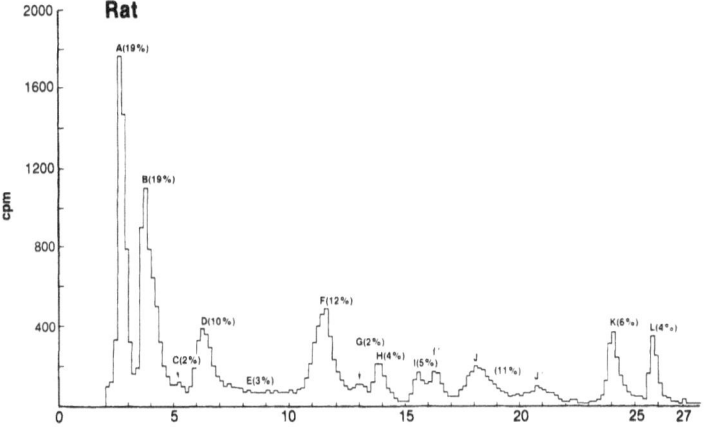

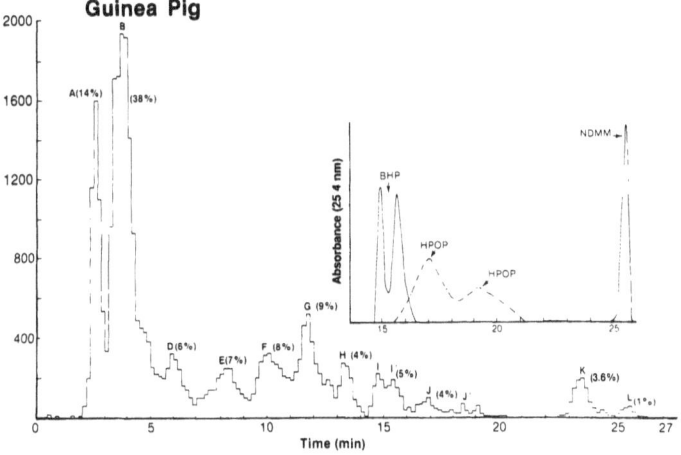

might be an indicator of a metabolic pathway peculiar to the hamster, and possibly related to the facility with which DMNM induces tumours of the pancreas ducts in hamsters (Reznik et al., 1978b).

Studies of the metabolism of DMNM carried out by Kokkinakis, Hollenberg and Scarpelli have shown that the initial β-oxidation in hamsters takes place very rapidly in the liver, and much more readily with the *cis* isomer than the *trans*, leading to formation of NHPOPA; the latter is also formed very rapidly from NBOPA by hamster liver *in vivo* or *in vitro* (Kokkinakis et al., 1984).

The further fate of NHPOPA, which is apparently a central metabolite of NDMM, NBOPA and NBHPA (Gingell et al., 1979), has been studied by Kokkinakis et al. (1985) and by this author, who examined the urinary metabolites of NHPOPA in rats. Kokkinakis identified metabolites that were a glucuronide and a sulphate ester of NHPOPA. These can be considered transport forms of this compound, which is strongly carcinogenic in its own right (Lijinsky et al., 1984c; Pour et al., 1979). It is probably not, as Pour claimed, the proximate carcinogen formed from NBOPA, DMNM and NBHPA, since the four compounds have such different tumourigenic effects in rats, qualitative and quantitative. In hamsters the relative potencies of the compounds, which produce essentially the same tumours, do not support NHPOPA as the proximate carcinogen, since it is less effective than NBOPA in this species (Lijinsky et al., 1984b).

The fact that the four nitrosamines (Fig. 4.11) all lead to methylation of DNA and RNA in liver and kidney (and probably other organs) of the rat and hamster, although not to the same extent (Lijinsky et al., 1988e),

Fig. 4.15. Metabolites of DMNM by α- or β-oxidation.

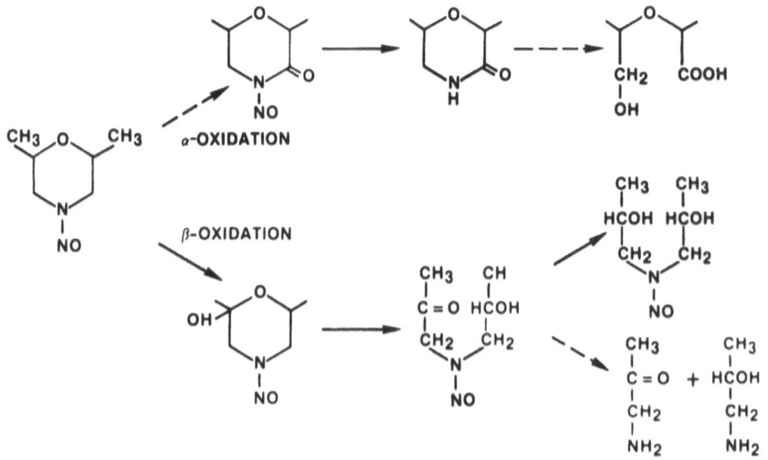

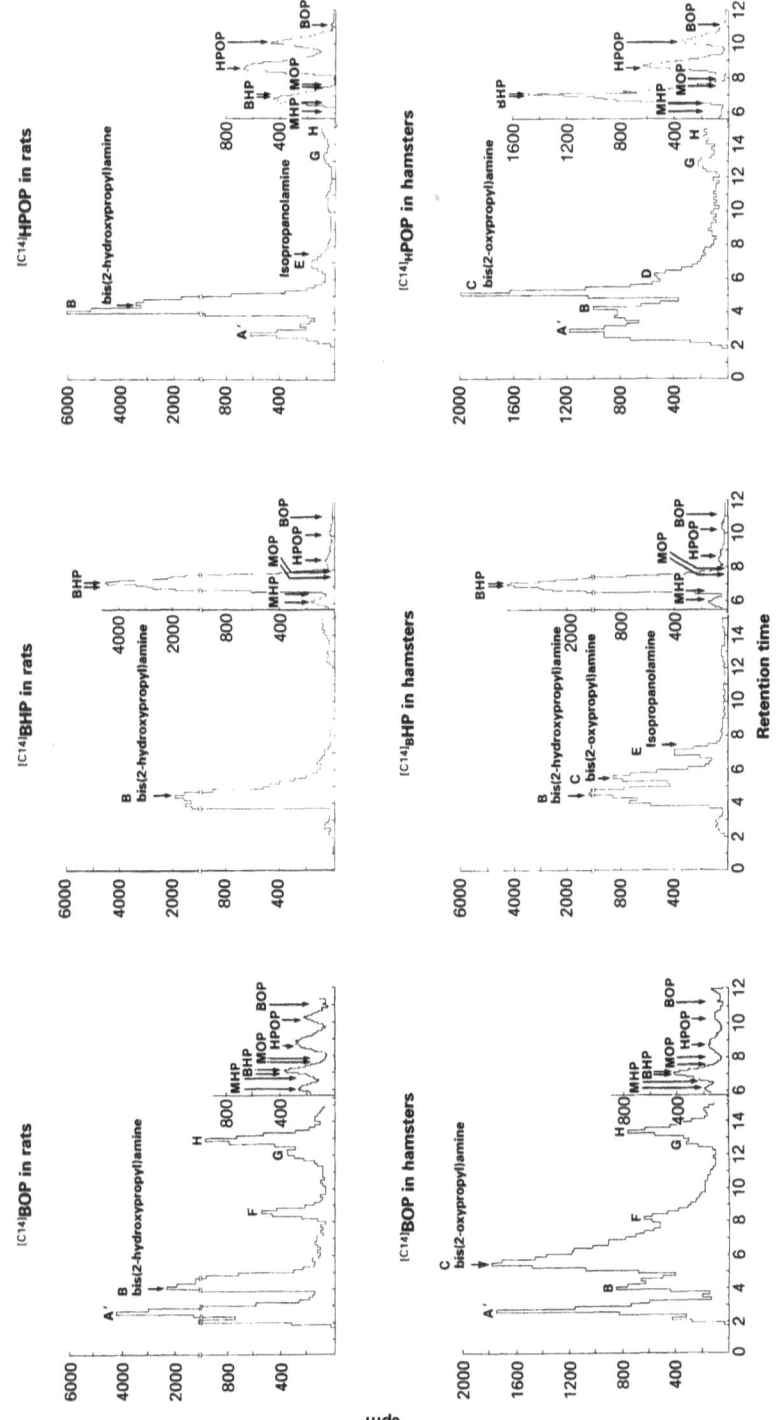

Fig. 4.16. Urinary metabolites of NBHPA, NBOPA and NHPOPA 24 h after oral administration in rats and hamsters.

suggest that some pathways of metabolism and activation are common between them. This makes the differences in carcinogenicity more difficult to understand, unless it is assumed that alkylation of DNA is not the key event in carcinogenesis by these compounds. Equally perplexing is the finding that DMNM is readily converted to a bacterial mutagen by rat liver microsomes (S_9 fraction), but it has not induced tumours in rat liver, whereas the other three nitrosamines are not mutagenic by rat liver activation, although all three have induced liver tumours in rats (Lijinsky et al., 1988e, f); all four compounds induce liver tumours in hamsters and are activated readily by hamster liver microsomes to bacterial mutagens (Andrews and Lijinsky, 1984).

The pattern of metabolites excreted in the urine of rats and hamsters given the three acyclic oxygenated nitrosamines (Fig. 4.11) show both similarities and differences (Fig. 4.16), the latter reflecting their different chemical structures and probably related to their different carcinogenic activities. A primary difference is that oxidation of DMNM and NHPOPA in the cyclic form probably occurs at the α-carbon, whereas this does not seem to take place to a large extent in the acyclic compounds. Evidence against α-oxidation of NBOPA comes from the absence of a significant deuterium isotope effect in carcinogenesis when NBOPA labelled with deuterium in the α-methylenes is compared with the unlabelled compound (Lijinsky et al., 1988f); the deuterium isotope effect in carcinogenesis by α-deuterated nitrosomorpholine is large (Lijinsky et al., 1976) and also in DMNM (Lijinsky et al., 1980f). A comparison of the carcinogenic effects of DMNM and the three metabolites suggests that dose-dependent differences in metabolism are responsible for the induction of tumours of liver as well as esophagus by NBHPA when given to rats by gavage, but only tumours of the esophagus when administered in drinking water. NBOPA induces liver tumours in female rats by gavage or given in drinking water, but tumours of lung, bladder and thyroid, not liver, in male rats by either dose regimen. NHPOPA seems to add the tumour responses of both analogous symmetrical nitrosamines, inducing in rats, whether administered by gavage or in drinking water, tumours of liver (in males or females), lung, esophagus, thyroid, nasal mucosa and bladder. These carcinogenesis results imply that the structure of the original nitrosamine plays the major role in determining the organs in which tumours arise, in addition to (or instead of) the metabolic formation of a methylating intermediate which methylates DNA (Lijinsky et al., 1988e; Thomas and Lijinsky, 1988).

The importance of the structure of the nitrosamine in determining target organs for carcinogenesis compels examination of the metabolism of these oxygenated nitrosamines, and here the urinary metabolites formed from them can provide insight. A smaller proportion of the dose of NBOPA was excreted in the urine than from NBHPA or NHPOPA, indicating that a greater proportion was metabolised within the body, leading to a somewhat greater exhalation of carbon dioxide (Table 4.4). In the hamster more of the dose of NHPOPA was excreted as CO_2 than in the rat, while the opposite was the case with NBOPA, again suggesting that metabolism of each compound was different in the rat and hamster, and that within one species metabolism of the two compounds was different. Concomitantly, more of the radioactive dose of NHPOPA was excreted in urine by the rat than by the hamster, and the reverse was the case for NBOPA. In total, a very similar proportion of the dose of each compound was excreted as CO_2 and in urine (approximately 60%) in either rats or hamsters.

Separation of the metabolites in the urine by HPLC showed the presence of a number of radiolabelled metabolites that were common among the three acyclic nitrosamines and DMNM. The proportions of various separated metabolites in the urine of rats and hamsters given 2 mg of NBOPA and NHPOPA, 20 mg of NBHPA or 5 mg of DMNM are shown in Table 4.5. Some of the unidentified metabolites might be the glucuronides and sulphate esters of NHPOPA and NBHPA reported by Kokkinakis et al. (1985) following administration of much larger doses of NHPOPA (16 mg per animal), and probably also were present in the urine of animals given DMNM.

Bis-(2-oxopropyl)amine and bis-(2-hydroxypropyl)amine are obviously formed through denitrosation of the corresponding nitrosamines, whereas *iso*propanolamine can be formed by oxidative elimination of the oxopropyl group of NHPOPA, followed by denitrosation, as suggested for the formation of methylamine as a urinary metabolite of nitrosodimethylamine (Keefer et al., 1987). Another possible product, 2-oxopropylamine (aminoacetone) was not identified, although it could be formed in a similar manner to *iso*propanolamine. Neither was hydroxypropyl-oxopropylamine identified. Methylnitroso-2-oxopropylamine (MNOPA) and methylnitroso-2-hydroxypropylamine (MNHPA) were also not detected in any of the urines, although Lawson et al. (1981 b) suggested that MNOPA was a proximate carcinogenic product formed *in vivo* from NBOPA; however, administration of MNOPA or MNHPA to rats led to excretion of the latter compound in the urine (Singer et al., 1981), so it should have been detected had this route from NBOPA been followed. The inter-relation of oxygenated metabolites of DMNM and methylnitroso-*n*-alkylamines is shown in Fig. 4.17.

Table 4.4. Metabolism of oxygenated propylnitrosamines

Compound	Dose	Animal	% of dose excreted in 6 h as CO_2	in urine
N-nitroso-bis-(2-oxopropyl)amine-1-^{14}C	2.5 mg/5 µCi	♀ Rat, 20 wk	17	19
		♂ Rat, 20 wk	17	9
		♂ Rat, 20 wk, castrated	11	5
		♂ Rat, 20 wk, estradiol	33	24
		♂ Rat, 20 wk, castrated + estradiol	28	26
	0.6 mg/2.5 µCi	♂ Rat, 4 wk	30	8
		♀ Rat, 4 wk	30	9
	2.3 mg/5.2 µCi	♂ Rat, 4 wk	22	13
		♀ Rat, 4 wk	26	10
		♂ Rat, 65 wk	33	9
		♂ Rat, 74 wk, castrated	38	12
		♀ Rat, 65 wk	29	11
	2 mg/13 µCi	♀ Hamster	12	19
(2-hydroxypropyl) (2-oxopropyl)amine-1-^{14}C	2 mg/11.5 µCi	♀ Rat	18	24
		♀ Hamster	14	33
bis-(2-hydroxypropyl)amine-1-^{14}C	20 mg/45 µCi	♀ Rat	2	45
		♂ Hamster	2	41
methyl-1-^{14}C-oxopropylamine	2 mg/5.5 µCi	♂ Rat	43	7
		♀ Rat	43	9
		♀ Hamster	22	9

Table 4.5. Metabolites of 2,6-dimethylnitrosomorpholine and related 2-oxygenated propylnitrosamines in urine (% of dose)

Compound	DMNM (2 mg)		NHPOPA (2.5 mg)		NBOPA (2.5 mg)		NBHPA (20 mg)	
Metabolite	Rat	Hamster	Rat	Hamster	Rat	Hamster	Rat	Hamster
CO_2 (24 h)	—	—	23	41	41	28	4	10
Urine								
NHPOPA	47	70	41	33	19	27	52	61
NBHPA	11	5	7.3	4.2	2.0	2.3	1.6	2.5
NBOPA	5	12	3.7	5.2	1.4	1.8	33	42
bis-(2-hydroxypropyl)amine	19	17	20	3.2	<1	<1	10	7.7
					3.4	3.7		
bis-(2-oxopropyl)amine	<1	13	0.4	3.8	0.1	4.1	—	4.7
Hydroxypropylamine	<1	4	1.3	—	0.3	0.8	—	2
Hydroxymethyl ethoxypropionic acid	19	16	—	—	—	—	—	—
dimethylmorpholine-3-one	6	5	—	—	—	—	—	—

142   *Metabolism and interactions*

The main conclusions of these studies, even though all of the metabolites in urine have not been identified, is that there are large and consistent differences between the results in rat and hamster. The presence of a major metabolite in hamster urine that is not found in rat urine [believed to be *bis*-(oxopropylamine)], suggests the possibility that the metabolism in hamsters leading to formation of this product might be related to induction of pancreas tumours in hamsters, but not rats.

Much has been discussed already about metabolism of nitrosamines by liver microsomes, and formation of the expected aldehydes. However, the aldehydes, other than formaldehyde, formed from the nitrosamines

Fig. 4.17. Common metabolites of DMNM and methylnitrosoalkylamines.

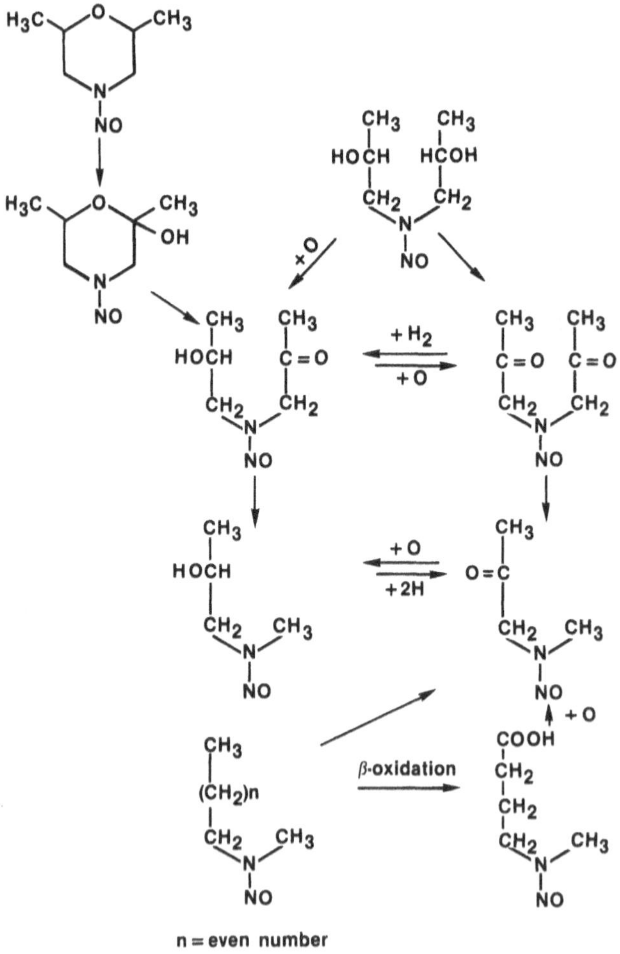

having alkyl groups bearing oxygen functions, such as MNHPA and MNOPA have not been identified. Hamster liver microsomes oxidise such compounds but rat liver microsomes do not. The metabolism of NBOPA and its metabolites leads to methylation of DNA and RNA in a number of organs, including liver, kidney, lung and pancreas, but tumours develop only in some organs in which methylation is found (Thomas and Lijinsky, 1988; Lijinsky, 1988b). It is possible that other types of alkylation of DNA (i.e. by alkyl groups other than methyl) occur at much lower levels, and this has been shown by Lawson and Nagel (1988) in hamster liver, but not in pancreas. However, hamsters develop tumours in both organs when treated with NBOPA. The existence of hydroxypropylated bases in liver DNA does indicate that some intermediate such as a 2-hydroxy-propyldiazonium ion is formed *in vivo*, although the significance of this in relation to the much greater extent of methylation of DNA is not known.

The mechanism by which a methylating agent is formed from NBOPA is not clear. Leung and Archer (1984) suggest that oxidation of one of the methylenes would give rise to an oxopropyldiazonium ion (and presumably pyruvaldehyde). The former would decompose according to some such scheme as shown in Fig. 4.18, which was derived from consideration of the decomposition of 2-oxopropylnitrosourea, eventually yielding diazomethane. The methylated bases in DNA produced from NBOPA were originally identified by comparison of their chromatographic properties with those of the model synthetic compounds (Lawson *et al.*, 1981 a). Recently, using deuterium-labelled NBOPA in both the methylenes and in the terminal methyl groups, it was shown that the alkylated base in the hydrolysates of the liver DNA treated with the compound had a mass of 167 mass units, corresponding to methyl-d_2-guanine, confirming unequivocally the original finding (Liberato *et al.*, 1989).

Michejda and Kupper offered an alternative mechanism by which a methylating agent could be formed from NBOPA, through the well known Baeyer-Villiger oxidation of ketones to esters. From NBOPA acetoxymethylnitroso-2-oxopropylamine would be formed (Fig. 4.19), which could hydrolyse to hydroxymethylnitroso-2-oxopropylamine and thence to formaldehyde and 2-oxopropyldiazonium hydroxide. This could then give rise by such a mechanism as Archer proposed to diazomethane, in agreement with the mass spectrometric finding that the methyl group comes from the α-methylene of NBOPA. The activation of MNOPA is less straightforward, since either alkyl chain of this asymmetric nitrosamine can be oxidised. If the methyl group is oxidised to formaldehyde this would release a 2-oxopropyldiazonium ion, and thence lead to diazomethane as already described. However, oxidation of the oxopropyl group at the α-methylene would give rise to a methyldiazonium ion and to pyruvic aldehyde. In spite of the induction of liver tumours in rats by MNOPA, however, Farrelly et al. (1988) could not demonstrate oxidation of this nitrosamine by rat liver microsomal enzymes, although it did occur with hamster enzymes and by rat hepatocytes. Farrelly concluded that the presence of oxygen on the β-carbon prevented metabolism by these enzymes. Extensive oxidation of MNOPA must take place, however, since methylation of rat liver DNA following administration to rats is similar

Fig. 4.18. Formation of 2-oxopropyldiazonium from oxopropylnitrosourea and an oxopropylnitrosamine.

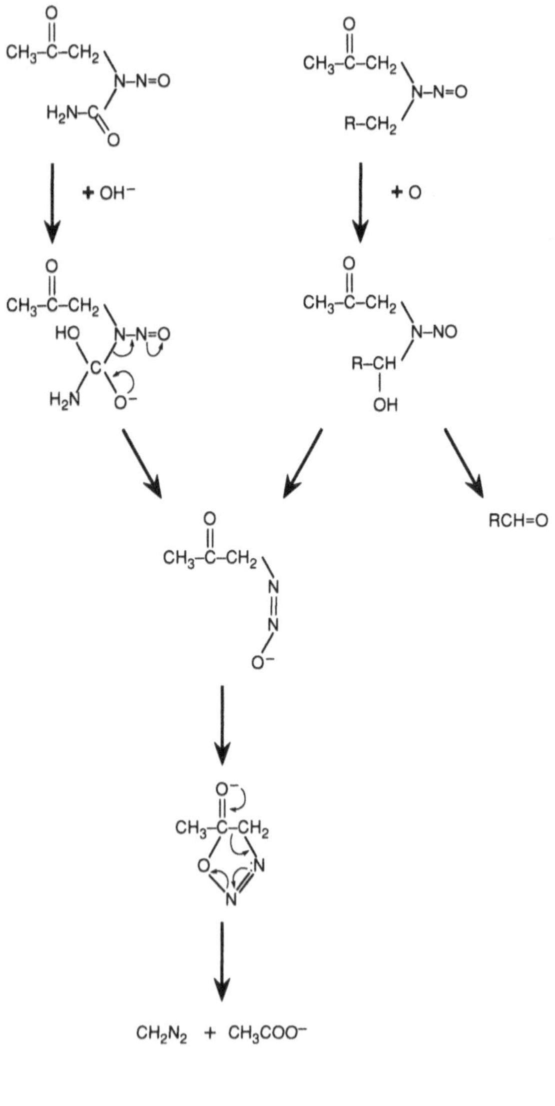

to or greater than that by an equimolar dose of NDMA. This leads to consideration of a Baeyer-Villiger oxidation to acetoxymethylnitrosomethylamine, and further, through the hydroxymethylnitrosomethylamine, to a methyldiazonium ion and formaldehyde.

When rats were given MNOPA labelled with deuterium in the methyl group and in the α-methylene, the methylguanine later isolated from hydrolysates of the liver DNA and RNA had a mass of 168, showing the presence of methyl-d_3-guanine (or CD_3-guanine); no methylguanine with a mass of 167 was detected, showing that all of the methylating agent came from the *N*-methyl group of MNOPA. This result supports the probability that all of the metabolism of MNOPA (and by inference of NBOPA) leading to formation of a methylating agent proceeds via a Baeyer-Villiger oxidation (Fig. 3.12, p. 82), and not through α-oxidation of the oxopropyl group (Fig. 4.18). Two problems at least remain. First, an enzyme system carrying out a Baeyer-Villiger oxidation has not been

Fig. 4.19. Baeyer-Villiger oxidation of NBOPA.

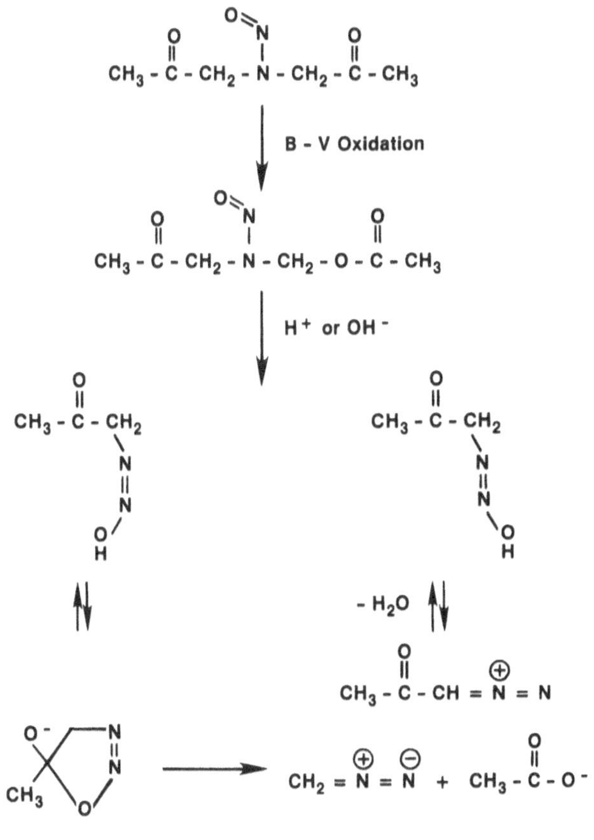

demonstrated in eukaryotes (although there is evidence for such in bacteria). Second, there is no direct information that methylation of DNA in liver (or other organs) by MNOPA is the mechanism by which tumours are induced.

4.7 Which enzymic activation leads to carcinogenesis?

One major problem in studying the metabolism of nitrosamines is deciding which system or complex of enzymes is the most appropriate in the context of carcinogenesis. Any system which selects particular enzymes, for example, mixed function oxidases as in microsomal preparations, risks omitting important enzymes in the cytoplasm which might supplement or complement the MFO's. This is, of course, why the usually conducted 'Ames' mutagenic assay does not inform us about mechanisms by which nitrosamines produce cancer. One cytosolic enzyme is alcohol dehydrogenase, which can oxidise hydroxyalkylnitrosamines to aldehydes, thereby transforming nitrosodiethanolamine to the acyclic form of 2-hydroxynitrosomorpholine. This, however, seems to be an excretory product, since it is a very weak carcinogen, although it is a directly acting mutagen.

It has not proved easy in animals, or *in vitro*, to study several metabolic pathways that might be concurrent, which makes determination of the balance of products from several pathways difficult. Yet this balance might be the most important factor in determining which organs develop tumours and which do not, following different doses and dose rates of a carcinogen. For example, this might be why there is such a large difference in carcinogenic effect between gavage treatment and drinking water treatment of rats with NDMA but not with MNEA. Similar, as discussed previously, are the differences between NBHPA and NBOPA by the two modes of administration. It is perhaps teleological to mention that it is not in the interest of an organism to metabolise a substance to form carcinogenic products, but the metabolism is by enzymes performing their normal function of detoxication of xenobiotics. Perhaps in some cases a close analogue (which happens to be carcinogenic) of an important metabolic intermediate or hormone is enzymically transformed along a normal pathway. This certainly seems to be the case with methylnitrosoalkylamines having even-numbered carbon chains. The larger members of the series appear not very susceptible to metabolism via the usual α-oxidation pathway, perhaps because they have such low solubility in water. As measured by the proportion of the dose that can be recovered as metabolites in the urine of rats, the major metabolism of these compounds is through a series of successive β-oxidations to form a series of methylnitrosaminoacids, ending with nitrososarcosine. The long chain nitrosamine simulates the alkyl chain of a fatty acid and is oxidised by appropriate enzymes (Knoop), perhaps in mitochondria, to the various products, the smallest of which are carcinogens (MNOPA, for example).

It is notable that the important product that is the ultimate or proximate carcinogen, in this case of bladder carcinogenesis, is a minor metabolite, the major metabolites being inactive or much less active. This raises the important point that our studies of the metabolism of carcinogenic nitrosamines usually focuses on identification and measurement of the major metabolites (which is easier), and there is less interest in the minor metabolites that might be formed. As mentioned previously it is almost certain that the major metabolites of NDMA, formaldehyde and the methyldiazonium ion, are not the products important in tumour induction, although their precursors might be, for example a stabilised form of the hydroxymethylnitrosamine, a further oxidation product the alkylnitrosoformamide, or the carbanion. Other possible carcinogenic intermediates are more difficult to imagine, but that might be due only to our limited chemical vision. It is certain that many compounds that give rise to a methyldiazonium ion as readily as NDMA (methylnitrosocyclohexylamine, for example) produce neither toxic changes in the liver of rats nor liver tumours. It is conceivable that the initial interaction in the target organ is with an enzyme that recognises a particular structure of the entire nitrosamine molecule, and that all subsequent biochemical and biological effects follow from this. Little or no attention has been paid to this possibility, partly because the concept of receptors has been restricted to hormonal or pharmacological receptors. However, there have been indications from the studies of structure–activity relations in carcinogenesis by alkylnitrosoureas and alkylnitrosocarbamates, which require no metabolic activation for mutagenic activity or alkylation of macromolecules, that particular cells in certain organs are responsive to particular chemical structures among these compounds. This Laboratory is beginning to investigate this matter.

The activation of a nitrosamine might be considered a series of steps involving distribution and partition of the administered compound into several compartments, in each of which metabolism takes place. But in only some of these compartments do the products have the necessary affinity for the critical cellular organelles, through disturbance of which the process of neoplastic transformation occurs. Some of these actions might well be co-operative. This might explain in another area of carcinogenesis why polynuclear hydrocarbons are such spectacularly potent carcinogens in mouse skin and less so in the skin of other species, yet are virtually without toxic or carcinogenic effects in liver and most other organs of adult mice and other species. The reactions and interactions must come together, and this can only happen in certain organs and cells; formation of an intermediate that alkylates DNA is certainly not enough, perhaps explaining why cancer cannot be produced in epithelial cells in a Petri dish. The concept of Foulds (1969) and others that cancer is a progressive process, probably more complicated than the vision of initiation followed by promotion suggested by Berenblum and Shubik (1947), indicates that the carcinogen continuously applied for a long period has a series of reactions and interactions subsequent to 'initiation'. Initiation itself might be a process involving a series of physiological and chemical interactions between the carcinogen and the host, including multiple reactions of the carcinogen or its several metabolites with cellular components, and probably cell replication.

The cyclic nitrosamines are metabolised with comparable efficiency to the acyclic nitrosamines, as measured by carbon dioxide excretion (Snyder et al., 1977), and metabolites have in several cases been identified and measured in urine, blood and several organs. Further, the pattern of distribution and elimination of cyclic nitrosamines which have been studied is similar to that of acyclic nitrosamines, concentrations increasing in the blood, etc. until one to two hours following treatment, then declining (Althoff et al., 1977b). In experiments in rats a small dose (26 μmol) of ^{14}C-MNEA led to a decline in specific radioactivity in the liver of 75% by 48 h, but in the kidney the decline was only 25%. There were some differences in uptake and distribution between the small NDMA and the larger nitrosodibutylamine and nitrosohexamethyleneimine, but this might be largely due to differences in the rate of absorption from the site of subcutaneous injection. The much slower absorption of the large water-insoluble compounds under these conditions is one reason for dissatisfaction with this mode of treatment, which requires administration of huge doses of, for example, methylnitrosododecylamine (Ketkar et al., 1981) in order to elicit tumours. These results, and others in similar experiments, suggest that the rate of metabolism of the various nitrosamines is quite similar, regardless of their cyclic or acyclic structure. Hydroxylated metabolites have been identified in the case of several acyclic compounds studied (Okada, 1984), but the role of most of them in carcinogenesis is usually not known. Certainly the more recent results of Wiessler (Richter et al., 1988a, b) and of Mirvish et al. (1985), have reinforced what was known previously, that most methyl or methylene groups in a nitrosamine can be enzymically oxidised in vivo. The proportions of the various hydroxylated products varies from one compound to another, and some greatly predominate, but the role of these hydroxylated derivatives in carcinogenesis is not at all clear.

Even with the simplest nitrosamines, such as methylnitrosoethylamine and nitrosodiethylamine some β-oxidation of the ethyl group of the nitrosamine occurs, although it has not been possible to detect the hydroxylated nitrosamine in the circulation or elsewhere in the body (Von Hofe et al., 1986). However, indirect evidence of formation of the hydroxyethylnitrosamines is the finding of hydroxyethylated porphyrin from NDEA-treated animals (White et al., 1983) and in DNA of rat liver from MNEA (Von Hofe et al., 1986b); the latter was present to the extent of only 2% of the ethylation provided by MNEA, however. It is difficult to believe that the hydroxyethylating agent could have a source other than the β-hydroxylated ethylnitrosamine (Streeter et al., 1989).

Suggestions that β-oxidation of the propyl group in propylnitrosamines occurred *in vivo* were first put forward by Krüger (1971), and these were borne out to some extent by the *in vitro* studies of Park and Archer (1978), who showed that nitrosodi-*n*-propylamine was oxidised to form both *n*-propyl and *iso*propyl products, possibly the respective diazonium ions; the *iso*propyl-diazonium compound would easily form a 2-hydroxypropyl (or isopropanol) product. In spite of much searching, particularly by the group at the Eppley Institute in Omaha, it has not been possible to demonstrate the formation of significant quantities of the 2-hydroxypropyl derivatives of propylnitrosamines in the bodies of rodents. It is unfortunate, therefore, that those investigators (Pour et al., 1974) imply or suggest in several publications that β-oxidised propylnitrosamines are formed *in vivo* and that they are important intermediates in the carcinogenic action of propylnitrosamines. The main reason for doubting this suggestion is that the carcinogenic properties (types of tumour produced) by propylnitrosamines are entirely different from those produced by the β-oxidised propylnitrosamines; the latter have the rare property of inducing pancreas duct cancers in hamsters (discussed elsewhere), which are not induced by nitrosodi-*n*-propylamine or methylnitroso-*n*-propylamine. Attention is rarely given to the equal possibility of gamma oxidation of the *n*-propyl group (to 3-hydroxypropyl-), probably because *N*-nitroso compounds containing a 3-hydroxypropyl group are weak and uninteresting carcinogens (or do not induce tumours at all) (Koepke et al., 1988a); nitroso-3-hydroxypropylurea is a weak mutagen (Lijinsky et al., 1987a) and one of the least potent carcinogens among the alkylnitrosoureas examined (Lijinsky and Kovatch, 1989b).

Although so many nitrosamines are oxidised to a variety of hydroxylated products *in vivo*, there is little evidence that the hydroxylated derivatives play an important role in carcinogenesis. Quite the contrary, most of the hydroxylated derivatives are less potent (some considerably less) than the parent nitrosamines, and only a handful show comparable potency with the parent. The hydroxylated derivatives probably play a role as excretory products since they are much more water soluble than the often insoluble parent nitrosamines; much of the dose administered is excreted unchanged in the urine, where some of them possibly act on the bladder mucosa following further oxidation, and give rise to bladder tumours. Hydroxylated nitrosamines of particular structures have this action in the bladder.

Oxidation of acyclic nitrosamines at several positions, leading to formation of a large variety of hydroxylated products, was first shown by Blattmann and Preussman (1974). The implication that such hydroxylation could take place at all non-hindered sites in an alkyl chain has been borne out in subsequent experiments by many investigators. These hydroxylated derivatives are not usually considered proximate carcinogens, since structure–activity studies, especially those with deuterium-labelled nitrosamines, show that oxidation of the carbon atom α- to the N-nitroso function is the key step in carcinogenesis in almost all cases.

The laboratory of Okada, in which the main thrust was the metabolism of nitrosodi-n-butylamine and other nitrosamines that are related to induction of bladder tumours, has reported the identification in rat urine of many of the possible permutations of hydroxylation of these large molecules (Okada, 1984). Wiessler and his colleagues have examined the numerous hydroxylated derivatives of nitrosodi-n-butylamine and nitrosodi-n-amylamine formed in rats under a variety of conditions, including bypass of the liver, and have suggested that the pharmacokinetics of these carcinogens depend on the relative activities of a variety of oxidising enzymes (Richter *et al.*, 1988 *a*, *b*). Mirvish and his colleagues have carried out an intense study of methylnitroso-n-amylamine, which is an esophageal carcinogen in rats, but induces liver and lung tumours in hamsters. These studies *in vivo* and *in vitro* in rats have shown that all of the possible positions are hydroxylated, and the relative proportions of the various derivatives have been measured. Attempts have been made to characterise the enzymes (in liver) responsible for the various hydroxylations; aldehydes resulting from further oxidation of the alcohols have also been identified in some cases (Ji *et al.*, 1989). Earlier studies focussed on the metabolites produced by the rat esophagus (Mirvish *et al.*, 1985), and a more wide-ranging study compared metabolism in the newborn and adult rat and hamster tissues. Excellent as these experiments were, they

Alkylation of DNA etc. 151

showed only small differences in metabolism between one system and another, although often large differences in enzymic activity, and shed little or no light on the reasons for the extraordinary organ and species specificity of these nitrosamines.

4.8 Alkylation of DNA and other macromolecules
4.8.1 Overview

It is commonly believed – and has been for decades – that the initial and most important reaction in tumour induction by *N*-nitroso compounds is conversion, directly or through metabolic activation, to alkylating agents which alkylate DNA. Alkylnitroso-ureas, -carbamates and -guanidines are unlikely to undergo metabolism, since they are very reactive and unstable and form an alkyldiazonium ion very rapidly. This same alkyldiazonium ion is formed metabolically from a nitrosamine containing the same alkyl group (Fig. 4.20) through oxidation of an α-methylene in the nitrosamine to an α-hydroxy derivative (which is suspected but has not been isolated). This is converted to a carbonyl compound (which can often be isolated and identified) and the remaining unstable monoalkylnitrosamine becomes an alkyldiazohydroxide, which also has not been isolated, but is a potent alkylating agent. Information about some of the steps in this process is circumstantial, but products of the reaction of the alkylating agent have been identified, including alcohols and nitrogen, as well as alkylated macromolecules, including DNA and proteins. These products do not account for all of the covalently bound radioactivity present after administration of a radio-labelled nitrosamine, and it is conceivable that other products play an important role in carcinogenesis.

Studies of the alkylation of DNA and RNA in animals by *N*-nitroso compounds began with Magee and Farber (1962) and in the early years produced several studies comparing alkylation by various agents with their carcinogenic effects. For example, Swann and Magee (1971) compared ethylation of nucleic acids by NDEA, ethylnitrosourea and

Fig. 4.20. Common reactant (alkyldiazonium ion) formed from alkylating *N*-nitroso compounds.

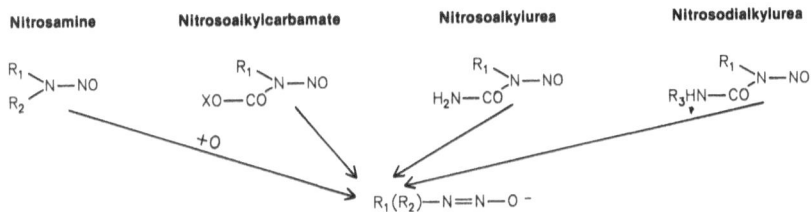

ethylmethanesulphonate in rats, and found considerable quantitative differences between them. The alkylated base measured was 7-ethylguanine, and the relation to the carcinogenic activity of the compounds was not convincing. Loveless (1969) in an incisive paper pointed out that the formation of 7-alkylguanosine in DNA, the product usually measured, would not produce structural changes in DNA that would lead to mispairing and mutation. On the other hand, alkylation at the enolic O^6 position of guanosine (O^6-ethylguanosine from ethylnitrosourea in the example discussed by Loveless) does lead to mispairing and can give rise to mutations. Extending that observation, Rajewsky and Goth between 1972 and 1974 examined O^6-ethylation of guanosine in DNA of brain and liver of rats treated with ENU, and reported that this alkylation was more stable in brain than in liver, leading to the proposal that slower repair of DNA damaged by alkylation in brain than in other organs (Goth and Rajewsky, 1974) was related to the induction of tumours of the brain in rats treated with ENU, which has not induced tumours in rat liver. These experiments opened up the large area of investigation of alkylation of DNA at mutation-prone sites, and variations in rates of repair of those lesions, in which N-nitroso compounds have played – and still play – an important role in this flourishing field.

Because most N-nitroso compounds are mutagenic to bacteria and in other test systems, the significant interaction with them or their metabolites in cells is believed to be formation of adducts with DNA in the nucleus by alkylation. This idea is supported by the finding that several N-nitroso compounds do form adducts with DNA following administration to animals. This does not, of course, *prove* that these adducts in DNA, or formation of them, cause the tumours that subsequently might arise, although conventional wisdom states that the mutational damage caused by such alkylation is essential for 'initiation' of tumours. This argument is difficult to sustain in face of the considerable number of potent carcinogenic nitrosamines, administration of which to animals has not led to identifiable covalent adducts with DNA, whether in organs in which the carcinogen induces tumours, or in those unaffected. Prime among these non-alkylating compounds are cyclic nitrosamines, of which a large number have been chronically tested in animals, and some of them have been used in the search for DNA adducts (in liver, for example).

The number of compounds that have been used to study the relation between alkylation of DNA and induction of tumours by N-nitroso compounds is quite small. There have been a vast number of studies using nitrosodimethylamine, a large number using nitrosodiethylamine, methylnitrosourea and ethylnitrosourea, and only a scattering of studies with other compounds. Even with the commonly used compounds there has been little unbiased investigation, paying attention to the precise organ- and species-specificity that characterises tumour induction by these compounds. That is, the focus has been on organs in which tumours are induced and on the species that is responsive, brain for ethylnitrosourea, mammary gland for methylnitrosourea, liver for nitrosodiethylamine and liver and kidney for nitrosodimethylamine. Little attention has been given to the use as controls of non-target organs or non-susceptible species, which would provide the most valid contrasts. Consequently, a lot of data has been gathered which is one-sided and might, therefore, be unimportant. Among the small number of other nitrosamines studied in this context are those inducing pancreas tumours in hamsters and esophageal tumours in rats, and in these cases there has been some attempt to relate lower levels or absence of alkylation of DNA in non-target organs to the absence of tumours.

It has proved difficult to study metabolism and DNA alkylation by N-nitroso compounds in such organs as esophagus, nasal mucosa and bladder, because of the paucity of epithelial tissue in those organs. Metabolism in esophagus and other tissues in organ culture has led to some success in identifying metabolites and in measuring alkylation of DNA (Hecht *et al.*, 1982; Castonguay *et al.*, 1983), using a small number of compounds having those organs as the targets. More difficult to rationalise have been the studies of Harris *et al.* (1979), using cultures of various human organs. He found, for example, that cultures of human esophagus and lung readily metabolised several nitrosamines in a similar manner to the same organs of rats, and that DNA adducts were similarly formed. The results of Montesano and Magee (1970) were similar. These results demonstrate that enzymes in those organs of rats and of humans were capable of metabolising these compounds, but it is not known whether the metabolic routes are the same or similar in both species, or in both organs. The metabolic products other than carbon dioxide were not usually measured. The major problem with these studies is that the target organs of these nitrosamines (or of any others) in humans is not known, and might never be known. Therefore, studies of metabolism of these nitrosamines in organs of the rat, hamster or mouse, in which the types of tumour induced are known, would be much more informative about mechanisms of carcinogenesis.

The well known organ specificity of N-nitroso compounds in different species, which is extensively discussed elsewhere, makes it perilous to extrapolate from observations of target organs in rats to those possible in humans. Such extrapolation is equally unwise for other types of carcinogen, for example some aromatic amines which induce bladder tumours in humans and dogs, but liver tumours in rats, although it is probable that a substance carcinogenic in animals is likely to pose an increased carcinogenic risk to humans. Even knowledge that a particular N-nitroso compound produces a particular level of alkylation of DNA in one or more human organs in culture must be greeted with limited enthusiasm. In the absence of information that relates such alkylation of DNA, qualitatively and quantitatively, to induction of tumours in that organ of humans, no conclusions can be drawn about the meaning of such results, even though rates of repair of those alkylated lesions are measured and compared with those in experimental animals. DNA repair has become a flag to wave to explain differences in carcinogenic effect of an N-nitroso compound between target and non-target organs, although it is not known with certainty that alkylation of DNA is involved in tumour induction, let alone repair of such alkylated damage. It is only recently that there has been a focus on specific, localised alkylation in specific regions of DNA, which logic requires be important, rather than the massive methylation produced, for example, by large doses of nitrosodimethylamine.

Even though most of the studies of alkylation of DNA by N-nitroso compounds during the past 25 years are, to be charitable, uninformative about mechanisms of carcinogenesis by these compounds, the volume of information on the subject demands some recounting.

4.8.2 Methylating agents

The first investigations of methylation of DNA by NDMA by Magee and Farber (1962) and of protein by Magee and Hultin (1962), both in rat liver, were novel and exciting, providing an insight into the biological activity of this seemingly inert solvent. These authors realised that their conditions for study differed from those in which the compound induced tumours, since a single large toxic dose was given to the rats. The system was one for detecting the formation of alkylating electrophilic intermediates, through observing and measuring alkylation of the selected nucleophiles DNA and RNA; many other nucleophilic components of cells could be – and were – alkylated, but these were rarely examined. Following administration of the very large toxic doses used in these early experiments, alkylation of DNA in the liver (the organ most frequently examined) was extensive. As much as 2% of all the guanine residues in the liver DNA of rats given 20 to 30 mg/kg body weight of NDMA was methylated, surely vastly more than was necessary to produce mutational damage at critical sites in the DNA; yet no liver tumours ever arose in rats given this treatment and allowed to survive. Mesenchymal tumours of the kidney, however, were frequently seen in those rats, although the extent of methylation of DNA in the rat kidneys was considerably less than that in liver; however the kidney contains many types of cell, some of which might be alkylated more than others. The pattern of methylation of liver and kidney DNA by NDMA in Syrian hamsters was very similar to that in rats, yet NDMA induces liver tumours in hamsters following few relatively large doses, but does not induce kidney tumours (Tomatis and Cefis, 1967; Lijinsky et al., 1987b).

In the beginning the alkylation considered most significant was at the tertiary nitrogen at the 7-position in guanine, and this was measured in DNA, but more commonly in the more abundant RNA. At that time the two nucleic acids were viewed equally, since they were only receptors for detecting electrophilic attack. On this basis considerable information was gained about the comparative effectiveness of various N-nitroso compounds in alkylating nucleic acids *in vivo*, although it was not easy to translate the results to the relative potency of the carcinogens. It was discovered quite early that the ability of cyclic nitrosamines (often very potent carcinogens) to form electrophiles that alkylate nucleic acids *in vivo* was miniscule, since no significant amount of nitrosamine-produced alkylation of nucleic acids was found (Lijinsky, 1976a; Lijinsky and Ross, 1969; Farrelly and Hecker, 1984) and no adduct has yet been characterised, except from nitrosopyrrolidine (Chung et al., 1989b). There were other carcinogenic nitrosamines that showed the same lack of alkylation of nucleic acids (such as methylnitrosoaniline), but the expense of preparing radiolabelled forms of the more complex nitrosamines limited the number of them that have been examined.

In contrast with methylnitrosoaniline, a number of other methylnitrosoalkylamines that induce esophageal tumours in rats can form a methyldiazonium ion through oxidation of the larger alkyl group; methylnitrosoaniline can form only a phenyldiazonium ion through oxidation of the methyl group. A study by Van Hofe et al. (1987) of methylation in the esophagus, lung, kidney and liver of rats by a series of methylnitrosoalkylamines showed that maximum methylation in the esophagus DNA, both N^7-MeG and O^6-MeG, was by methylnitroso-n-butylamine and methylnitroso-n-amylamine, the two most powerful esophageal carcinogens among the group (Table 4.6). However, the almost equally potent methylnitroso-n-propylamine produced only a low level of DNA methylation in the rat esophagus, while methylnitroso-n-hexylamine and higher homologues did not produce detectable levels of esophageal DNA methylation, although they were quite effective esophageal carcinogens. The question arises 'how do methylnitrosoalkylamines induce esophageal tumours in rats, if they do not methylate

Table 4.6. *Methylation of DNA in rats by methylnitrosoalkylamines 6 h after 0.1 mmol/kg by gavage*

		μmol N^7- or O^6-MeG/mol G			
		Liver		Esophagus	
	Tumours	N^7-MeG	O^6-MeG	N^7-MeG	O^6-MeG
Methylnitroso-					
methylamine	Liver	3600	350	—	—
ethylamine	Liver	2500	250	—	—
n-propylamine	Esophagus	1200	100	80	10
n-butylamine	Esophagus	600	50	800	80
n-amylamine	Esophagus	100	—	600	60
n-hexylamine	Liver, Esophagus	500	50	—	—
n-heptylamine	Liver	600	50	—	—
n-octylamine	Liver	300	tr*	—	—
n-nonylamine	Liver	300	tr	—	—
n-decylamine	Bladder	200	tr	—	—
n-undecylamine	Liver	200	tr	—	—
n-dodecylamine	Bladder	100	tr	—	—

* tr = Trace.

esophageal DNA?'. A corollary is that a very small amount of methylation is necessary, if at the right region of the DNA, and most of the methylation by those nitrosamines that produce extensive methylation is irrelevant. Nevertheless, it is only possible to measure alkylation that is within the level of sensitivity of our methods, and it is fruitless to speculate whether in other cases there are undetectable levels of methylation or no methylation at all.

It seems likely that there are enzymes in the rat esophagus which can oxidise methylalkylnitrosamines to more proximate carcinogens, including alkylating agents, and that these give rise to esophageal tumours. However, these esophageal enzymes appear unable to activate nitrosodimethylamine or methylnitrosoalkylamines with alkyl chains longer than nine carbon atoms, and with little ability to activate methylnitrosoethylamine, which induces tumours of the rat esophagus only at high doses; it does not give rise to detectable methylation of esophageal DNA (Von Hofe et al., 1987). Nitrosodiethylamine is a potent inducer of tumours in the rat esophagus but there has been no report of ethylation of DNA in the rat esophagus by this nitrosamine. Neither has there been investigation of DNA alkylation in the esophagus of a species, such as the Syrian hamster, in which nitrosamines do not induce esophageal tumours, although it can properly be assumed that the hamster esophagus lacks the necessary activating enzymes.

Comparison of the methylation of DNA in other organs by methylalkylnitrosamines suggests that there is not a close correlation with induction of tumours in those organs. For example, DNA methylation in rat liver has been measured with the entire series from C-1 to C-12, and is greatest by C-1 (nitrosodimethylamine) and less by C-2 and still less by C-3 (methylnitroso-n-propylamine, which does not induce liver tumours). As the alkyl chain lengthens there is not much change in the level of DNA methylation in rat liver, although the C-4 and C-5 compounds have not induced liver tumours in rats, while C-6, C-7, C-8 and C-9 have. The still larger molecules produce quite small levels of methylation, extremely small in the form of O^6-methylation of guanine, and only the odd-numbered C-11 compound induced liver tumours in rats, not the C-10, C-12 or C-14 compounds, although all three induced bladder tumours (Lijinsky et al., 1981c).

In the lung and kidney, levels of methylation of DNA were much lower than in liver, and showed a pattern that was not readily interpretable in relation to tumour induction. In lung, for example, NDMA produced extensive methylation (O^6-MeG and N^7-MeG) and this nitrosamine induced a high incidence of lung tumours when given to rats by gavage. However, methylnitrosoethylamine, which induced almost as high an incidence of lung tumours (again by gavage) produced much less N^7-MeG and almost no O^6-MeG in lung DNA under these conditions (Von Hofe et al., 1987). In rat lung methylnitroso-n-propylamine, -n-butylamine and -n-amylamine produced higher levels of N^7-MeG than MNEA, and some O^6-MeG, although none of them induced lung tumours in rats. The C-6 to C-9 compounds gave rise to no detectable methylation of rat lung DNA, either N^7-MeG or O^6-MeG, although all four compounds induced a high incidence of lung tumours in rats. The lack of correlation between induction of lung tumours and methylation of rat lung DNA is particularly striking among the methylnitrosoalkylamines, although conclusions are constrained (in either direction) by the multitude of cell types in the lung.

In the case of the rat kidney, only NDMA, MNEA, methylnitroso-n-propylamine and methylnitroso-n-butylamine gave rise to any detectable methylation of DNA, both N^7-MeG and O^6-MeG declining from the smaller to the larger molecules. On the other hand, only NDMA induced tumours of the kidney in rats, and even the closely similar nitrosamine in other respects, MNEA, failed to do so. The lack of correlation between DNA alkylation and tumour induction was less striking, and less important, in the rat kidney than in the rat lung for the methylnitroso-n-alkylamine series. It was unfortunate that among the series of methyl-

Table 4.7. *Ethylation of nucleic acids* in vivo *(6 h post-treatment)*

Compound	Dose (μmol/animal)	Species and organ	Ethylated bases (pmol/mg DNA)		Tumours
			N^7-EtG	O^6-EtG	
Ethylnitrosourea	21	Rat liver	4.5	1.2	−
		Rat kidney	1.9	0.4	−
		Rat lung	3.6	0.4	+
		Rat spleen	2.9	ND*	−
		Rat colon	4.6	0.9	+
	21	Hamster liver	2.9	ND	−
		Hamster brain	6.7	1.4	−
Nitroso-diethylamine	20	Rat liver	5.3	1.8	+
	20	Hamster liver	14	5	+
Ethylnitroso-methylamine	17	Rat liver	3.3	1.4	+
		Rat kidney	1	0.2	−
		Rat lung	1.5	ND	+
		Rat spleen	1.3	ND	−
		Rat pancreas	2.6	0.6	−
		Rat colon	2.5	0.5	−
	17	Hamster liver	3.9	1.2	+
		Hamster kidney	1.4	0.3	−
		Hamster lung	2.9	0.4	−

* ND = None detected.

nitrosoalkylamines studied by Kleihues it was not possible to examine DNA alkylation in the bladder of rats, in which many of them gave rise to tumours, because of the very small amount of epithelial tissue obtained from the bladder of a rat.

Most of this series of nitrosamines have been given chronically to hamsters and produced tumours of the liver (but not in the esophagus) and usually in the lungs. Exceptions were NDMA, MNEA and methylnitroso-*n*-propylamine. There has not yet been a systematic study of methylation by this series of nitrosamines in the various organs of the hamster, but NDMA and MNEA give rise to as much or more DNA methylation in hamster liver, lung and kidney as in those organs of rats, yet no tumours of lung or kidney have been induced in the hamster by those two nitrosamines.

It is probably too much to expect a process that leads to induction of neoplasia in only a handful of cells (since it is believed that clones of tumour cells arise from few transformed cells in an organ) would be reflected in such a gross extent of DNA damage as to be detected and measured easily by chemical means. Nevertheless, that is what has been sought by myriads of investigators worldwide during the past three decades. It is also not surprising that great extents of alkylation of DNA in several organs were found on administration of some carcinogenic nitrosamines, nor that very small or zero levels of DNA alkylation were found with other, equally carcinogenic, nitrosamines, for example, cyclic nitrosamines. Indirect evidence that such adducts might be formed, probably transiently, has been obtained by observing that pretreatment with some of them reduces the activity of the DNA-repair enzyme, O^6-methylguanine-transferase (Pegg and Lijinsky, 1984).

4.8.3 Ethylating agents

Among compounds that are ethylating agents there has been considerably less investigation of their ability to alkylate DNA, in relation to organ-specific carcinogenesis. One reason is the much smaller extent of ethylation compared with methylation by the analogous methylnitroso compound, often as much as two orders of magnitude less by the ethyl compound. Why this difference exists is not clear, but might be due to the greater lability of the ethyldiazonium ion compared with the methyldiazonium ion in water, as well as steric constraints imposed by the larger ethyl group. In spite of the greater potency of nitrosodiethylamine in inducing liver tumours in rats, the extent of DNA ethylation (O^6-EtG and N^7-EtG) by NDEA is more than 50 times lower than is the comparable methylation by NDMA (Von Hofe et al., 1986a). The ratio of O^6-EtG to N^7-EtG is, however, about 1:3, relatively more in favour of the mutagenic O^6-guanine lesion than is the case with methylating agents. In hamsters NDEA is considerably less potent in inducing liver tumours than is NDMA, yet the relative effectiveness of the two nitrosamines in alkylating liver DNA is the same in both rats and hamsters. Indeed, the extent of alkylation by the two nitrosamines in the several tissues examined in rats and in hamsters seems unrelated to the species and dependent only on the chemistry and biochemistry of the nitrosamines; clearly their carcinogenic effects depend very much on which species is treated with the nitrosamines.

In confirmation of this, ethylation of DNA by methylnitrosoethylamine in rats and hamsters is very similar to that by NDEA (Table 4.7), and appears to follow the chemistry of the nitrosamine, whereas the carcinogenic effects of MNEA are appreciably different from those of either NDEA or NDMA, although basically similar in inducing liver tumours. Perplexingly, in view of the large differences in carcinogenic activity produced by replacing hydrogen in the α-methylene of MNEA with deuterium (which greatly increased its potency in inducing liver tumours in rats), there was virtually no difference in the extent of ethylation of rat liver DNA between the deuterated and undeuterated compound. This reinforces the evidence that metabolism and formation of an alkylating agent from a nitrosamine follow different dynamics from those reactions leading to induction of tumours.

Ethylation of DNA by ENU is considerably lower than is methylation by MNU, and their relative extents are similar to those from NDEA and NDMA. However, the relative carcinogenic potencies of the two alkylnitrosoureas are more similar than are those of the two nitrosamines, again suggesting that DNA alkylation is not the most important mediator of carcinogenesis by *N*-nitroso compounds. There have been several studies comparing methylation and ethylation by the same type of *N*-nitroso compound, and usually the correlation with carcinogenesis is fragile, if it exists at all (e.g. Lickhachev *et al.*, 1983b). They reported that, in hamsters as in rats, O^6-alkylation of guanine in the brain by ENU or MNU was more stable (i.e. less rapidly repaired) than in liver, but neither alkylnitrosourea induced brain tumours in hamsters, as they do in rats. Unfortunately there have been few studies of alkylation by ethylnitroso compounds, so that it is unwise to draw sweeping conclusions. The small extent of ethylation of DNA makes the fluorescence method of measuring alkylguanines less applicable, because the sensitivity of this method is no greater for ethylguanines than for methylguanines. Equally unfortunately the synthesis of radiolabelled ethylnitroso compounds is more difficult and much more expensive than for the corresponding methylnitroso compounds. The few studies that have been conducted, however, have shown that if DNA alkylation is related to carcinogenesis, ethylation is quantitatively much more effective than is methylation.

4.8.4 Other alkylating agents

There has been little study of alkylation by the more complex alkylnitroso compounds, such as propyl-, butyl-, hydroxypropyl-, etc. However, there is some evidence that there is 2-hydroxypropylation, as well as methylation in the liver of hamsters (Nagel *et al.*, 1987), but not in hamster pancreas following treatment with NBOPA, which induces pancreas tumours in hamsters, but not in rats. This is only fragmentary evidence, but it emphasises that focussing, as has been the overwhelming pattern of research in this area, on *methylation* of nucleic acids might have been misplaced. Some of the early reports of methylation might have been artefacts, due to use of radiolabelled compounds contaminated with traces of methylating impurities.

No doubt one of the reasons for the attention given to methylation of nucleic acids, is that many complex nitrosamines are extensively metabolised to smaller and simpler molecules, which are able to form methylating agents. For example, the series of nitrosopropylamine derivatives having oxygen functions on the β-carbon, all give rise through metabolism to some methylating agent (Lijinsky et al., 1988e; Liberato et al., 1989), resulting in methylation of DNA being the most common form of nucleic acid alkylation *in vivo*. Similarly, the methylnitroso-*n*-alkylamines that induce bladder tumours are metabolised by β-oxidation of the alkyl chain to form, eventually, methylnitroso-2-oxopropylamine, which is as effective a methylating agent for DNA as is NDMA (Lijinsky, 1988b). It is increasingly likely that methylation of DNA is simply too extensive to be meaningful in induction of tumours, at least to justify measuring methylation of DNA as an index of the carcinogenic potency or effectiveness of a particular treatment. It is certainly possible that specific alkylation at particular sites in DNA of certain organs is related to induction of tumours, which appear much later. But there is no way in which chemical measurement of such methylation can be helpful, and it can be left to the methods of molecular biology for resolution. Much of the plethora of work in this area must be irrelevant.

4.8.5 *DNA alkylation in relation to cancer*

The question of the relation of alkylation of DNA to tumour induction is pertinent to the directly acting alkylnitrosoureas and alkylnitrosocarbamates, which do not require the intervention of metabolising enzymes, unlike nitrosamines. Although the alkylnitrosamides (as they can generically be termed) are relatively unstable chemically, compared with nitrosamines, they become quite well distributed around the body following administration to animals, and probably enter every available compartment. They react, therefore, with nucleophiles in a fairly random manner, including the macromolecules DNA, RNA and proteins; the specific activities of these isolated from animals treated with a radiolabelled N-nitroso compound are usually similar, soluble protein being lowest. Alkylation of nucleic acids can be detected and measured in many organs and tissues, but the compounds are very selective in the organs in which they induce tumours. It is surprising that enzymic conversion of nitrosamines to eventually excreted carbon dioxide is as rapid as conversion of analogous alkylnitrosoureas to excreted carbon dioxide (Table 4.1), suggesting that the enzymic requirement is not necessarily limiting in the metabolic pathways of the nitrosamines.

It is unlikely that alkylnitrosoureas and similar directly acting N-nitroso compounds have to be metabolised in order to effect carcinogenesis. But their spontaneous breakdown to an alkylating diazonium ion (for example) does not explain their organ- and species-specific carcinogenicity, even if it occurs at different rates for different compounds. Dose–response studies with these compounds (described in Chapter 7) reveal that the target organs remain the same for a given compound – but often differ markedly from other compounds giving rise to the same alkylating agent – at higher or lower doses. This suggests a specificity for the particular structure of interest. For example, considering methylnitrosourea, methylnitrosourethane and methylnitrosonitroguanidine, oral administration of the urea induces in rats forestomach tumours and tumours of the nervous system; the second induces only tumours of the forestomach; the third induces tumours of the glandular stomach. If the carcinogenicity of the corresponding ethyl compounds is compared, ethylnitrosourea induces a large variety of tumours, including those in the forestomach, nervous system, mammary gland, uterus, lung, mesothelioma (in males), Zymbal gland, intestine, colon and bladder; the carbamate induces only tumours of the forestomach, and the guanidine glandular stomach tumours. It seems that the carbamate and guanidine structures constrict the carcinogenic action to the two types of epithelium in the stomach, regardless of the nature of the alkyl group, whereas the urea structure does not restrict carcinogenic activity in this way. Studies of DNA alkylation by these directly acting compounds have not been comprehensive. To the contrary, they have been limited to a few organs. However, in some studies by the author, it has become apparent that uptake of the nitroso-carbamate and the corresponding nitroso-urea are similar, as are distribution in various organs and alkylation of DNA in them, yet they have precise and distinct tumour-inducing effects in particular organs. This argues against alkylation of DNA as the key event in tumour induction by these compounds, although such alkylation might in most cases occur and be necessary. Chemical quantification and characterisation of alkylation of DNA by these compounds does not, therefore, seem a worthwhile pursuit if the goal is to elucidate the mechanisms of carcinogenesis in relation to the biological specificity of carcinogens. Some of the results are useful, however, because they shed light on dosimetry of exposure of animals (including man) and of organs within them to particular N-nitroso compounds, DNA in this case being the receptor molecule for measurement of the extent of interaction.

Comparison of the alkylation of DNA in the liver of rats by a number of methylating agents was undertaken as long as 30 years ago. In animals there was a clear distinction between the type and extent of alkylation by dimethyl sulphate and methylmethanesulphonate, which are at best very weak carcinogens, and the powerful carcinogens nitrosodimethylamine and methylnitrosourea. The latter nitroso compound, like the former, produced such extensive methylation of rat liver DNA that mass spectrometric methods could be applied to identification of the methylated bases in the DNA (Lijinsky et al., 1968, 1972c). At the time it was not known how close numerically the extent of methylation by these two nitroso compounds was, but more recently it has been shown that the differences are not great, although NDMA induces liver tumours readily in rats and hamsters, whereas methylnitrosourea does not. As discussed in another chapter, the failure of methylnitrosourea (and most alkylnitrosoureas) to induce liver tumours in rats is not due to some inherent resistance of liver cells to alkylnitrosourea-induced neoplastic transformation, since a number of alkylnitrosoureas (hydroxyethylnitrosoethylurea, hydroxyethylnitrosochloroethylurea, phenylethylnitrosourea) have induced liver tumours in rats, although none has done so in hamsters. It has been claimed, correctly, that O^6-alkylation of guanine in DNA by methyl- and ethyl-nitrosourea is more easily repaired by methyltransferases in the liver (Pegg, 1983b) than in the brain (Goth and Rajewsky, 1974). Therefore, the liver cells can recover from the mutational damage, while the cells of the nervous system do not; hence the alkylnitrosoureas induce tumours of the nervous system in the rat, but not liver tumours. This hypothesis was questioned at the time (Singer, 1975; Lijinsky, 1976), and has become no more acceptable and valid since. Since O^6-guanine alkylation is produced in a similar way and to similar extents in rat (and hamster) liver, even by low doses of dialkylnitrosamines, as well as by alkylnitrosoureas, the activity of the repair enzymes and the rate of DNA repair would not logically be very different. Yet the alkylation remaining following repair is sufficient to give rise to tumours when the alkylating agent is the nitrosamine, but not when it is the alkylnitrosourea; this is obviously an untenable argument.

4.8.6 Unresolved discrepancies in the DNA alkylation/cancer relation

The conclusion is that, while there is a role for DNA alkylation and consequent mutagenic changes in DNA in carcinogenesis by N-nitroso compounds, other actions or reactions of the compound are of overwhelming importance in induction of tumours. Their nature is unknown, but they seem to depend on features other than formation of an alkylating agent, which is often the same structure from compounds with different target organs.

These conclusions are borne out by comparing the pattern of alkylation of DNA brought about by a number of N-nitroso compounds that would be expected chemically to form the same alkylating (methylating) agent.

Table 4.8. Methylation of DNA by alkylating carcinogens in vivo (6 h post-treatment)

Compound	Dose (μmol/animal)	Species	Organ	Methylation of DNA (pmol/mg DNA)		Tumours
				N^7-MeG	O^6-MeG	
Nitrosodimethylamine	18	♂ Rat	Liver	380	23	+
			Kidney	51	3	+
		♂ Hamster	Liver	420	29	+
Azoxymethane	27	♂ Rat	Liver	740	48	−
			Kidney	21	2	+
			Colon	19	1	+
	27	♂ Hamster	Liver	1200	110	+
Methylnitrosoethylamine	20	♂ Rat	Liver	270	24	+
			Kidney	44	5	−
			Spleen	84	8	−
			Lung	22	5	+
	20	♂ Hamster	Liver	1000	71	+
			Kidney	40	5	−
			Lung	30	5	−

Methylnitrosourea	20	♂ Rat	Liver	103	11	—
			Kidney	111	16	—
			Lung	54	5	—
			Brain	23	2	+
			Spleen	180	14	—
	20	♂ Hamster	Liver	130	16	—
			Kidney	108	11	—
			Lung	22	3	—
Methylnitrosoethylurea	20	♂ Rat	Liver	54	3	—
			Lung	21	2	+
			Brain	17	2	+
Methylnitroso-oxopropylamine	42	♂ Rat	Liver	2400	280	—
Nitrosobis-(2-oxopropyl)amine	16	♂ Rat	Liver	55	7	—
			Kidney	24	3	—
	16	♀ Rat	Liver	230	18	+

The differences (Table 4.8) are not large, in spite of the variety of different tumours induced by the respective compounds. In these experiments an attempt has been made to administer similar molar doses of the various compounds to animals, to make comparisons easier. Past experiments, of which there are a great many in the literature, have often employed large, acutely toxic doses, that introduce the issue of widespread cellular damage and cell death, which might modify measurements of alkylation. Furthermore, reliance on acute toxicity as a benchmark for dosing, makes it difficult to compare results between one compound and another because, as in tumour-induction studies, the straightforward chemistry is confused with the seldom studied and imperfectly understood pharmacology. Therefore, while these numerous previous studies can be recounted, they are difficult to fit into a pattern related to carcinogenesis. They are presented here for completeness, and not necessarily for relevance. Many of them are duplicative, involving the same compound or compounds in slightly different conditions, all purporting to confirm the same point; such a narrow focus is neither helpful nor economic of research resources.

The recent studies of Barbacid and his group showed that following administration of methylnitrosourea to rats about seven weeks old, in relatively high doses, mammary carcinomas later arose in high incidence. This has been related to the methylation of DNA in cells of the mammary gland at specific sites (in the 12th codon), thereby inducing expression of an H-ras oncogene (Zarbl et al., 1985). The very sophisticated elaboration of this finding, including tranfection of fibroblasts with DNA from the mammary tumours and induction of sarcomas in nude mice recipients, has produced the hypothesis that initiation of the mammary tumours is through activation of the oncogene. However, cause and effect have not been proved in this or any similar case. On the other hand, the age of the rat at which this carcinogenic effect appears following a single treatment with MNU is critical, and other treatment regimens in rats with this carcinogen do not result in mammary tumours, but in tumours of the nervous system (Lijinsky, 1989).

Ethylnitrosourea and many other alkylnitrosoureas also induce mammary carcinomas in rats, and in circumstances in which methylnitrosoureas do not, such as continuous treatment with low doses (Lijinsky and Kovatch, 1989b; Lijinsky et al., 1989b). It is a pity, therefore, that studies of site-specific alkylation of DNA in the mammary gland of rats by ethyl- and higher alkylnitrosoureas, comparable to those with MNU, have not been carried out. The extent of alkylation by these other alkylnitrosoureas would undoubtedly be less than by MNU, but it would be interesting to discover whether the same site is alkylated, resulting in expression of the same oncogene, which could be related to induction of the same carcinoma. Whether the results were the same or different, they would further illuminate the problem of chemical carcinogenesis, by providing more than a one-point evaluation of the hypothesis. This is, of course, a chemist's view of how the problem should be investigated.

Alkylation of DNA by a number of directly acting N-nitroso compounds other than alkylnitrosoureas has been little studied, excepting MNNG. Kleihues and Wiestler (1984) showed that methylation of DNA in the stomach mucosa of rats (in which MNNG induces tumours) is strongly affected by thiols, whereas MNU is not. It is also likely that alkylnitrosocarbamates are modified in their alkylating ability by thiols, which is suggested as the reason for their limited carcinogenic activity when given orally; that is, most of the compound is expected to decompose in the stomach wall, and little remains to enter the circulation and alkylate macromolecules in internal organs. However, there have been few studies of alkylation by alkylnitrosocarbamates or by alkylnitrosonitroguanidines, probably because their carcinogenic properties are less interesting than those of alkylnitrosoureas. Nevertheless, such studies as have been done with these two classes of directly acting alkylating agents show that they do enter the circulation in significant quantity and alkylate DNA in the liver of rats or hamsters as effectively as do the corresponding alkylnitrosoureas.

For example, in the case of hydroxyethylating compounds, listed in Table 4.9, equimolar doses of hydroxyethylnitrosourea, nitrosooxazolidone and hydroxyethylnitrosoethylurea lead to approximately the same extent of DNA alkylation in the liver (Fig. 4.21), and considerably greater than after treatment with larger doses of nitrosodiethanolamine (Farrelly et al., 1987b), which gives rise to liver tumours at high incidence in rats. It is cautionary, when trying to relate DNA alkylation by N-nitroso compounds to organ-specific induction of tumours, that hydroxyethylation of DNA as O^6-hydroxyethylguanosine by hydroxyethylnitrosourea, which does not induce liver tumours in rats, is much greater (40 μmol vs. 4 μmol HEdG/mol dG) than by hydroxyethylnitrosoethylurea or hydroxyethylnitrosochloroethylurea, which do induce liver tumours in rats (Ludeke et al., 1991a). Equimolar doses of the three compounds were given to rats by gavage, followed six hours later by isolation of DNA from liver and other organs. There is similar hydroxyethylation of nucleic acids in the liver of hamsters treated with these directly acting compounds (Lijinsky, 1988b), although none of them, nor other alkylnitrosoureas, have induced tumours in hamster liver.

Table 4.9. *Hydroxyethylation of DNA in vivo (6 h post-treatment)*

Compound	Dose (μmol/animal)	Species and organ	Alkylated bases (pmol/mg DNA*)		Tumours
			N^7-HEG	O^6-HEG	
(a)					
Hydroxyethyl-nitrosourea	22	Rat liver	20	7	−
		Hamster liver	24	7	−
Hydroxyethyl-nitrosoethylurea	22	Rat liver	6	2	+
		Hamster liver	19	10	−
Nitroso-oxazolidone	22	Rat liver	13	5	−
		Hamster liver	12	4	−
Nitroso-diethanolamine	370	Rat liver	14	7	+

Compound	Dose (mmol/kg)	Species and organ	μmol/mol dG** O^6-HEdG	Tumours
(b)				
Hydroxyethyl-nitrosourea	0.38	Rat liver	37	−
		Rat kidney	24	−
		Rat brain	6	−
		Rat lung	18	+
Hydroxyethyl-nitrosoethylurea	0.38	Rat liver	4	+
		Rat kidney	3	−
		Rat brain	2	+
		Rat lung	3	+
Hydroxyethyl-nitroso-chlorethylurea	0.38	Rat liver	3	+
		Rat kidney	2	+
		Rat brain	2	−
		Rat lung	2	−

* Measured by radioassay of purine bases.
** Measured by immunofluorescence of nucleosides.

The directly acting N-nitroso compounds seem to alkylate nucleic acids in a variety of organs (Fong et al., 1990), as would be predicted from their chemical structure and reactivity, but this does not relate to their predilection for inducing tumours in precisely targeted organs in each species. It is impossible to conclude, therefore, that alkylation of DNA by these compounds has the only direct bearing on the later appearance of tumours in particular organs; this is also true of nitrosamines. The solution to the riddle of organ- and species-specific carcinogenesis seems to lie in physicochemical properties of N-nitroso compounds or of particular metabolites, the nature of which we must continue to seek.

Alkylation of DNA etc. 169

4.8.7 Does DNA repair (of alkylated mutagenic lesions) have a role in carcinogenesis by N-nitroso compounds?

Much has been said and written, and numerous investigations during the past 30 years have focussed on the production of mutations through alkylation of DNA in animals. Based on the classical studies of the relation of DNA repair – or lack of it – to the well known susceptibility of people with the genetic disorder *Xeroderma Pigmentosum* to induction of skin cancer by sunlight, much has been made of the possibility that differences in organ- and species-specific tumour induction by carcinogens could be due to differences in efficiency of repair of lesions in DNA between different cell types and locations. The report by Goth and Rajewsky (1974) that ethylation of DNA was repaired more rapidly in rat liver than in the brain of rats treated with ethylnitrosourea was very stimulating and has almost assumed the position of dogma, in spite of considerable evidence that the stability of the alkylated products is organ-specific across species, but the induction of tumours is not.

Among the cant that is common in this area of research it is possible to find a few investigations designed to question the hypothesis, rather than to accept it and prove its correctness. For example, the reports that

Fig. 4.21. Hydroxyethylation by NDELA, hydroxyethylnitrosourea and nitrosooxazolidone. Hydroxypropylation by NBHPA, etc.

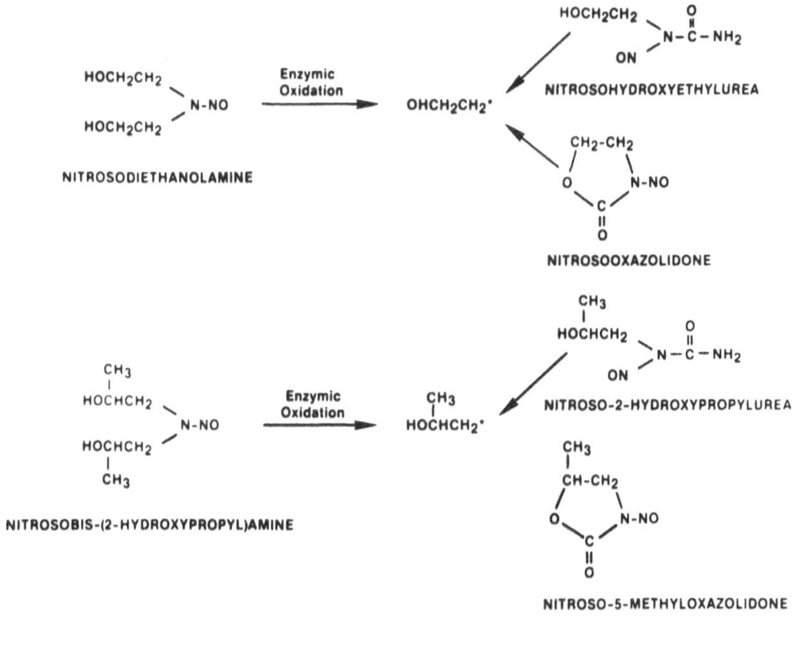

Table 4.10. Stability of methylation of DNA in rat liver produced by N-nitroso compounds (20 µmol)

Compound	Species	Highest methylation (h) (µmol/mol G)		At 48 h		% drop		Liver tumours
		O^6-MeG	N^7-MeG	O^6-MeG	N^7-MeG	O^6	N^7	
Nitrosodimethylamine	Rat	700	3500 (6)	120	1700	83	51	+
Methylnitrosoethylamine	Hamster	350	1600 (6)	230	740	34	54	+
Methylnitrosoethylamine	Rat ♀	500	2700 (6)	200	1700	60	37	+
	Rat ♂	160	900	23	485	86	47	+
Nitrosobis-(2-oxopropyl)amine	Hamster	510	2500 (6)	330	1800	35	28	+
Methylnitrosourea	Hamster	60	570 (3)	19	230	68	60	−
Methylnitrosourethane	Rat	43	470 (6)	6	220	87	53	−
Azoxymethane	Rat	500	3100 (6)	130	1700	74	45	−

ethylation of DNA by ethylnitrosourea was similarly stable in the brain of rats, hamsters and gerbils and unstable in the liver of those species, would be expected to inspire some reservations when ethylnitrosourea induced brain tumours in rats, but not in hamsters or gerbils. There is no logic in assuming that methylation or ethylation of DNA in the same cells (almost certainly produced by the same alkylating diazonium ion, regardless of the structure of the compounds giving rise to it, directly or metabolically) would be repaired at different rates. Yet there is nothing in common in the types of tumours induced, respectively, by ethylnitrosourea and nitrosodiethylamine in rats or hamsters. It is easy in Science, as in other spheres of human activity, to find what one is looking for. But there has begun a questioning of the conventional wisdom, and the broadly acting carcinogenic N-nitroso compounds afford an opportunity to put the hypothesis to the test.

A first consideration is whether the great variety of compounds that could theoretically give rise to the same alkylating intermediate do so, and whether they produce, *in vivo*, a similar extent of the particular alkylation of DNA, as theory would also imply. Within broad limits this seems to be the case, even when, as with nitrosobis-(2-oxopropyl)amine formation of a methylating agent was not predicted, but was discovered (Lawson *et al.*, 1981 *a*).

Recently Koepke *et al.* (1988 *b*) investigated the relative stability of methylation and hydroxyethylation of guanine in DNA in the liver of rats that had been treated with a single dose of methylnitrosohydroxyethylamine, which induces tumours of the liver and esophagus. They reported that hydroxyethylation and methylation occurred in liver DNA in a dose-dependent manner, the same in males and females except at the highest doses, at which alkylation was greater in females; methylnitrosohydroxyethylamine induced more liver tumours in female rats than in males. As expected, N^7-guanine alkylation was fairly stable over 96 h observation, whereas loss of O^6-alkylguanines (methyl- and hydroxyethyl-) was rapid and similar for both methyl and hydroxyethyl. Since the activity of the alkyltransferase responsible for the DNA repair (Pegg *et al.*, 1984) is much slower for hydroxyethylguanine than for methylguanine, an alternative mechanism of removal of O^6-hydroxyethylguanine is indicated. It must be emphasised that the relative contribution of methylation and hydroxyethylation of DNA, if either, and their repair to the induction of tumours in rats is quite unknown.

Similar studies with a number of methylating N-nitroso compounds, some of which do, and others do not, induce liver tumours in rats or hamsters are in progress, and most are incomplete. Equimolar small doses (approximately 20 micromol per animal) produce no detectable toxic effects in the liver within 48 h following treatment, but all of the compounds formed DNA methylated at N^7 and O^6 of guanine in amounts readily measured by the fluorescence method (Herron and Shank, 1979). There were some quantitative differences in the effects produced, the extent of methylation being lower from methylnitrosourea and methylnitrosourethane than by NDMA, MNEA and azoxymethane. However, as would be expected on chemical grounds, there is no large difference in stability of O^6-methylguanine in liver DNA (Table 4.10) regardless of the chemical structure of the methylating agent administered to the animal. This was equally true of rats and hamsters, although the stability of the methylated DNA in hamster liver tended to be greater than in rat liver. Therefore, the reason for the failure of the orally administered directly acting alkylnitrosamides to induce tumours of the liver in rats or hamsters is not because they do not alkylate liver DNA *in vivo*, nor because such alkylation is repaired more rapidly than that produced by analogous nitrosamines which do induce tumours of the liver in those species. These results do not support a role for differences in rates of DNA repair, or differences in DNA alkylation, as an explanation for the differences between structurally analogous *N*-nitroso compounds in the organs in which they induce tumours in various species.

5

Toxicity of N-nitroso compounds

5.1 Introduction

That the toxicity of nitrosodimethylamine was discovered by observing its effects – albeit inadvertently – in humans, is not entirely unusual, although it is a long time since the effects of new materials in humans were tested by exposing humans to them. Freund's report (1937) of a gross exposure of workers to NDMA went unnoticed until a similar exposure almost 20 years later (Barnes and Magee, 1954). It is obviously not a new phenomenon that the 'old' literature is overlooked or ignored. More recently deliberate poisonings of people with nitrosodimethylamine have been reported, one in Germany (Fussgaenger and Ditschunheit, 1980), the other in Omaha, Nebraska (Herron and Shank, 1980). In the latter incident the nitrosamine was added to lemonade, causing severe liver necrosis in several people who drank it, and death of some. In the case of Ulm, Germany, the nitrosamine was incorporated in a dish of blackberries. The immediate response in the early 1950s was to examine the toxicity of the material in experimental animals (Barnes and Magee, 1954).

In rats an estimate was made of LD_{50} (median lethal dose) as 30 to 40 mg/kg body weight of NDMA in a single intraperitoneal dose, based on the effect, lethal or not, of a range of doses the highest of which caused death of all of the animals (Weil, 1952); the rats develop acute centrilobular necrosis in the liver, and some die, but liver regeneration occurs in those that survive more than two or three days. It is interesting and important that some of the surviving animals develop mesenchymal kidney tumours (studied in detail by Hard and Butler, 1970), but not liver tumours, although there is very little toxic damage to the kidneys. The failure of these rats to develop any liver tumours, which do not appear even following several successive relatively large doses of NDMA (although kidney and lung tumours developed in those rats) is one of the lasting conundrums of nitrosamine toxicology.

As with most toxins, the precise reason for the damage to the cells and for their death, and for the death of the animals, is not known. It is fairly certain, however, that NDMA must be metabolised to exert its toxic effects. Organisms which are apparently unable to metabolise it are not harmed and can live in NDMA solutions (bacteria, for example). Isolated cells that lack this metabolic capacity, fibroblasts, for example, are unaffected by NDMA, as has been noticed in the various fibroblast transformation assays used to detect carcinogenic potential, and described in Chapter 6.

It seems likely that oxidation of the nitrosamine to the α-hydroxy derivative is an important step in toxicity of NDMA, and this is a reaction which certainly takes place in liver cells. The product decomposes spontaneously to formaldehyde and a methylating agent, believed to be the methyldiazonium ion; both products are toxic, and both can be produced directly from several other precursors which do not show the same pattern of toxicity in liver as does NDMA. For example, methylnitrosourea forms a methyldiazonium ion directly (and is a powerful directly acting methylating agent), but its toxicity is low. Early studies by Smuckler and others (Lijinsky *et al.*, 1972*c*; Druckrey *et al.*, 1967) show little toxicity to the liver by even relatively large doses of MNU or ENU administered intravenously. Whereas 20–30 mg/kg body weight of NDMA produces extensive liver necrosis in rats (and in hamsters), leading to death in some animals, a similar intravenous dose of MNU has very slight effects in the liver; when repeated over the course of 10 to 20 weeks tumours of the uterus and, less commonly, of the mammary gland, lung and nervous system eventually appear.

It was known more than 20 years ago, and corroborated in many experiments since, that MNU and other simple alkylnitrosoureas do not induce liver toxicity or liver tumours in intact adult animals. Therefore, it is unlikely that the methyldiazonium ion arising from metabolism of NDMA contributes at all to the toxic effects in the liver. Furthermore, methylnitrosoethylamine is less toxic to rat liver than NDMA, although it gives rise to just as much methyldiazonium ion, as measured by methylation of liver DNA (Lijinsky, 1988*b*). The same is true of the other product, formaldehyde, which has not been associated with liver toxicity in the many toxicological studies that have been carried out with it. Some other metabolite of NDMA, therefore, appears to be responsible for its liver toxicity, perhaps the presumed initial product of oxidation, hydroxymethylnitrosomethylamine, or a further oxidation product, methylnitrosoformamide. The former compound is very unstable and has not been isolated, although acetate and phosphate esters have been prepared. These are directly acting mutagens, but it has not been noted that they produced liver damage, severe or mild, when given to rats (Berman et al., 1979). Methylnitrosoformamide is highly mutagenic and more toxic to rats (3 mg/kg body weight I.P. was fatal) than acetoxy-NDMA. Ethylnitrosoformamide is much less toxic (140 mg/kg I.P. was fatal to rats) than its methyl homologue, and also less toxic than the isomer methylnitrosoacetamide (oral LD_{50} in rats, 20 mg/kg).

5.2 Toxicity of nitrosodimethylamine

The first report of experimentally induced toxicity of NDMA (Barnes and Magee, 1954) was very thorough, describing effects in several species. The LD_{50} in rats was estimated as 26.5 mg/kg body weight, rats given a single oral dose of this size dying between two and six days after. The abdomen was often swollen and contained bloody fluid (as much as 30 mL) at death. The liver was swollen, dark and mottled. The effects of similar doses delivered by I.V., I.P. or S.C. injection in rats were similar, but percutaneous application of 100 mg/kg had no obvious effect even when repeated on four successive days. It is known now that nitrosamines are absorbed readily through the skin (Lijinsky et al., 1981a), so it is probable that the lack of toxic effects of topically applied NDMA in the 1954 study was due to failure of much of the dose to be absorbed, because of evaporation, for example. In mice and rabbits given 15 to 20 mg/kg NDMA death was more rapid than in rats, usually within 24 h, but the liver damage appeared less severe, with less blood in the fluid exudate in the peritoneal cavity. Guinea pigs were less susceptible than rats, dying between four and eight days following 25 to 50 mg/kg NDMA, with bloody peritoneal exudate and congested, mottled liver, while dogs given 20 mg/kg NDMA died within 24 h with a congested liver, but little fluid exudate in the peritoneum. Chronic administration of lower doses of NDMA to rats (100 or 200 ppm in the diet) led to death of the animals between 30 days at the high dose and 90 days at the lower dose, following massive reduction in weight compared with controls, and showing extreme emaciation; a continuous dose of 50 ppm in the diet had no particular effect, although the rats did develop liver tumours much later. Histologically the livers of the rats given single doses or continuous doses of NDMA showed varying degrees of centrilobular necrosis, similar in some ways to the effects of carbon tetrachloride (which does not cause haemorrhage); in the feeding experiments there was fibrosis. The kidneys were normal. The liver lesions in the dog were similar to those in rats. It

Table 5.1. *Toxicity of NDMA and NDEA in various species*

Species	LD_{50} (mg/kg) and Toxic effect	
	NDMA	NDEA
Mouse	20 Liver necrosis	280
Rat	30 Liver necrosis	220
Syrian Hamster	30 Liver necrosis	250
Chinese Hamster	18 Liver necrosis	230
European Hamster	35 Liver necrosis	250
Rabbit	15 Liver necrosis	NT*
Guinea pig	25 Liver necrosis	190
Mink	7 Liver necrosis	NT
Dog	50 Gut haemorrhage	NT
Trout	1800	NT
Newt	> 16000	NT

*NT = not tested.

was notable that in the periportal region there were areas of apparently undamaged liver parenchyma. In rats, guinea pigs and dogs given single doses of NDMA there was considerable bleeding into the gut, indicating that the toxic effects of this nitrosamine were not confined to the liver. The toxic effects of NDMA in various species are summarised in Table 5.1.

In 1964 there was a report by Koppang (1964) of toxic liver injury in sheep and cattle which had been observed since 1961 in Norway. These animals had been fed herring meal as part of the diet; the remainder of the diet was vegetable material. When feeding of herring meal ceased the sick animals often recovered. In animals that died there was extensive centrilobular necrosis, haemorrhage into the peritoneal cavity and into the gastrointestinal tract and fluid exudate in the abdominal cavity, effects reminiscent of those of nitrosodimethylamine in rats. Investigations of the toxins present in the herring meal by Ender and Ceh (1968) revealed nitrosodimethylamine as one of the components, not aflatoxins as in peanut meal that caused liver necrosis in domestic animals. The herring meal had been treated with a solution of sodium nitrite as a preservative, but it was not clear how NDMA could form to such an extent in neutral or alkaline conditions, as opposed to the acidic conditions usually considered necessary for nitrosation. Nitrosatable secondary and tertiary amines, such as dimethylamine, trimethylamine and its oxide are probably formed through bacterial action in the fish meal.

The uniform pattern of toxicity of NDMA across a large variety of species, including humans, suggests that the metabolic activation of the nitrosamine to toxic products proceeds in a similar way in those species. In contrast, the carcinogenic effects of even this simplest nitrosamine are often different among several species. These differences suggest that the metabolic products of NDMA producing the toxic effects are not necessarily the same as those causing tumour formation.

5.3 Toxicity of N-nitroso compounds in several species

Most of the studies of toxicity of N-nitroso compounds have been carried out in rats, the same species in which most of the carcinogenicity studies have been conducted. While it is not true that the many strains of rat have exactly the same response to the toxic effects of N-nitroso compounds – and there are definitely differences between strains in response to their carcinogenic effects, in the relative responses of various organs, for example – it seems to be the case that inter-strain differences are considerably smaller than inter-species differences. The error in comparing the effects of several compounds measured in different laboratories on 'rats', therefore, is likely to be small.

The first attempt to systematise the toxic (and carcinogenic) effects of N-nitroso compounds in experimental animals was by Druckrey, Preussmann and their colleagues, summarised in their landmark review of effects in rats (Druckrey et al., 1967). The same year a more biologically orientated and broader review by Magee and Barnes (1967) was published. The information was mostly complementary between the two, but there was some overlap in the areas of toxicity and carcinogenicity of N-nitroso compounds. There is a school of thought which considers carcinogenesis only a particular example of chronic toxicity (Shubik and Sicé, 1956), but for the present discussion 'toxicity' refers to acute or relatively short-term effects, and will centre on effects in animals.

In Table 5.2 are listed the LD_{50} values (calculated according to Weil, 1952) of the N-nitroso compounds in cases in which acute toxicity tests have been performed. Because of the design of the carcinogenesis tests conducted in Druckrey's laboratory – and continued by investigators with the same viewpoint – the investigations of the N-nitroso compounds carried out in the early days included determination of the LD_{50}. Considering those acyclic nitrosamines that are believed to be metabolised to methylating intermediates, including a methyldiazonium ion, a large series of methylnitrosoalkylamines have been studied. Apart from the prototype, NDMA, the most toxic of the series are methylnitrosovinylamine and methylnitrosobenzylamine, both inducing liver toxicity and both similar in toxic potency to NDMA. Implicit is the probability that accompanying formation of the methylating agent is oxidation of the other alkyl group to a carbonyl compound (ketene and benzaldehyde in this case). Ketene is extremely reactive and, if formed, would immediately

Table 5.2. *Acute toxicity of N-nitroso compounds*

Compound	Species	Route	LD$_{50}$ (mg/kg)	Relative carcinogenicity
Symmetrical Nitrosamines				
N-nitroso-				
dimethylamine	Rat	Oral, I.P., Inhal.	30	+ + + +
	Hamster	S.C.	30	+ + + +
	Mouse	I.P.	20	+ + + +
	Guinea pig	I.P.	16	+ + + +
diethylamine	Rat	Oral, I.V.	280	+ + + +
	Hamster	S.C.	250	+ + + +
	Guinea pig	I.P.	190	+ + + +
	European Hamster	S.C.	220	+ + +
di-*n*-propylamine	Rat	Oral	480	+ + + +
	Hamster	S.C.	600	+ + + +
di-*iso*propylamine	Rat	Oral	850	+ +
diallylamine	Rat	Oral	800	+
	Hamster	S.C.	1300	+ +
di-*n*-butylamine	Rat	Oral, S.C., I.P.	1200	+ + +
	Hamster	Oral	2150	+ + +
di-*iso*butylamine	Hamster	S.C.	5600	+ +
di-*n*-amylamine	Rat	Oral, S.C.	1750, 3000	+
dicyclohexylamine	Rat	Oral	5000	–
diphenylamine	Rat	Oral	3000	+
dibenzylamine	Rat	Oral	900	–
iminodiacetonitrile	Rat	Oral	> 7500	+ +
diethanolamine	Hamster	S.C.	11300	+ +

diethanolamine diacetate	Rat	Oral	5000	+++
bis-2-hydroxypropylamine	Rat	S.C.	500	++++
bis-2-oxopropylamine	Rat	S.C., Oral	100	++++
	Hamster	S.C.	100	+++++
bis-2-oxobutylamine	Hamster	S.C.	280	+++

Asymmetric Nitrosamines

N-nitroso-				
methyl-methoxyamine	Rat	I.V.	130	—
methyl-acetoxymethylamine	Rat	I.P.	26	++++
ethyl-ethoxyamine	Rat	Oral	1000	++
methylethylamine	Rat	Oral	90	+++++
methylvinylamine	Rat	Oral	24	+++++
methyl-chloroethylamine	Rat	Oral	22	+++++
methyl-cyanomethylamine	Rat	Oral	45	+++++
methyl-n-propylamine	Hamster	S.C.	500	+++++
methyl-n-butylamine	Rat	Oral	130	+++++
methyl-tert-butylamine	Rat	Oral	700	—
methylallylamine	Rat	Oral, I.V.	320	+++
methyl-methoxymethylamine	Rat	Oral	700	++++
methyl-1-methoxyethylamine	Rat	Oral	240	+++++
methyl-2-oxopropylamine	Rat	S.C., Oral	35	+++++
	Hamster	S.C.	35	+++++
sarcosine	Rat	Oral, I.V.	5000	+++
sarcosine ethyl ester	Rat	Oral, S.C.	>5000	+++
methylamylamine	Rat	Oral, I.P.	120	++++
methylcyclohexylamine	Rat	Oral	130, 30	++++
	Hamster	Oral	170	++
	Mouse	Oral	60	
methyl-n-heptylamine	Rat	S.C.	420	++++
methylphenylamine	Rat	Oral	200	+++
	Hamster	Oral	150	++

Table 5.2 (cont.)

Compound	Species	Route	LD_{50} (mg/kg)	Relative carcinogenicity
methyl-2-aminopyridine	Rat	Oral	60	+++
methyl-3-aminopyridine	Rat	Oral	10	–
methyl-4-aminopyridine	Rat	Oral	200	–
methylbenzylamine	Rat	I.P., Oral	5, 18	++++
methyl-2-phenylethylamine	Rat	Oral	48	++++
methylamino-benzaldehyde	Rat	Oral	2000	–
methylamino-1,1-dimethyl-3-butanone	Rat	Oral	2100	–
methyl-dodecylamine	Hamster	Oral	1200	+++
	European Hamster	S.C.	3600	+++
vinylethylamine	Hamster	S.C.	110	++++
	Rat	S.C.	88	++++
ethyl-methoxymethylamine	Rat	Oral	540	++++
ethyl-1-methoxyethylamine	Rat	Oral	1000	++++
ethyl-ethanolamine	Rat	Oral	> 7500	+++
ethyl-isopropylamine	Rat	Oral	1100	++++
ethyl-n-butylamine	Rat	Oral, I.V.	380	+++
ethyl-tert-butylamine	Rat	Oral	1600	–
ethyl-4-picolylamine	Rat	Oral, I.V.	40	++++
propyl-hydroxypropylamine	Hamster	S.C.	1500	++++
propyl-oxopropylamine	Hamster	S.C.	1200	++++
hydroxypropyl-oxopropylamine	Hamster	S.C.	380	++++
propyl-1-acetoxypropylamine	Hamster	S.C.	500	+++

Compound	Species	Route	Dose	Rating
propyl-1-methoxypropylamine	Hamster	S.C.	580	+++
butyl-4-hydroxybutylamine	Rat	Oral	1800	+++
butyl-amylamine	Rat	S.C.	2500	++
oxopropyl-2-oxobutylamine	Hamster	S.C.	170	+++
dinitroso-N,N'-dimethylethylenediamine	Rat	Oral	150	+++

Cyclic Nitrosamines

Compound	Species	Route	Dose	Rating
N-nitroso-				
azetidine	Mouse	Oral	> 1600	++
	Rat	Oral	> 1600	++
oxazolidine	Rat	Oral	1500	+
tetrahydro-oxazine (1, 2)	Rat	Oral	830	+++
tetrahydro-oxazine (1, 3)	Rat	Oral	600	++
dihydro-oxazine (1, 2)	Rat	Oral	900	+++
pyrrolidine	Rat	Oral, I.P.	900, 650	+++
piperidine	Rat	Oral, S.C., I.P.	200, 100, 85	++++
	Hamster	S.C.	300	+++
	Mouse	S.C.	70	+++
hexamethyleneimine	Hamster	S.C.	140	++++
	Rat	Oral	340	+++
	Mouse	S.C.	70	++++
heptamethyleneimine	Rat	Oral	280	++++
	European Hamster	S.C.	275	+++
octamethyleneimine	Rat	Oral	570	+++
morpholine	Rat	Oral, I.V.	280, 100	++++
	Hamster	Oral, S.C.	1050, 500	+++
thiomorpholine	Rat	Oral	800	+++
2-methylpiperidine (S+ or R−)	Rat	Oral	600	+++
2,6-dimethylmorpholine	Rat	S.C.	430	++++
	Hamster	Oral	370	++++

Table 5.2 (cont.)

Compound	Species	Route	LD$_{50}$ (mg/kg)	Relative carcinogenicity
imidazolidone	Mouse	S.C.	370	+++
	European Hamster	Oral, S.C.	1400, 850	+++
	Guinea pig	Oral	280	++++
1-piperazine	Rat	S.C.	250	+++
N-methylpiperazine	Rat	Oral	2300	–
N-carbethoxypiperazine	Rat	Oral	1000	++
nornicotine	Rat	S.C.	400	+++
anabasine	Rat	S.C.	1000	+++
indoline	Rat	S.C.	>1000	++
trimethylhydrazine	Rat	Oral	320	–
Dinitrosopiperazine	Rat	Oral	95	+++
Dinitroso-2,6-dimethylpiperazine	Rat	Oral, S.C.	160	+++
	Hamster	S.C.	375	+++
	European Hamster	S.C.	230	++
Trinitrosotrimethylenetriamine	Rat	Oral	160	–
Nitrosamides				
N-nitroso-				
methylurea	Rat	Oral, I.V.	180, 110	++++
	Hamster	S.C.	70	++++
ethylurea	Rat	Oral, S.C., I.V.	300, 240, 240	++++
benzylurea	Rat	Oral, S.C.	550	++
methylacetamide	Rat	Oral	20	+++
phenylurea	Rat	S.C.	150	++
dimethylurea	Rat	Oral, I.V.	280	+++
trimethylurea	Rat	Oral, I.V.	250	+++

methyldiethylurea	European Hamster	S.C.	280	+++
	Hamster	S.C.	280	+++
n-butylurea	Rat	S.C.	1200	+++
methylurethane	Rat	Oral, I.V.	240, 4	++++
ethylurethane	Rat	I.V.	>330	+++
carbaryl	Rat	Oral	>1500	++
methylnitroguanidine	Rat	Oral, S.C.	100, 420	+++
cimetidine	Rat	Oral	1800	–
propyl-propionamide	Hamster	S.C.	310	+++
streptozotocin	Mouse	Oral	260	+++

react, whereas benzaldehyde is quite a stable product, unlikely itself to have serious toxic effects (LD_{50} in rats 1.3 g/kg).

Methylnitrosocyclohexylamine is very toxic, as is methylnitrosophenylethylamine. The remaining methylnitrosoalkylamines examined were less toxic than NDMA, even methylnitrosoethylamine having more than double the LD_{50} of NDMA. The LD_{50} increased as the length of the alkyl chain increased, and was approximately 120 mg/kg for methylnitrosoamylamine and 400 mg/kg for methylnitroso-n-heptylamine. Methylnitrosoallylamine was considerably less toxic than NDMA, in contrast to methylnitrosovinylamine. Ethylnitrosovinylamine was less toxic than its methyl analogue, in conformity with a pattern of lower toxicity of ethylnitroso compounds, compared with their methyl analogues; for example nitrosodiethylamine was less toxic than methylnitrosoethylamine. The toxicity of the symmetrical nitrosodialkylamines fell rapidly as the size of the alkyl group increased, the LD_{50} being 280 mg/kg for NDEA, 480 mg/kg for nitrosodi-n-propylamine (higher for nitrosodi*iso*propyl- and -diallyl-amine), 1200 mg/kg for nitrosodi-n-butylamine, 3000 mg/kg for nitrosodiamylamine and 5000 mg/kg for nitrosodicyclohexylamine. The larger molecules are, of course, much less water-soluble and possibly more difficult to oxidise than the smaller ones, although the contrast between the high toxicity of methylnitrosocyclohexylamine and nitrosodicyclohexylamine is more striking than that between methylnitrosobenzylamine and nitrosodibenzylamine; it is notable that neither nitrosodicyclohexylamine nor nitrosodibenzylamine is carcinogenic, although nitrosodiphenylamine is (Cardy *et al.*, 1979). The smallest acyclic nitrosamines are the most carcinogenic, and carcinogenic potency seems to decrease, in general, as the molecular weight increases.

Many of these compounds are toxic to the liver or lung of rats, which is the cause of death, whether or not the nitrosamine induces liver or lung tumours on chronic administration. There is apparently a difference between those interactions of the nitrosamine which cause toxic effects in the liver or lung of rats, and those which give rise to tumours, although neither chain of events is well understood.

It is conceivable – even likely – that oxidation at the α-carbon of an acyclic nitrosamine produces the interactive intermediate which is responsible for the toxicity of the compound, possibly through formation of an alkylating agent. This is also possible for cyclic nitrosamines, although in those cases the nature of the product is different from those formed from acyclic alkylating agents. Nevertheless, the toxicity of cyclic nitrosamines is comparable with that of the acyclic ones. There is a difference in that several cyclic nitrosamines have a profound effect on the central nervous system, producing convulsions and brain oedema in rats at the higher doses used for determination of LD_{50} values. This was true for nitrosopyrrolidine, nitrosopiperidine and nitrosohexamethyleneimine (Lee and Lijinsky, 1966). Larger cyclic nitrosamines, nitrosoheptamethyleneimine and nitrosooctamethyleneimine also produced convulsions when given to rats at high doses, although not consistently (i.e. not in every treated rat), as it is with the smaller molecules. In order to obtain a measure of the acute toxicity of these compounds through metabolic cellular damage in organs such as liver or lung, rats were anesthetised prior to treatment with the convulsant cyclic nitrosamines, in which case they survived immediate death, but usually revived immediately from the anesthesia. Centrilobular necrosis in the liver was commonly observed within three days following the treatment, and frequently lung toxicity also. The smallest cyclic nitrosamine, nitrosoazetidine had such low toxicity that an LD_{50} could not be determined (Lijinsky et al., 1967); only two of five rats died (with some liver damage) after receiving 1.6 g/kg of nitrosoazetidine.

Among cyclic nitrosamines the smallest molecules were the least toxic, and larger ones were increasingly toxic up to ring sizes of six or seven carbons, above which toxicity declined; nitrosododecamethyleneimine showed low toxicity. This is the opposite of the pattern of acyclic nitrosamines, among which the smaller molecules tended to be the most toxic. The parallel of toxicity with carcinogenicity of cyclic nitrosamines is apparent, because nitrosoazetidine and nitrosopyrrolidine are considerably weaker carcinogens in rats than are the larger nitrosopiperidine, nitrosomorpholine, nitrosohexamethyleneimine and nitrosoheptamethyleneimine. Dinitrosopiperazine is similar in toxicity to nitrosopiperidine, whereas the much weaker carcinogen 1-nitroso-4-methylpiperazine is also much less toxic than dinitrosopiperazine. The relation between toxicity and chemical structure among cyclic nitrosamines is quite different from acyclic nitrosamines, suggesting that whatever similarities exist between the two types in metabolic activation, they are circumscribed and modified by other parallel or subsequent reactions.

5.4 Toxicity of nitrosamines with substituted alkyl groups

At least in rats, the toxicity of nitrosamines is greatly reduced by oxygen substitution, especially hydroxylation. For example, nitrosodiethanolamine and ethylnitrosoethanolamine are much less toxic than nitrosodiethylamine, the LD_{50} being almost 50 times greater for the hydroxylated compounds. Indeed it has been difficult to establish an LD_{50} for nitrosodiethanolamine (Druckrey et al., 1967; Lijinsky et al., 1980c).

The differences in carcinogenic potency are not so large, although the hydroxyethylated compounds are less potent than NDEA. The bladder carcinogen butylnitroso-4-hydroxybutylamine is also less toxic than nitrosodi-*n*-butylamine; nitrosobis-(2-hydroxypropyl)amine and nitrosobis-(2-oxopropyl)amine are less toxic than nitrosodi-*n*-propylamine. Nitrosamines containing carboxy substituents, such as nitrososarcosine, invariably have slight toxicity, although some are, nevertheless, carcinogenic. Many of these compounds are metabolites of nitrosamines and might represent detoxication products.

In contrast, both halogen- and cyano-substituted nitrosamines are as toxic or more toxic than their unsubstituted parents. For example, methylnitroso-2-chloroethylamine is considerably more toxic than methylnitrosoethylamine, and methylnitroso-cyanomethylamine (methylnitrosaminoacetonitrile) is equally as toxic as nitrosodimethylamine. Electronic effects probably play an important role in toxicity by these compounds. Nitrosobis-(2-chloroethyl)amine is a very toxic compound, as is bischloroethylamine itself, whereas nitrosiminodiacetonitrile is considerably less toxic than NDMA. The corresponding acid, nitrosimino-diacetic acid is almost non-toxic when given orally to rats (Lijinsky *et al.*, 1973*b*).

Since so many of the acyclic nitrosamines are metabolised to, among other products, the methyldiazonium ion, which methylates nucleophiles it encounters, this alkylating moiety is unlikely to be the agent mainly responsible for the toxic effects of these compounds. While, under suitable conditions, the effectiveness of methylnitrosoalkylamines as methylating agents is very comparable, their toxicity varies enormously. This suggests that formation of a methyldiazonium ion, and by analogy an ethyldiazonium ion from the ethyl compounds, has little relevance to toxicity, and might not be involved at all.

An additional reason for reaching this conclusion is that alkylnitrosoureas and alkylnitrosocarbamates, which give rise directly to alkyldiazonium ions, have relatively low toxicity for the liver, although they alkylate macromolecules as readily as the corresponding nitrosamines *in vivo*. For example, methyl- and ethyl-nitrosourea and methyl- and ethylnitrosourethane have higher LD_{50} values than the corresponding dialkylnitrosamines, even (except in the case of methylnitrosourethane) when administered to rats intravenously – this to reduce decomposition of these unstable compounds. Death of the animals given these directly acting alkylating agents orally is usually due to ulceration of the gastrointestinal tract, particularly the stomach. It is of interest that dimethylnitrosourea and trimethylnitrosourea are little different in toxicity from MNU, although the former compounds are enormously more chemically stable than the latter.

Comparison of the toxic effects of directly acting alkylnitrosoureas with toxicity of nitrosamines of similar structure (the latter being more toxic) suggests that the metabolic activation of the nitrosamines is responsible for their strong toxicity. The differences in chemical properties between the two classes of N-nitroso compound, rather than the similarities, seem to give a clue to the mechanisms by which they induce toxicity, much as the same or similar differences underlie the very different array of organs and tissues in which they induce tumours. For example, alkylnitrosoureas do not induce tumours of the esophagus and nasal mucosa in rats, whereas these are among the most common targets of carcinogenic nitrosamines.

If the formation of an alkyldiazonium ion, directly or through metabolism, is not the mechanism by which they exert their toxic effects, a review of the relation between acute toxicity and chemical structure provides no coherent alternative mechanism. Perhaps the most perplexing of these effects is the liver toxicity of the simple nitrosamine, NDMA, for it is difficult to conceive an alternative to its conversion to a methylating agent as its mode of toxic action. Oxidation of the methyl group to form a hydroxymethyl-derivative, which is not isolable, could result in interaction with some cellular component, so as to disrupt its normal function. The acetate ester of this compound, α-acetoxy-dimethylnitrosamine, which could easily hydrolyse *in vivo*, is quite toxic, although not especially to the liver. Nitrosodimethylamine is not very toxic to organs of the rat other than liver, although it is activated to a methylating agent in lung and kidney, in which it produces extensive methylation of DNA and other macromolecules.

Another possibility is the reactive carbanion formed by removal of a proton from the methyl group of NDMA, which is very facile in the presence of base, as shown by Keefer and Fodor (1970). This is the basis for easy preparation of deuterium- or tritium-labelled nitrosamines by exchange with D_2O or tritiated water in the presence of alkali (Singer and Lijinsky, 1979). The carbanion would react with electrophiles, which are less available than nucleophiles in a cell, so that the ramifications of formation of this product are difficult to predict. Also possible is the formation of an aldehyde from hydroxy-NDMA by further oxidation; the product methylnitrosoformamide is highly toxic and bears some resemblance to pyruvic aldehyde. Although the alcohol could be esterified, for example by sulphate or phosphate, it is difficult to see why these products would be the proximate toxic agents, since the acetate ester seems not to have the appropriate activity. As Druckrey *et al.* (1967) showed, following speculation many years ago, analogues of NDMA such as dimethylformamide or dimethylhydrazine or dimethylsulphoxide do not show similar toxicity to NDMA. This leaves the apparently simple problem open.

Koepke *et al.* (1979) have suggested that an oxadiazolium ion might be an intermediate in methylation of macromolecules, and presumably could be the intermediate in toxicity of many nitrosamines, but not NDMA. Among them are methylnitrosoethylamine and its higher homologues, as well as the much less toxic methylnitrosoethanolamine (Koepke *et al.*, 1988). A majority of nitrosamines are quite toxic to the liver of rats and hamsters (and mice), whether cyclic or acyclic, and it is tempting to believe that the mechanisms by which they are converted into toxic agents are similar, especially since they are assumed to be activated by liver enzymes. Many of them undergo an extensive series of degradative reactions before they are excreted or join the carbon pool of the body (e.g. methylnitrosoalkylamines – Lijinsky, 1983*b*), but the toxic effects are probably a result of reactions that are quite immediate. If products at the end of the reaction chain were the toxic agents of interest, one would expect all of the nitrosamines of a series, such as methylnitrosoalkylamines, to have similar toxicity, whereas they vary very widely, from small molecules that are very toxic to large molecules that are weakly toxic.

For those who view the relationship between toxicity and carcinogenic potency as very close it must be an obstacle to explain why so many nitrosamines acutely toxic to the liver of experimental animals do not induce liver tumours by chronic treatment of those same animals. The failure to induce liver tumours is probably not a matter of dose, for example with methylnitroso-*n*-propylamine and methylnitroso-*n*-butylamine, toxic to the liver and lungs in both rats and hamsters, but inducing liver and lung tumours only in hamsters (Lijinsky and Kovatch, 1988*b*). Nitrosodiethylamine is much less toxic to the liver than is NDMA, but the former is a more potent carcinogen in rats, less potent in hamsters, and induces liver tumours in both species (Lijinsky *et al.*, 1987*b*).

Comparing NDMA with its non-nitrosamine isomer azoxymethane, both are very toxic to the liver in both rats and hamsters, and both are potent liver carcinogens in hamsters. But in rats azoxymethane is considerably weaker than NDMA in inducing liver tumours when administered in drinking water (Lijinsky and Kovatch, 1989*a*), while when given in pulsed doses by gavage NDMA induces fewer liver tumours and azoxymethane none; both compounds induce a high incidence of kidney tumours in rats by this regimen (Lijinsky *et al.*, 1987*b*), as they do also when injected into the bladder (intravesicular administration).

The ethylnitrosoalkylamines are usually less toxic than the corresponding methylnitroso-compounds, although not always less carcinogenic. This indicates that those reactions occurring in the ethyl group form less toxic products than the analogous reactions of the methyl group. This applies equally to the pair (methyl- and ethyl-) of nitrosamines derived from 'vinylamine' and from hydroxyethylamine. Methylnitrosochloroethylamine is little different in toxicity from nitrosobis-(chloroethyl)amine, but in these cases the toxic effect of the haloalkyl group might well be more important than the activating properties of the N-nitroso group. The relatively low toxicity of the nitrosamines with hydroxylated alkyl groups is no doubt partially due to their hydrophilicity, which leads to excretion of some, and occasionally most, of the dose unchanged, for example nitrosodiethanolamine and nitrosobis-(2-hydroxypropyl)amine (Farrelly et al., 1987b). Both of these are at least partially oxidised to more toxic carbonyl compounds. However, it is not clear why methylnitrosoethanolamine is not oxidised to a toxic intermediate in its passage through the liver, much as is methylnitrosoethylamine.

Even if the toxic intermediates formed metabolically from the nitrosamines were known, it might not lead to complete elucidation of the mechanisms of toxic injury to the liver of rats, for example. Hard and Butler (1970) have explored the toxic action of NDMA in rat kidney and have used the well known finding of McLean that protein starvation of rats decreased the toxicity of NDMA (increased the LD_{50}), possibly by decreasing the synthesis of detoxifying enzymes (McLean and Verschuuren, 1969). Nitrosomorpholine has similar toxic effects to NDMA in many ways, and has been extensively investigated by Bannasch and Massner (1976), although the metabolism and activation of these two compounds would be expected to differ markedly. Other unexplored facets of the toxicity of nitrosamines include the convulsant effects in rats of several cyclic nitrosamines, nitroso-pyrrolidine, -piperidine, -hexamethyleneimine and -heptamethyleneimine, but specifically excluding nitrosomorpholine. Ring size and shape are not the sole determinants of this toxic effect, although much larger nitrosamines, nitrosododecamethyleneimine, and smaller, nitrosoazetidine, not only are not convulsants, but have very low toxicity overall. At this juncture it seems that neither the mechanism of toxicity by cyclic nitrosamines nor their mechanisms of carcinogenesis are close to being understood.

5.5   Toxicity by different routes of administration

There is great variation in the toxicity of N-nitroso compounds, which is made more complex by the variety of routes by which they were administered to determine their LD_{50}. By the oral route, using gavage for example, the rate of absorption and entry into the circulation is quite similar for most of the compounds, regardless of their lipophilicity. However, when the compounds are given by S.C. injection, or by I.P. administration the rate at which the compound enters the circulation depends greatly on its solubility in water and partition ratio between the solvent in which it is administered, often a natural fat, and water. I.V. injection, although seldom employed, ensures very rapid distribution of the administered compound and is useful for unstable alkylnitrosoureas (Lijinsky *et al.*, 1972c) or alkylnitrosourethanes (Druckrey *et al.*, 1967). Methylnitrosododecylamine has a high molecular weight (for a nitrosamine) and is only slightly soluble in water. Its toxicity by S.C. injection was very low, thereby permitting administration of very large quantities, as a proportion of the LD_{50} so commonly used in some laboratories, in the chronic carcinogenicity study. This was wasteful of compound, most of which probably remained at the site of injection, and led to an artificially low measure of the carcinogenic potency of this nitrosamine (Ketkar *et al.*, 1981). In the pioneering experiments of Druckrey and Preussmann, most of the treatments used the oral route of administration, and their toxicity data are comparable within themselves and with those in many other laboratories.

5.6 Subchronic toxicity

A distinction must be drawn between acute toxicity, or death which follows shortly after administration of one or few treatments, and subchronic toxicity, or death which follows administration of several relatively small doses during a period of several weeks, but too short a time for tumours to eventuate. It is surprising that there is considerable variation in this effect between nitrosamines of closely similar structures, which, in this case, were administered at approximately equimolar doses. The causes of death also frequently differ between closely similar compounds. However, for a given compound the acute toxic cause of death and the subchronic toxic cause of death are usually the same, and the animals are simply able to survive the smaller doses for a longer time, perhaps even partially repairing the damage between treatments. Examination of the results show no obvious relation between subchronic toxicity, as measured by the size of the dose leading to death from toxic effects, and the carcinogenic potency of the compound; the target of the toxic effects is usually the liver, even by the many compounds that at lower dose rates induce tumours in other organs but not in the liver. There does not seem, therefore, to be a close connection between the mechanisms of production of toxic lesions and mechanisms of induction of tumours; but there might be similarities in the chemistry of activation producing the two effects.

The simplest nitrosamine, NDMA, has a single dose LD_{50} of 20 to 30 mg/kg body weight in rats, and also for mice, hamsters, guinea pigs and dogs; doses of 10 mg/kg twice a week to male or female rats were not cumulatively lethal until tumours appeared (Lijinsky et al., 1987b), although 15 mg/kg twice a week to female rats led to death of animals after eight weeks. In hamsters, three doses of 12 to 18 mg/kg of NDMA during five weeks did not cause death (Tomatis and Cefis, 1967), although 15 mg/kg once a week led to death of some animals in four to six weeks (the survivors developed tumours much later, Lijinsky et al., 1987b); 7.5 mg/kg per week caused no early deaths of hamsters. These results suggest that the animals were able to repair NDMA-induced liver damage between treatments, at least partially. The toxic effects of the methylating agent azoxymethane were closely similar to those of its isomer NDMA, in both rats and hamsters.

5.7 Carcinogenic and toxic potency

Nitrosodiethylamine was considerably less toxic than its lower homologue (NDMA), with the LD_{50} approximately 200 mg/kg body weight in both rats and hamsters, also causing centrilobular necrosis in the liver. The LD_{50} of methylnitrosoethylamine, which can be considered a hybrid of NDMA and NDEA, was 90 mg/kg body weight in rats, logically midway between those of the two symmetrical nitrosamines. Neither NDEA nor MNEA as multiple small doses to either rats or hamsters caused any deaths before the appearance of tumours. While this might seem quite logical in hamsters, in which species NDMA is a more potent carcinogen than either MNEA or NDEA, it is not so straightforward in rats, in which NDEA is a much more potent carcinogen than NDMA, although MNEA is less potent. Nevertheless, among these three prototypic nitrosamines there is no indication of any connection between acute toxicity and carcinogenicity; there are many more examples, some of which follow.

In the case of many N-nitroso compounds that have been tested for carcinogenicity there was no prior determination of acute toxicity, but a considerable number of determinations have been made, notably by Druckrey et al. (1967). The range of potency as toxins is much wider than the range of carcinogenic potency, in general, ranging in rats from 18 mg/kg for methylnitrosobenzylamine to 7500 mg/kg for ethylnitrosohydroxyethylamine and to nitrosodiethanolamine, which is virtually not toxic at all, but nevertheless carcinogenic. There are many noncarcinogenic nitrosamines, of course, which are also not measurably toxic. Alkylnitrosoureas and alkylnitrosocarbamates, although directly acting alkylating agents, are generally much less acutely toxic than many dialkylnitrosamines which must be activated to form an alkylating agent. This indicates that the alkylating agent formed in common by the nitrosamines and the directly acting compounds is not the toxic agent causing death of animals following a single treatment. Cyclic nitrosamines are neither the most, nor the least, toxic among nitrosamines of all classes; formation of an alkylating agent from cyclic nitrosamines has not been convincingly demonstrated, yet they are carcinogenic and toxic compounds, not qualitatively dissimilar in their actions from acyclic nitrosamines.

The single dose LD_{50} values of a number of N-nitroso compounds in rats and other species are shown in Table 5.2, and large differences can be seen. The cumulative toxicity of a number of N-nitroso compounds that have been administered in multiple small doses shows some variation from the single dose effects, and large variations from carcinogenic effectiveness. These variations again underline the divergence between events leading to toxicity and events leading to tumour induction. Several compounds containing halogenated alkyl groups, such as chloroethylnitrosoureas, cause fatal nephropathy and sclerosis in the kidneys, but in smaller doses seldom induce kidney tumours. As related above, many nitrosamines cause liver toxicity, but not by any means do all induce liver tumours. Toxic effects in other organs are not common – although histopathological examination of animals that died without tumours was not always thorough – but they have been seen in lung, esophagus and gastrointestinal tract of animals treated with compounds which after prolonged exposure sometimes give rise to tumours in those organs. It is notable, and unexpected, that alkylnitrosoureas, which are very potent carcinogens, are usually only mildly toxic, although they are potent alkylating agents. The toxic effects of N-nitroso compounds are not directly related to their capacity to form alkylating agents, nor to their carcinogenic action.

5.8 Structure–activity relations in toxicity and mechanisms

Considering methylnitrosoalkylamines that induce tumours in the rat esophagus and which methylate nucleic acids in rat esophagus (Von Hofe *et al.*, 1987; Labuc and Archer, 1982), and also in rat liver in which they usually do not induce tumours, there is a large variation in toxicity from compound to compound. For example, while methylnitrosoethylamine is less toxic than NDMA, methylnitroso-*n*-butylamine

(MNBA) is very toxic, so that a single gavage dose of approximately 50 mg/kg or two successive doses of 25 mg/kg cause death of all males so treated. When male rats were given 12 mg/kg body weight of MNBA twice a week by gavage, four of 20 animals died after four weeks of treatment, but the remainder survived continuing treatment to eventually develop tumours of the esophagus. The animals that died early had severe congestion of the lungs and some showed hyperplasia in the liver, but none of the surviving animals developed tumours of either the lungs or liver. Female rats given 16 mg/kg body weight of MNBA by gavage twice a week showed a similar response; five of 20 animals died during the first four weeks, following which continued treatment caused no additional deaths before appearance of tumours. The lungs and liver were again the target of the toxic effects, but no tumours appeared in these organs.

The effects of MNBA given to rats in drinking water were similar to those by gavage. Half of the male rats given 4 mg/week (approximately 15 mg/kg body weight) died within two weeks; two of 20 males given 8 mg/kg died after one week, the remainder surviving the treatment and developing tumours. Two of 20 female rats given 10 mg/kg MNBA died after one week and the remainder survived to develop tumours of the esophagus. Parallel groups of male and female rats were treated with MNBA labelled with deuterium in the methyl group or in the α-methylene of the butyl group, and the methyl-d_3 compound was more toxic than the unlabelled compound, causing death of eight of 20 males at 8 mg/kg and 14 of 20 females at 10 mg/kg within two weeks. On the other hand the butyl-d_2 compound at the same doses caused no early deaths of either male or female rats, which all died of esophageal tumours after 20 weeks treatment. At higher doses (20 mg/kg/week) the d_2 compound caused no early deaths. These results suggest that the toxic product is formed by α-oxidation of the butyl group, not by oxidation of the methyl group; however, the nature of the toxic metabolite is not known, but is not likely to be the methyldiazonium ion, for reasons discussed above.

In contrast with the toxic effects of methylnitroso-n-butylamine, an equimolar dose of methylnitroso-n-propylamine, 1.4 mg or approximately 7 mg/kg per week, caused no early deaths of female rats given the compound in drinking water. The next higher homologue, methylnitroso-n-amylamine was not tested in comparable conditions, but according to Druckrey *et al.* (1967) is less toxic than methylnitrosoethylamine. The still higher homologues methylnitroso-n-hexylamine, -heptylamine, and larger were even less toxic and caused no early deaths in the course of treatments that led to induction of tumours, using considerably higher doses than those of MNBA. Therefore, MNBA joins a number of other methylnitrosoalkylamines in being exceptionally toxic, but not more potently carcinogenic than several less-toxic compounds of similar structure.

Among the very toxic compounds are methylnitrosobenzylamine, methylnitrosocyclohexylamine and methylnitroso-2-phenylethylamine, the reported acute single dose LD_{50} levels for which are 18, 30 and 48 mg/kg, respectively, in rats (Druckrey et al., 1967). In subchronic studies with methylnitrosobenzylamine, one week's treatment in drinking water with 2.5 mg (12 mg/kg/week) caused several male rats to die with severe lung toxicity and haemorrhage, esophagitis and pancreas hyperplasia, whereas half of that dose was tolerated until the animals died with esophageal tumours after 20 weeks. An equimolar dose of methylnitrosophenylethylamine, approximately 14 mg/kg/week was well tolerated by male rats, and caused no early deaths. However, 12 mg/week (60 mg/kg/week) in drinking water caused death after one week of 11 of 20 male rats, although after a four-week interval survivors accepted continued treatment until death with esophageal tumours. The cause of early death was toxicity in the lungs and liver. Equimolar treatment with methyl-d_3-nitrosophenylethylamine caused early death of fewer male rats than the unlabelled compound, seven of 20 dying after one week of treatment. This was at variance with the effects of deuterium labelling on the toxicity of methylnitroso-n-butylamine, in which the methyl-d_3-compound was more toxic than the unlabelled compound, and suggests that the route of metabolism leading to formation of the toxic product can be different for nitrosamines of different structure. Methylnitrosocyclohexylamine was not toxic after administration at 25 mg/kg/week in drinking water to male rats, whereas methylnitrosoneopentylamine caused death of 3 of 20 male rats after one week of treatment in drinking water with 22 mg/kg/week; half of that dose was well tolerated until tumours of the esophagus appeared after 20 weeks.

It appears that the high toxicity of methylnitrosoalkylamines is related to particular structures of the alkyl group, but that the entire molecule must be involved in the toxic effect of whatever important metabolite is formed; that is, the presence of the methyl group is essential for the toxic effect, since the corresponding N-ethyl compounds are less toxic than the N-methylnitrosoalkylamines. Also in sharp contrast, the nitrosodialkylamines corresponding to the larger alkyl groups, for example, nitrosodi-n-butylamine, nitrosodi-n-amylamine and nitrosodibenzylamine are all very weakly toxic (in contrast with the potent toxicity of the N-methyl analogues), and the first two are weaker carcinogens than the N-methyl analogues, while the last is reported as non-carcinogenic (Druckrey et al., 1967). These results again suggest that formation of a methyldiazonium ion following metabolic fragmentation of the methylnitrosoalkylamine is not the mechanism by which these compounds exert their toxic effect. Methylnitrosophenylamine is quite toxic, but cannot form a methyldiazonium ion.

Although, in general, the presence of oxygen functions in the alkyl chain of a nitrosamine conspicuously lowers toxicity (but not necessarily carcinogenic potency), this is not always the case. For example, 35 mg/kg of methylnitroso-2-hydroxypropylamine given twice a week by gavage led to death of female rats after four weeks of treatment, whereas a slightly lower dose of methylnitroso-2-oxopropylamine, 25 mg/kg, in male rats caused no early deaths.

5.9 Toxicity of cyclic nitrosamines and carcinogenicity

The toxicity of acyclic nitrosamines tends to be greater among those of lower molecular weight, but this is not so among cyclic nitrosamines. For example, it was not possible to establish an LD_{50} for the simplest cyclic nitrosamine, nitrosoazetidine, which was nevertheless quite carcinogenic (Lijinsky et al., 1967), and the toxicity of the next higher homologue, nitrosopyrrolidine, is quite low, 900 mg/kg (Druckrey et al., 1967). The next higher homologue, nitrosopiperidine, was considerably more toxic with a single (oral) dose LD_{50} of 200 mg/kg (Druckrey et al., 1967). An oxygen-containing analogue of nitrosopiperidine, nitrosomorpholine, had an acute oral LD_{50} of 320 mg/kg, but the analogue containing sulphur, nitrosothiomorpholine, was less toxic (LD_{50}, 800 mg/kg) and the nitrogen-containing analogue, 1-nitrosopiperazine, was even less toxic, with an LD_{50} of 2260 mg/kg (Garcia et al., 1970); dinitrosopiperazine was more toxic than nitrosopiperidine (LD_{50}, 160 mg/kg, Druckrey et al., 1967).

Higher carbocyclic analogues of nitrosopiperidine, nitrosohexamethyleneimine, nitrosoheptamethyleneimine and nitrosooctamethyleneimine had slightly lower toxicities, 336, 283 and 566 mg/kg by gavage, respectively (Goodall et al., 1968; Lijinsky et al., 1969). Light anesthesia prevented death from convulsions of rats treated with the larger cyclic nitrosamines. Those that died within one to three days had lung congestion and haemorrhage and/or centrilobular necrosis of the liver. The C-7 and C-8 compounds induced only tumours of the lung and esophagus, although they were strongly toxic to the liver in large single doses, while the C-6 compound induced tumours of the liver and esophagus. The lack of tumourigenesis in the liver by these compounds that are so toxic to the liver, distinguishes the routes of activation to toxic and carcinogenic metabolites of the cyclic nitrosamines, as has already been noted among acyclic nitrosamines. The target organ of carcinogenicity of these cyclic nitrosamines was not necessarily related to the target organs of their toxic effect, since nitrosopiperidine, nitrosothiomorpholine and dinitrosopiperazine did not induce lung tumours in rats, but dinitrosopiperazine did give rise to liver tumours. Since covalent DNA adducts of cyclic nitrosamines have seldom been detected, the implication of DNA adduct formation in either toxicity or carcinogenicity by cyclic nitrosamines is not yet proved.

In subchronic studies in rats cyclic nitrosamines did not cause early deaths, with the exception of nitrosoheptamethyleneimine at 40–50 mg/kg/week in drinking water, which caused death of two of 20 male rats after four weeks of treatment. No early deaths occurred in female rats given the same treatment, and there was no significant effect in male or female rats that had chronic murine pneumonia (Schreiber et al., 1972) and were given the same nitrosoheptamethyleneimine treatment. Cumulative toxicity by cyclic nitrosamines seemed to be less frequent than by their acyclic counterparts, although in carcinogenic potency the two classes were quite comparable. Even the very potent carcinogen nitroso-2,6-dimethylmorpholine given by gavage twice a week to male rats (50 mg/kg) or to female rats (80 mg/kg) did not cause early death, although animals died with esophageal tumours beginning at week 11. Nitroso-3,4-epoxypiperidine by gavage twice a week at approximately 40 mg/kg caused death of female rats without tumours, but with lesions of the kidneys, liver and pancreas, beginning at eight weeks and lasting until week 15; a lower dose, 15 mg/kg/twice a week, caused death of eight of 20 female rats between the fourth and fifth week, following which continued treatment led to appearance of tumours (liver, esophagus, nasal mucosa) beginning at week 24. The increased toxicity can be attributed to the strongly electrophilic epoxide structure.

5.10 Hamsters: toxicity of nitrosamines with oxygenated alkyl groups

In hamsters, all of which were treated by gavage, the many acyclic nitrosamines studied were not more toxic than to rats. Even methylnitroso-n-butylamine, which was lethally toxic to rats at 20 mg/kg body weight did not cause early death of hamsters after administration of 40 to 50 mg/kg/week; no others of the methylnitrosoalkylamine series, beginning with methylnitrosoethylamine, caused early death of hamsters at equimolar doses. However, acyclic nitrosamines containing a 2-oxy- or 2-hydroxy-propyl group were toxic to Syrian hamsters, as has been reported by Pour and his associates following single treatment, usually by S.C. injection (Pour et al., 1977, 1979). Treatment of hamsters with nitrosobis-(2-oxopropyl)amine, 50 mg/kg once a week caused death of seven of 20 animals after eight weeks from liver necrosis and hepatitis, and pancreatitis; surviving animals later developed liver tumours and tumours of the pancreas ducts (Lijinsky et al., 1984b). Lower doses of the nitrosamine (20 mg/kg/week) caused few early deaths, at 11 and 12 weeks, and continued treatment led to death from liver and pancreas tumours after week 24. No early deaths occurred of hamsters treated with 50 mg/kg/week of nitroso-(2-hydroxypropyl)(2-oxopropyl)amine, a common metabolite of NBOPA and nitroso-2,6-dimethylmorpholine; all three compounds induce tumours of the liver and pancreas on prolonged administration to hamsters, and with similar effectiveness. The same dose of nitrosodimethylmorpholine does not lead to early deaths of hamsters, either. Therefore, the exceptional toxicity of NBOPA probably involves biochemical mechanisms which are unrelated to those by which these compounds induce tumours of the liver and pancreas in hamsters.

Methylnitroso-2-oxopropylamine at 20 mg/kg by gavage once a week caused death of five of 20 female hamsters between the third and seventh week, with severe damage to the liver, inflammation and haemorrhage in the lungs, but at most mild effects in the pancreas; surviving animals developed liver tumours, but no tumours of the pancreas or lungs (Lijinsky and Kovatch, 1985a). The analogue, methylnitroso-2-hydroxypropylamine given to hamsters by gavage at 50 mg/kg/week cause death of six of 20 animals between week three and week six; liver toxicity and hepatitis were severe, there were some toxic effects in the lung, and toxicity to the pancreas was mild, yet most of the surviving hamsters later developed tumours of both the liver and pancreas, as well as lung tumours.

5.11 Toxicity of alkylnitrosoureas and alkylnitrosocarbamates

The toxicity of alkylnitrosoureas and other alkylnitrosamides is usually low, less, for example, than that of nitrosamines that would be expected to give rise to the same alkylating agent through metabolism as the alkylnitrosamide does directly. The alkylnitrosoureas would be expected to interact directly with a number of nucleophiles in cells, and yet their toxic effect is small. This suggests either that they readily decompose to innocuous products, which is unlikely, or that their structures do not lend themselves to formation of intermediates which produce toxic changes in cells, as do nitrosamines. Methylnitrosourea and ethylnitrosourea in doses of 80 and 240 mg/kg respectively, injected into the hepatic portal vein of rats produced zero or minimal liver damage, as observed by electron microscopy (Lijinsky et al., 1972c). In chronic studies alkylnitrosoureas caused early death of rats only when they contained halogenated alkyl groups.

On the other hand, some alkylnitrosocarbamates were quite toxic to rats. Methylnitrosourethane (methylnitroso-ethylcarbamate) at 10 mg/kg per week by gavage caused death within two weeks of six of 20 female rats concurrently given otherwise nontoxic treatments with 3 mg/kg methylnitrosoethylamine and 2.5 mg/kg dinitroso-2,6-dimethylpiperazine (Lijinsky et al., 1983c). A number of N-nitroso derivatives of N-methylcarbamate insecticides also led to early death of rats given the compounds by gavage in relatively small doses (Lijinsky and Schmähl, 1978). Nitrosoaldicarb was the most toxic, causing death of eight of 12 rats following two weekly treatments with 20 mg/kg, and to death of seven of 12 rats following two to nine weeks of treatment with half that dose. Nitrosocarbofuran at 25 mg/kg/week caused death of nine of 12 female rats in the second week of treatment; at half the dose only one rat died in the 10th week. Nitrosomethomyl at 18 mg/kg/week was much less toxic, and led to death of one rat at week two, and three additional animals died between week seven and nine, of the 12 rats treated. The remaining alkylnitrosocarbamates, including nitrosocarbaryl, caused no early deaths of rats when administered at similar doses by gavage, although all had very similar carcinogenic potencies (inducing tumours of the forestomach). The reason for the exceptional toxicity of nitrosocarbofuran and nitrosoaldicarb is not known.

Intravesicular administration of small doses of methylnitrosourethane (1.4 mg or two or three doses of 0.4 mg) caused severe inflammation of the bladder in rats and nephropathy, pyelonephritis and other toxic lesions in the kidneys; several rats died within a few days. The toxic effects were similar, but much less severe, following repeated intravesicular administration of 1.6 mg of ethylnitrosourethane, and no rats died. On the other hand, repeated intravesicular treatments of rats with 1 mg of methylnitrosourea produced much less severe toxic injury to the bladder or kidneys and no rats died. The cyclic alkylnitrosocarbamate nitrosooxazolidone in repeated intravesicular doses of 2.4 mg caused toxic damage to the bladder and kidneys of rats similar to that by ethylnitrosourethane and some animals died after several weeks.

5.12 Toxicity of haloalkyl nitrosoureas

The halogenated alkylnitrosoureas were frequently toxic even at low dose rates. For example, 2-fluoroethylnitrosourea at 10 mg/kg orally caused death of seven of 12 male rats after a single treatment (Lijinsky and Kovatch, 1989b). At one half of that dose twice a week some animals died by week 14, when treatment stopped; remaining animals died steadily until week 32, all without tumours. At 5 mg/kg, twice a week, all of 12 female rats had died, beginning at week eight, and none had tumours. In females receiving a lower dose, 2.5 mg/kg twice a week, the rate of death was slower, beginning at week 16; all animals were dead at week 44 and none had tumours. A group of males receiving a still lower dose, 1.7 mg/kg twice a week, showed good survival with no early deaths, and there was a small incidence of induced tumours. The main cause of early death of these rats was toxic damage to the kidneys, nephropathy and glomerulosclerosis, which was seen also in animals that survived. This effect on the kidneys was a common feature of rats treated with nitrosoureas containing halogenated alkyl groups, although kidney tumours were never seen in those rats that survived the early toxic effects. On the other hand, kidney tumours in rats were induced by several compounds that did not have particularly severe toxic effects in the kidneys. 2-Fluoroethylnitrosourea also was toxic to mice, causing severe skin damage and death after application of a 0.04 M solution in acetone to the skin (Lijinsky and Reuber, 1988); lower doses produced skin tumours.

Among other nephrotoxic alkylnitrosoureas were chloroethylnitroso-hydroxyethylurea and chloroethylnitroso-2-hydroxypropylurea which, after six weeks of treatment by gavage with approximately 15 mg/kg in females and 11 mg/kg in males, both twice a week, caused death of, respectively, 13 females and 11 females; the remainder died within the next two or three weeks. The males treated with either compound survived better, but all died before week 30, without tumours (Lijinsky et al., 1986a). At lower doses, 7 mg/kg in females and 4.5 mg/kg in males twice a week by gavage, all of the females died by week 18, at which time all of the males were still alive; most subsequently died by week 30, but three survived to week 46. Deaths caused by the hydroxyethyl compound tended to be earlier than those caused by the hydroxypropyl compound, and at 4 mg/kg twice a week female rats given the former were dead by week 21 (beginning at week 16), whereas females given the latter were not all dead until week 41. None of the rats treated with the chloro-ethylnitrosoureas by gavage died with tumours, even those that survived to 40 weeks or more, and it was particularly noticeable that none had kidney tumours. When given to rats in drinking water – a gentler treatment – chloroethylnitrosohydroxypropylurea at a dose of 14 mg/kg/week to females and 9 mg/kg/week to males did not cause death of females until week 15, but they all died by week 40 without tumours; all but two males survived to week 60 and they had a significant incidence of tumours of the liver and lungs. The commonly used cancer therapeutic agents BCNU and CCNU are both chloroethylnitrosoureas. Both are strongly cytotoxic and the margin between their acutely toxic dose and their therapeutic dose is small. In rats they caused death from renal toxicity and delayed bone marrow toxicity, which was also the common toxic effect in humans. The LD_{50} in rats (single dose) was, BCNU 30 mg/kg, CCNU 70 mg/kg. Chloroethylnitrosohydroxyethylurea (HECNU) is a promising cancer therapeutic agent (Eisenbrand *et al.*, 1976), less toxic than BCNU. The analogous 2-hydroxypropylurea is still less toxic, and is also a potential anti-cancer drug (Zeller *et al.*, 1989).

Several chlorinated nitrosotrialkylureas (Fig. 5.1) were synthesised and tested for carcinogenicity in rats by gavage administration and all were very toxic. They were chlorinated derivatives of nitrosotrialkylureas that had induced high incidences of mammary tumours and tumours of the nervous system in rats (Lijinsky *et al.*, 1980*b*). Chloroethylnitrosodimethylurea (Lijinsky *et al.*, 1979*a*) given at 14 mg/kg twice a week caused death of all male rats between weeks four and eight. At 7 mg/kg twice a week, 10 of 15 rats died between weeks eight and nine. At 1.7 mg/kg twice a week survival was good to week 54, after which animals began to die with lung tumours and tumours of the forestomach. The diethyl analogue was studied only at one dose level, 2 mg/kg, twice a week which caused no early deaths; few of these rats developed tumours, however. Methylnitrosobis-(2-chloroethyl)urea was toxic at 25 mg/kg twice a week, although this compound did not have the chloroethyl group neighbouring the N-nitroso function and was not, therefore, a chloroethylating agent; two of 15 female rats died after 1 week. At 5 mg/kg twice a week, six of 15 rats died between weeks seven and 15. A still lower dose, 1.5 mg/kg twice a week, caused no early deaths from toxic effects of the compound, but neither did a significant number of tumours develop.

5.1 Structures of chloroalkylnitrosoureas.

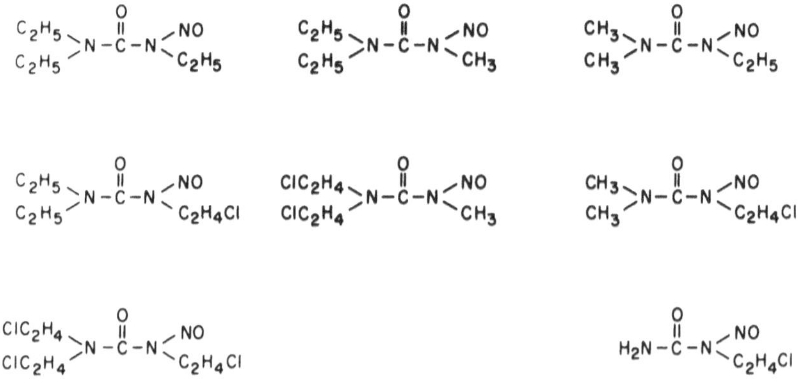

Nitrosotris-(2-chloroethyl)urea at 6 mg/kg twice a week caused death of 10 of 15 female rats between week three and 15, and a still lower dose of 1.5 mg/kg twice a week caused death of half the rats between weeks 12 and 18; at neither dose was there a significant incidence of tumours in long-lived survivors. It is apparent that the low doses necessitated by the more toxic compounds prevented delivery of sufficient compound to cause tumours within the abbreviated lifetime of the rats, almost all of which in these experiments were dead before week 90. It is obvious, however, that none of these compounds can be considered non-carcinogenic based on such restricted chronic studies, which is a sign of the ease with which under particular experimental conditions a carcinogen might have no apparent tumourigenic effect.

5.13 Conclusions

The conclusions that can be drawn from the studies of toxicity of so many N-nitroso compounds are that there is often no discernible pattern of toxicity among a group of structurally similar compounds, and that the targets of the toxic effects cannot be predicted with assurance for a given species (except possibly for halogenated compounds). There is no quantitative relation between toxic potency and carcinogenic potency of N-nitroso compounds, even among those that are closely related structurally. More importantly, the organ in which toxic effects are manifest is often, perhaps usually, not the same as that in which the nitroso compound induces tumours on chronic administration. It follows that the hyperplasia subsequent to cell death in an organ does not lead to tumour formation unless the toxic effect was produced by a carcinogen, and not inevitably even in that case (cf. the effects of large toxic doses of NDMA in rat liver). The evidence points to a probable distinction between those products or metabolites that are responsible for the toxic effects and those that are responsible for induction of tumours.

6

Mutagenesis and cell transformation by N-nitroso compounds

6.1 Introduction

Probably the first chemical mutagen in common use was the N-nitroso compound MNNG (methylnitrosonitroguanidine), which was prepared for use as an explosive, but was found to be a mutagen (Mandell and Greenberg, 1960). It was known at the same time that mustard gas and some nitrogen mustards were mutagens, and were alkylating agents. The availability of chemical mutagens was a useful adjunct to the X-rays used by Müller and his colleagues to study mutagenesis. MNNG has become the most widely used agent for mutagenesis and transforming activity in cells. It has been less useful as a mutagen for germ cells of multicellular organisms and is limited in its carcinogenic activity, compared with other directly acting N-nitroso compounds. However, it has given rise to tumours of the glandular stomach in several species, that are rarely induced by other carcinogens.

The mutagenic activity of MNNG was believed to be due to its activity as a methylating agent – and this is probably the case. The formation of diazomethane as the intermediate was considered likely, and attention turned to other similar methylating agents, such as methylnitrosourea (MNU). The ensuing studies were quite productive, although it was shown shortly, using deuterium-labelled MNNG, that methylation of DNA in bacteria did not involve formation of diazomethane as an intermediate (Süssmuth et al., 1972). This was also shown for methylation of DNA in rats by MNU (Lijinsky et al., 1972c).

Alkylnitrosamides are fairly well established as 'classical' mutagens, and a large part of the conventional wisdom about mechanisms of mutagenesis has derived from studies of the biological effects of these compounds. Included has been the identification of ethylnitrosourea (ENU) as a 'supermutagen' in the germ cells of mice (Russell et al., 1979).

In contrast, MNU is not in the same class as a mouse mutagen. The reason for the differences is an important issue, because there is no consistent difference between ethylating and methylating agents as mutagens or carcinogens in a variety of organisms and test systems.

Since alkylnitrosamides are directly acting compounds, giving rise without need for metabolic activation to powerful alkylating agents (believed to be alkyldiazonium ions), they are an ideal series of compounds in which to examine the effect of variations in chemical structure on mutagenic activity. Other than having different alkyl groups, with a variety of sizes and functional groups, they have similar physicochemical and chemical properties, at least among variants of a given type, such as alkylnitrosoureas, alkylnitrosocarbamate esters and alkylnitrosoguanidines. The difference between methylnitrosourea and ethylnitrosourea as mutagens for germ cells of mice is unexpected, as has been mentioned. Equally unexpected is the lack of mutagenic activity in mouse germ cells of ethylnitrosourethane (Russell and Montgomery, 1982). Ethylnitrosourethane is an equally effective alkylating agent with its alkylnitrosourea analogue, and the reason for its lack of mutagenic activity in this mammal is possibly due to failure to reach the germ cells of the mice, although it reaches other organs in animals and alkylates macromolecules therein. The differences cannot be due to lack of mutagenic activity since, as will be discussed later, alkylnitrosocarbamate esters are more potent mutagens in bacteria and other single-cell systems than the alkylnitrosoguanidines, and much more potent than the corresponding alkylnitrosoureas (Lijinsky and Andrews, 1979); this is especially true of the N-nitroso derivatives of N-alkylcarbamate insecticides, among which activities in bacteria of 10^5 to 10^6 revertants per micromol are not uncommon (Lijinsky and Elespuru, 1976).

One conclusion is that study of mutagenesis in bacteria, for example, tells us little about mutagenesis in mammals. Among alkylnitrosoureas, ENU is one of the least potent in bacterial systems, much less potent in *Salmonella typhimurium* than MNU (the difference being about two orders of magnitude). Presumably there is something special about ethylnitrosourea in its affinity for the germ cells of mice, perhaps an affinity for a particular receptor for the ethylnitrosourea molecule, as distinct from a receptor for the ethyldiazonium ion or similar ethylating agent released by decomposition (solvolysis) of this compound; there is no good chemical reason why ethylnitrosourethane (which forms the same ethylating intermediate) should have less access than ENU to the germ cells of mice. It is a pity, therefore, that the specific locus assay for mutagenesis in mice, which is definitive, is so expensive and demanding of animals and human effort. Otherwise it would be simple to test the idea that the ethylnitrosourea structure is the necessary attribute of supermutagenicity, by testing the analogous diethylnitrosourea, ethylnitrosomethylurea, ethylnitrosodimethylurea or triethylnitrosourea in the same system. This has been possible in other mutagenic systems (Lijinsky et al., 1987a), and in systems to measure carcinogenic potency in animals, which is also expensive. This theme in relation to carcinogenesis by N-nitroso compounds will be discussed further, that is, the relationship between the structure of the molecule and its carcinogenic activity, independent of the formation of an alkyldiazonium intermediate.

The most popular system in which to measure the mutagenic activity of substances is the *Salmonella typhimurium* histidine reversion assay, which was developed and brought into use by B. N. Ames. In the beginning this system was touted as a predictor of carcinogenic activity of substances and as a convenient and inexpensive substitute for chronic toxicity testing in animals. Unfortunately this claim was based on inadequate understanding of carcinogenesis, which was responsible for the assumption that induction of cancer by an agent was through mutagenic effects following damage of some kind to DNA in cells. This was – and is – one of several feasible mechanisms of carcinogenesis, although not proved. The claim that all carcinogens act through mutagenic mechanisms was unsupported at the time, and has appeared even dogmatic in face of the subsequent findings that not few, but many, carcinogens are not mutagenic. Some of these agents have been stubbornly inactive in mutagenesis systems of various kinds, and even in several systems for detecting transforming activity *in vitro*, which do not necessarily require mutagenesis to produce a positive effect. Many *in vitro* mutagenesis systems have been developed during the past two decades and their effectiveness has been compared, including validation of one against another. These developments have been in response to the insistence that known carcinogens be proved to be mutagenic. There has been relatively little interest in proving that all mutagens are carcinogens – and no reason to do so.

The reason for this digression is because so many N-nitroso compounds have been well tested in systems to measure carcinogenesis and in those to measure mutagenesis, and many have failed to produce correlations between one activity and the other, even qualitatively. Of course there has never been a quantitative relationship between them, and the complexity of a multicellular organism defies comparison with a unicellular organism or cells in culture.

6.2 Activation systems for nitrosamines

In the beginning there was concern that the two most common nitrosamines studied, nitrosodimethylamine and nitrosodiethylamine, were not mutagenic to *Salmonella* in the 'plate' test (McCann et al., 1975), when activated by rat liver microsomal fraction, even though both compounds are powerful liver carcinogens in rats. Heinrich Malling (1971) had shown that incubation of the nitrosamine with mouse liver enzymes generated mutagenic substances. One of the difficulties lay in the conduct of the standard assay, but if the nitrosamine was preincubated with the bacteria in suspension, an adequate number of mutated colonies was seen when the suspension was plated and incubated (Yahagi et al., 1977). Nevertheless, the number of mutations produced was in no way commensurate with the potency of these compounds as liver carcinogens in rats compared with other more mutagenic nitrosamines.

These findings led to a focus on metabolism and activation of the nitrosamine, rather than on the nature of the mutagenic intermediate formed. This has been a continuing diversion, and rather begs the question whether mutagenesis and carcinogenesis by these agents are the same mechanistically or are perhaps even unrelated to one another. In the latter case, the metabolism by 'microsomal' enzymes which is extensively investigated is interesting biochemically, but does not necessarily have any bearing on the mechanisms by which these compounds cause toxicity or induce tumours (which themselves appear to be unrelated phenomena). It is notable that half a century of study of the toxicity of the very simple molecule carbon tetrachloride has not led to a complete understanding of how it kills cells. To those familiar with toxicology it is naive to assume that toxins kill cells by damaging their DNA, or even nucleoprotein, since in non-dividing differentiated cells DNA is a passenger; can the nuclear DNA in non-mammalian erythrocytes be essential, when erythrocytes of mammals can do perfectly well without a nucleus? Many toxins seem to cause death of cells through disruption of their organisation, particularly of membranes, and some of the classic toxic chemicals inhibit or prevent the action of enzymes crucial to cell homeostasis (e.g. iodoacetic acid, investigated by Peters in the historic period of Biochemistry).

Prival and other investigators have studied enzymes which activate nitrosamines to mutagens. Their findings indicate that the standard mixed-function oxidases are perhaps not the most important nitrosamine-activating enzymes in rat liver, although they metabolise and activate to mutagens a variety of carcinogens, including polynuclear hydrocarbons and aromatic amines. With these, too, there are discrepancies between enzymes involved in mutagenic activation, and those seemingly related to their carcinogenic action (Bigger et al., 1980). Many nitrosamines which are carcinogenic to rats – even inducing tumours in rat liver – stubbornly fail to be activated to bacterial mutagens by rat liver 'microsomal' enzymes (commonly used as the so-called S_9 fraction). There have been several reports that the principal enzymes responsible for such activation of nitrosodimethylamine are not microsomal, since highly purified microsomes were incompetent in this activation (Prival and Mitchell, 1981). This suggests that unsuspected cytosolic enzymes or a required factor are responsible for activation of this nitrosamine, but perhaps not of others, which are equally well activated to mutagens by crude S_9 fraction and by microsomes purified by centrifugation (Prival and Mitchell, 1981).

There are patterns of mutagenesis related to the chemical structure of nitrosamines, as there are structurally related patterns of carcinogenesis. These patterns are to some extent dependent on whether rat or hamster liver enzymes are used for activation (Table 6.1). Hamster liver enzymes are invariably more effective in activating nitrosamines, as measured by the number of revertants formed from a given amount of nitrosamine in the presence of a standard amount of microsomal protein. Nitrosamines bearing an oxygen on the β-carbon of an alkyl chain, for example, nitrosobis-(2-oxopropyl)amine, are not activated by rat liver S_9 (Andrews and Lijinsky, 1984), and were shown to be mutagens only with hamster liver S_9 activation. What role these hamster enzymes play in metabolising all nitrosamines is not known, nor whether those hamster enzymes are involved in nitrosamine carcinogenesis in hamsters, but this seems an important area for study. It has also been suggested (Prival and Mitchell, 1981) that rat liver contains a factor inhibitory to the formation of mutagens by mouse or hamster liver microsomes.

An additional complication is the finding by Malling, dating back to the early days of chemical mutagenesis, that the effectiveness with which rat liver microsomal fraction activates nitrosodimethylamine to mutagens drops as the microsomes are purified, suggesting that enzymes other than mixed-function oxidases are important. These disparate findings do not lend credence to the assumption that understanding mechanisms of nitrosamine carcinogenesis will come through examination of bacterial mutagenesis by these compounds.

The matter is clarified a little by the observation made many years ago that hamster liver microsomes (or S_9 fraction) were more effective in activating many types of carcinogen, including nitrosamines, to bacterial mutagens than rat liver enzymes (Bartsch et al., 1976; Raineri et al., 1981). This, tangentially, is another weakening of the link between mutagenesis and carcinogenesis, since many of these carcinogenic compounds are more

Table 6.1. *Mutagenicity of nitrosamines*

Compound	Activation, liver S$_9$ (R-Rat, H-Hamster, A-Aroclor, P-Phenobarbital)	Number of revertants per micromol	Carcinogenicity and principal target (NT = not tested)
Acyclic Nitrosamines			
N-nitroso-			
dimethylamine	RA	—	+ + + + Liver
	HA	27	+ + + + Liver
acetoxydimethylamine	—	2000	+ + + + Lung, Intestine
diethylamine	RA	—	+ + + + Esophagus, Liver
	HA	61	+ + + Liver
di-*n*-propylamine	RA	10	+ + + + Liver, Esophagus
	HA	1130	+ + + + Lung, Nasal
di-*iso*propylamine	RA	0	+ + Nasal, Liver
	HA	100	NT
di-allylamine	RA	22	—
	HA	450	+ + Nasal
di-*n*-butylamine	RA	21	+ + + Liver, Lung, Bladder
	HA	950	+ + + Lung, Bladder
di-*iso*butylamine	RA	95	+ + Nasal
	HA	44	+ + Nasal
di-*sec*-butylamine	RP	0	—
di-*n*-octylamine	RA	30	—
dicyclohexylamine	RA	0	—
diphenylamine	RA	0	+ Bladder
	HA	0	NT
dibenzylamine	RA	0	—
acetoxydibenzylamine	—	600	+ + Forestomach, Liver

Table 6.1 (cont.)

Compound	Activation, liver S_g (R-Rat, H-Hamster, A-Aroclor, P-Phenobarbital)	Number of revertants per micromol	Carcinogenicity and principal target (NT = not tested)
diethanolamine	RA	0	++ Liver
	HA	0	+ Nasal
iminodiacetonitrile	RP	0	—
iminodipropionitrile	RP	0	—
bis-(2-chloroethyl)amine	—	1050	+++ Forestomach
bis-(2-methoxyethyl)amine	RP	0	+++ Liver
bis-(2-ethoxyethyl)amine	RP	10	++ Liver
bis-(2,2-diethoxyethyl)amine	RP	0	—
bis-trifluorethylamine	RP	0	—
bis-(2-chloropropyl)amine	—	550	++ Forestomach
bis-(2-oxopropyl)amine	RA	0	++++ Lung, Liver
	HA	25	++++ Liver, Pancreas
bis-(2-hydroxypropyl)amine	RA	0	++ Esophagus, Liver
	HA	34	++ Liver, Pancreas
methylmethoxyamine	RA	13	NT
methylethylamine	RA	0	+++ Liver
	HA	0	+++ Liver
methyl-n-propylamine	RA	6	++++ Esophagus
	HA	100	+++ Liver, Nasal
methyl-isopropylamine	RP	0	NT
methyl-n-butylamine	RA	6	+++ Esophagus
	HA	140	+++ Liver, Nasal
methyl-n-pentylamine	RA	16	+++ Esophagus
	HA	780	+++ Lung, Liver
methyl-n-hexylamine	RA	50	+++ Esophagus, Liver
	HA	670	+++ Liver, Lung, Bladder

methyl-n-heptylamine	RA	300	+ + Liver, Lung
	HA	660	+ + Lung, Liver
methyl-n-octylamine	RA	1030	+ + Liver, Lung, Bladder
	HA	720	+ + Liver, Lung
methyl-n-nonylamine	RA	2490	+ + Liver, Lung
methyl-n-decylamine	RA	1920	+ + Lung, Bladder
methyl-n-undecylamine	RA	6200	+ + Liver, Lung
methyl-n-dodecylamine	RA	1230	+ + Bladder
	HA	430	+ + Bladder
methyl-n-tridecylamine	RA	870	NT
	HA	1910	NT
methyl-n-tetradecylamine	RA	0	+ + Bladder
	HA	770	NT
methyl-n-octadecylamine	RA	0	NT
	HA	0	NT
methyl-neopentylamine	RA	0	+ + + Esophagus
	HA	0	NT
methyl-trifluoroethylamine	RA	0	+ + + Esophagus
	HA	13	NT
methyl-hydroxyethylamine	RA	0	+ + + Liver, Esophagus
	HA	180	NT
methyl-2-hydroxypropylamine	RA	0	+ + + Esophagus, Nasal
	HA	1020	+ + Liver, Pancreas
methyl-3-hydroxypropylamine	RA	0	+ + Liver, Lung
	HA	85	NT
methyl-2,3-dihydroxypropylamine	RA	16	+ + + Esophagus
	HA	97	+ + + + Nasal
methyl-2-oxopropylamine	RA	0	+ + + Esophagus
	HA	18	+ + + + Liver, Nasal
methyl-3-carboxypropylamine	RA	0	+ + Bladder
	HA	0	NT

Table 6.1 (cont.)

Compound	Activation, liver S_9 (R-Rat, H-Hamster, A-Aroclor, P-Phenobarbital)	Number of revertants per micromol	Carcinogenicity and principal target (NT = not tested)
methylamino-3-pyridylbutanone (NNK)	RA	69	+++ Liver, Lung
	HA	2 200	+++ Trachea, Nasal
methylamino-3-pyridylbutanol (NNA1)	RA	0	+++ Lung
	HA	5 500	NT
methylcyclohexylamine	RA	0	++++ Esophagus
	HA	16	+ Lung, Liver
methylaniline	RA	0	+++ Esophagus
	HA	0	+ Spleen, Liver
methyl-4-fluoroaniline	RA	0	++ Esophagus
	HA	0	NT
methyl-4-nitroaniline	RA	63	+ Liver
	HA	42	NT
methyl-2-aminopyridine	RA	90	++ Esophagus
methyl-3-aminopyridine	RA	0	—
methyl-4-aminopyridine	RA	0	—
methyl-2,2-dimethylethylenediamine	RA	0	+++ Esophagus
	HA	39	NT
methylbenzylamine	RA	0	+++ Esophagus
	HA	39	NT
methyl-4-methylbenzylamine	RA	8	+++ Esophagus
	HA	21	NT
methyl-4-cyanobenzylamine	RA	49	NT
	HA	370	NT
methyl-4-nitrobenzylamine	RA	16	NT
	HA	120	NT

methyl-4-fluorobenzylamine	RA	10	NT
	HA	42	NT
methyl-4-chlorobenzylamine	RA	93	NT
	HA	940	NT
ethyl-n-butylamine	RP	52	+++ Esophagus
n-propyl-n-butylamine	RP	230	NT
methyl-2-phenylethylamine	RA	3600	++++ Esophagus
phenylbenzylamine	RA	0	+ Esophagus
	HA	0	NT
ethanolhydroxypropylamine	RA	0	++ Liver, Esophagus
	HA	0	+++ Nasal
ethyl-ethanolamine	RA	17	+++ Liver, Kidney
ethanoloxopropylamine	RA	180	++ Liver
	HA	300	++++ Liver, Pancreas
dihydroxypropylethanolamine	RA	0	++ Liver
	HA	160	+ Trachea
allylethanolamine	RA	290	++ Nasal, Liver
	HA	1170	NT
hydroxypropyloxopropylamine	RA	19	+++ Liver, Esophagus
	HA	1700	+++++ Liver, Pancreas
trihydroxydipropylamine	RA	41	+++ Esophagus
	HA	280	+++ Forestomach
dihydroxypropyl-2-oxopropylamine	RA	0	++++ Esophagus
	HA	170	++++ Forestomach, Pancreas
2,3-dihydroxypropylallylamine	RA	0	++++ Esophagus, Nasal
	HA	35	+++ Nasal
allylhydroxypropylamine	RA	230	++++ Liver, Nasal
	HA	5000	++ Nasal
allyl-2-oxopropylamine	RA	350	++ Liver
	HA	11000	+++ Nasal

Table 6.1 (cont.)

Compound	Activation, liver S_9 (R-Rat, H-Hamster, A-Aroclor, P-Phenobarbital)	Number of revertants per micromol	Carcinogenicity and principal target (NT = not tested)
n-butyl-4-hydroxybutylamine	RA	24	+++ Bladder
	HA	33	++ Bladder
n-butyl-3-carboxypropylamine	RA	0	++ Bladder
	HA	0	NT
trimethylurea	RP	140	+++ Brain
1-methyl-3,3-diethylurea	RA	0	+++ Brain
	HA	5 100	+++ Spleen
triethylurea	RP	22	+++ Mammary gland
1-ethyl-3,3-dimethylurea	RA	0	++ Brain, Mammary gland
	HA	10	NT
Cyclic Nitrosamines			
N-nitroso-			
azetidine	RA	11	++ Liver
pyrrolidine	RA	23	++ Liver
	HA	440	++ Lung
pyrroline	RA	0	++ Liver
	HA	17	NT
nornicotine	RA	11	++ Esophagus, Nasal
2,5-dimethylpyrrolidine	RP	0	—
3-hydroxypyrrolidine	RP	0	++ Liver
proline	RP	0	—
hydroxyproline	RP	0	+ Spleen
3,4-dibromopyrrolidine	RP	0	NT
3,4-dichloropyrrolidine	RP	210	+++ Esophagus
1,3-oxazolidine	RA	48	+++ Liver

Compound	Method	Value	Rating	Organ
2-methyl-1,3-oxazolidine	HA	85	+++	Liver
	RA	14	++	Liver
5-methyl-1,3-oxazolidine	HA	370	NT	
	RA	67	++	Liver
thiazolidine	HA	1900	+++	Liver
	RA	0	—	
	HA	0	NT	
piperidine	RA, RP	80, 18	+++	Esophagus
2-methylpiperidine	RA, RP	33, 29	+++	Liver, Esophagus
3-methylpiperidine	RA, RP	35, 48	+++	Esophagus
4-methylpiperidine	RA, RP	35, 85	+++	Esophagus
2,6-dimethylpiperidine	RA	0, 0	—	
2,2,6,6-tetramethylpiperidine	RA	0		
3,5-dimethylpiperidine	RA	43	+++	Esophagus
3,5-dimethylpiperidine – *cis*	RA	41	+++	Esophagus
3,5-dimethylpiperidine – *trans*	RA	77	+++	Esophagus
4-*tert*-butylpiperidine	RA	0	++	Esophagus
	HA	45	NT	
4-cyclohexylpiperidine	RA	0	—	
	HA	0		
4-phenylpiperidine	RA, RP	146	+++	Liver, Esophagus
3-hydroxypiperidine	RA, RP	39, 22	+++	Esophagus, Liver
4-hydroxypiperidine	RA, RP	17, 10	+++	Liver, Nasal
4-piperidone	RA, RP	10, 51	+++	Liver, Nasal
3-methyl-4-piperidone	RA	15	+++	Esophagus
3,4,-epoxypiperidine	—	74	+++	Liver
Δ²-tetrahydropyridine	—	182	+++	Esophagus
Δ³-tetrahydropyridine	RA	225	+++	Liver, Esophagus
Guvacoline	RA	31	—	
	HA	57	NT	
2-carboxypiperidine	RA	0	—	
3-carboxypiperidine	RA	0	—	

Table 6.1 (*cont.*)

Compound	Activation, liver S_9 (R-Rat, H-Hamster, A-Aroclor, P-Phenobarbital)	Number of revertants per micromol	Carcinogenicity and principal target (NT = not tested)
4-carboxypiperidine	RA	0	—
methylphenidate	RA	0	—
	HA	0	NT
3-chloropiperidine	RA, RP	120, 210	+ + + + Esophagus
4-chloropiperidine	RA, RP	41, 60	+ + + + Esophagus
3,4-dichloropiperidine	RA, RP	55, 30	+ + + + Esophagus
3,4-dibromopiperidine	RA, RP	140, 190	+ + + + Esophagus
morpholine	RA, RP	10, 38	+ + + + Liver
	HA	1160	+ + + + Nasal
2-methylmorpholine	RA	16	+ + + + Liver, Esophagus
	HA	7300	+ + + + Nasal, Liver
2,6-dimethylmorpholine	RA	17	+ + + + Esophagus
	HA	3900	+ + + + Liver, Pancreas
2,6-dimethylmorpholine – *cis*	RA	7	+ + + + Esophagus
	HA	7000	+ + + + Liver, Pancreas
2,6-dimethylmorpholine – *trans*	RA	12	+ + + + Esophagus
	HA	3300	+ + Liver, Pancreas
2-phenyl-3-methylmorpholine	RP	12	—
	HA	23	NT
2-hydroxymorpholine	—	30	
	RA	22	+ + Liver, Lung
	HA	87	NT
tetrahydro-1,3-oxazine	RA	10	+ + + Liver
	HA	350	NT

Compound			Organs
thiomorpholine	RA	0	+++ Esophagus
	HA	0	NT
dithiazine	RA	0	—
	HA	0	NT
thialdine	RA	0	+++ Esophagus, Liver
	HA	0	NT
piperazine	RP	0	—
N-methylpiperazine	RA	0	+ Nasal
	HA	2600	NT
3,5-dimethylpiperazine	RA	0	+++ Thymus
	HA	960	NT
3,4,5,-trimethylpiperazine	RA	0	+++ Thymus
	HA	1630	+++ Lung, Forestomach
4-acetyl-3,5-dimethylpiperazine	RA	43	+++ Esophagus
4-benzoyl-3,5-dimethylpiperazine	RA	0	+++ Liver, Forestomach
	HA	10	NT
hexamethyleneimine	RA	1020	+++ Liver, Esophagus
heptamethyleneimine	RA	650	++++ Esophagus, Lung
octamethyleneimine	RA	1250	+++ Esophagus, Lung
	HA	7300	NT
dodecamethyleneimine	RA	740	++ Liver
Dinitroso-			
piperazine	RA	29	++ Liver, Nasal
	HA	460	NT
2-methylpiperazine	RP	47	+++ Esophagus
2,5-dimethylpiperazine	RA	240	+++ Esophagus
2,6-dimethylpiperazine	RP	70	+++ Esophagus
2,3,5,6-tetramethylpiperazine	RP	0	—
homopiperazine	RP	270	+++ Esophagus

Table 6.2. *Rat carcinogens activated to mutagens by hamster liver S_9 but not by rat liver S_9*

	Target organ for tumour induction	
Compound	in rats	in hamsters
N-Nitroso-		
di-*iso*propylamine	Liver	NT*
bis-2-hydroxypropylamine	Liver, Lung	Liver, Pancreas
bis-2-oxopropylamine	Liver, Lung	Liver, pancreas
methyl-tetradecylamine	Bladder	NT
methyl-trifluoroethylamine	Esophagus	NT
methyl-hydroxyethylamine	Liver	NT
methyl-2-hydroxypropylamine	Esophagus	Liver, Pancreas
methyl-3-hydroxypropylamine	Liver	NT
methyl-2-oxopropylamine	Liver, Esophagus	Liver, Pancreas
methylcyclohexylamine	Esophagus	Lung, Liver
methyldimethylethylenediamine	Esophagus	Liver
methylbenzylamine	Esophagus	NT
dihydroxypropylethanolamine	Liver	Trachea
dihydroxypropyloxopropylamine	Esophagus	Pancreas
dihydroxypropylallylamine	Esophagus	Nasal cavity
3-pyrroline	Liver	NT
4-*tert*-butylpiperidine	Esophagus	NT
N-methylpiperazine	Nasal cavity	NT
3,5-dimethylpiperazine	Thymus	NT
3,4,5-trimethylpiperazine	Thymus	Lung
4-benzoyl-dimethylpiperazine	Liver	NT
methyldiethylurea	Nervous system	Spleen
ethyldimethylurea	Mammary gland	NT

* NT = Not tested.

potent, not less potent (Table 6.2), in rats than in hamsters (Lijinsky, 1987). An important group of nitrosamines which are more mutagenic with hamster liver S_9 activation than with rat liver S_9 activation are those with an oxygen function (hydroxyl or carbonyl) on the 2-carbon atom of the alkyl chain, for example nitrosobis-(2-oxopropyl)amine, nitrosobis-(2-hydroxypropyl)amine or methylnitroso-2-oxopropylamine. An explanation of this lack of activity with rat liver S_9 activation might lie in the resistance of the so-called β-oxygenated nitrosamines to oxidation by microsomal enzymes of rat liver, as demonstrated by Farrelly *et al.* (1984). They are metabolised by rat liver parenchymal cells, as must be assumed if they induce tumours in the liver of rats.

A variety of nitrosamines do not fit easily into the category of those that are activated to mutagens by rat liver microsomes and are metabolised by rat liver microsomes, or those that are not activated to mutagens by rat liver microsomes and are not metabolised by rat liver microsomes. These misfits include an important anomaly, methylnitrosoethylamine, which has been reported as non-mutagenic (Rao, 1984; Andrews and Lijinsky, 1984), but mutagenic when incubated at lower pH (Guttenplan, 1984), although it is a powerful carcinogen in both rats and hamsters, and its principal target is the liver in both species (Lijinsky et al., 1987b). Another compound, less important in this regard than methylnitrosoethylamine, is methylnitrosocyclohexylamine, which does not induce tumours in the liver of rats, although it is a very potent esophageal carcinogen in this species, and does induce a small incidence of liver tumours in hamsters (Lijinsky and Kovatch, 1988b). Methylnitrosocyclohexylamine has not been mutagenic with rat liver and was marginally so with hamster liver activation. Whether or not enzyme preparations from the esophagus would activate this nitrosamine to a bacterial mutagen is not known, although indications from such experiments with other nitrosamines carcinogenic for the esophagus of rats are uncertain (Scanlan et al., 1980).

An analogue of methylnitrosocyclohexylamine, methylnitrosoaniline (methylnitrosophenylamine) is also non-mutagenic to bacteria, with or without microsomal activation by enzymes from rat or hamster liver. This is to some extent understandable, since methylnitrosoaniline cannot give rise to a simple alkylating agent, because it possesses only one oxidisable carbon atom, that of the methyl group. When this is oxidised (which occurs readily, as shown by Kroeger-Koepke et al. (1981)), there remains a monophenylnitrosamine, which can give rise to phenyldiazonium ion, readily detectable by coupling with a phenol to form an azo dye. The phenyldiazonium ion is quite a different type of alkylating agent from, say, a methyldiazonium ion, and cannot readily form a phenyl adduct with a base in DNA. Indeed, phenylating agents are usually active in the frameshift strains of Salmonella typhimurium, as is, for example, phenylnitrosourea (Lijinsky and Andrews, 1979). The non-mutagenicity of methylnitrosoaniline is another example of the divergence between mutagenicity and carcinogenicity among this group of compounds, since methylnitrosoaniline is carcinogenic in rats and, like methylnitroso-cyclohexylamine, induces only tumours of the esophagus (Goodall et al., 1970). In hamsters both compounds are very weakly carcinogenic, inducing small incidences of tumours of the spleen, lungs or liver after a very long latent period (Lijinsky and Kovatch, 1988b). It is pertinent that phenylnitrosobenzylamine also induces tumours of the esophagus in rats, but is non-mutagenic, and this compound can also give rise through α-oxidation to a phenyldiazonium ion. These compounds form a consistent set in their carcinogenic effects in rats, and their lack of mutagenicity is also consistent.

A pair of closely related derivatives of methylnitrosoaniline that has been examined are the 4-fluoro- and 4-nitro- derivatives (Fig. 6.1). The latter is a much weaker carcinogen than the former, and induces only a small incidence of liver tumours in rats, whereas the fluoro-derivative is almost as potent as methylnitrosoaniline in inducing tumours of the esophagus (Kroeger-Koepke et al., 1983). In contrast, methylnitroso-4-fluoraniline is not mutagenic, whereas methylnitroso-4-nitroaniline is mutagenic to *Salmonella* without microsomal activation, probably through reduction of the nitro group by bacterial enzymes, as is common with aromatic nitro compounds. It is not likely that the mechanism of mutagenesis by this compound in bacteria tells us anything about the mechanisms of carcinogenesis among this group of aromatic nitrosamines.

Among the most important nitrosamines encountered in the environment is nitrosodiethanolamine, and its homologue nitrosobis-(2-hydroxypropyl)amine, which as mentioned previously, are found in synthetic cutting oils and in cosmetics, sometimes at quite high concentrations. Both compounds have in common with many other oxygenated alkyl nitrosamines a resistance to conversion by rat liver microsomal enzymes to products mutagenic to bacteria. Many of these compounds can be detected as mutagens using microsomal enzymes prepared from hamster liver. However, nitrosodiethanolamine is a marked exception, having failed to be mutagenic in any system, including *Drosophila*, although it did produce genomic changes in somatic cells

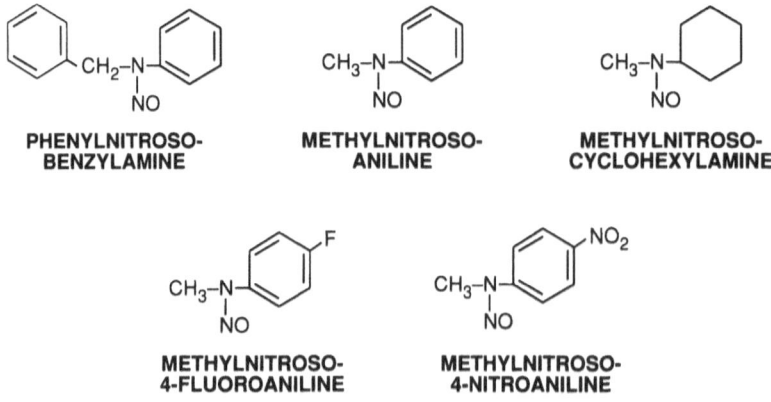

Fig. 6.1. Phenylnitrosobenzylamine, methylnitrosoaniline, methylnitroso-cyclohexylamine, methylnitroso-fluoroaniline and methylnitroso-4-nitroaniline.

(Fahmy and Fahmy, 1984). The liver of rats is a prime target of this nitrosamine, which does not induce liver tumours in hamsters, but tumours of the nasal mucosa. Tumours of the nasal mucosa, and of the esophagus and kidneys are also induced in rats by nitrosodiethanolamine, which cannot be considered a carcinogen with narrow range of activity as explanation of its failure to demonstrate mutagenic activity. Some other non-mutagenic carcinogens, such as the antihistamine methapyrilene, are less troublesome in this regard, because it (and others) induces only liver tumours and only in rats (Lijinsky *et al.*, 1980c).

An enormous amount of effort has been devoted to demonstrating – or attempting to – that the apparently non-mutagenic carcinogens are, in fact, weakly mutagenic. Some of these studies involve using abnormal physiological conditions, such as lowered pH (Guttenplan, 1980) and are not entirely convincing. It might be better to conclude that there are mechanisms of carcinogenesis which do not involve direct structural changes in cellular DNA, and therefore are not revealed by studying the mutagenic effects of the carcinogens in *in vitro* systems.

6.3 Structure of *N*-nitroso compounds related to mutagenesis

Of much greater interest are systematic studies of mutagenesis by *N*-nitroso compounds, and relating this to what we can observe of their chemical reactions within cells. Although the relationship between mutagenic activity and carcinogenic activity has been of considerable interest a unified and consistent pattern between the two biological actions of the *N*-nitroso compounds has not emerged. If some other compounds with very similar chemical reactivity, such as azoxyalkanes and dialkyltriazenes, are brought into the picture, the confusion increases. One of the long-standing problems is the often close similarity between the cyclic nitrosamines and their acyclic analogues both in mutagenicity and carcinogenicity, but the failure to detect chemical and biochemical reactions in common between the two types of nitrosamine. Many inferences have been made following studies of the effects of small changes in chemical structure (such as the effect of methyl groups in various positions which can block particular sites of oxidation) on both mutagenesis and carcinogenesis, which are often affected in parallel. However, in only one case have adducts of a cyclic nitrosamine or its metabolites to DNA *in vivo* been convincingly demonstrated (Chung *et al.*, 1989b). In contrast, the finding of simple DNA adducts of methyl- and ethyl-nitroso compounds has been a mainstay of the current hypotheses of mechanisms of chemical carcinogenesis. Some investigators have devoted much time to the study of metabolism of cyclic nitrosamines and they have speculated on the nature of possible adducts with DNA; however, except in the case of nitrosopyrrolidine such adducts have not been identified, (Hecht et al., 1981; Farrelly and Hecker, 1984). Earlier, rather crude studies of the problem by this author produced negative results (i.e. no identified adducts) which may not be definitive (Lijinsky, 1976a).

The lack of correlation is extended by the existence of carcinogens which alkylate DNA in animals in the organs in which they induce tumours, and yet are not mutagenic when activated by enzyme preparations, even by cells, of those organs. A ready example is methylnitrosoethylamine, which is a potent carcinogen for rat liver, yet is not activated to a bacterial mutagen by liver enzymes from this species, although methylation and ethylation of DNA in the liver (and other organs) is extensive (Lijinsky, 1988b). Therefore, the rat liver cells are able to activate methylnitrosoethylamine to alkylating agents *in vivo*, but this activation does not take place, or is different, *in vitro*. This defies logic, if we accept the premise that formation of agents which alkylate DNA is the mechanism by which these carcinogens operate.

Similarly, azoxymethane and other azoxyalkanes are potent carcinogens, similar in action and potency with the nitrosamines with which they are isomeric. They are equally effective as the nitrosamines as alkylating agents for DNA in liver and other organs, yet are not activated to bacterial mutagens by rat or hamster liver enzyme preparations (Lijinsky et al., 1985a). It appears that the reason for the failure of rat or hamster liver microsomal enzymes to activate azoxyalkanes to bacterial mutagens, in spite of the fact that they induce liver tumours when given to rats in drinking water and liver tumours in hamsters by gavage, is because azoxyalkanes are not activated by microsomal enzymes. Instead, they must be metabolised by other enzymes in rat liver which are responsible for formation of the alkylating agents, which alkylate DNA. Whether or not this alkylation is involved in whole or in part in the induction of tumours by azoxyalkanes is a separate question, which has not yet been answered.

Examination of the relation between chemical structure of N-nitroso compounds and their mutagenic activity has shown remarkable consistencies, if the relationship to carcinogenesis is omitted. It is very likely that mutagenesis by nitrosamines comes about through oxidation at the carbon α- to the N-nitroso function followed by reactions that lead to products interacting with DNA. Among these might be the alpha-hydroxyalkylnitrosamine, a further oxidation product, the corresponding alkylnitrosamide or an alkyldiazonium ion derived from them. The alpha-hydroxyalkylnitrosamines are unstable, but are directly acting mutagens

Table 6.3. *Mutagenicity of directly acting N-nitroso compounds in Salmonella typhimurium (strain TA 1535) – plate assay*

Compound	No. of revertants per micromol	Carcinogenicity and principal target in rats
(A) Nitrosoalkylamides		
N-nitroso-		
methylformamide	55 000	+ + + + Forestomach
methylacetamide	160 000	+ + + Forestomach
ethylformamide	8 000	NT
(B) Nitrosoureas		
1-nitroso-		
methylurea	4 000	+ + + + Nervous system
ethylurea	100	+ + + + Mammary gland, Lung
n-propylurea	200	+ + + Mammary gland, Thymus
*iso*propylurea	90	NT
allylurea	560	+ + + Mammary gland, Lung
n-butylurea	140	+ + + Mammary gland, Lung
iso-butylurea	730	+ + Duodenum
sec-butylurea	390	NT
n-amylurea	530	+ + + Lung, Uterus
n-hexylurea	1 200	+ + + Lung, Uterus
undecylurea	320	NT
tridecylurea	5	NT
phenylurea	660	+ + Sarcoma at injection site
cyclohexylurea	52	NT
benzylurea	11	+ + Forestomach
phenylethylurea	4 600	+ + + Liver
hydroxyethylurea	2 400	+ + + Lung, Colon
methoxyethylurea	2 100	+ + + Mammary, Lung

Table 6.3 (cont.)

Compound	No. of revertants per micromol	Carcinogenicity and principal target in rats
2-hydroxypropylurea	3100	+ + + + Thymus
3-hydroxypropylurea	220	+ + Stomach, Liver
2-oxopropylurea	2200	+ + + Kidney
fluorethylurea	2700	+ + + Stomach
dimethylaminoethylurea	40000	NT
1,3-dimethylurea	73	+ + Nervous system
1-methyl-3-ethylurea	86	+ + + Nervous system
1,3-diethylurea	34	+ + + + Mammary gland, Lung
1-ethyl-3-methylurea	34	+ + + + Mammary gland, Lung
1-ethyl-3-hydroxyethylurea	180	+ + + Lung, Nervous system
1-ethyl-3-oxopropylurea	66	+ + + Lung, Mammary gland
1-hydroxyethyl-3-ethylurea	2900	+ + + Liver, Lung
1-hydroxyethyl-3-chloroethylurea	1900	+ + Liver, Kidney
1-chloroethyl-3-hydroxyethylurea	330	NT
1-chloroethyl-3-hydroxypropylurea	500	+ + Lung
1,3-chloroethylurea (BCNU)	240	+ + Lung
1-hydroxypropyl-3-chloroethylurea	3800	+ + Liver, Lung
1-oxopropyl-3-chloroethylurea	250	+ + + Liver, Nervous system
1-oxopropyl-3-ethylurea	21	NT
1-dimethylaminoethyl-3-ethylurea	2000	+ + Mammary gland, Tongue
(C) Nitrosocarbamates N-nitroso-		
methylurethane	4500	+ + + + Forestomach
ethylurethane	1800	+ + + + Forestomach
methylphenylcarbamate	63000	+ + + Forestomach

carbaryl	33 000	+++	Forestomach
methomyl	1 000 000	+++	Forestomach
aldicarb	27 000	+++	Forestomach
bux-ten	116 000	+++	Forestomach
landrin	17 000	+++	Forestomach
carbofuran	9 500	+++	Forestomach
baygon	88 000	+++	Forestomach
oxazolidone	9 300	++	Forestomach
5-methyloxazolidone	5 200	++	Forestomach
tetrahydrooxazone	5 300	NT	
(D) Nitrosoguanidines			
N-nitroso-methylnitroguanidine	12 000	+++	Glandular stomach
ethylnitroguanidine	5 800	+++	Glandular stomach
cimetidine	3 600	–	

*NT = Not tested.

when tested as the esters (e.g., acetoxymethylnitrosomethylamine). The more stable alkylnitrosamides, such as methylnitrosoformamide (Huisgen and Reimlinger, 1956) from nitrosodimethylamine, and ethylnitrosoformamide or methylnitrosoacetamide from methylnitrosoethylamine, are very potent bacterial mutagens (Table 6.3) and might well typify the proximate mutagens formed metabolically from nitrosamines by mammalian enzymes. Nitrosolactams are the equivalent derivatives from cyclic nitrosamines, but have been little studied. Nitrosamines of similar structure (e.g. methylnitrosoalkylamines) probably exert their mutagenic effects by a similar process, provided that they are susceptible to metabolism by the same enzymes. Much of the information has been obtained with the *Salmonella typhimurium* system developed by Ames and his collaborators. Information from other systems, including mammalian cells *in vitro*, and a variety of other bacterial systems, support the conclusions from the analysis of the *Salmonella* data.

6.4 Mutagenesis by directly acting *N*-nitroso compounds

Let us first consider the series of simple directly acting *N*-nitroso compounds, including alkylnitrosoureas, alkylnitrosocarbamates and alkylnitrosonitroguanidines (Table 6.3) (Fig. 6.2). The last are something of an anomaly, since they bear little relation to any naturally occurring compound or any common commercial product. The nitro substituent is essential for stability of these compounds, and other stabilising substituents would also serve. Attempts have been made to prepare and isolate alkylnitrosoguanidines, but all have failed. It is, of course, possible to

Fig. 6.2. Methylnitrosonitroguanidine, methylnitroso-guanidine, methylnitrosocarbamate ester and methylnitrosourea.

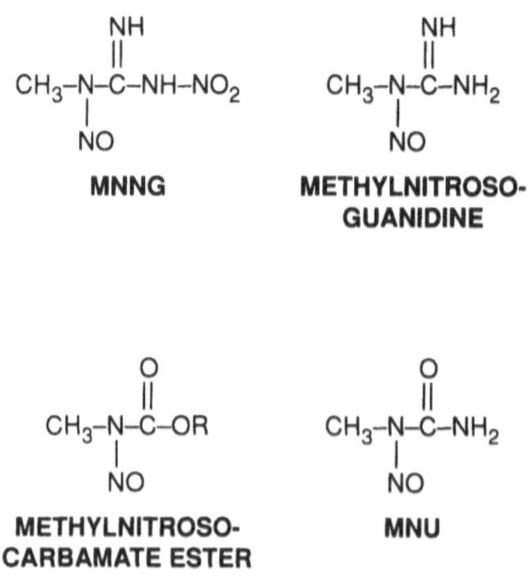

$$CH_3-\underset{\underset{NO}{|}}{N}-\overset{\overset{NH}{||}}{C}-NH-NO_2$$
MNNG

$$CH_3-\underset{\underset{NO}{|}}{N}-\overset{\overset{NH}{||}}{C}-NH_2$$
METHYLNITROSO-GUANIDINE

$$CH_3-\underset{\underset{NO}{|}}{N}-\overset{\overset{O}{||}}{C}-OR$$
METHYLNITROSO-CARBAMATE ESTER

$$CH_3-\underset{\underset{NO}{|}}{N}-\overset{\overset{O}{||}}{C}-NH_2$$
MNU

nitrosate N-alkylguanidines, but the products appear to be unstable. In unpublished studies, this author and R.K. Elespuru nitrosated N-methylguanidine in aqueous solution, neutralised it with sodium bicarbonate and detected strong mutagenic activity in the solution, comparable with that of an equimolar solution of methylnitroso-N'-nitroguanidine. Although this result suggested that almost quantitative conversion of methylguanidine into its N-nitroso derivative had occurred, no means of extracting the nitroso derivative (a strong base) could be found. Furthermore, evaporation of the aqueous solution to dryness led to destruction of mutagenic activity, so that extraction of the residue with a solvent in which the inorganic materials were insoluble was also not effective. It is instructive that feeding to rats of methylguanidine together with sodium nitrite led to no increase in tumour incidence (Lijinsky and Taylor, 1977c; Sander and Schweinsberg, 1972), suggesting that insufficient methylnitrosoguanidine was formed *in vivo* to give rise to tumours, or that the nitroso derivative was not carcinogenic under these conditions. Similarly, feeding the natural guanidine derivative arginine together with nitrite to rats also failed to result in increased tumour incidence (Lijinsky and Taylor, 1977c). It is not known whether nitrosation of arginine results in a mutagenic product.

The compound first mentioned in this Chapter, MNNG, serves well as a surrogate for mutagenic alkylnitrosoguanidines. Ethylnitrosonitroguanidine and some higher homologues are also mutagenic, although less potent than the methyl compound, in *Salmonella* or other bacterial systems. Alkylnitrosocarbamate esters, such as methylnitrosourethane, are also potent mutagens, often more effective than the corresponding nitrosonitroguanidines. Among homologues, ethylnitrosourethane is somewhat less potent than the methyl analogue, although both are more potent mutagens to *Salmonella* than the corresponding alkylnitrosonitroguanidines.

Many N-methyl-N-carbamic acid esters are heavily used pesticides, notably insecticides such as carbaryl (1-naphthyl ester); these readily form N-nitroso derivatives (Fig. 3.7). They are widely used because they are cholinesterase inhibitors very toxic to insects, but of relatively low toxicity to mammals (unlike some other insecticides such as organophosphorus compounds, which are highly toxic to mammals). Because of the likelihood of human exposure to these alkylcarbamate esters, which can be nitrosated in the environment or *in vivo*, their N-nitroso derivatives have been extensively investigated. They are particularly potent bacterial mutagens, differing considerably in potency from one to another, although the 'business end' giving rise to the alkylating agent presumably responsible for this mutagenic activity is the same.

It is possible that the reasons for the differences between them in activity, including differences from the parent compound methylnitrosourethane, could be due to differences in uptake by the bacteria (possibly related to liposolubility). Indeed, Elespuru (1976) showed that differences in uptake at least partially explained the differences in mutagenic activity in *Hemophilus influenzae* between methylnitrosourea, MNNG and nitrosocarbaryl, all of which are assumed to give rise spontaneously to the same alkylating agent, the methyldiazonium ion. It is reasonable to assume that the same alkyldiazonium ion, whatever its source, would produce very similar patterns of mutagenic alkylations in the bacterial DNA.

The nitrosated alkylcarbamates differ greatly in mutagenic activity, perhaps due in part to differences in stability. Nitrosomethomyl must certainly qualify as one of the most potent bacterial mutagens. It is interesting, therefore, that the great differences in mutagenic potency among this group of compounds are not matched by differences in carcinogenic potency, which all lie within a small range in rats (Lijinsky and Schmähl, 1978). Considering the wide variation in mutagenic potency among alkylnitrosoureas, from low to quite high, it is notable that all of the alkylnitrosocarbamates are very potent. This suggests that even in bacteria factors other than formation of a simple methylating or other alkylating agent are instrumental in determining mutagenic potency. However, in the case of the cyclic alkylnitrosocarbamate, nitrosooxazolidone, its mutagenic activity is very similar to that of the analogous alkylnitrosourea, 2-hydroxyethylnitrosourea, which can give rise to the same alkylating group. This is also true of its homologue, nitroso-5-methyloxazolidone, a potent bacterial mutagen comparable with 2-hydroxypropylnitrosourea, which can form the same 2-hydroxypropyldiazonium ion (Fig. 4.21). On the other hand, 2-hydroxypropylnitrosourea and nitroso-5-methyloxazolidone have vastly different carcinogenic effects in rats when administered orally, as do hydroxyethylnitrosourea and nitrosooxazolidone (Lijinsky and Reuber, 1983*b*).

The differences among these compounds in mutagenic potency might be partly due to differences in uptake into the target cells, but it is not plausible that such differences among compounds of very similar physical chemical properties (i.e. liposolubility, partition coefficient, and solubility) would be significant. Rather it appears that differences in chemical reactivity among them is a better explanation. That, in turn, requires large differences among them in rates of decomposition into the alkyldiazonium ion (or other alkylating agent), or a direct interaction with the target macromolecule, presumably DNA. It remains enigmatic that nitrosomethomyl is so much more mutagenic than methylnitrosourethane, while many of the other bulkier and less water-soluble alkylnitrosocarbamates, such as methylnitrosophenylcarbamate, nitrosolandrin and nitrosocarbofuran are intermediate in mutagenic potency.

Another enigma is the difference in mutagenicity of two orders of magnitude between methylnitrosourea and ethylnitrosourea (not paralleled by differences in carcinogenic potencies between them), in comparison with the very similar mutagenic (and carcinogenic) potency between methylnitrosourethane and ethylnitrosourethane, MNNG and ENNG. These contradictions are difficult to reconcile with the methyl compounds acting solely through formation of a methyldiazonium ion, and the ethyl compounds acting through formation of an ethyldiazonium ion. The conclusion is inescapable that the facile explanation of the biological actions of these simple alkylating agents is incorrect, that is that they all form by an S_n1 mechanism an alkyldiazonium ion (or similar identical product) which then reacts with a target macromolecule, thereby producing a mutation or initiating neoplastic transformation. On the other hand, in carcinogenesis there could be direct interaction of the administered alkylnitrosamide with some receptor site, in DNA or elsewhere; the affinity of this receptor for the alkylating agent could depend greatly on the structure of the latter. The three types of alkylnitrosamide differ greatly in their target organs in animals. It is conceivable that, the alkylnitrosamide having bound to the receptor, there could be an intramolecular rearrangement with elimination, leaving the alkyl group on the DNA or other macromolecule, which could, if appropriate, produce profound changes in its function. Few experiments have been designed to investigate this possibility (Buckley, 1987).

6.5 Mutagenic nitrosamines requiring metabolic activation

Unlike alkylnitrosoureas and other alkylnitrosoamides, there are few nitrosamines which are mutagenic without requiring metabolic activation. Among them are some unsaturated nitrosamines, such as allylnitrosamines and nitrosodehydropiperidines, that are presumably converted by bacterial enzymes to epoxides, which are directly acting mutagens on the bacterial DNA. Similarly, methylnitroso-4-nitroaniline is mutagenic to bacteria without metabolic activation, presumably through typical bacterial reduction of an aromatic nitro group to a hydroxylamine which forms adducts directly with the bacterial DNA, causing mutations; unsubstituted methylnitrosoaniline is not mutagenic even with metabolic activation.

In Table 6.1 are listed the numbers of mutant (revertant) colonies produced per micromol of nitrosamine, calculated from dose–response measurements. The standard conditions for the Ames assay using *Salmonella* as prescribed by McCann *et al.* (1975) for the plate test were employed, with a 9000 × g supernatant prepared from male Sprague-Dawley rat liver or male Syrian golden hamster liver. The amount of compound on a plate usually was one to 1000 μg; larger quantities (five or 10 mg) of refractory compounds were used.

There is considerable chemical similarity between cyclic nitrosamines and acyclic nitrosamines and similarity in their mutagenic capacities, but important differences in the chemistry of the major products of metabolic activation of the two types in animals, although their carcinogenic action is often similar. It is easy to understand that α-oxidation of acyclic nitrosamines would give rise to a carbonyl compound and to an unstable monoalkylnitrosamine; the same would be formed by nitrosation of a primary amine, and would, in turn, form an alkyldiazonium ion, which is an alkylating agent and mutagen. Most cyclic nitrosamines have a different fate when metabolised (see Chapter 4). In these cases α-oxidation again gives rise to a carbonyl compound following ring opening. The unstable nitrosaminoaldehyde resulting could form a diazoaldehyde, which could alkylate macromolecules. Although Chung and Hecht (1985) and others have produced alkylated nucleotide products through synthetic reactions, there has been no success in isolating such products from the DNA of animals treated with cyclic nitrosamines, other than nitrosopyrrolidine (Chung *et al.*, 1989*a*). The question of the role of such adducts present at very low levels in carcinogenesis or mutagenesis by cyclic nitrosamines remains, therefore, moot.

Leaving aside the mechanism of mutagenesis by cyclic nitrosamines, the action of acyclic nitrosamines seems easier to understand. There is a pattern of mutagenesis that relates to some extent chemical structure with mutagenic activity. The need for metabolic activation of nitrosamines means that there are at least two steps in bacterial mutagenesis by these compounds, one activation and the other interaction of the product with DNA, either of which can determine mutagenic potency. Considering only activation by crude S_9 fraction from rat or hamster liver, there are variations depending on whether enzymes have been induced by phenobarbital, Aroclor, or other inducers, or whether enzymes from uninduced animals are used. Because different enzymes are involved in activation of different nitrosamines, comparison of the mutagenicity of one nitrosamine with another is less easy to assess, because the activated mutagenic products are not always known. Other obstacles in the way of straightforward comparisons of mutagenesis among nitrosamines are the effects of pH of the medium in which incubation occurs (Guttenplan, 1980), and whether liquid preincubation or mutagenesis on the agar plate is used.

6.5.1 Methylnitrosoalkylamines

The three simplest nitrosamines, NDMA, MNEA and NDEA are not mutagenic in the standard plate assay with rat liver S_9, although they are very potent rat liver carcinogens. Two of them, NDMA and NDEA, but not MNEA, are mutagenic to *Salmonella* TA 1535 following liquid preincubation, although they are not very potent even with this modification (Rao *et al*., 1979). Why MNEA, which possesses characteristics of both NDMA and NDEA is not mutagenic under these conditions is an enigma. It is not because MNEA is not metabolised by microsomal enzymes; Farrelly *et al*. (1982) and others (Lai *et al*., 1980) have shown that both formaldehyde and acetaldehyde are formed by microsomal metabolism of MNEA *in vitro*, as would be predicted. It is logical to assume that, concomitantly, the appropriate alkyldiazonium ion, ethyl- or methyl-, would be formed and would alkylate DNA (in bacteria, if they were present), as they do when arising from microsomal metabolism of NDMA or NDEA. If this is not so – and it appears not to be – then our proposed singular mechanism of mutagenesis through alkylation of bacterial DNA by an alkyldiazonium ion must be re-evaluated.

Not only is methylnitrosoethylamine an outlier, because of its lack of mutagenicity in the standard microbial assay, but the powerful carcinogenic homologues, methylnitroso-*n*-propylamine, methylnitroso-*n*-butylamine and methylnitroso-*n*-pentylamine, are very weakly or nonmutagenic with rat liver microsomal activation. Of course, they do not cause liver cancer in rats, although they do in hamsters, and they are mutagenic with hamster liver microsomal activation. It could be claimed then, that there is a parallel between liver enzyme-induced mutagenesis and liver carcinogenesis induced by a nitrosamine. However, not only would that cast doubt on the usefulness of the *Salmonella* mutagenesis assay, which uses (usually successfully) rat liver enzymes to demonstrate the mutagenicity of many carcinogens which do not induce tumours in rat liver, but would negate the findings that many nitrosamines which are not rat liver carcinogens give positive results in the bacterial mutagenesis assay. Among them are nitrosopiperidine, nitrosoheptamethyleneimine, dinitrosohomopiperazine and methylnitroso-2-phenylethylamine. These noncongruent findings suggest that activating pathways which convert nitrosamines to bacterial mutagens are not necessarily the same as those converting them to carcinogens. A further complication is that liver from male rats is usually assumed to yield a microsomal fraction more effective in activating carcinogens to mutagenic derivatives than that from female rats, although some nitrosamines induce liver tumours in female rats but not in male rats. This suggests that the capacity of the liver enzymes to convert the nitrosamine to mutagens *in vitro* is not directly related to its activation *in vivo* to a carcinogenic product.

As shown in Table 6.1, among methylnitrosoalkylamines there is a pattern of increasing mutagenicity (number of revertants per micromol of nitrosamine) with increasing molecular size or length of the alkyl chain, if the simplest compound, nitrosodimethylamine is excluded; it is something of an anomaly. The mutagenicity of the nitrosamines with the longer alkyl chains is high, but unexpectedly drops to virtually zero with methylnitroso-*n*-octadecylamine is also non-mutagenic, but this nitrosamine has not been tested for carcinogenic activity. It is probable that the confirmation that there is a size limit for mutagenesis in this system, methylnitroso-*n*-octadecylamine is also non-mutagenic, but this nitrosamine has not been tested for carcinogenic activity. It is probable that the conversion of these asymmetric nitrosamines to methylating agents is the mechanism by which they become mutagenic, since it is difficult to believe that the large alkyldiazonium ion formed by oxidation of the methyl group would be the mutagenic agent. Among the alkylnitrosoureas, those with longer alkyl groups are usually less mutagenic – sometimes much less mutagenic – than those with short alkyl groups. Among the methylnitroso-*n*-alkylamines, however, mutagenic activity increases in the opposite direction from short to long. Increasing uptake into cells, due to great liposolubility, cannot be responsible for both. Therefore, it seems likely that the methyl group is the common alkylating agent responsible for the mutagenic activity of this group of compounds or that current notions of the mechanism of mutagenesis by these compounds are incorrect. The differences in potency could be due to differences in an alternative mechanism of activation of the nitrosamine, to a form in which the whole molecule is the DNA-interacting mutagenic moiety. This could also apply to the cyclic nitrosamines, which must act by some such mechanism, and are equally mutagenic with their acyclic counterparts. Such an explanation would provide unity for the types of compound that are so similar chemically, and in mutagenic and carcinogenic activity.

6.5.2 *Alkylnitrosamines with oxygen substituents*

In considering the large group of nitrosamines which have substituents in the alkyl groups, there is much less uniformity in the relation of mutagenic activity to chemical structure than among the unsubstituted compounds. Those asymmetric nitrosamines containing a

β-oxidised alkyl group, such as methylnitroso-2-oxopropylamine (MNOPA) and methylnitroso-2-hydroxypropylamine are usually not mutagenic when activated by rat liver microsomes, but are by hamster liver microsomes. The results of measuring the mutagenic properties of the methylnitrosoalkylamines with substituents in the alkyl chain are shown in Table 6.4. With rat liver enzyme activation there is no striking pattern relating chemical structure with mutagenic activity. Methylnitroso-2-hydroxypropylamine and methylnitroso-2,3-dihydroxypropylamine (Fig. 6.3) were weakly mutagenic with rat liver microsome activation, and the ketone, MNOPA was not mutagenic, whereas with hamster liver microsome activation the last was minimally mutagenic but the two alcohols were quite potent. This is inverse to the order of their carcinogenic activity, in both species. Methylnitroso-2-hydroxyethylamine and methylnitroso-3-hydroxypropylamine were non-mutagenic with rat liver microsome activation, but mutagenic with hamster liver microsome activation, while methylnitroso-3-carboxypropylamine was not mutagenic when activated by rat or hamster liver enzymes.

Presumably, with these and other methylnitrosamines, oxidation of the methyl group did not lead to formation of a mutagenic intermediate, and oxidation of the α-methylene of the other alkyl group was precluded, so that formation of a methyldiazonium ion was minimal or zero. Thus methylnitrosoneopentylamine, methylnitrosotrifluorethylamine, methylnitrosocyclohexylamine, methylnitrosobenzylamine (Fig. 6.4) and methylnitroso*iso*propylamine were non-mutagenic. However, methylnitroso-2-phenylethylamine was a potent mutagen under the same conditions, although it was not more potent a carcinogen in rats than the others, and, like them, did not induce tumours of the liver, but only tumours of the esophagus or nasal mucosa. It appears that the ethylene group was essential for activation to a mutagen, again suggesting that the character of the whole nitrosamine molecule was the key to activation to a mutagenic form.

To continue further in this vein, since the methylnitroso function did not seem vital, we can consider a group of nitrosamines containing a variety of alkyl groups, many with oxygen substituents, others with fluorine, and some unsaturated (which causes large changes in structurally related mutagenesis).

6.5.3 Symmetrical nitrosamines

Nitrosodi-*iso*propylamine is not mutagenic with rat liver microsome activation, presumably because access of the activating enzyme to the α-carbons is sterically hindered; it is, however, a weak, but definite, carcinogen in rats and is a mutagen with hamster liver microsome

Table 6.4. *Mutagenicity of methylnitrosoalkylamines with substituents*

Compound	Activation (R Rat H Hamster)	Mutagenicity (revertants/μmol)	Carcinogenesis target organ
Methylnitroso-ethylamine	R	0	Liver, Lung
	H	0	Liver
hydroxyethylamine	R	0	Liver, Esophagus
	H	180	ND
trifluorethylamine	R	0	Esophagus
	H	13	ND
3-hydroxypropylamine	R	0	Liver, Lung
	H	85	ND
2-hydroxypropylamine	R	0	Esophagus, Nasal
	H	1020	Liver, Pancreas
2,3-dihydroxypropylamine	R	16	Esophagus
	H	97	Nasal
2-oxopropylamine	R	0	Esophagus
	H	18	Liver, Nasal
3-carboxypropylamine	R	0	Bladder
	H	0	ND
cyclohexylamine	R	0	Esophagus
	H	16	Lung, Liver
phenylamine	R	0	Esophagus
	H	0	Liver, Spleen

4-fluorophenylamine	R	0	Esophagus
	H	0	ND
4-nitrophenylamine	R	63	Liver
	H	42	ND
benzylamine	R	0	Esophagus
	H	39	ND
2-phenylethylamine	R	3600	Esophagus
2-aminopyridine	R	90	Esophagus
3-aminopyridine	R	0	
4-aminopyridine	R	0	—
amino-3-pyridylbutanone	R	69	Liver, Lung
	H	2200	Trachea, Nasal, Lung
amino-3-pyridylbutanol	R	0	Lung, Pancreas
	H	5500	ND
N',N'-dimethylethylenediamine	R	0	Esophagus
	H	39	Liver

ND = Not done.

activation. For the same reason nitrosodi-*sec*-butylamine is not mutagenic, but is not carcinogenic in rats. Nitrosodi-*iso*butylamine is as mutagenic as nitrosodi-*n*-butylamine, although the former is a weaker carcinogen and induces only tumours of the nasal mucosa in rats. Although the carcinogenicity of nitrosodi-*n*-pentylamine and nitrosodi-*n*-hexylamine have been examined, there is no report of the mutagenicity of any of the many isomers of these two nitrosamines. Nitrosodi-*n*-heptylamine has not been tested for mutagenicity or carcinogenicity, but nitrosodi-*n*-octylamine has been tested for both and is mutagenic, but not carcinogenic, to rats. It is possible that higher homologues than nitrosodioctylamine would be mutagenic, although virtually certain that they would not be carcinogenic, because of their molecular size. Nitrosodiphenylamine, nitrosodicyclohexylamine and nitrosodibenzylamine (Fig. 6.5) are not mutagenic, the first because it cannot be activated by α-oxidation, the latter two because they might be too large to have access to the activating enzymes. This possibility is fortified by the finding

Fig. 6.3. Methylnitroso-3-hydroxypropylamine, methylnitroso-2-hydroxypropylamine, methylnitroso-2-oxopropylamine, methylnitroso-2,3-dihydroxypropylamine, methylnitrosohydroxyethylamine, and methylnitroso-3-carboxypropylamine. These compounds are shown in the order stated.

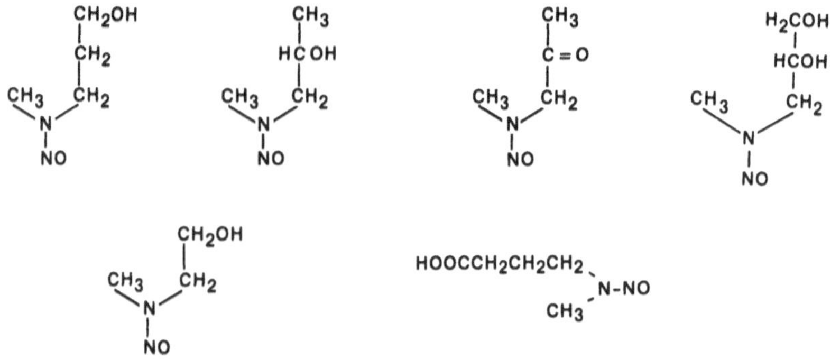

Fig. 6.4. Methylnitroso-phenylamine, -benzylamine, -phenylethylamine, -neopentylamine, and -trifluoroethylamine.

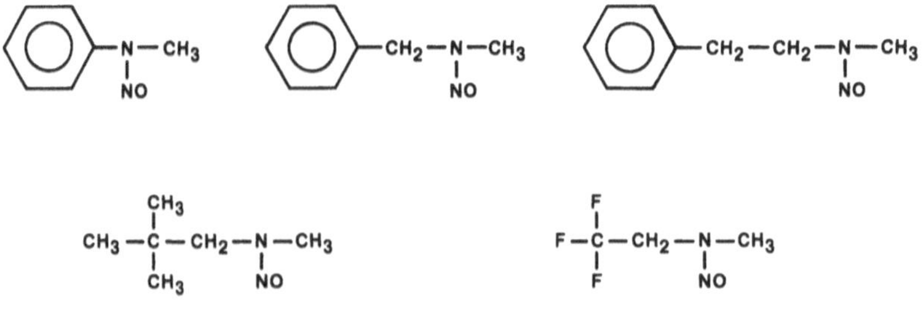

that α-acetoxydibenzylnitrosamine is quite mutagenic, and is the ester of the preactivated form of dibenzylnitrosamine (Lyle *et al.*, 1983). Molecular size, therefore, appears to be an important limiting factor in activation of symmetrical dialkylnitrosamines to mutagens. In this context it is interesting that the largest cyclic nitrosamine examined, nitrosododecamethyleneimine, is less mutagenic than a smaller homologue, nitrosooctamethyleneimine; the maximal size for a mutagenic cyclic nitrosamine is probably not much larger than the C-12 compound.

Among symmetrical dialkylnitrosamines containing substituents on the alkyl groups, many are not mutagenic with rat liver microsomal activation (although they are carcinogenic in rats), but often are mutagenic with hamster liver microsomal activation. Among them are nitrosobis-(2-hydroxypropyl)amine and nitrosobis-(2-oxopropyl)amine. Neither is a very potent mutagen, and there is little difference in potency between them, although the ketone is a much more potent carcinogen than the alcohol, in both rats and hamsters; both of these nitrosamines induce liver tumours (among others) in both species, and thus present the same dilemma discussed elsewhere in this Chapter. The lower homologue of **NBHPA**, nitrosodiethanolamine, is not mutagenic to bacteria with activation by either rat or hamster liver microsomes; it induces liver tumours in rats, but not in hamsters. Farrelly *et al.* (1984) have shown that nitrosodiethanolamine is not metabolised by liver microsomes, which is certainly one reason for the lack of mutagenicity. On the other hand, nitrosodiethanolamine is oxidised in rat liver cells *in vivo*, since it produces hydroxyethylation of liver nucleic acids, albeit to a very small extent. Nevertheless, the absence of mutagenic activity from this nitrosamine is

Fig. 6.5. Nitrosodiphenylamine, nitrosodicyclohexylamine, nitrosodibenzylamine, nitroso-α-acetoxydibenzylamine.

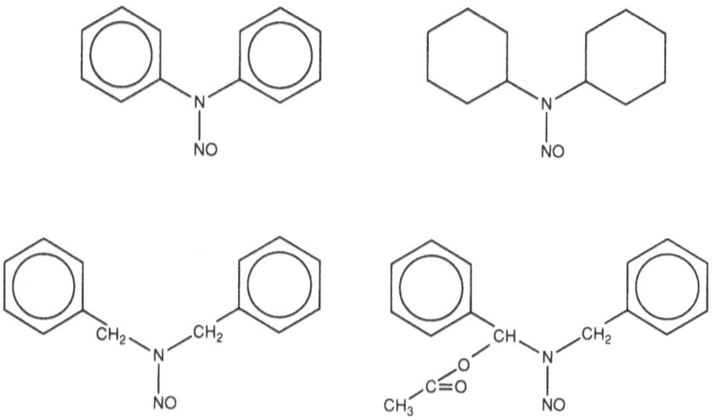

perplexing, since the liver of the rat is its principal target organ. Nitrosodiallylamine is much more mutagenic in the presence of hamster liver microsomes than in the presence of rat liver microsomes, like so many nitrosamines, and it is carcinogenic to the nasal mucosa of hamster, not in liver, and it was reported non-carcinogenic to rats (Druckrey et al., 1967); a more recent report (Althoff et al., 1977a) suggests that it is a very weak carcinogen in rats. Nitrosodiallylamine is readily oxidised by rat liver enzymes to form acrolein, together with other products (Farrelly et al., 1984), among which some would be expected to alkylate rat liver DNA. These products of activation are assumed to be responsible for the mutagenesis of this nitrosamine in bacteria.

Several derivatives of nitrosodiethanolamine have been examined, including the methyl and ethyl ethers (Fig. 6.6), both of which are potent liver carcinogens in rats, the ethyl ether (nitrosobis-(2-ethoxyethyl)amine) being somewhat less effective than the methyl ether. However, as mutagens in the presence of rat liver microsomes, the methyl ether resembles nitrosodiethanolamine, and is not mutagenic, while the ethyl ether is similar in potency to nitrosodiethylamine, which it to some extent resembles. It is difficult to believe that the methyl ether is not oxidised by rat liver microsomes, while the ethyl ether is metabolised, although this has not been investigated. If α-oxidation occurred, the methyl ether would give rise to a methoxyethyldiazonium ion, which would be a powerful mutagen, to judge by the potent mutagenicity to *Salmonella* of methoxyethylnitrosourea (Lijinsky et al., 1987a). These inconsistencies regarding the mutagenicity and carcinogenicity of nitrosamines suggest that the basic premises on which the correlations rest are not correct.

Another derivative of nitrosodiethanolamine, nitrosobis-(2,2-diethoxy-ethyl)amine is not carcinogenic and this compound, too, is not mutagenic in the presence of rat liver microsomes. This nitrosamine can be considered the diethylacetal of the compound derived from oxidation of nitrosodiethanolamine, nitrosobis-(2-oxoethyl)amine, the dialdehyde of nitrosodiethanolamine. This aldehyde and the corresponding hemiacetal are likely products of oxidation of nitrosodiethanolamine by alcohol dehydrogenase. There is no doubt that the diethylacetal would be easily hydrolysed to the aldehyde, which must therefore be considered an inactive nitrosamine. It is quite probable that the aldehyde function prevents oxidation of the α-methylene of the nitrosamine. In the same vein, nitrosiminodipropionitrile, or nitrosobis-2-cyanoethylamine, is not mutagenic under the usual conditions of testing, and it is not carcinogenic; the probable reason is that the electronegative cyano group prevents oxidation of the α-methylene of the nitrosamine. Nitrosiminodiacetonitrile, the lower homologue of nitrosiminodipropionitrile, is likewise neither mutagenic nor carcinogenic.

A further example of the effect of electronegative substituents on the activation of nitrosamines to mutagens is the inactivity of nitrosobistrifluoroethylamine (Fig. 6.7), which is neither mutagenic nor carcinogenic (Preussmann et al., 1981). This compound is not oxidised at the α-methylenes and is biologically inactive. In contrast, methylnitrosotrifluoroethylamine is a very potent carcinogen, presumably because the methyl group is oxidised to form the alkylating trifluoroethyldiazonium ion, which can alkylate DNA. However, methylnitrosotrifluoroethylamine is not converted to a mutagen by rat liver microsomes, although it is weakly mutagenic with hamster liver activation.

The effects of hamster liver microsomes in activating some of these compounds refractory to the effect of rat liver enzymes are often profound, and lead to classification of the compound as a mutagen. Other than a means of optimising the assay, it is difficult to understand how use

Fig. 6.6. Nitrosodiallylamine, nitrosobis-(2-hydroxypropyl)amine, nitrososodiethanolamine, its ethyl- and methyl-ethers, and methoxyethylnitrosourea.

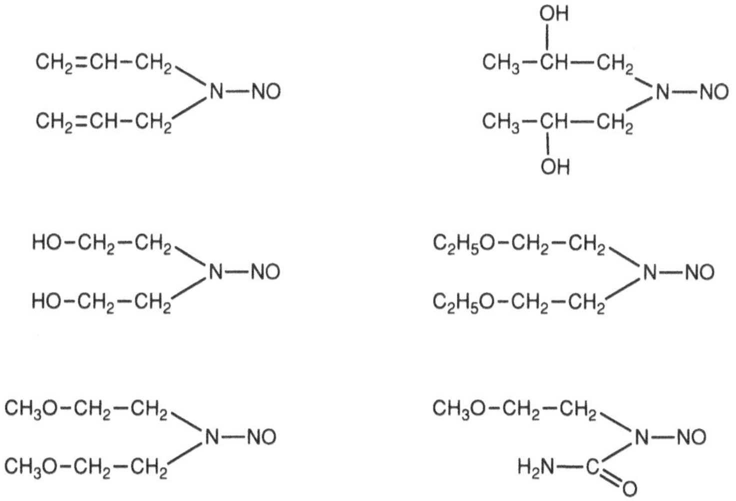

Fig. 6.7. Nitrosodiethylamine and nitrosobis-(trifluoroethyl)amine.

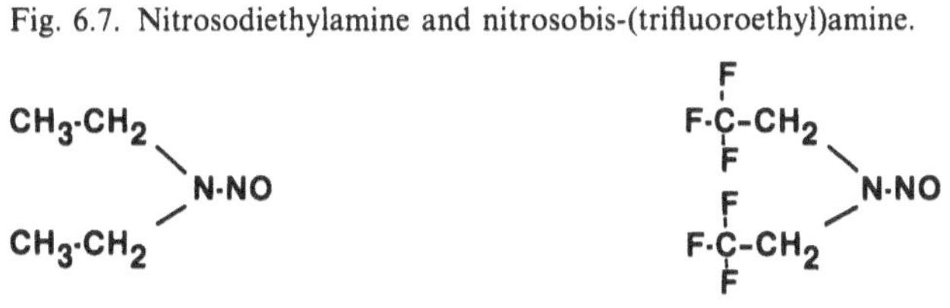

Table 6.5. *Carcinogenic nitrosamines which are not mutagenic with rat or hamster liver S_9*

	Target organ for tumours	
Compound	in rats	in hamsters
N-nitroso-		
diphenylamine	Bladder	NT
diethanolamine	Liver	Nasal cavity
bis-(2-methoxyethyl)amine	Liver	NT
methylethylamine	Liver	Liver
methylneopentylamine	Esophagus	NT
sarcosine	Esophagus	NT
methyl-3-carboxypropylamine	Bladder	NT
methylphenylamine	Esophagus	Liver, Spleen
methyl-4-fluorophenylamine	Esophagus	NT
phenylbenzylamine	Esophagus	NT
ethanol-hydroxypropylamine	Liver, Esophagus	Nasal cavity
thiomorpholine	Esophagus	NT
thialdine	Liver, Esophagus	NT

NT = Not tested.

of enzymes from multiple species helps to understand mechanisms of mutagenesis or carcinogenesis by these compounds. Similarly, the use of enzyme preparations from organs other than liver to activate a particular carcinogen to a mutagen, is often hailed as a triumph, when the organ in question (such as kidney, lung or nasal mucosa) is the target for the carcinogenic action of the compound. What is usually lacking from such studies is the extension to a broader group of compounds, some of which do and some do not induce tumours in that particular organ. These controls providing negative as well as positive correlations would help convince skeptics that there is a relation, and that the results are not fortuitous and the compounds selected merely to support a prejudice.

Examples of studies of mutagenesis with microsomes from organs other than liver include those from esophagus and lung (Scanlan *et al.*, 1980; Bartsch *et al.*, 1975; Mori *et al.*, 1986; Camus *et al.*, 1976). Whatever the findings, the question is still begged why so many nitrosamines are activated to mutagens by rat liver microsomal enzymes, yet do not induce tumours in rat liver.

6.6 Mutagenic nitrosamines which are not carcinogens

To complement the carcinogenic nitrosamines which are not mutagenic (Table 6.5), there are a number of nitrosamines which are mutagenic, but not carcinogenic. This raises questions not only of

Table 6.6. *Mutagenic* N-*nitroso compounds which are not carcinogens*

> N-nitroso-
> di-n-octylamine
> methyl-methoxyamine
> methyl-4-nitrophenylamine
> hydroxyproline
> guvacoline
> phenmetrazine (2-phenyl-3-methylmorpholine)
> cimetidine

numerical predictability of carcinogenic properties, but also reflects on our understanding of mechanisms of carcinogenesis. If a compound is mutagenic, why is it not carcinogenic?

Among this group of non-correlants is nitroso-N,O-dimethylhydroxylamine (nitroso-N-methoxy-N-methylamine) which is mutagenic (and a prophage inducer), but not carcinogenic. Its higher homologue, nitroso-N,O-diethylhydroxylamine (Fig. 6.8) is a fairly potent carcinogen. Another such compound is nitrosoguvacoline, and a list of others is in Table 6.6. It is possible that nitrosodi-n-octylamine is simply too large to enter mammalian cells, but not, apparently, too large to be activated by liver enzymes; methylnitroso-n-octylamine is quite carcinogenic. The same size restriction might apply to nitrosomethylphenidate, a nitrosamino ester, of which several and their parent acids – e.g. nitrosoproline – are not carcinogenic; nitrosophenmetrazine, nitroso-4-cyclohexylpiperidine and nitrosodiallylamine allow of no easy explanation for their

Fig. 6.8. Methylnitrosomethoxyamine, ethylnitrosoethoxyamine, nitrosoguvacoline and nitroso-oxazines, with their carcinogenicity stated.

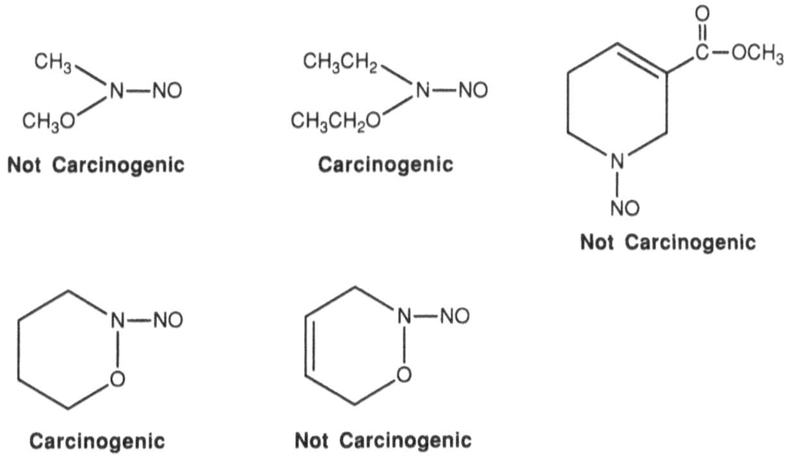

mutagenicity but lack of carcinogenicity. It can only be surmised that there are sound chemical reasons why they are not carcinogenic, when closely similar nitrosamines are carcinogenic, sometimes quite potently. For example, nitroso-2-methylpiperidine and nitroso-2-methyloxazolidine are carcinogenic, as is nitroso-4-phenylpiperidine. There might be special reasons, related to biological denitrosation (Jensen, 1983), for the non-carcinogenicity of nitrosocimetidine.

6.7 Prophage induction activity of *N*-nitroso compounds

The propensity of many *N*-nitroso compounds to give rise to alkylating agents which can damage DNA makes them good candidates for bacteriophage induction assays. These can indicate which compounds are likely to act through DNA-alkylating mechanisms which lead to mutagenesis. The colourimetric assay, developed by Elespuru and Yarmolinsky (1979), which measures the colour formed by production of the enzyme β-galactosidase as a result of cleavage of the lambda repressor has been used to good effect in studying *N*-nitroso compounds. The nitrosamines require metabolic activation which the bacteria carrying the phage do not provide and, of course, this imposes the usual limitations already discussed on the usefulness of the assay for predicting mechanisms of carcinogenesis.

However, among directly acting *N*-nitroso compounds, especially alkylnitrosoureas, the assay has proved very useful (Table 6.7). It has been possible to rank many of these compounds according to the magnitude of their effect in the assay, and this has correlated to some extent with their carcinogenic potency. At least this was so for the monoalkylnitrosoureas, such as the homologous series, methyl, ethyl, propyl, etc. Unfortunately, none of the dialkylnitrosoureas, which are much more stable than the monoalkylureas, were active in this assay (Lijinsky *et al.*, 1987a). This is contrary to their relative activity as carcinogens in animals, which compares well with the corresponding monoalkylnitrosoureas. For example, diethylnitrosourea is equally potent in inducing mammary tumours in rats with ethylnitrosourea, but the latter induces prophage and diethylnitrosourea does not.

Among the compounds which need to be activated, many of the nitrosamines which are inactive in the *Salmonella* assay give good positive results in the prophage induction assay, which uses hamster liver microsomal fraction (S_9). For example, methylnitrosoethylamine, methylnitroso-*n*-propylamine and methylnitroso-*n*-butylamine give positive responses similar to those of nitrosodiethylamine (in concordance with their relative carcinogenic activity). Methylnitrosocyclohexylamine, methylnitrosobenzylamine and methylnitrosoaniline (Fig. 6.1) are also

Table 6.7. *Lambda prophage induction by alkylnitrosamides in a uvrB lexA3 strain of* E. coli

	Peak response	
Compound	Enzyme units	Concentration (μg/ml)
N-nitroso-		
methylurea	125	200
ethylurea	400	250
allylurea	60	250
n-butylurea	85	250
n-amylurea	160	250
n-hexylurea	90	250
2-phenylethylurea	90	150
oxazolidone	375	30
hydroxyethylurea	230	250
methoxyethylurea	235	250
fluoroethylurea	235	250
2-hydroxypropylurea	100	250
3-hydroxypropylurea	400	250
5-methyl-oxazolidone	250	40
carbaryl	200	5
MNNG	175	10
ENNG	160	25
Control	40	0

positive in the prophage induction assay (Elespuru, 1984). However, several nitrosamines, such as nitrosobis-(2-oxopropyl)amine and nitrosobis-(2-hydroxypropyl)amine, which are mutagenic to *Salmonella* with hamster liver enzyme activation, were inactive in the prophage induction assay. The discrepancies make the assay unreliable as a guide to the mechanism of carcinogenesis by nitrosamines, but so are all other short-term *in vitro* assays.

6.8 Diffusibility of products of metabolic activation – a limitation on detectable mutagenic activity

The usual method of determining mutagenicity of a compound is by incubation with an activating system, enzyme preparation or cells, together with the test organism. The success of the assay depends on the diffusion of the mutagenic product of the activation from the site of formation to the surface, and then to the DNA of the target organism. In many cases the nature of the reactive metabolite is not known, but can be guessed. It is probable that many of them, while able to diffuse to other parts of the cell in which they are produced and damage key cellular components, might not be sufficiently stable to pass into a target cell and exert toxic damage. This might partially explain the otherwise perplexing failure of some compounds to be activated to mutagens by microsomal enzymes in suspension, and the commonly observed large differences in mutagenic activity of a compound between laboratories and even in the same laboratory at different times. There is, unfortunately, no way to improve this situation or to reduce the variation from experiment to experiment, which makes quantitative mutagenicity measurements of dubious value – and less reliable than quantitative carcinogenicity, itself not at all precise.

Attempts to determine the diffusibility of unstable reactive metabolites, for example phenyldiazonium ions, have been only moderately successful (Gold and Hines 1984). For the much less stable α-hydroxynitrosamines there has been little insight into their diffusibility into cells and to particular cellular targets.

6.9 Transformation and mutagenesis in cultured cells
6.9.1 *Mutagenesis in mammalian cells* in vitro

In addition to mutagenesis in micro-organisms and DNA damage in bacteriophage as a measure of the genotoxicity and potential carcinogenicity of substances such as *N*-nitroso compounds, it has become fashionable to use cells in culture as detectors of genetic damaging agents. There has been dissatisfaction with micro-organisms as being too simple and unrepresentative of the complexity of differentiated cells of higher organisms which, of course, are the cells capable of undergoing neoplasia. It is usually assumed that genetic damage is necessary for induction of neoplasia and cancer in animals, although that is not the only important event in the process induced by carcinogenic agents. There is increasing evidence that many carcinogens induce tumours through a mechanism not involving direct damage to DNA, for example plasticisers, hypolipidaemic agents, hormones, methapyrilene, etc. (Lijinsky, 1990c). Nevertheless, agents that produce DNA damage seem to be more likely to be carcinogens than those that do not, and this makes investigation of their activity in metazoan cells in culture important. However, the fibroblasts which are usually grown in culture are, alas, not noticeably better at activating nitrosamines than are bacteria, so that external activating systems, such as rat liver S_9 fractions, have to be employed, with all of their attendant problems and disadvantages. It would be much preferable to use cells as targets that can themselves metabolise and activate the test compound. However, it is very difficult to grow the type of cell that has the metabolic capacity, notably epithelial cells, and is susceptible to mutagenesis in culture. Partially transformed cells which grow well in culture, but lack the array of activating enzymes in the liver, such as mouse lymphoma cells (Clive et al., 1972) are an uneasy compromise.

There is possibly a link between the susceptibility of the cells that are commonly the target of carcinogens in animals (usually epithelial in origin) and the difficulty of growing those same cells in culture. The compromise presently used is a combination of cells that grow readily in culture, usually mesenchymal and often partially mutated or transformed, together with a rat liver S_9 fraction or with rat or hamster hepatocytes. Use of the subcellular fraction or hepatocytes was predicated on the supposition that whatever was the reactive intermediate formed would be sufficiently stable to pass from the site of formation into the target cell and to its nucleus, in order to mutate the DNA. It is quite possible, but not often considered, that the reactive intermediates involved in carcinogenesis by the test substances might be very unstable, while more stable reactive metabolites can reach the cell's nucleus, but not represent the carcinogenic intermediate. This is a problem inherent in all short-term assays used as surrogates for chronic toxicity assays in animals, and does not diminish their usefulness for stochastic purposes.

A significant impediment in using such *in vitro* assays for studying mechanisms of carcinogenic action is that they do not use or require the same cells as are the target of carcinogenic action. Most N-nitroso compounds do not have the liver of rats as their targets. Others do not have the male rat liver as target, although they might induce tumours in female rats (for example, nitrosobis-(2-oxopropyl)amine (Lijinsky et al., 1988f). In those cases the reactive mutagenic products formed from the compound by male rat liver enzymes are unlikely to be those responsible for its induction of tumours. It would seem a flight of fancy, impelled by faith rather than science, to imagine that such experiments provide insight into mechanisms of carcinogenesis. However, a large number of N-nitroso compounds have been tested for mutagenic or transforming activity and the results compared with carcinogenesis tests in animals, with the failings common to all such efforts.

Plants and plant cells have been used as target cells for detecting mutagenic or transforming activity of N-nitroso compounds. Similar problems arise here as with animal cells, in that the requisite activating systems are not present in the target cells, and must be supplied externally, with all of the caveats this involves. Unfortunately, the number of N-nitroso compounds that have been examined in plant systems (Gichner and Veleminsky, 1986) is small, limited to MNU and ENU, MNNG and a few nitrosamines. Although much has been discovered about the mutagenicity of these compounds in the plant cells, the small scope of the studies has not allowed any generalisations about the relation of mutagenic activity to chemical structure and biochemical reactivity of the compounds.

Both reverse mutations and forward mutations have been studied, and there has seemed to be some technical advantage in using reversions. Among the two approaches to measuring mutagenic activity there does not seem to be a large difference in relative potency among a group of related N-nitroso compounds. That is, the comparisons are consistent with formation of similar (or the same) intermediates which interact with DNA and lead to mutagenic events in a similar fashion.

Among the most commonly used mammalian cells for detection of mutagenic activity are hamster V79 cells, much used by Huberman and his associates (Huberman and Sachs, 1974). A large number of nitrosamines have been examined in this system, and most of the carcinogens have been mutagenic, when rat hepatocytes are incorporated as an activating system. However, there is not a good quantitative correlation between mutagenicity in this system and carcinogenicity, a similar situation to that with bacterial mutagenesis. Few directly acting N-nitroso compounds have been studied in the V79 cell system, so that the correlation between mutagenicity and carcinogenicity is not clear even when metabolic activation is not necessary.

The genetic markers used by Jones *et al.* (1981) in Chinese hamster V79 cells have included thioguanine resistance and ouabain resistance. The expression of potency used by Huberman is D_{10c}, or the concentration in nM producing a ten-fold increase in mutant frequency over controls. The linear portion of the dose–response curves was used for calculation. Using this measure there was a considerable range of potency among the thirty or so nitrosamines that have been tested in this system, rat hepatocyte activation being used. Many of the nitrosamines occupied a similar position on the scale of activity for both markers, but some were different. For example, the most effective nitrosamine in this system was methylnitroso-n-butylamine, at both markers, with nitrosodiethylamine and nitrosodimethylamine of similar potency at the ouabain (O) marker, but not at the thioguanine (T) marker. Of somewhat lesser potency were methylnitroso-n-propylamine, methylnitrosophenylethylamine, nitrosodi-n-propylamine, methylnitrosodiethylurea, nitrosopiperidine and nitrosohexamethyleneimine. It is notable that only two of these moderately potent mutagens are liver carcinogens in rats, although rat liver microsomes are efficient in activating them to mutagens for V79 cells. This anomaly has been commented on elsewhere.

The remaining nitrosamines were less-effective mutagens than those already mentioned, and their potency was similar. They include a number of nitrosamines which are potent liver carcinogens in rats, as well as those with other rat organs as targets. They are methylnitrosoethylamine, methylnitroso-*n*-undecylamine, nitrosopyrrolidine, nitroso-4-phenylpiperidine, nitrosodi-*n*-butylamine, nitrosomorpholine, dinitrosopiperazine, nitrosobis-(2-hydroxypropyl)amine, nitroso-2,6-dimethylmorpholine, nitrosoazetidine and nitrosoheptamethyleneimine. Among the inactive compounds were 1-nitrosopiperazine, nitrosodiallylamine, methylnitroso-*tert*-butylamine, nitrosodiphenylamine and nitrosododecamethyleneimine. The first three of these were also non-carcinogenic, but the latter two were carcinogenic to rats, the last even inducing liver tumours. The correlation between mutagenicity to V79 cells and carcinogenicity in rats was semiquantitative, but the clustering of most of the compounds in the region of low mutagenic activity made the correlation artificially close; in the outlying parts of the correlation, the quantitative relationship was not good (potent carcinogens weakly mutagenic, and potent mutagens are weakly carcinogenic), and some carcinogens failed to show any mutagenic activity in the assay. The results are another indication that the mechanisms of mutagenicity and tumour induction by nitrosamines are not the same, and might not be even similar.

Langenbach (1986) has evaluated the relationship between structure of a number of dialkylnitrosamines and their mutagenic activity to V79 cells, noting that nitrosamines containing a carbonyl group were more active than those containing an hydroxyl, compared with their unsubstituted analogues. Furthermore, he noted (Langenbach *et al.*, 1980) that for the oxygen-bearing nitrosamines Syrian hamster liver S_9 fraction activated them to bacterial mutagens (rat liver S_9 was ineffective) less readily than hamster hepatocytes activated them to mutagens for V79 cells. There was a better correlation of results in the latter system with carcinogenicity in hamsters. These results have a bearing on the greater effectiveness as carcinogens of these nitrosamines in hamsters than in rats.

Meaningful quantitative relations would hardly be expected, and none has been convincingly shown, so the assays remain indicators of the likelihood that a substance is a carcinogen, or not, but nothing more firm or definite. The greatest obstacle to accepting the results of such tests as anything more, is the large number of *N*-nitroso compounds mutagenic to cells *in vitro* following rat liver enzyme activation that do not induce tumours in rat liver; many of them, but not all, induce tumours in hamster liver, and almost invariably hamster liver enzymes (or cells) are considerably more active than their rat liver equivalents in activating nitrosamines to mammalian cell mutagens or transforming agents. With microsomal or hepatocyte activation many nitrosamines known to be carcinogens are also mutagenic to these mammalian cells, but there is no discrimination in mutagenic potency which is related to carcinogenic potency. Quite the contrary, several potent carcinogens are only weakly mutagenic in this system.

Ho et al. (1984) studied the mutagenic effects of N-nitroso compounds in Chinese Hamster embryo cells (CHO-K_1-BH_4) at the hypoxanthine-guanine phosphoribosyl transferase (HGPRT) locus, a commonly used system for detecting mutagenic activity. As with so many other assays using fibroblasts, the CHO cells were quite unresponsive to nitrosamines unless a system were incorporated to metabolically activate them. Such systems included injection of both target cells and the nitrosamines into athymic nude mice (Hsie et al., 1978), or using the rat liver S_9 fraction in cell culture. These added systems, of course, complicated interpretation of the results and particularly comparisons of potency of several nitrosamines. Nevertheless, usable results were obtained, although the number of compounds examined was rather small. The compounds used included nitrosamines which readily form alkylating intermediates and those which do not. The former were nitrosodimethylamine and nitrosodiethylamine, both of which were mutagenic as well as carcinogenic. The latter included nitrosomorpholine, nitrosopyrrolidine and dinitrosopiperazine, which were carcinogenic and mutagenic but much weaker in this system than the two acyclic nitrosodialkylamines; nitroso-4-methylpiperazine was carcinogenic but not mutagenic; nitroso-3-methyl-2-phenylmorpholine and nitroso-2,5-dimethylpyrrolidine were neither mutagenic nor carcinogenic.

The directly acting carcinogens are, of course, mutagenic in mammalian cells in culture, but there is again no relation between mutagenic and carcinogenic potency. Alkylnitrosocarbamates are much more potent mutagens than alkylnitrosoureas, but are no more effective carcinogens. The apparent disparity results from using a simplified system to simulate the complexity of an animals, together with the assumption that tumours arise only through mutational changes in the DNA of target cells.

6.9.2 Cell transformation in vitro

Cell transformation assays do not necessarily involve induction of mutagenic events in DNA by the agent, a possibility recognised by Barrett et al. (1984). The common transformation assays seem to be variants on the system introduced by DiPaolo and Donovan (1967) and others. These assays used fibroblasts from animals, or sometimes from humans (children's foreskin, for example), including rat or hamster embryo cells, which can be quick-frozen and used at a later date with little loss of viability. There has been little success in transforming epithelial cells *in vitro* with chemicals, mainly because of the difficulty of growing epithelial cells in culture; multiple replication of the transformed cells is necessary to produce colonies, which are characterised by their changed morphology (Pienta, 1980).

The disadvantage of fibroblasts – the other side of the relative ease of growing them in culture, because they are relatively undifferentiated – is that they lack many of the enzymes present in specialised, differentiated epithelial cells, from which most cancers arise. The lack of these enzymes prevents the fibroblasts from being transformed by other than directly acting compounds, which are a small proportion of carcinogens. To study compounds that are not directly acting it is necessary to provide an activating system, with all of the incongruities that comprises, as has been discussed elsewhere. The common activating systems are crude S_9 fractions from male Sprague-Dawley rats or, occasionally, from hamsters, and hepatocytes, usually from rats but also from hamsters. The test compounds must be incubated with the activating system and the target cells of the test. This makes the tests no less useful for predicting probability of carcinogenic potential, but renders them less useful for probing mechanisms of carcinogenesis.

Directly acting alkylnitrosoureas, alkylnitrosoguanidines and alkylnitrosocarbamates are excellent – even classic – transforming agents. However, some nitrosamines not likely to be mutagens are transforming agents in the usual assays introduced by DiPaolo and Donovan (1967) and developed by Pienta (1980), for example, nitrosodiphenylamine.

In a series of studies using Syrian hamster embryo cells, Pienta (1980) found several *N*-nitroso compounds to be transforming agents, without need of metabolic activation. Nitrosodiethylamine was not active, although nitrosodimethylamine showed some activity, at a dose of 1 μg/ml of medium. Ethylnitrosourea and MNNG were both quite active, MNNG at a concentration of only 0.05 μg/ml. However, it was surprising that nitrosopiperidine, which like other nitrosamines is assumed to require metabolic activation, gave a positive response at 0.1 μg/ml. The target cells are considered incapable of metabolising nitrosamines to mutagenic intermediates, and these results suggest that *in vitro* transformation in this system might not always be a mutagenic process. Additional support for this alternative is the positive result with nitrosodiphenylamine which both theoretically and practically cannot form a mutagenic alkylating agent, yet is a directly acting transforming agent in the hamster embryo cell system and is a carcinogen, inducing bladder tumours in rats (Cardy et al., 1979).

Another variant of the DiPaolo/Pienta embryo cell transformation system is that used by Quarles et al. (1975) in which pregnant hamsters are given the compound and foetal cell cultures are prepared from the transplacentally exposed foetuses, which are grown in culture and examined for the incidence of cell transformation. This allows host-mediated activation of the test compound. The transformed cells were repeatedly subcultured and colonies were finally injected (10^6 cells) S.C. into athymic nude mice, which were then observed for development of tumours (fibrosarcomas), which usually appeared within three to six weeks. Growth in soft agar was one criterion of transformation of the foetal cells.

All of the solvents used and a number of non-carcinogens were negative in the assay, but a variety of carcinogens were positive, including benzo[a]pyrene, urethane, 4-nitroquinoline-N-oxide and several N-nitroso compounds, which produced transformed colonies, that grew in soft agar and gave rise to fibrosarcomas in nude mice. Nitrosodiethylamine gave a strong response, and it is a compound that must be activated metabolically; it is likely that other nitrosamines would show a similar response in this system. In view of the response to the nitrosamine, it is not surprising that a strong directly acting alkylating agent β-propiolactone produced a large response in the assay.

Methylnitrosourethane, also a directly acting alkylating agent, produced a large number of transformed colonies, and was about as effective as urethane, which in animals is a much weaker carcinogen. Methylnitrosourethane is a prototype of the N-nitroso derivatives of a number of N-methylcarbamate insecticides. The N-nitroso derivatives of these insecticides were potent carcinogens in rats, although none of the parent insecticides was carcinogenic. The N-nitroso derivatives gave many transformed colonies in the transplacental assay, and all showed similar activity, including their effectiveness in producing colonies that grew into tumours in nude mice. The compounds were the N-nitroso derivatives of carbaryl, carbofuran, baygon, landrin, methomyl, Bux-ten and aldicarb (Fig. 3.7). The similarity in their activity in this assay (Quarles et al., 1979) resembles the similarity in their carcinogenic effectiveness in rats, in which they induced tumours of the forestomach (mainly carcinomas) when given by gavage (Lijinsky and Schmähl, 1978). They also were similar in the extent to which they interacted with and caused damage to human DNA in vitro; nitroso-aldicarb was the most active (Blevins et al., 1977). However, the results were in sharp contrast with their activities as bacterial mutagens, in which the variation was several orders of magnitude in Salmonella (Lijinsky and Andrews, 1979) and in E. coli (Lijinsky and Elespuru, 1976).

The variations in activity in the bacterial assays might be due (at least partly) to differences in the rate of uptake of the several compounds into the bacteria. This was shown, for example, by Elespuru (1976) to explain the large differences in mutagenic activity to *Hemophilus influenzae*, between methylnitrosourea and nitrosocarbaryl. The differences in uptake of these and other carcinogens into a whole animal are probably small, especially when the administration of the compound is by mouth, and therefore less dependent on physical properties, such as solubility, partition and diffusibility, than when administered by S.C. or I.P. injection. This is the reason discussed elsewhere, for the unreliability of S.C. injection of compounds as a means of comparison of toxic or carcinogenic potency.

6.9.3 Other systems

In another assay, by Ho *et al.* (1984), in which the compound is tested for its ability to induce sister-chromatid-exchange (SCE) in human leucocytes, a larger number of nitrosamines was evaluated than in the CHO/HGPRT assay. Metabolic activation with rat liver S_9 was incorporated in these assays, and was required for activity. The match of SCE with carcinogenicity was not particularly good. Several non-carcinogenic nitrosamines produced SCE, while some carcinogenic nitrosamines did not. Two carboxylic acid derivatives of nitrosopiperidine, nitrosopipecolic acid and nitrosonipecotic acid, induced SCE in leucocytes, but are not carcinogenic, while the parent compound is a potent carcinogen which also induced SCE; it is conceivable that the two nitrosamino acids release nitrous acid in the system, and that this is the agent inducing SCE. The potently carcinogenic derivatives of nitrosopiperidine, 3-methyl-, 4-chloro-, 3,4,-dichloro- and 3-hydroxy-nitrosopiperidine were positive in the SCE assay, but the equally potent carcinogen nitroso-4-piperidone was inactive (although it was mutagenic to *Salmonella*). The carcinogenic dinitrosopiperazine induced SCE in leucocytes and the non-carcinogenic 1-nitrosopiperazine did not, but neither did the potent carcinogen 2-methyldinitrosopiperazine; the nitroso derivative of the naturally occurring amine guvacoline (in betel nut) was mutagenic and induced SCE, but was not carcinogenic.

6.10 Conclusions

The general defects of short-term assays as detectors of carcinogenic potential have been discussed by many reviewers (Tennant *et al.*, 1987; Zeiger, 1987; Piegorsch and Hoel, 1988), and the findings with *N*-nitroso compounds support their reservations, even though many *N*-

nitroso compounds are directly acting alkylating agents. Many short-term assays in addition to those described above have been used to assess the activity of N-nitroso compounds against their known carcinogenic activities, and all have had failings similar to the simple bacterial mutagenesis assays, of which those using *Salmonella typhimurium* are as good as any. In a few cases, with a great deal of effort, some carcinogens that failed to respond in the bacterial assays have given weak positive responses in some specialised assays, such as yeast mutagenesis (Mehta and Von Borstel, 1984) and mutation of lymphocytes (mouse lymphoma cells) in culture (Clive *et al.*, 1972), but these have not provided assurance that screening assays of those types will provide a better 'score' in evaluating compounds as carcinogens or non-carcinogens than the more generally used assays. Furthermore, the less-usual assays provide no additional insight into mechanisms of carcinogenesis by N-nitroso compounds or other types of carcinogen and, in fairness, they were not designed to do so.

The stochastic relation between the activities in mutagenic assays and capacity to induce tumours might, however, reveal the existence of activation pathways leading to formation of a number of reactive intermediates producing various biological effects. This, in turn, suggests the usefulness of the mutagenic assays for indicating the potential of substances to be carcinogens but with the constraint that some carcinogens, perhaps activated by different mechanisms, will fail to be detected in the short-term assays.

7
Structure–activity relations in carcinogenesis by *N*-nitroso compounds

7.1 Early developments

The first report of Magee and Barnes (1956) of the carcinogenic activity of nitrosodimethylamine, which induced liver tumours in rats, caused a small stir, but had no great impact. Scientists interested in this field of research probably shared the feelings of J. M. Barnes that the compound was an interesting curiosity, but of no great importance or significance. As a water-soluble carcinogen it was a rarity, but that it induced liver tumours in rats – along with so many other carcinogens – reduced interest, although these tumours were highly vascular hemangiosarcomas, not hepatocellular neoplasms. The early studies of the metabolism of nitrosodimethylamine undertaken by the staff of the Medical Research Council Toxicology Research Unit at Carshalton, England, of which J. M. Barnes was director and P. N. Magee a senior staff member and pathologist put nitrosamines 'on the map'. These studies, by Heath and others, have been described in other chapters, and laid the foundation for much of our present understanding of carcinogenesis by *N*-nitroso compounds in particular, but also of carcinogenesis in general.

An early discovery was that nitrosodimethylamine and methylnitroso-*n*-butylamine were metabolised in a similar way, producing carbon dioxide and an aldehyde (Heath and Dutton, 1958). This led to conjecture about the fate of the remainder of the molecule, and to the idea of P. D. Lawley and P. N. Magee that nitrosodimethylamine might give rise to an alkylating agent, more precisely a methylating agent. The structure of DNA had been recently determined by Watson and Crick, and the idea was gaining ground that the origins of cancer as a mutation in somatic cells, proposed decades earlier by Boveri (1914), might be through alkylation of DNA. Such an hypothesis was much more plausible than others then current.

That hypothesis focussed attention on those aspects of the chemistry of nitrosamines that favoured formation of an alkylating agent. The exciting discoveries of NDMA-induced protein and nucleic acid methylation in rat liver by Magee and Hultin (1962) and Magee and Farber (1962) paved the way for most of the research on the biological effects of *N*-nitroso compounds that has occurred since. These findings inspired the study of carcinogenesis by some close relatives of NDMA, such as its homologue nitrosodiethylamine, first studied by Preussmann and Schmähl in Freiburg and later the subject of the first dose–response study of a nitrosamine (Druckrey *et al.*, 1963); NDEA was a more potent carcinogen in rats than NDMA and had the unusual effect of inducing tumours in the esophagus (a very unusual tumour in experimental animals) in addition to the common liver tumours. NDEA could plausibly, of course, give rise to an ethylating agent as easily as NDMA forms a methylating agent, and this was soon demonstrated. In those early days alkylation of DNA was measured at the N-7 position of guanine, which was the most prominent alkylated base found in DNA *in vivo* or *in vitro*.

The common thread connecting the studies of nitrosamines of various structures was the close connection between the induction of tumours and a chemical structure permitting formation of an alkylating agent which could alkylate DNA; this was the motivation for the grand studies of Druckrey and his co-workers (Druckrey *et al.*, 1967). The precise nature of the alkylating agent was – and still is – not known; carbonium ions, carbocations and alkyldiazonium ions have been suggested in turn. An early discovery was that methylnitroso-*n*-butylamine could be metabolised to either a methylating agent or to an *n*-butylating agent, and this nitrosamine was a powerful carcinogen in rats (Magee and Lee, 1963). In contrast, its isomer methylnitroso-*tert*-butylamine could be oxidised only in the methyl group, not in the *tert*-butyl group, and so could not give rise to a methylating agent; it could form a *tert*-butylating agent which, however, is sterically constrained from interacting with nucleophilic sites on DNA. In congruence with this chemical constraint methylnitroso-*tert*-butylamine is not carcinogenic. It also has low toxicity compared with methylnitroso-*n*-butylamine. Thus began a series of studies exploring the relationship between chemical structure and carcinogenic activity (as well as mutagenic activity) which still continues, and which is the subject of this Chapter. It is probable that the catalogue of information so far produced constitutes the most comprehensive set of its kind, and can be reasonably

Early developments

expected to provide insight into mechanisms of carcinogenesis by these agents.

The early studies in this area by Magee and his associates provided the solid base for all of the subsequent investigations, and also the most important insights into molecular mechanisms of their carcinogenic action, most of which have stood the test of time. However, few compounds were included in these early investigations by Magee. The major effort was in Freiburg by H. Druckrey and the chemist R. Preussmann, who prepared a large number and great variety of N-nitroso compounds for study as carcinogens; a number of other scientists were collaborators. Experiments with more than 60 compounds were conducted during the decade from 1957 to 1967, results of which were published in a large number of reports, culminating in a monumental review in 1967 (Druckrey *et al.*, 1967), in which the results were summarised.

That review has proven of enormous value and is rich in information about the chemistry of the N-nitroso compounds and descriptions of the morphology of the tumours which they induced in rats, many of them seen for the first time in experimental animals. However, comparisons of the potency of one compound with another can be made only in the broadest terms from the results presented, because the compounds were administered by several routes and in doses which varied quite widely, instead of at more chemically logical equimolar doses. The basis for choosing a dose for each compound was the acute toxic LD_{50}, the dose calculated (according to Weil, 1952) to cause death of half of the animals so treated within a few days of a single application of the compound. Fixed proportions of the LD_{50} dose were then given to groups of animals at intervals during many weeks, by S.C., I.V. or I.P. injection, or in drinking water, until the animals died of tumours induced, or until the end of their lives if no tumours were induced by the treatment. This procedure ensured that a dose of the compound was administered which was close to the maximum dose that could be tolerated by the animals without substantial adverse effects in the short term, although tumours did usually eventuate. There was no suggestion in this design that carcinogenic effectiveness and acute toxicity of the compounds were related (which clearly they are not), although this was the interpretation placed on the results by successor investigators less thoughtful than Druckrey and his colleagues.

The consequences have sometimes been that a compound has been reinvestigated, and in some instances the original report of inactivity has been reversed, and a significant incidence of tumours found. For example, experiments with nitrosodiallylamine, previously reported inactive

(Druckrey et al., 1967), showed that tumours of the nasal cavity were induced in rats (Pour et al., 1985). It is not possible to find in this report what dose was received by the rats (only mg/kg is stated, but no time parameter) or the total dose, but it seems to be larger than reported by Druckrey. Another compound, nitrosodiphenylamine, has considerable environmental importance since it was a widely used chemical in the manufacture of rubber (IARC, 1982a, Monograph No. 28), and also theoretical importance because it is a classical non-alkylating, non-mutagenic nitrosamine (Ames et al., 1973). However, it gave rise to bladder tumours when fed to rats at high doses, but its effect in mice was minimal (Cardy et al., 1979).

Methylnitroso-n-heptylamine, also reported as non-carcinogenic in the Druckrey compendium, was later shown to be a carcinogen of reasonable potency (Lijinsky et al., 1983). The reason for this discrepancy can be traced to the fact that the later experiment was conducted by oral administration of methylnitrosoheptylamine, whereas the earlier treatment used larger doses, but given by S.C. injection (Druckrey et al., 1967). This underlines the disadvantage of S.C. injection of water-insoluble compounds to study their carcinogenic effects, because they escape only slowly from the site of injection, whereas substances administered orally are absorbed into the circulation within a few hours. This endorses the need to test compounds for carcinogenicity at more than one dose rate and, preferably, using more than one route of administration, especially when the result is an unexpected lack of carcinogenic activity. It is much easier to conduct an experiment that shows a carcinogen failing to induce tumours than it is to demonstrate experimentally that a non-carcinogen produces a positive tumour response; a positive finding in carcinogenesis testing therefore outweighs any number of negative results, unless the former can be definitively related to a flaw in the experiment.

Another byproduct of the design of Druckrey's studies was the assignment of nitrosodiethanolamine to the class of 'weak' carcinogens (Druckrey et al., 1967). This compound was found to be a widely distributed trace contaminant of synthetic cutting oils and of cosmetics and toiletries, to which millions of people are frequently exposed (see Chapter 2). In tobacco (Hoffmann et al., 1982), its presence was noticed when interest developed in the environmental distribution of N-nitroso compounds, and sophisticated and sensitive methods of analysis for these compounds were developed. Nitrosodiethanolamine is a compound of very low toxicity, so that the determination of an LD_{50} is difficult; the animals could almost bathe in the compound without noticeable immediate damage. In the chronic experiment reported by Druckrey this low toxicity led to administration of a dose of NDELA so large that it amounted to one-third of the weight of a rat by the time they began to die from the liver tumours induced, which was less than a year from the beginning of treatment.

A more recent investigation of NDELA ignored the low toxicity of the compound and simply used a series of reasonable, but much lower than formerly, doses of nitrosodiethanolamine (Lijinsky et al., 1980c). Even these lower dose rates led to quite rapid induction of liver tumours in high incidence, within six months, and a wide-ranging dose–response study (Lijinsky and Reuber, 1984b; Lijinsky and Kovatch, 1985b) showed that even the lowest dose applied gave rise to a detectable tumour response in the rats. Once again, mice were much less susceptible and hamsters responded only with tumours of the nasal mucosa (Hoffmann et al., 1983; Lijinsky et al., 1988e), and also seemed less susceptible than rats to NDELA. Thus it was established that nitrosodiethanolamine is a carcinogen of considerable potency, of the same order as many other nitrosamines, and induces in rats a variety of tumours in addition to those of liver, in the esophagus, kidney, tongue and nasal mucosa. Since nitrosodiethanolamine is absorbed through the skin (Lethco et al., 1982; Lijinsky et al., 1981a), exposure to it poses an obvious increased cancer risk to humans.

Amid the comprehensiveness of the structures of N-nitroso compounds examined by Druckrey and his colleagues, which ranged from the simplest dialkylnitrosamines to complex sugar-containing structures (Guttner et al., 1971) and those with aromatic rings, and from the simple methylnitrosourea to dinitroso derivatives of dicarboxylic acid dialkylamides, some truisms were apparent, and were stated by the authors. For example, it seemed that symmetrical dialkylnitrosamines favoured the liver as a target organ for carcinogenesis in rats, while asymmetric nitrosamines favoured the esophagus. Within a restricted group of compounds this was largely true, but deeper examination, including studies of several dose rates and various routes of administration, showed that these simple rules did not apply broadly. In particular, as we have learned more about the metabolism, often quite complex, of these molecules, explanations for variations in target organ between compounds of similar structures are indicated.

7.2 Alkylation of DNA in relation to carcinogenesis

Nitrosamines with the N-nitroso group adjacent to an aromatic ring, which would not be expected to form a DNA-alkylating metabolite, were usually weak carcinogens or non-carcinogens, and nitrosamines containing carboxylic acid residues were also weak carcinogens or were inactive. The most common target of nitrosamines in rats was the esophagus while alkylnitrosoureas had a predilection for the nervous system, which was not a target of any nitrosamine. These early conclusions have been borne out by the results of more recent studies, which have also shown that cyclic nitrosamines and their acyclic counterparts often have quite similar carcinogenic action, although the former have rarely been shown to produce detectable alkylation of DNA *in vivo*, while the latter almost invariably do. The early studies in the author's laboratory were devoted to trying to identify alkylated bases in DNA of rats treated with several cyclic nitrosamines (Lijinsky and Ross, 1969), without success. Although it is likely that the strongly mutagenic cyclic nitrosamines do interact with DNA in target organs, perhaps transiently, demonstration of this has so far eluded investigators, except in one case (Chung *et al.*, 1989*b*). Therefore, the original impetus for the structure–activity studies, to confirm that induction of tumours by *N*-nitroso compounds was mediated by formation of intermediates which alkylated DNA, partly succeeded (acyclic dialkylnitrosamines) and partly failed (cyclic nitrosamines and alkylarylnitrosamines, the latter being incapable of forming an alkylating agent). This is discussed in the following sections and in Chapter 4.

7.2.1 *Methylnitrosoalkylamines and dialkylnitrosamines*

Symmetrical nitrosamines (Fig. 7.1), unlike their asymmetric counterparts, offer only one alkyl group to activating and metabolising enzymes; one would expect their carcinogenic effects to be somewhat limited, and so they are. Their carcinogenic potency falls rapidly with increasing molecular size, which is not true of methylnitrosoalkylamines. In hamsters nitrosodimethylamine is the most potent symmetrical nitrosamine; in rats nitrosodiethylamine is more potent than nitrosodimethylamine, probably because the rat esophagus is so sensitive to the former, but not to the latter. In rats nitrosodi-*n*-propylamine induces liver and esophageal tumours and is weaker than its diethyl- or dimethyl-homologues; nitrosodi-*iso*propylamine is a very weak carcinogen, as also is nitrosodiallylamine, their action being almost limited to the nasal mucosa. Nitrosodi-*n*-butylamine is still weaker than its lower homologues, but is unusual in inducing bladder tumours in several species, as well as tumours of the forestomach, lung and liver in rats; its interesting spectrum of tumours has made its complicated metabolism the most exhaustively studied among nitrosamines. The symmetrical isomers of nitrosodibutylamine, di-*iso*butyl- and di-*sec*-butyl-, are respectively much weaker and

Fig. 7.1. Symmetrical nitrosamines (a) Nitroso-dimethyl, -diethyl, -di-*n*-propyl, -di-*iso*propyl, -di-*n*-butyl, -di-*n*-octyl-amine; (b) Other alkyl groups.

(a)

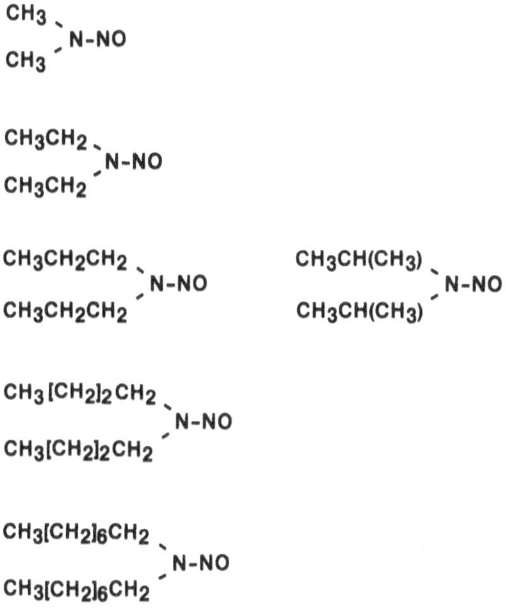

$$CH_3\text{-}N(CH_3)\text{-}N\text{=}O$$

$$(CH_3CH_2)_2N\text{-}NO$$

$$(CH_3CH_2CH_2)_2N\text{-}NO \qquad (CH_3CH(CH_3))_2N\text{-}NO$$

$$(CH_3[CH_2]_2CH_2)_2N\text{-}NO$$

$$(CH_3[CH_2]_6CH_2)_2N\text{-}NO$$

(b) $ON\text{-}NR_2$

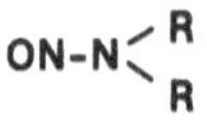

Compound	R
ALLYL	$CH_2\text{-}CH=CH_2$
sec-BUTYL	$CH(CH_3)(CH_2CH_3)$
iso-BUTYL	$CH_2\text{-}CH(CH_3)_2$
PHENYL	C_6H_5
HYDROXYETHYL	$CH_2\text{-}CH_2OH$
METHOXYETHYL	$CH_2\text{-}CH_2OCH_3$
ETHOXYETHYL	$CH_2\text{-}CH_2OCH_2\text{-}CH_3$
2,2-DIETHOXYETHYL	$CH_2\text{-}CH(OCH_2CH_3)_2$
TRIFLUOROETHYL	CH_2CF_3
CYANOMETHYL	CH_2CN
CYANOETHYL	CH_2CH_2CN
CARBOXYMETHYL	$CH_2\text{-}COOH$
2-HYDROXYPROPYL	$CH_2\text{-}CH(OH)\text{-}CH_3$
2-OXOPROPYL	$CH_2\text{-}CO\text{-}CH_3$

inactive, illustrating the powerful effect of substituents that, perhaps, block metabolism on the α-methylene which might be an essential reaction in carcinogenesis. Nitrosodi-*tert*-butylamine has not been prepared, but it would probably be non-carcinogenic.

Nitrosodi-*n*-amylamine is the only one of the large number of isomers that has been tested for carcinogenicity. It is a weak liver carcinogen in rats and induced no tumours of the esophagus, although methylnitroso-*n*-amylamine is a very powerful esophageal carcinogen in rats. Among larger symmetrical nitrosamines that have been examined, nitrosodi-*n*-octylamine is inactive, as are nitrosodicyclohexylamine and nitrosodibenzylamine. In these large molecules the limits of absorption or activation, or both, seem to have been reached or exceeded.

The picture presented by nitrosodialkylamines bearing substituents in the alkyl groups is more complex than for the unsubstituted compounds. Nitrosobis-(trifluoroethyl)amine is inactive, as are the bis-cyanoalkylnitrosamines, bis-carboxyalkylnitrosamines and the diethylacetal of nitrosaminobis-acetaldehyde; the analogous alkylnitrosomethylamines, when they have been studied, are potent carcinogens, usually targeting the esophagus in rats. Symmetrical nitrosamines with hydroxylated alkyl groups are weak carcinogens, but much more potent when converted to ethers. The unusual ketone, nitrosobis-(2-oxopropyl)amine is a potent carcinogen, inducing lung, thyroid, bladder or liver tumours in rats, and pancreas tumours in hamsters. Nitrosamines with chlorinated alkyl groups are very toxic and have moderate carcinogenic activity. The asymmetric alkylnitrosomethylamine analogues of these symmetric nitrosamines are all potent inducers of esophageal tumours in rats. The importance of the methylnitrosamino grouping in orientating the nitrosamine towards the rat esophagus will be discussed later; the same compounds seem less potent in the hamster, and those of them that have been examined target the hamster liver.

Most methylnitrosoalkylamines of whatever structure proved to be carcinogenic, although not always inducing tumours of the esophagus in rats, as was suggested by Druckrey *et al.* (1967). In fact, neither nitrosodimethylamine, the first member of the homologous series, nor methylnitrosoethylamine, the second, is primarily an esophageal carcinogen in rats, although the latter compound has induced tumours of the rat esophagus at high doses (Lijinsky and Reuber, 1981*b*) or when the terminal methyl of the ethyl group is fully deuterated (Lijinsky *et al.*, 1982*e*), in addition to liver tumours. The next higher homologues, however, had the esophagus of the rat as their primary target and induced few other tumours. Methylnitroso-*n*-propylamine, methylnitroso-*n*-butylamine and methylnitroso-*n*-amylamine were carcinogens of very comparable – and very high – potency, and induced tumours in the rat esophagus whether given orally (in drinking water or food or by gavage) or by S.C. or I.V. injection, or by intravesicular administration into the bladder (Lijinsky, 1990*d*); there were also tumours of the forestomach and tongue. This illustrated powerfully that nitrosamines are primarily systemic carcinogens, the local action of which is of minor importance. The next homologue, methylnitroso-*n*-hexylamine, induces tumours of the esophagus in rats, but also tumours of the liver and lung (Lijinsky *et al.*, 1983*e*). Both the *n*-butyl and *n*-hexyl compounds were more potent carcinogens when given in drinking water than by gavage. The incidence of esophageal tumours induced by methylnitroso-*n*-hexylamine was very high, whether given by gavage or in drinking water, but there were no tumours of the liver or lung following drinking-water treatment. This illustrates the importance of pharmacokinetics in determining which organs are the targets for tumour induction.

Neither methylnitroso-*n*-butylamine nor methylnitroso-*n*-hexylamine induced tumours of the esophagus in Syrian hamsters, and instead induced tumours of the liver, lung and nasal mucosa, as well as forestomach. Thus far tumours of the esophagus in hamsters have been extremely rare, only isolated animals having them. No nitrosamine has been associated with a significant incidence of esophageal tumours in hamsters, and it appears that this organ, so sensitive to nitrosamines in the rat, is refractory to carcinogenesis in the hamster. This profound difference between the two species is of great importance in understanding carcinogenesis. A partial explanation might lie in the low capacity of the hamster esophagus for metabolising nitrosamines, compared with the rat esophagus, which readily metabolises methylnitrosoalkylamines (Ludeke *et al.*, 1991*b*) but has a low capacity for metabolising nitrosodimethylamine (Von Hofe *et al.*, 1987) which, in turn, does not induce tumours in the rat esophagus.

This leaves unanswered the question why there is such a difference in enzymic constitution between these two rather similar species, and why such a difference is also evident between rats and mice (Mehta *et al.*, 1984), in the latter of which, however, nitrosamines do induce tumours of the esophagus. The sensitivity of the rat esophagus to very small doses of the alkylating nitrosamines is so great (Von Hofe *et al.*, 1987), that even low levels of activating enzyme would be expected to produce enough alkylating product to initiate tumour development; but apparently not in the hamster. There is a similar refractoriness on the part of the guinea pig and rabbit esophagus to induction of tumours by nitrosamines, although other organs are susceptible. Nitrosodiethylamine has induced tumours in the esophagus of the cat, in which very few nitrosamines have been tested, so that the breadth of the response of that organ of the cat to the variety of nitrosamine structures that have been examined in other species is not known. Nevertheless, it seems that induction of tumours in the esophagus depends on the species to a large extent, and the presence or absence of particular activating enzymes does not wholly explain the interspecies differences. From the response of the rat esophagus to particular nitrosamine structures, falling in a relatively narrow range of molecular size and shape of molecules that induce tumours, it seems that there are receptors for which these molecules have an affinity in the esophagus of the rat, but are absent from the esophagus of insusceptible species; the nature of these receptors is not known, although some inferences might be drawn about their nature.

7.2.2 Cyclic and acyclic nitrosamines in the rat esophagus

Regarding susceptibility of the esophagus of the rat to carcinogenesis by nitrosamines, this does not extend to alkylnitrosoureas or other alkylnitrosamides. Although the rat esophagus is so exquisitely sensitive to the action of the nitrosamines, presumably acting as alkylating agents, the directly acting alkylnitrosoureas are apparently incapable of inducing tumours in the rat esophagus. Methylnitrosourea has been shown to alkylate DNA in the rat esophagus (Qin et al., 1990). However, even when given in drinking water ethylnitrosourea or diethylnitrosourea have not induced tumours in the rat esophagus (while inducing tumours in many other organs, including the forestomach), although they form directly the same alkylating agent, probably an ethyldiazonium ion, as is formed from nitrosodiethylamine by metabolism; nitrosodiethylamine is a very potent esophageal carcinogen in rats, producing a significant response after administration of as little as two milligrams during the lifetime of the rat (Lijinsky et al., 1981b).

Similarly for nitrosodi-n-propylamine and n-propylnitrosourea, nitrosodi-n-butylamine and n-butylnitrosourea, the formation of an alkylating agent in the esophagus of the rat seems less important than whether the alkylating agent is formed directly or through metabolic activation of a nitrosamine. Nitrosodimethylamine is also an anomaly, since it does not give rise to esophageal tumours in rats, neither does it methylate DNA in the rat esophagus (Von Hofe et al., 1987). The simple methylnitroso-n-alkylamines (-propyl, -butyl, -amyl, etc.) (Fig. 7.2), very similar in structure to nitrosodimethylamine, are powerful inducers of esophageal tumours in rats and methylate DNA in the esophagus; methylnitroso-

Alkylation of DNA 261

ethylamine methylates esophageal DNA in rats to a very small extent and induces some esophageal tumours.

These results suggest that whatever the receptor in the rat esophagus, it has affinity only for nitrosamines and not for nitrosamides; also the affinity is shown only by nitrosamines whose molecular dimensions lie within certain limits, which exclude nitrosodimethylamine and, to some extent, methylnitrosoethylamine. The largest molecules inducing tumours in the rat esophagus are methylnitroso-n-heptylamine and methylnitroso-n-octylamine, which only act in this organ when they are given in drinking water. Perhaps their physical properties of low water solubility and high lipid partition prevent their reaching the esophagus epithelium in sufficient concentration to induce tumours when they are administered parenterally.

Fig. 7.2. Some methylnitroso-n-alkylamines and their carcinogenic properties.

Structure	RELATIVE CARCINOGENIC POTENCY	RAT TARGET ORGAN
ON–N(CH$_3$)(CH$_2$CH$_3$)	+++	LIVER
ON–N(CH$_3$)[(CH$_2$)$_2$CH$_3$]	++++	ESOPHAGUS
ON–N(CH$_3$)[(CH$_2$)$_3$CH$_3$]	++++	ESOPHAGUS
ON–N(CH$_3$)[(CH$_2$)$_4$CH$_3$]	++++	ESOPHAGUS
ON–N(CH$_3$)[(CH$_2$)$_5$CH$_3$]	++++	ESOPHAGUS, LIVER
ON–N(CH$_3$)[(CH$_2$)$_6$CH$_3$]	+++	LIVER, LUNG
ON–N(CH$_3$)[(CH$_2$)$_7$CH$_3$]	+++	LIVER, BLADDER
ON–N(CH$_3$)[(CH$_2$)$_8$CH$_3$]	+++	LIVER
ON–N(CH$_3$)[(CH$_2$)$_9$CH$_3$]	++	BLADDER
ON–N(CH$_3$)[(CH$_2$)$_{10}$CH$_3$]	++	LIVER, LUNG
ON–N(CH$_3$)[(CH$_2$)$_{11}$CH$_3$]	++	BLADDER
ON–N(CH$_3$)[(CH$_2$)$_{13}$CH$_3$]	++	BLADDER

262 *Structure–activity relations*

These required molecular dimensions include methylnitroso-2-phenylethylamine, methylnitrosobenzylamine and methylnitrosoaniline. They also comprise a large number of cyclic nitrosamines (Fig. 7.3) – with a variety of substituents – including derivatives of nitrosopiperidine, nitrosomorpholine, nitrosohexamethyleneimine, nitrosoheptamethyleneimine, nitrosooctamethyleneimine and dinitrosopiperazine. The very large nitrosododecamethyleneimine (Goodall and Lijinsky, 1985), the small nitrosoazetidine (Lijinsky and Taylor, 1977*b*) and nitrosopyrrolidine are not included (although some derivatives of nitrosopyrrolidine induce tumours of the esophagus in rats: Lijinsky and Taylor, 1976*d*).

In assessing the carcinogenicity of cyclic nitrosamines, the idea was proposed (Lyle *et al.*, 1976) that those cyclic nitrosamines which formed a rigid chair were more likely to induce tumours of the esophagus in rats, and therefore were more potent esophageal carcinogens, than those that formed more flexible configurations. This certainly appeared to be true of several derivatives of nitrosopiperidine, which was itself a potent esophageal carcinogen in rats. For example, 4-*tert*-butyl-nitrosopiperidine, in spite of its bulk, induced esophageal tumours, while 4-phenylnitrosopiperidine (Fig. 7.3a) induced tumours of the esophagus and of the liver in rats, but 4-cyclohexylnitrosopiperidine was not carcinogenic

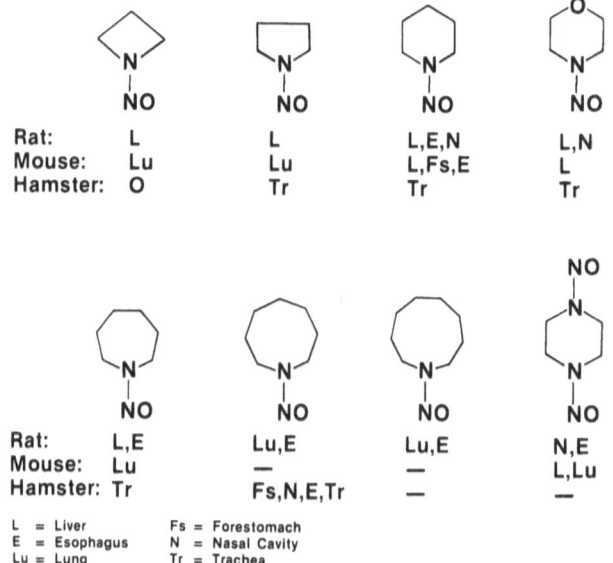

Fig. 7.3. Cyclic nitrosamines: nitroso-azetidine, -pyrrolidine, -piperidine, -morpholine, -hexamethyleneimine, -heptamethyleneimine, -octamethyleneimine, dinitrosopiperazine.

Rat:	L	L	L,E,N	L,N
Mouse:	Lu	Lu	L,Fs,E	L
Hamster:	O	Tr	Tr	Tr

Rat:	L,E	Lu,E	Lu,E	N,E
Mouse:	Lu	–	–	L,Lu
Hamster:	Tr	Fs,N,E,Tr	–	–

L = Liver   Fs = Forestomach
E = Esophagus N = Nasal Cavity
Lu = Lung   Tr = Trachea

(Lijinsky et al., 1981d). On the other hand, 4-methylnitrosopiperidine was not notably different in carcinogenicity from nitrosopiperidine, or from its 3-methyl isomer, all inducing tumours of the esophagus in rats (Lijinsky and Taylor, 1975a). Also in support of the hypothesis, 2-methyldinitrosopiperazine was considerably more potent in inducing esophageal tumours in rats than was dinitrosopiperazine itself, and 2,5-dimethyldinitrosopiperazine was also more potent, while 2,6-dimethyldinitrosopiperazine was more effective than the 2-methyl derivative (Lijinsky and Taylor, 1975g).

The hypothesis, however, lost cohesion when the two isomers, *cis* and *trans*, of 2,6-dimethylnitrosomorpholine were examined. The *cis* isomer, which is a rigid chair (Fig. 4.12), was considerably less potent in inducing esophageal tumours in rats than was the flexible *trans* isomer (Lijinsky and Reuber, 1980c). The same was true of the *cis* and *trans* isomers of 3,5-dimethylnitrosopiperidine, the flexible *trans* being a more potent carcinogen than the rigid chair *cis* isomer (Lijinsky et al., 1982d). In hydrogen-deuterium exchange experiments it was shown that the flexible *trans* form of these two β-dimethylnitrosamines exchanged the hydrogens α- to the *N*-nitroso function much more readily than did the rigid *cis* isomers, probably a consequence of the flexibility of the molecule, and hence formed an α-carbanion more readily (Singer and Lijinsky, 1979; Lijinsky et al., 1980f). Whether or not formation of an α-carbanion has

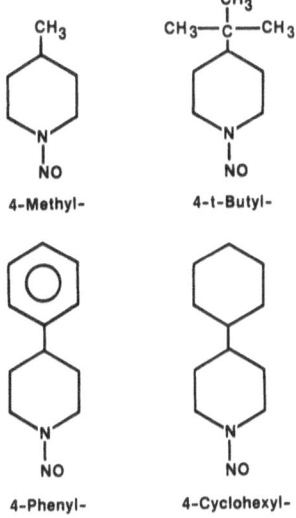

Fig. 7.3a. 4-Methyl-, 4-*tert*-butyl-, 4-phenyl-, and 4-cyclohexyl-nitrosopiperidine.

any role in carcinogenicity by these compounds is a matter of conjecture, but these results point to a fundamental observation about carcinogenic nitrosamines, to wit, that they must normally have at least one carbon atom α- to the N-nitroso function free and unhindered. Blockage of this position decreases or eliminates carcinogenic activity.

7.3 Alkylnitrosamides (alkylnitroso-ureas, -carbamates and -guanidines)

This group of N-nitroso compounds are, in general, unstable and reactive. They require no activation to form adducts with biological materials in the body. They are directly acting mutagens in a variety of systems, including micro-organisms, and are presumed to be directly acting carcinogens. Their carcinogenic actions (i.e. the number of organs in which they induce tumours in animals) are very diverse, and they give rise to several types of tumour in animals that have close analogues in man, and in some cases they provide the best or only animal models for particular human tumours. They are, therefore, very useful for studying the morphogenesis and pathogenesis of such tumours, although no satisfying etiological link has been proposed between exposure to directly acting N-nitroso compounds and any human cancer.

The largest group of these compounds which has been studied is the

Fig. 7.4. Alkylnitrosoureas.

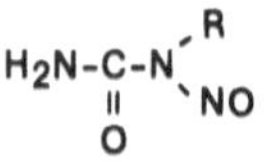

$$H_2N-C(=O)-N(R)(NO)$$

$R = CH_3-$	METHYL
$R = CH_3CH_2-$	ETHYL
$R = CH_3CH_2CH_2-$	n-PROPYL
$R = CH_3[CH_2]_2CH_2-$	n-BUTYL
$R = CH_2=CHCH_2-$	ALLYL
$R = CH_3[CH_2]_3CH_2-$	n-AMYL
$R = CH_3[CH_2]_4CH_2-$	n-HEXYL
$R = CH_3[CH_2]_9CH_2-$	n-UNDECYL
$R = CH_3[CH_2]_{11}CH_2-$	n-TRIDECYL
$R = FCH_2CH_2-$	FLUOROETHYL
$R = HOCH_2CH_2-$	HYDROXYETHYL
$R = CH_3CH(OH)CH_2-$	2-HYDROXYPROPYL
$R = HOCH_2CH_2CH_2-$	3-HYDROXYPROPYL
$R = CH_3CH(CH_3)-$	iso-PROPYL
$R = C_6H_5-$	PHENYL
$R = C_6H_{11}-$	CYCLOHEXYL
$R = C_6H_5CH_2-$	BENZYL
$R = C_6H_5CH_2CH_2-$	PHENYLETHYL

alkylnitrosoureas. The structures of most of those for which information on mutagenicity or carcinogenicity is available, are shown in Fig. 7.4. The structures of the analogous alkylnitrosocarbamates and alkylnitrosoguanidines are shown in Fig. 7.5.

The surprises in the results of administering alkylnitrosoureas and other directly acting N-nitroso compounds to animals are not the number of types of tumour they induce, but the many types of tumour they do not induce. We know that they, like the water-soluble nitrosamines, are rapidly and widely distributed in the animal body; this in spite of the limited stability of many of them under physiological conditions. They reach every organ and tissue, as they are known to give rise to tumours in the skin and nervous system, for example, when administered by mouth or by S.C. injection. Yet few of them induce tumours in the liver or kidney, which have an excellent blood supply (and, furthermore, are susceptible to tumour induction by a number of nitrosamines which have to be activated) and none of them gives rise to tumours of the esophagus in rats or mice, which is a prime target organ of many nitrosamines, perhaps of a majority of them. These discrepancies are the most vexing problems in N-nitroso compound carcinogenesis, to which there has been no satisfactory solution from chemistry, biochemistry, physiology or molecular biology. If there is light at the end of the tunnel, it might lie in the area of receptor biochemistry, but a description of the biological effects of alkylnitrosamides, and a comparison with those of structurally related nitrosamines must first be recounted, because any plausible hypothesis to explain the biological effects of N-nitroso compounds must take into account those findings.

Most of the directly acting N-nitroso compounds produce skin tumours

Fig. 7.5. Alkylnitrosoguanidines, alkylnitrosocarbamates and trialkylnitrosoureas.

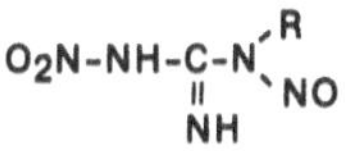

$O_2N-NH-C(=NH)-N(R)(NO)$

(a) $R = CH_3-$
(b) $R = CH_3CH_2-$

$R_2O-C(=O)-N(R_1)(NO)$

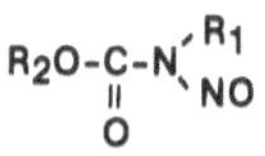

(a) $R_1 = CH_3-$; $R_2 = CH_3CH_2-$
(b) $R_1 = CH_3CH_2-$; $R_2 = CH_3CH_2-$
(c) $R_1 = CH_3-$; $R_2 = $ 1-NAPHTHYL ($C_{10}H_7$)

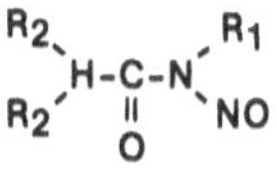

$R_2(R_2')H-C(=O)-N(R_1)(NO)$

(a) $R_1 = R_2 = CH_3-$
(b) $R_1 = R_2 = CH_3CH_2-$
(c) $R_1 = CH_3-$; $R_2 = CH_3CH_2-$

after multiple applications to mice, but they differed in effectiveness (Lijinsky and Winter, 1981). A few of them did not induce skin tumours, including methyl- and ethyl-nitrosourethane, allylnitrosourea, cyclohexyl-nitrosourea and phenylethylnitrosourea, although the last was among the most potent bacterial mutagens of the alkylnitrosoureas examined. Many of these directly acting N-nitroso compounds, when painted on mouse skin, gave rise to a variety of tumours of internal organs, including those compounds which failed to induce skin tumours (Lijinsky and Reuber, 1988). There is no satisfactory explanation for the lack of skin tumourigenesis by some of these compounds, which are neither the largest nor the smallest molecules, suggesting that physical properties are not an explanation. Nor can the absence of thiols which accelerate nitrosamide decomposition be the reason for the failure of the alkylnitrosourethanes to be active, since nitrosocarbaryl does induce skin tumours, albeit weakly, while the cyclic nitrosocarbamates, nitrosooxazolidone and 5-methylnitrosooxazolidone, are quite potent skin carcinogens. Furthermore, the alkylnitrosonitroguanidines are also potent mouse skin carcinogens (Lijinsky, 1982), and they, too, are susceptible to accelerated decomposition by thiols (Kleihues and Wiestler, 1984). It is possible that, as in tumour induction in other organs by N-nitroso compounds, the directly acting N-nitroso compounds interact with cellular receptors in mouse skin for which those compounds that are not skin carcinogens do not have affinity.

Alkylnitrosamides – directly acting xenobiotics – elicit a broad tumour spectrum in rats, but a very narrow one in hamsters. In spite of the formation of supposedly the same proximate carcinogen (perhaps an alkyldiazonium ion) from a nitrosamine through enzymic activation and from the analogous alkylnitrosamide directly, the tumours induced by the two classes of N-nitroso compound are usually quite different (Tables 7.1 and 7.2). Many more alkylnitrosoureas have been studied than alkyl-nitrosocarbamates or alkylnitrosoguanidines, which also differ in tumour response between the three classes even if containing the same alkyl group – an enigma in itself. Therefore, comparison will be of alkylnitrosoureas with the analogous dialkylnitrosamine. The first four in the series, methyl-, ethyl-, n-propyl- and n-butyl- illustrate the conundrum. All four symmetrical nitrosamines induce liver tumours in rats, and all except nitrosodimethylamine induce tumours of the esophagus in rats. None of the alkylnitrosoureas induces tumours of liver or esophagus in rats, but instead methylnitrosourea gives rise to nervous system tumours and the other three homologues induce a large variety of tumours. These include tumours of mammary gland, Zymbal's gland, uterus, duodenum and

Table 7.1. *Tumours induced by nitrosamines in rats and hamsters* (%)

Tumour site	No. of compounds tested in	
	Rats (130 nitrosamines)	Hamsters (41 nitrosamines)
Liver	57 (44)	26 (63)
Lung	28 (22)	18 (44)
Kidney	8 (6)	—
Esophagus	66 (51)	—
Nasal mucosa	49 (38)	25 (61)
Bladder	11 (8)	3 (7)
Tongue	25 (19)	—
Forestomach	25 (19)	15 (37)
Pancreas	—	11 (27)
Thyroid – follicular cell	6 (5)	—
Trachea	11 (8)	11 (27)
Thymus	2 (2)	—
Intestine	1 (1)	—
Colon	2 (2)	1 (2)
Spleen	2 (2)	—
No tumours	24 (18)	—

intestine, thyroid gland, and mesotheliomas and thymus lymphoma, in addition to tumours of the nervous system, none of which are induced by the analogous nitrosamines. A different set of incongruities is seen in hamsters treated with the same members of the two classes of N-nitroso compounds. Nitrosodi-n-butylamine has induced bladder tumours in both rats and hamsters (and, indeed, in other species in which it has been tested), but no bladder tumours have appeared after treatment with n-butylnitrosourea.

There are few tumour types that have been produced in common by dialkylnitrosamines and alkylnitrosoureas; they include those of lung and forestomach and, occasionally, follicular cell tumours of the thyroid gland. Other alkylnitrosoureas containing substituted alkyl groups sometimes induce tumours in addition to those listed above, including osteosarcomas and liver tumours. These differences in target organ specificity are summarised for a number of nitrosamines and analogous alkylnitrosoureas in Table 7.3.

As already mentioned, there are large differences in the types of tumour induced between alkylnitrosoureas and the analogous alkylnitrosocarbamates and alkylnitrosoguanidines. They are locally acting, inducing tumours of the stomach when administered orally, in the forestomach and in the glandular stomach, especially by the nitrosoguanidines, but

Table 7.2. *Tumours induced by nitrosamides in rats and hamsters (%)*

Tumour site	No. of compounds tested in	
	Rats (50 nitrosamides)	Hamsters (17 nitrosamides)
Liver	10 (20)	3 (18)
Lung	26 (52)	—
Kidney	7 (14)	—
Bladder	8 (16)	—
Tongue	5 (10)	—
Forestomach	33 (66)	15 (88)
Gland. stomach	4 (8)	1 (6)
Thyroid – follicular cell	10 (20)	—
Thyroid – C. cell	3 (6)	—
Trachea	—	1 (6)
Thymus	9 (18)	—
Mammary gland	15 (30)	1 (6)
Uterus/Cervix	19 (38)	4 (23)
Duodenum	7 (14)	1 (6)
Intestine	8 (16)	—
Colon	12 (24)	2 (12)
Zymbal gland	11 (22)	—
Nervous system	16 (32)	1 (6)
Mesothelioma	15 (30)	—
Skin	9 (18)	—
Spleen	3 (6)	14 (82)
No tumours	1 (2)	—

alkylnitrosoureas are mainly systemic carcinogens. There are some differences in stability between them (see Chapter 3), but they are similar in their pathways of decomposition and in the products of solvolysis, so their chemistry that is obvious to us does not explain the large differences in carcinogenic effect between them. The nitrosamines are, of course, quite different chemically from the alkylnitrosamides, and the presence in the former of a second alkyl group on the amine nitrogen appears to make them very different from the latter in affinity for particular target cells. Since we do not know what the receptors for these compounds might be, or where they are, it is not surprising that studies conducted *in vitro* with cells or cell fragments separated from the organ and the organisms have failed to give much insight into the mechanisms by which N-nitroso compounds cause cancer.

Table 7.3. *Tumours induced in rats by alkylnitrosoureas and analogous symmetrical nitrosamines*

Alkyl group	Nitrosourea*	Dialkylnitrosamine
Methyl	Nervous system	Liver, Kidney, Lung
Ethyl	Mammary gland, Mesothelioma, Lung, Uterus, Nervous system, Zymbal gland	Liver, Esophagus
n-Propyl	Thymus, Mammary Gland, Intestine, Zymbal gland	Liver, Esophagus, Nasal
n-Butyl	Mammary gland, Mesothelioma, Lung, Uterus, Colon	Liver, Lung, Bladder
n-Amyl	Lung, Uterus, Mesothelioma, Mammary gland, Zymbal gland	Liver
n-Octyl	Nervous system	—
Allyl	Lung, Thymus, Mammary gland, Uterus, Zymbal gland, Colon	(Nasal)
Hydroxyethyl	Lung, Thyroid, Intestine, Zymbal gland, Mammary gland, Lung	Liver, Nasal, Esophagus
Methoxyethyl	Thyroid, Intestine, Zymbal gland, Lung	Liver, Nasal
2-Hydroxypropyl	Thymus (lymphoma), Lung	Esophagus, Nasal, Lung, Liver
2-Oxopropyl	Kidney, Lung, Mesothelioma	Lung, Liver, Bladder, Thyroid
Phenyl	Subcutaneous sarcoma	Bladder
Benzyl	Forestomach	—
iso-Butyl	Duodenum	Nasal, Lung

*Most induced high incidences of forestomach tumours by gavage.

7.4 Importance of α-oxidation in nitrosamine carcinogenesis

7.4.1 Effects of replacement of hydrogen with deuterium

The work of Magee with methylnitroso-*n*-butylamine and methylnitroso-*tert*-butylamine (Magee and Lee, 1963) pointed out the difference between the toxic and carcinogenic former compound and the latter non-carcinogenic compound as lying in the inability of the *tertiary*-butyl nitrosamine to generate a methylating agent. This was related to the lack of oxidation of the fully substituted *tert*-butyl group, compared with the ready oxidation of the 1-methylene of the *n*-butyl group. Later studies by Druckrey's group reinforced the concept that α-oxidation was important in activation of carcinogenic nitrosamines. Nitrosodi-*iso*propylamine, which can be considered α-methylated nitrosodiethylamine, is much less carcinogenic than NDEA, but does induce tumours in rats (Druckrey *et al.*, 1967; Lijinsky and Taylor, 1979*b*), in spite of the considerable hindrance to oxidation presented by the α-methyl groups. Also, and more significantly, methylnitrosophenylamine (methylnitrosoaniline) is a carcinogen of reasonable potency in rats (Goodall *et al.*, 1968), although it resembles methylnitroso-*tert*-butylamine in incapacity to form a methylating agent. Its mechanism of carcinogenesis is enigmatic, although by oxidation of the methyl group it could form a phenyldiazonium ion, for example; the question then arises why the hindered *tert*-butyl nitrosamine could not also form an alkylating agent through oxidation of the methyl group. This is not, however, the only obstacle to acceptance of this simple mechanism of nitrosamine carcinogenesis.

The role of α-oxidation in activation of nitrosamines to carcinogenic forms became clearer following studies of the effect of substitution of deuterium for hydrogen at the α-carbon of nitrosamines. After discovery by this author and L. Keefer that nitrosopyrrolidine labelled with deuterium at the α-methylenes by deutero-reduction of succinimide (followed by nitrosation) lost deuterium when it was distilled from base, the tendency of nitrosamines to form α-carbanions was reported (Keefer and Fodor, 1970). This property of nitrosamines was later developed by Seebach as an important route of organic synthesis (Seebach and Enders, 1975). The susceptibility to exchange of α-hydrogen atoms in nitrosamines led to speculation that removal of the α-hydrogen by carbon–hydrogen cleavage might be a rate-limiting step in activation of these molecules (presumably carried out by mixed function oxidases or other enzymes). In that case, because of the greater strength of the carbon–deuterium bond, nitrosamines with deuterium in the α-methylene groups would be expected to have a lessened carcinogenic activity, compared with the non-deuterated compound.

Importance of α-oxidation

Fig. 7.6. Carcinogenesis by deuterium-labelled nitrosamines in rats.

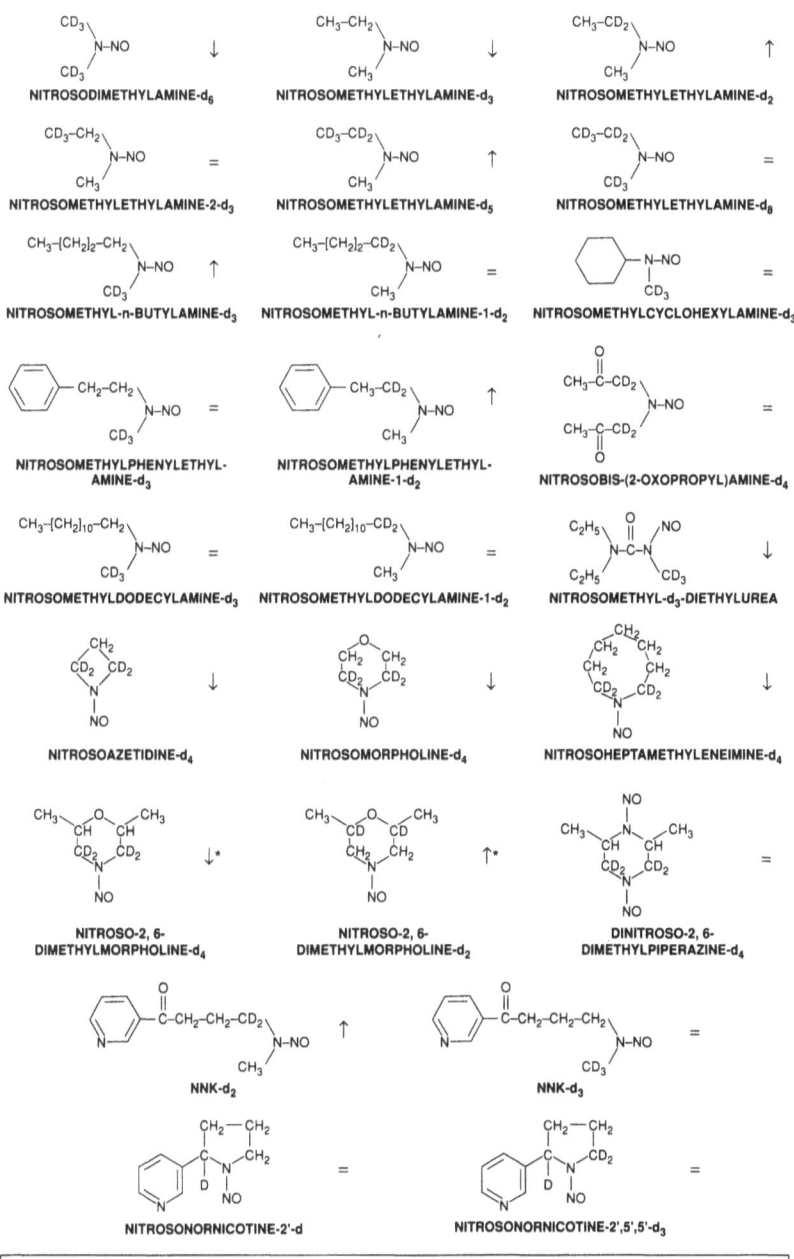

And so it was, for most of the deuterium-labelled nitrosamines studied (Fig. 7.6). Nitrosodimethylamine-d_6 was a less potent carcinogen than NDMA (Keefer et al., 1973), which could be attributed to the lower rate of formation of the methylating agent formed by oxidation; however, there does not seem to be a significant lowering of the extent of methylation of DNA by the deuterium-labelled compound compared with the unlabelled compound (Oshiro et al., 1975). Nitrosomorpholine labelled with deuterium at the α-positions to the N-nitroso function was reduced five-fold in carcinogenic activity compared with nitrosomorpholine when equimolar doses were given to rats in drinking water (Lijinsky et al., 1976), and in Syrian hamsters when given by gavage; liver tumours were predominant in the rat, and nasal cavity tumours in the hamster (Lijinsky et al., 1984a). The results indicate that, as with nitrosodimethylamine, nitrosomorpholine is oxidised at one of the carbon atoms α- to the N-nitroso function as the first – and rate-limiting – step in activation to a proximate carcinogenic form. However, whereas NDMA is known to give rise to an intermediate that alkylates DNA, no such alkylated adduct of DNA arising from nitrosomorpholine has been demonstrated; transient formation of such an adduct can be inferred from the demonstration that nitrosomorpholine is a mutagen to bacteria or cells in culture, following suitable metabolic activation (Rao, 1984; Jones et al., 1981). It is not easy to draw conclusions about possible mechanisms of carcinogenesis by these two nitrosamines from these somewhat disparate results, although they do point to one or more products resulting from α-oxidation of nitrosamines being important in carcinogenesis, whether or not they give rise also to intermediates which form stable adducts with DNA.

Similar comparisons have been made between several other N-nitroso compounds and their deuterated counterparts. Among a number of directly acting alkylnitrosoureas, nitrosocarbamates and nitrosoguanidines containing methyl or ethyl groups, replacement of hydrogen with deuterium in the alkyl group did not lead to any significant change in carcinogenic activity, whether tested by repeated skin painting on mice (Lijinsky, 1982) or by gavage administration to rats (Lijinsky and Reuber, 1982a). These experiments indicate that oxidation of carbon–hydrogen bonds in the alkyl group is not part of the mechanism of carcinogenesis by these compounds. Instead, the results support the idea that the directly acting N-nitroso compounds induce tumours by direct transfer of the unchanged alkyl group.

The carcinogenic potency of the α-deuterated cyclic nitrosamine nitrosoheptamethyleneimine was reduced compared with the unlabelled compound in rats; tumours of the esophagus were the major effect of this carcinogen (Lijinsky et al., 1982b). There was only a small apparent reduction in carcinogenic activity between α-deuterium-labelled nitrosoazetidine and the unlabelled compound (Lijinsky and Taylor, 1977b). When 2,6-dimethyldinitrosopiperazine was labelled with deuterium in the α- positions, there was no significant difference in potency (measured by the median week of death from tumours or in incidence of tumours of the esophagus) between the labelled and unlabelled compound; two dose rates, a factor of five apart, were used in this experiment, the results of which differed from those with the previously discussed esophageal carcinogen nitrosoheptamethyleneimine.

2,6-Dimethylnitrosomorpholine, structurally similar to 2,6-dimethyldinitrosopiperazine, also induced tumours of the esophagus in rats, but showed markedly different behaviour when labelled with deuterium in the α-position to the N-nitroso function. The deuterium-labelled morpholine derivative was significantly less carcinogenic than the unlabelled compound in rats (Lijinsky et al., 1980f). This suggested that there are differences between these two structurally similar compounds in the mechanism by which they are activated to carcinogenic forms. It is of some relevance that labelling of 2,6-dimethylnitrosomorpholine with deuterium in the methines at the β-positions increased the potency of the compound as an esophageal carcinogen in rats, by reducing oxidation at the β-positions and thereby enhancing oxidation at the α-positions. This competition between methylene or methine groups at various positions in a nitrosamine molecule for oxidative enzymes has often been reported, as has the tendency for each possible position in the molecule to be oxidised (Blattmann and Preussmann, 1974). It is possible that the tendency of the β-carbon to be oxidised is enhanced in α-deuterium-labelled dimethylnitrosomorpholine but not in dimethyldinitrosopiperazine, or that the second nitroso group in the latter changes the activity of the molecule so as to blur the difference between the α-deuterated compound and the same without deuterium.

An unexpected and striking finding about the effects of deuteration on carcinogenesis by dimethylnitrosomorpholine was that in hamsters the effects were precisely the reverse of those in rats. That is, the α-deuterium-labelled compound was more potent and the β-deuterium-labelled compound was less potent than the unlabelled compound (Rao et al., 1981); in hamsters DMNM induces tumours of the liver, lung, nasal mucosa and of the pancreas ducts (Lijinsky et al., 1982f), but dimethyldinitrosopiperazine induces no pancreas tumours (Lijinsky et al., 1983f). Gingell et al. (1976) demonstrated that β-oxidation is a favoured reaction in metabolism of DMNM in hamsters, leading to formation of

2-hydroxypropylnitroso-2-oxopropylamine, which has different carcinogenic properties. In rats β-oxidation of DMNM also occurs, with formation of much the same pattern of metabolites as is seen in the hamster (Underwood and Lijinsky, 1982). The inference from these results is that the pathways leading to formation of tumours in the two species are different, perhaps accounting in part for the very different types of tumour induced by this compound in the two species. A reasonable explanation is that β-oxidation is important in causing the tumours found in the hamster, but is not important in the genesis of the esophageal tumours found in the rat. As will be described elsewhere there is an association between nitrosamines containing a propyl group oxygenated on the 2-carbon (or convertible into a molecule having this structure) and the induction of pancreas tumours in hamsters.

Although most nitrosamines are believed to be activated by oxidation at the α-carbon, in by no means all of them is it the rate-limiting step as revealed by the effects of replacement of hydrogen by deuterium. For example, in the case of one of the 2-oxygenated propylnitrosamines already mentioned as inducing pancreas tumours in hamsters, nitrosobis-(2-oxopropyl)amine (NBOPA), the derivative having deuterium in the α-methylene groups did not differ in carcinogenic potency from the unlabelled compound (Lijinsky et al., 1988f). This was so whether treatment was of rats by gavage or in drinking water; in the drinking water experiment there was a small, but statistically significant, decrease in mortality rate of males given NBOPA-d_4 compared with the unlabelled compound; it suggests a possible contribution of α-oxidation to carcinogenesis by this compound, although not in the liver, since few liver tumours were seen in male rats.

Other compounds which showed little or no change in carcinogenic effectiveness when α-hydrogens were replaced by deuterium were the bladder carcinogen methylnitroso-n-dodecylamine and the rat esophageal carcinogens methylnitrosocyclohexylamine, methylnitroso-n-butylamine and methylnitroso-2-phenylethylamine. The carcinogenicity of methylnitrosododecylamine was unaffected by deuterium substitution, whether at the α-methylene or in the N-methyl group. The insolubility of this compound in water made it necessary to administer it by gavage, and this pulsed dosing might obscure a small deuterium isotope effect on carcinogenesis. Nevertheless, it would be incorrect to assume that α-oxidation at either the methyl- or the dodecyl-group is a rate-limiting step in carcinogenesis by this compound. This is not an unreasonable assumption, since we know that extensive chain shortening by β-oxidations of the dodecyl chain occurs, leading to formation of the proximate bladder carcinogen, methylnitroso-2-oxopropylamine (Lijinsky, 1983b; Singer et al., 1981). Methyl-d_3-nitrosocyclohexylamine did not differ in carcinogenic potency from the unlabelled compound, given to rats in drinking water at two dose rates (Lijinsky and Reuber, 1980b), suggesting that oxidation of the methyl group was not a rate-limiting step in carcinogenesis. The comparable derivative labelled with deuterium in the methine of the cyclohexyl group was not tested, and this is more likely to be rate-limiting, since methylation of DNA in the rat esophagus was shown by several methylnitrosoalkylamines that induce esophageal tumours (Von Hofe et al., 1987).

Supporting this to some extent are the results of comparing the carcinogenicity of methylnitroso-*n*-butylamine labelled with deuterium in the methyl or in the α-methylene. In this case, the methyl-labelled compound was somewhat more potent than the unlabelled compound, indicating that suppression of oxidation of the methyl group can, *pari passu*, increase the rate at which the enzymes oxidise the α-methylene of the butyl group, which leads to increased formation of the methylating agent (Lijinsky et al., 1980d). However, methylnitrosobutylamine with deuterium in the α-methylene was not less carcinogenic than the unlabelled compound in the same experiment, as was expected; this could be due to limitations in the sensitivity of the system.

Another potent esophageal carcinogen in which the effect of deuterium substitution has been investigated is methylnitroso-2-phenylethylamine. In this case deuteration of the methyl group did not change the potency of the compound, whereas deuterium in the α-methylene led to an increase in carcinogenic potency (Lijinsky et al., 1982c). This effect is contrary to that with methylnitrosobutylamine and indicates that a reduction in oxidation of the phenylethyl increases carcinogenic potency, which is in turn dependent on increased oxidation of the methyl group, which would lead to formation of a phenylethylating agent. It is notable that 2-phenylethylnitrosourea (which also produces a phenylethylating agent) is a powerful bacterial mutagen, much more so than ethylnitrosourea (see Chapter 6), so that phenylethylation can be considered a mutagenic lesion. The results suggest that a single mechanism of induction of esophageal tumours in rats by nitrosamines (such as methylation of DNA) does not operate universally.

1-Methylnitroso-3,3-diethylurea induces tumours of the central nervous system in rats when given by oral administration (Lijinsky et al., 1980b), as do methylnitrosourea and 1,3-dimethylnitrosourea (Druckrey et al., 1967). The latter two compounds are directly acting alkylating agents, and directly acting bacterial mutagens, whereas the trialkylnitrosourea requires metabolic activation for mutagenesis (Lijinsky and Andrews, 1983) and, presumably, for carcinogenesis. As a probe of the mechanism of action of this compound *in vivo*, hydrogen in the methyl group of methylnitrosodiethylurea was replaced with deuterium, and the compound was given to rats at equimolar doses with the unlabelled compound. The deuterium-labelled derivative was less carcinogenic than the unlabelled compound, indicating that oxidation of the methyl group was a rate-limiting step in carcinogenesis; the corollary is that methylation by methylnitrosodiethylurea is not its mechanism of inducing tumours, since oxidation of the methyl group is involved, perhaps in the manner described by Kolar and Carubelli (1979) for a dimethyl-aryltriazene, which formed a conjugate through an N-methyl group.

Hecht has also studied the effects of deuterium substitution on carcinogenesis by tobacco-specific nitrosamines, and has used deuterium-labelled nitrosamines to study their mechanisms of action (Hecht *et al.*, 1981). In one experiment nitrosonornicotine was labelled with deuterium in the 2 (methine) position of the pyrrolidine ring and in the 5 (methylene) position. Neither the 2-deuterated compound nor that labelled in both 2 and 5 positions showed any significant difference in carcinogenic effectiveness from the unlabelled compound.

Of the studies of the effect of deuterium substitution on carcinogenic activity of N-nitroso compounds, the most thorough has been with methylnitrosoethylamine. The number of possible sites for deuterium substitution is any or all of three carbon atoms, in the methyl group or at the 1-carbon or 2-carbon of the ethyl group. Because this nitrosamine is asymmetric, and can give rise to either a methylating agent or an ethylating agent, or both, deuterium substitution in one position or another has a variety of effects on carcinogenicity (Lijinsky and Reuber, 1980a). Deuterium in the methyl group reduces carcinogenic potency, as measured by a decrease in the rate of mortality of rats given the deuterated compound compared with those given an equimolar dose of the unlabelled compound. On the other hand, deuterium in the α-methylene of the ethyl group greatly increased carcinogenic potency, increasing the rate of mortality from tumours (median week of death from tumours 37 versus 63) and increasing the incidence of liver tumours considerably (Table 7.4). The previously discussed competition for oxidising enzymes between the methyl group and the α-methylene underlies the deuterium isotope effect, and indicates that suppression of oxidation of the ethyl group (Fig. 7.7) – which leads to formation of a methylating agent – increases oxidation of the methyl group – which leads to formation of an ethylating agent. Therefore, it can be concluded that in rat liver, ethylation of DNA by

Table 7.4. *Induction of tumours in 20 male rats by deuterated methylnitrosoethylamines in drinking water*

Compound	Concentration (mg/L)	Median week of death	No. of animals with tumours of		
			Liver	Esophagus	Nasal
MNEA	30	63	9	1	4
	6	108	3	0	0
MNEA-d_3	30	81	9	1	1
	6	102	4	0	0
MNEA-d_2	30	37	16	1	0
	6	91	9	0	0
MNE(2-d_3)A	30	70	20	14	0
	6	99	6	0	0
MNEA-d_5	30	29	20	5	0
	6	85	12	0	0
MNEA-d_8	30	80	13	7	1
	6	107	5	0	0

MNEA is more important in inducing tumours than is methylation by this compound (Lijinsky and Reuber, 1980a). To some extent this assumption is supported by the fact that NDEA is a considerably more potent carcinogen in rats than is NDMA (Lijinsky and Reuber, 1981b).

A further interesting complication is introduced when the 2-methyl of the ethyl group of MNEA is labelled with deuterium. In that case, MNEA-d_5, in which the ethyl group is entirely labelled with deuterium, is even more potent than MNEA-d_2 (Table 7.4); the median week of death from tumours was 29 (MNEA-d_5) versus 37 (MNEA-d_2) versus 63

Fig. 7.7. Effect of deuterium labelling on carcinogenicity of MNEA.

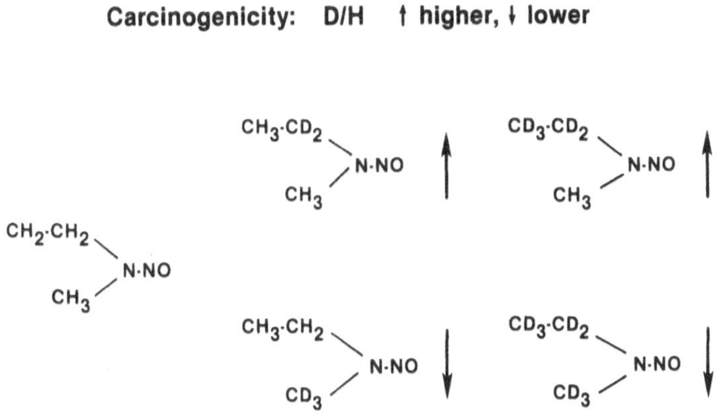

Carcinogenicity: D/H ↑ higher, ↓ lower

(MNEA), and the incidence of liver tumours was 100% (Lijinsky and Reuber, 1980a). In addition there was a significant incidence of tumours of the esophagus, which are induced by MNEA itself only at much higher dose rates (Lijinsky and Reuber, 1981b). The effect of replacing the methyl group in MNEA-d_5 with a trideuteromethyl group (as in MNEA-d_8) is to reduce the carcinogenic potency considerably (median week of death 80 versus 29); the incidence of liver tumours was also decreased, but there was a considerable incidence of tumours of the esophagus, not significantly different from that induced by MNEA-d_5. The effect of deuterium labelling of the N-methyl group in reducing the carcinogenic potency of MNEA therefore applies equally to MNEA containing a fully deuterated ethyl group. The presence of deuterium on both α-carbons seems to cancel out, and the overpowering effect of the deuterium in the α-methylene shows itself. However, the effect of the deuterium in the 2-methyl of the ethyl group in causing the compound to induce esophageal tumours is unexplained by the effects of methylation or ethylation of DNA, respectively.

This effect was reproduced in an experiment in which only the 2-methyl of the ethyl group was labelled with deuterium (Lijinsky et al., 1982e). In that study there was little difference in mortality rate between the male rats given the deuterated compound and those given the unlabelled MNEA, at either of two dose rates, but the incidence of liver tumours was twice as great in both groups given the deuterated compound. However, the biggest difference was that 14 rats (70%) at the higher dose of the labelled compound had tumours of the esophagus (compared with one rat in the comparable MNEA group; at the lower dose no esophageal tumours were seen).

The meaning of these results is not entirely clear, although they seem to indicate that oxidation of the terminal methyl of the ethyl group in MNEA is related to induction of liver tumours, and blockage of this oxidation (by deuterium substitution) leads to induction of esophageal tumours. In support of this, methylnitroso-n-propylamine, which can be considered a derivative of MNEA in which one of the hydrogens of the terminal methyl is replaced by a methyl group, induces esophageal tumours exclusively, and no liver tumours (Lijinsky et al., 1983d). It might be expected, if this hypothesis is correct, that hydroxyethylation of DNA in the liver by MNEA is important in induction of liver tumours in rats, while methylation or ethylation is important in induction of tumours of the esophagus. This explanation is too simple, since hydroxyethylation does not seem to be an important effect of MNEA in rat liver (Von Hofe et al., 1986b), nor does oxidation of MNEA to methylnitroso-2-

hydroxyethylamine take place to other than a small extent in rats *in vivo*. Also, the latter compound is a less potent carcinogen than MNEA (Koepke *et al.*, 1988*a*) and induces liver tumours but few esophageal tumours. It is probable that the complex interplay of enzymes that oxidise the three carbon atoms in MNEA is the key to the explanation of the effects of deuterium substitution on carcinogenesis by MNEA. We know too little about the enzymes involved, especially those other than the mixed function oxidases which, surprisingly, do not convert MNEA *in vitro* to mutagenic metabolites. This nitrosamine is inactive in the 'Ames' assay, even using preincubation with the rat liver microsomal fraction, which causes the otherwise inactive methylnitroso-*n*-propylamine to be mutagenic (Rao, 1984). The possibility that products of oxidation of MNEA, other than alkyldiazonium ions arising from α-hydroxylation, are involved in carcinogenesis cannot be overlooked.

The most perplexing aspect of the deuterium-labelling experiments with nitrosamines is that the carcinogenesis results are often clear-cut, suggesting that a single rate-limiting step controls the outcome of the complex series of events resulting in tumours that kill the animal. Yet there are only small effects of the presence of deuterium on alkylation of DNA *in vivo* (Swann *et al.*, 1983), which is supposed to be the significant initial event in carcinogenesis. Deuterium labelling of MNEA in the methyl group and in the ethyl group has profound effects on carcinogenesis, yet deuterium has no appreciable effect on either methylation or ethylation of DNA compared with the unlabelled compound (Lijinsky, 1988*b*). This in spite of the effect observed by Keefer and his colleagues (Mico *et al.*, 1985) of deuterium labelling of NDMA on the rate of metabolism. This leads to the conclusion that overall metabolism is not directly connected with induction of tumours, but minor pathways, not readily measurable, are the important ones in leading to neoplastic transformation and tumour induction.

A side-effect of the results of replacement of hydrogen with deuterium on carcinogenic activity of nitrosamines is the need for care in evaluating studies of metabolism of nitrosamines conducted with tritium-labelled compounds. The carbon–tritium bond is, like the carbon–deuterium bond, stronger than the carbon–hydrogen bond, therefore needing more energy to break the former. So, in any study of metabolism using nitrosamines labelled with tritium and therefore containing methyl or methylene groups with a small proportion of molecules having a carbon–tritium bond in one position or another, the carbon–tritium bond will tend to be cleaved much more slowly than the carbon–hydrogen bonds which are more abundant. Therefore, retention or loss of the tritium label gives no indication of the rate of metabolism of the compound in those positions at which metabolism takes place. Studies of $^{14}C/^3H$ ratios are misleading in such experiments, although they have been used; the published results should be evaluated with particular care.

7.4.2 Effects of replacement of α-hydrogens by other groups

As mentioned previously, replacement of α-hydrogens in the potent carcinogen nitrosodiethylamine with methyl groups gives rise to nitrosodi-*iso*propylamine, which is a very weak carcinogen. On the other hand, replacement of α-hydrogens in nitrosodimethylamine (NDMA) by methyl groups gives rise to nitrosodiethylamine (NDEA), which is considerably more potent in rats than is NDMA, and the target organs change from liver, kidney and lung, to liver and esophagus (Lijinsky *et al.*, 1987*b*; Lijinsky and Kovatch, 1989*a*). Any theory attempting to explain the increase in carcinogenicity of NDEA compared with NDMA is complicated – even confounded – by the finding that in Syrian hamsters NDEA is a considerably weaker carcinogen than is NDMA, and both induce mainly liver tumours, with additionally some nasal mucosal tumours in the case of NDEA (Lijinsky *et al.*, 1987*b*). If alkylation of DNA in liver is the main mechanism by which the compounds induce tumours, then ethylation is more effective in the rat liver than is methylation, but methylation is more effective in hamster liver than is ethylation; the extent of methylation or ethylation, respectively, is very similar in both species (Lijinsky, 1988*b*), indicating that the mechanism of metabolic activation to the respective alkylating agents is similar in both species. The disparate extents of DNA alkylation and of potency in inducing liver tumours suggest that there are other essential actions of each of the carcinogens in order to give rise to visible tumours. In the absence of information about these other processes it is improper to label them 'promoting' effects, although they are important in the progression of transformed cells to tumours. These aspects of carcinogenesis will be discussed elsewhere, but it is important to note that substantial DNA alkylation by these compounds was measured in organs in which they do not induce tumours (hamster kidney and lungs by NDMA, rat kidney by MNEA) and that, of course, most of the affected cells in the liver do not progress to tumours.

In parallel with the weak carcinogenicity of nitrosodi-*iso*propylamine, its homologue nitrosodi-*sec*-butylamine, which can be considered nitroso-di-*n*-propylamine (a potent inducer of liver and esophagus tumours in rats) in which the α-hydrogens are replaced by methyl groups (Fig. 7.1), was not carcinogenic (Lijinsky and Taylor, 1979*b*). Again, it seems that blockage of oxidation of the α-carbon in these nitrosamines (nitrosodiethylamine and nitrosodi-*n*-propylamine) reduces or eliminates carcinogenic activity. Methyl substitution on the 2-carbon of nitrosodi-*n*-propylamine, to form nitrosodi-*iso*butylamine, also resulted in sharply reduced carcinogenic potency and an orientation to induce tumours of the nasal mucosa in rats, as opposed to liver and esophagus tumours (Lijinsky and Taylor, 1979*b*). Nitrosodi-*n*-butylamine, the compound which is nitrosodi-*n*-propylamine with methyl substituents in the 3-carbon, also is a weaker carcinogen than the latter, and induces in rats (and in hamsters) tumours of the bladder, as well as tumours of the liver, lung and forestomach in rats (Lijinsky and Reuber, 1983*a*). The results of treatment of rats or Syrian hamsters with *N*-nitroso compounds of many structures, by several modes of administration, mainly from the author's laboratory, are listed in Tables 7.5a and 7.5b.

In cyclic nitrosamines, as in acyclic ones, methyl substitution on the α-carbons also reduces – often drastically – or eliminates carcinogenic activity. For example, 2,5-dimethylnitrosopyrrolidine induced no tumours after administration to rats of doses equimolar with those of nitrosopyrrolidine itself that induced 100% incidence of liver tumours (Lijinsky and Taylor, 1976*d*). Similarly, 2,6-dimethylnitrosopiperidine (Fig. 7.8) induced no tumours in rats in contrast with the induction of 100% incidence of esophageal tumours by an equimolar dose of its parent, nitrosopiperidine (Lijinsky and Taylor, 1975*a*). 2,2,6,6-Tetramethylnitrosopiperidine was also non-carcinogenic, as would be expected. 2-Methylnitrosopiperidine was less potent than nitrosopiperidine, and induced liver tumours as well as tumours of the esophagus in rats, perhaps because the animals lived somewhat longer and had time to develop visible liver tumours; however, by no means can this be assumed to be the reason for the appearance of liver tumours. Rats treated with many other esophageal carcinogens do not usually develop liver tumours, no matter how long they live.

The question must be asked why replacement of a single hydrogen atom with methyl at one or both α-carbons of nitrosopiperidine reduces or eliminates carcinogenic activity? There is some evidence that the position of the nitroso group in the two rotamers of an asymmetric nitrosamine is an important determinant of enzymic activation leading to carcinogenesis (Chapter 4). It is probable that one of the rotamers of nitroso-2-methylpiperidine is less readily oxidised than the opposite rotamer, and the number of appropriately orientated molecules presented to the enzyme is reduced in the equilibrium mixture. In the α-dimethyl derivatives of nitrosopiperidine and nitrosopyrrolidine it is likely that access to the enzyme is simply too hindered. The two stereoisomers of 2-methylnitroso-

Table 7.5a. *Carcinogenic effects of N-nitroso compounds in standard chronic experiments – comparative effects in rats and Syrian hamsters*

	Dose (μmol/wk)	Total dose per animal† (mmol)	Treatment (Gavage, G; Drinking water, W)	Median time of death (weeks)	% of treated animals with tumours of each type
Acyclic nitrosamines					
N-nitroso-					
dimethylamine	45	1.3	W – rat ♀	31	Liver 100
	18	0.5	W – rat ♀	79	Liver 85, Lung 10
	7	0.22	W – rat ♀	91	Liver 70
	54	1.6	G – rat ♂	45	Lung 80, Kidney 50, Liver 50, Nasal 15
	27	0.8	G – rat ♀	59	Kidney 83, Lung 75, Liver 42
	54	0.38	G – rat ♀	62	Kidney 100, Lung 77, Liver 62, Thyroid 15
	54	1.6	I.U. – rat ♀	59	Kidney 100, Lung 92, Thyroid 17
	20	0.08	G – hamster ♂	43	Liver 78, Nasal 14
	10	0.2	G – hamster ♂	29	Liver 60
acetoxymethyl-methylamine			*I.P. – rat	18	Intestine, Nervous system, Lung
diethylamine	110	1.9	W – rat ♀	18	Liver 100, Esophagus 90, Forestomach 20
	44	1.0	W – rat ♀	26	Esophagus 95, Liver 60
	70	2.0	W – rat ♀	25	Liver 100, Esophagus 17
	70	2.0	W – rat ♂	30	Liver 100, Esophagus 100
	18	0.5	W – rat ♀	35	Esophagus 100
	7	0.2	W – rat ♀	45	Esophagus 100
	2.7	0.08	W – rat ♀	100	Esophagus 90, Liver 25

	1.1	0.07	W – rat ♀	102	Esophagus 85, Liver 25
	0.4	0.05	W – rat ♀	97	Esophagus 70, Liver 35
	49	1.0	G – rat ♂	36	Liver 100, Nasal 88, Esophagus 69, Kidney 15
	49	1.0	G – rat ♀	29	Liver 100, Kidney tubules 38, Esophagus 25
	24	0.7	G – rat ♀	33	Liver 92, Nasal 83, Esophagus 50
	49	1.2	I.U. – rat ♀	31	Liver 92, Esophagus 67, Nasal 33
	20	0.5	G – hamster ♂	47	Liver 85, Nasal 30
	10	0.2	G – hamster ♂	69	Liver 30, Trachea 30, Nasal 20
di-*n*-propylamine	35	1.0	W – rat ♂	31	Esophagus 100, Forestomach 60, Tongue 30
	68	2.0	G – rat ♂	34	Liver 100, Esophagus 60, Nasal 70, Lung 40
	34	1.0	G – rat ♀	54	Liver 75, Nasal 67, Esophagus 33
di-*iso*-propylamine	460	18	W – rat ♂	45	Nasal 83, Liver 20
	70	3.5	W – rat ♂	56	Nasal 53
diallylamine			*W – rat		0
di-*n*-butylamine	68	2.0	G – rat ♂	94	Liver 60, Lung 65, Forestomach 50, Bladder 30
di-*sec*-butylamine	70	3.5	W – rat ♀	—	0
di-*iso*-butylamine	280	14	W – rat ♀	55	Nasal 70, Trachea 45, Lung 30
	70	2.1	W – rat ♀	102	Lung 27
di-*n*-octylamine	220	11	G – rat ♂		0
	220	11	G – rat ♀		0
methylethylamine	170	5	W – rat ♂	31	Liver 100, Esophagus 35
	34	1	W – rat ♂	63	Liver 45, Nasal 20
	7	0.2	W – rat ♂	105	Liver 30
	52	1.6	G – rat ♂	38	Liver, 95, Nasal 50, Lung 44
	52	1.4	G – rat ♀	30	Liver 100, Lung 25

Table 7.5a (cont.)

	Dose (μmol/wk)	Total dose per animal† (mmol)	Treatment (Gavage, G; Drinking water, W)	Median time of death (weeks)	% of treated animals with tumours of each type
	52	1.6	I.U. – rat ♀	54	Liver 92, Lung 17, Kidney 25
	20	0.5	G – hamster ♂	35	Liver 95, Nasal 55, Pancreas 10, Trachea 10
	10	0.2	G – hamster ♂	81	Liver 89, Nasal 22
	10	0.15	G – hamster ♂	70	Liver 70, Nasal 20, Colon 20
methylhydroxyethylamine	51	2.0	G – rat ♂	45	Liver 35, Esophagus 20, Nasal 25
	51	2.0	G – rat ♀	40	Liver 70, Esophagus 50
methyldimethylamino-ethylamine	110	2.2	W – rat ♂	28	Esophagus 100, Nasal 67
	110	2.2	W – rat ♀	28	Esophagus 100, Nasal 50
	110	2.2	G – rat ♂	29	Esophagus 100, Nasal 75
	110	2.2	G – rat ♀	26	Esophagus 100, Nasal 42
	110	2.5	I.U. – rat ♀	36	Esophagus 100, Nasal 33
	55	3.8	G – hamster ♂	78	Liver 25, Nasal 33
	55	3.8	G – hamster ♀	93	Nasal 67, Lung 33, Colon 17
methyl-n-propylamine	14	0.3	W – rat ♀	28	Esophagus 100, Forestomach 35, Pharynx 35
	21	0.6	G – hamster ♂	53	Liver 83, Nasal 92, Lung 17
	21	0.6	G – hamster ♀	52	Liver 50, Nasal 83
methyl-n-butylamine	14	0.28	W – rat ♂	27	Esophagus 90, Forestomach 35, Tongue 40
	14	0.28	W – rat ♀	24	Esophagus 75, Forestomach 35, Pharynx 35

	5	0.12	W – rat ♂	30	Esophagus 100, Forestomach 25, Tongue 25
	41	0.8	G – rat ♂	26	Esophagus 100, Forestomach 50, Nasal 50, Tongue 30
	41	0.8	G – rat ♀	20	Esophagus 80, Nasal 20
	52	0.9	I.U. – rat ♀	22	Esophagus 100, Liver 25
	43	1.0	G – hamster ♂	26	Liver 75, Forestomach 75, Nasal 50, Lung 33
	43	1.0	G – hamster ♀	25	Nasal 75, Forestomach 50, Lung 33, Liver 17
	22	0.65	G – hamster ♂	52	Liver 75, Nasal 75, Forestomach 58, Lung 25
	22	0.65	G – hamster ♀	39	Nasal 92, Liver 50, Forestomach 42, Lung 17
methyl-*tert*-butylamine			*W – rat		0
methyl-*n*-amylamine			*W – rat		Esophagus
	43	1.1	G – hamster ♂	39	Forestomach 90, Lung 40, Nasal 20
	43	1.1	G – hamster ♀	31	Forestomach 91, Nasal 100, Lung 82, Liver 36
methylneopentylamine	17	0.4	W – rat ♂	32	Esophagus 80
methyl-*n*-hexylamine	35	0.7	W – rat ♂	25	Esophagus 100, Tongue 45, Forestomach 20
	220	3.5	G – rat ♂	17	Esophagus 95, Liver 55, Lung 65, Trachea 30
	86	2.4	G – rat ♂	31	Esophagus 100, Liver 100, Lung 45
	36	1.2	G – rat ♂	63	Lung 90, Liver 80, Esophagus 50
	53	1.6	I.U. – rat ♀	65	Lung 100, Uterus 67, Bladder 22
	43	2.1	G – hamster ♂	53	Forestomach 91, Liver 73, Lung 55, Bladder 45, Nasal 45

Table 7.5a (cont.)

	Dose (μmol/wk)	Total dose per animal† (mmol)	Treatment (Gavage, G; Drinking water, W)	Median time of death (weeks)	% of treated animals with tumours of each type
	43	2.1	G – hamster ♀	56	Forestomach 83, Lung 42, Bladder 42, Liver 67, Nasal 42
methyl-*n*-heptylamine	89	3.0	W – rat ♂	42	Esophagus 95, Lung 65, Liver 30
	220	5.4	G – rat ♂	27	Liver 100, Trachea 60, Lung 45
	89	2.7	G – rat ♂	51	Liver 100, Lung 100
	43	1.8	G – hamster ♂	45	Lung 100, Liver 83, Nasal 67, Forestomach 67
	43	1.8	G – hamster ♀	64	Lung 80, Liver 80, Forestomach 80, Nasal 40
methyl-*n*-octylamine	100	5	W – rat ♂	56	Esophagus 100, Lung 70, Bladder 10
	220	6.5	G – rat ♂	35	Liver 85, Bladder 70, Lung 60
	43	1.8	G – hamster ♂	52	Lung 42, Liver 75, Bladder 33, Forestomach 67, Nasal 17
	43	1.8	G – hamster ♀	57	Liver 75, Lung 50, Nasal 25, Bladder 25, Forestomach 25
methyl-*n*-nonylamine	220	6.5	G – rat ♂	43	Liver 90, Lung 60
methyl-*n*-decylamine	216	6.5	G – rat ♂	81	Bladder 85, Lung 40
methyl-*n*-undecylamine	215	6.5	G – rat ♂	48	Liver 90, Lung 80
methyl-*n*-dodecylamine	210	6.5	G – rat ♂	47	Bladder 67, Lung 33
	210	6.5	G – rat ♀	49	Bladder 100
	105	3.3	G – rat ♂	101	Bladder 95, Lung 25, Liver 15

Compound					
methyl-*n*-tetradecylamine	105	3.2	G – rat ♀	53	Bladder 85, Lung 60
	39	1.1	G – hamster ♂	58	Bladder 87, Forestomach 40
	43	1.3	G – hamster ♀	54	Bladder 87, Lung 40
	220	6.5	G – rat ♂	84	Bladder 100, Lung 15
methylphenylamine	37	1.8	W – rat ♂	78	Esophagus 85, Forestomach 20
	37	1.8	W – rat ♀	90	Esophagus 80, Forestomach 15
	37	1.8	G – hamster ♀	61	Spleen 28, Liver 32
methyl-4-fluorophenylamine	37	1.8	W – rat ♂	96	Esophagus 90, Liver 25
methyl-4-nitrophenylamine	37	1.8	W – rat ♂	114	Liver 35
methylcyclohexylamine	35	1.1	W – rat ♀	29	Esophagus 95
	9	0.26	W – rat ♀	34	Esophagus 95
	3.5	0.1	W – rat ♀	58	Esophagus 95
	18	0.9	G – hamster ♂	103	Lung 25
	18	0.9	G – hamster ♀	96	Liver 25
methylbenzylamine	9	0.2	W – rat ♂	24	Esophagus 85, Tongue 35
methyl-2-phenylethylamine	70	1.5	W – rat ♂	20	Esophagus 95, Forestomach 55
	17	0.5	W – rat ♂	31	Esophagus 95, Forestomach 65
	6	0.19	W – rat ♂	33	Esophagus 95, Forestomach 40
	2	0.06	W – rat ♂	110	Esophagus 80, Forestomach 80, Tongue 30
	0.7	0.02	W – rat ♂	111	Esophagus 20, Forestomach 40
diphenylamine	4200	420	Food – rat ♂		Bladder 40
	2800	280	Food – rat ♀		Bladder 86
phenylbenzylamine	2100	140	Food – rat ♂	69	Esophagus 50, Spleen 50, Forestomach 20
methyltrifluoroethylamine	44	1.2	W – rat ♂	36	Esophagus 100, Forestomach 35, Tongue 20
	18	0.5	W – rat ♂	94	Esophagus 90, Nasal 15
methyl-2-hydroxypropylamine	85	1.9	W – rat ♀	29	Esophagus 100, Lung 75

Table 7.5a (cont.)

	Dose (μmol/wk)	Total dose per animal† (mmol)	Treatment (Gavage, G; Drinking water, W)	Median time of death (weeks)	% of treated animals with tumours of each type
	21	0.6	W – rat ♂	59	Esophagus 45, Nasal 90, Lung 25, Liver 10
	21	0.6	W – rat ♀	50	Esophagus 75, Nasal 95, Lung 30, Liver 25
	85	0.4	G – rat ♀	40	Kidney 100, Esophagus 88, Lung 75, Thyroid 63
	34	0.6	G – rat ♂	32	Esophagus 100, Liver 17, Nasal 17
	34	0.6	G – rat ♀	24	Esophagus 100
	34	1.0	I.U. – rat ♀	83	Bladder 67, Nasal 67, Esophagus 25
	17	0.4	G – hamster ♀	18	Liver 65, Pancreas 45, Lung 45, Nasal 15
	8	0.18	G – hamster ♀	25	Nasal 90, Liver 35, Lung 20
methyl-3-hydroxypropylamine	51	4.0	G – rat ♂	100	Liver 40, Lung 35
	51	4.0	G – rat ♀	100	Liver 60, Lung 35
methyl-2-oxopropylamine	86	1.8	W – rat ♂	23	Esophagus 100, Trachea 55, Nasal 25, Liver 10
	86	1.8	W – rat ♀	24	Esophagus 95, Liver 75, Nasal, 65, Trachea 30
	22	0.65	W – rat ♂	40	Esophagus 70, Forestomach 20, Tongue 20
	22	0.65	W – rat ♀	33	Esophagus 85, Liver 10

Compound				
	110	1.7	G – rat ♂	14	Esophagus 60, Nasal 20, Trachea 10
	86	0.9	G – rat ♂	22	Esophagus 92, Kidney pelvis 17
	34	0.55	G – rat ♂	22	Esophagus 100, Trachea 15
	34	0.55	G – rat ♀	22	Esophagus 100
	17	0.5	G – rat ♂	42	Esophagus 92
	34	1.0	I.U. – rat ♀	43	Bladder 75, Nasal 58
	9	0.2	G – hamster ♀	26	Liver 80, Nasal 95
	17	0.24	G – hamster ♀	22	Liver 83, Nasal 67
methyl-2,3-dihydroxypropylamine	85	3.4	W – rat ♀	44	Nasal 75, Esophagus 75, Tongue 15
	22	0.9	W – rat ♀	84	Liver 45, Lung 40, Esophagus 20, Nasal 20
methylamino-3-pyridylbutanone (NNK)	18	0.7	G – hamster ♀	66	Nasal 50, Forestomach 10
			*S.C. – rat		Lung, Liver, Nasal
	51	1.3	G – rat ♀	42	Liver 83, Lung 25, Nasal 50
	51	1.5	I.U. – rat ♀	70	Nasal 50, Lung 42, Liver 33
	1.1	0.10	G – hamster ♂	100	Liver
			*S.C. – hamster		Trachea, Nasal, Lung
			*W – rat		Lung, Pancreas
methylamino-3-pyridylbutanol	1.1	0.10	G – hamster		0
methylmethoxyamine	280	14	W – rat ♂		0
	280	14	W – rat ♀		0
methyl-3-carboxpropylamine	205	12	W – rat ♂	66	Bladder 95, Liver 10
	205	12	W – rat ♀	66	Bladder 100
	410	11.5	W – rat ♀	52	Bladder 90
	205	6.2	I.U. – rat ♀	55	Bladder 75, Kidney pelvis 33
diethanolamine	1870	84	W – rat ♂	58	Liver 100, Nasal 95, Kidney 25, Esophagus 25

Table 7.5a (cont.)

	Dose (μmol/wk)	Total dose per animal† (mmol)	Treatment (Gavage, G; Drinking water, W)	Median time of death (weeks)	% of treated animals with tumours of each type
	1870	84	W – rat ♀	47	Liver 100, Nasal 15, Kidney 10
	750	37	W – rat ♂	69	Liver 100, Nasal 65, Kidney 10
	750	37	W – rat ♀	61	Liver 100, Nasal 70, Kidney 20
	300	22	W – rat ♂	86	Liver 100, Nasal 56
	300	22	W – rat ♀	78	Liver 100, Nasal 31
	300	15	W – rat ♂	90	Liver 100, Nasal 31
	300	15	W – rat ♀	84	Liver 100, Nasal 31
	120	6	W – rat ♂	102	Liver 70
	120	6	W – rat ♀	115	Liver 100, Kidney 11
	112	5.6	W – rat ♀	101	Liver 75
	48	4.8	W – rat ♂	108	Liver 55, Kidney 10
	48	4.8	W – rat ♀	113	Liver 70
	21	2.1	W – rat ♂	106	Liver 15
	21	2.1	W – rat ♀	110	Liver 25
	750	22	G – rat ♂	40	Nasal 100, Liver 60, Esophagus 50, Tongue 20
	600	14	G – hamster ♀	35	Nasal 90
bis-(2-methoxyethyl)amine	70	3.5	W – rat ♀	63	Liver 100, Nasal 33
bis-(2-ethoxyethyl)amine	70	3.5	W – rat ♀	92	Liver 67
bis-(2,2-diethoxyethyl)amine	70	3.5	W – rat ♀		0
ethanol-2-hydroxypropylamine	135	6.7	W – rat ♀	59	Liver 85, Esophagus 25, Nasal 30
	123	3.5	G – hamster ♀	35	Nasal 80, Liver 20, Pancreas 15, Lung 20

Compound				Tumor sites
ethanol-2-oxopropylamine	34	1.7	W – rat ♀	113	Liver 35, Tongue 10
	42	1.9	G – hamster ♀	56	Liver 80, Pancreas 50, Forestomach 10
ethanol-2,3-dihydroxypropylamine	134	6.7	W – rat ♀	120	Liver 50
	34	3.5	W – rat ♀	119	Liver 40, Forestomach 10
	120	5.5	G – hamster ♀	67	Trachea 15
allylethanolamine	135	6.7	W – rat ♀	54	Nasal 95, Liver 30, Esophagus 15
allyl-2-hydroxypropylamine	140	6.9	W – rat ♀	50	Nasal 85, Liver 70, Esophagus 30
	125	3.8	G – hamster ♀	35	Nasal 85, Trachea 10
allyl-2,3-dihydroxypropylamine	140	6.9	W – rat ♀	47	Esophagus 85, Nasal 80
	125	4.4	G – hamster ♀	36	Nasal 65, Trachea 35, Forestomach 15
allyl-2-oxopropylamine	34	1.7	W – rat ♀	74	Liver 80, Esophagus 35
	42	1.7	G – hamster ♀	50	Nasal 60, Liver 25
2-oxopropyl-2-hydroxypropylamine	140	2.9	W – rat ♀	24	Esophagus 95, Liver 70, Lung 45
	34	1.4	W – rat ♀	41	Esophagus 90, Liver 80, Lung 65, Nasal 30
	34	1.55	W – rat ♂	69	Esophagus 50, Liver 45, Lung 100, Colon 30, Nasal 25, Thyroid 25
	31	1.1	G – rat ♂	45	Esophagus 100, Lung 100, Nasal 90, Liver 30, Thyroid 30, Colon 20
	75	2.2	G – rat ♂	34	Esophagus 100, Lung 100, Nasal 80, Thyroid 80, Kidney pelvis 50, Bladder 30
	31	1.1	I.U. – rat ♀	48	Bladder 17, Liver 17
	38	0.8	G – hamster ♀	27	Liver 80, Pancreas 70, Lung 10
	34	0.8	W – rat ♀	26	Esophagus 100, Forestomach 45, Tongue 40, Lung 25
2-oxopropyl-2,3-dihydroxypropylamine	8	0.26	W – rat ♀	49	Esophagus 85, Forestomach 75, Tongue 70, Lung 15

Table 7.5a (cont.)

	Dose (µmol/wk)	Total dose per animal† (mmol)	Treatment (Gavage, G; Drinking water, W)	Median time of death (weeks)	% of treated animals with tumours of each type
	38	1.3	G – hamster ♀	38	Forestomach 75, Pancreas 25, Liver 10
2-hydroxypropyl-2,3-dihydroxypropylamine	136	3.5	W – rat ♀	27	Esophagus 75, Forestomach 10
	34	0.8	W – rat ♀	24	Esophagus 90, Forestomach 55, Tongue 35
	8	0.3	W – rat ♀	63	Esophagus 90, Tongue 75, Forestomach 50
	37	1.5	G – hamster ♀	57	Forestomach 65, Pancreas 10
bis-(2,2,2-trifluoroethyl)amine	18	0.5	W – rat ♀		0
bis-(2-chloroethyl)amine	13	0.4	G – rat ♀	86	Forestomach 83
bis-(2-chloropropyl)amine	68	1.4	G – rat ♀	55	Forestomach 10, Liver 10
bis-(2-cyanoethyl)amine	70	3.5	W – rat ♀		0
bis-(2-cyanomethyl)amine			*W – rat		0
iminodiacetic acid	1200	90	W – rat ♂		0
	1200	90	W – rat ♀		0
bis-(2-hydroxypropyl)amine	140	5.7	W – rat ♀	42	Esophagus 80, Nasal 35
	67	3.3	W – rat ♀	74	Nasal 40, Liver 20
	250	7.4	G – rat ♂	39	Esophagus 95, Nasal 75, Lung 65, Thyroid 50, Liver 35
	120	4	G – hamster ♀	32	Nasal 80, Pancreas 50, Lung 30, Liver 20

Compound	Dose	Animal	N	Tumor sites
bis-(2-oxopropyl)amine	35	W – rat ♂	60	Lung 100, Thyroid 70, Liver 20, Kidney 25
	35	W – rat ♀	61	Liver 90, Lung 75
	76	G – rat ♂	33	Lung 100, Thyroid 90, Bladder 80, Kidney pelvis 50
	32	G – rat ♂	44	Lung 100, Thyroid 90, Bladder 60
	32	G – rat ♀	59	Liver 92, Lung 83, Thyroid 25
	32	I.U. – rat ♀	87	Bladder 83, Lung 50, Kidney 42, Liver 17
	15	G – hamster ♀	25	Liver 100, Pancreas 50, Lung 20
	38	G – hamster ♀	22	Liver 100, Pancreas 36
1-acetoxydibenzylamine	37	G – rat ♀	120	Forestomach 45, Liver 25, Lung 25
Cyclic nitrosamines				
N-nitroso-				
azetidine	200	W – rat ♀	53	Liver 100
	67	W – rat ♂	96	Liver 30
	120	G – hamster ♂	64	Liver 35, Nasal 10, Forestomach
pyrrolidine	90	W – rat ♀	101	Liver 100
	200	W – rat ♂	80	Liver 93
	200	W – rat ♀	75	Liver 93
3-hydroxypyrrolidine		*W – rat		Liver
2,5-dimethylpyrrolidine	200	W – rat ♂		0
	200	W – rat ♀		0
normicotine	200	W – rat ♀	43	Nasal 100
		*S.C. – hamster		Trachea, Nasal
proline	1000	W – rat ♂		0
4-hydroxyproline	1000	W – rat ♂	108	Spleen 15
3,4-dichloropyrrolidine	60	W – rat ♀	44	Esophagus 93

Table 7.5a (cont.)

	Dose (μmol/wk)	Total dose per animal† (mmol)	Treatment (Gavage, G; Drinking water, W)	Median time of death (weeks)	% of treated animals with tumours of each type
3-pyrroline	100	6	W – rat ♂	97	Liver 13, Nasal 13
	100	6	W – rat ♀	73	Liver 53
piperidine	88	3.9	W – rat ♂	38	Nasal 93, Esophagus 87, Liver 13
	88	3.9	W – rat ♀	35	Nasal 100, Esophagus 67, Liver 27
	88	2.5	W – rat ♀	30	Esophagus 95, Forestomach 10
2-methylpiperidine	88	4.4	W – rat ♂	80	Esophagus 80, Nasal 33
	88	4.4	W – rat ♀	77	Liver 47, Esophagus 40, Nasal 27
3-methylpiperidine	88	4.4	W – rat ♂	55	Esophagus 93, Nasal 33
	88	4.4	W – rat ♀	52	Esophagus 93, Nasal 47
4-methylpiperidine	88	3.5	W – rat ♂	39	Esophagus 100, Nasal 80
	88	3.5	W – rat ♀	42	Esophagus 93, Nasal 67, Lung 13
3,5-dimethylpiperidine (82% cis, 16% trans)					
cis	88	4.4	W – rat ♀	61	Esophagus 80, Forestomach 20
	72	3.6	W – rat ♀	90	Esophagus 85, Forestomach 40, Tongue 30
trans	14	0.7	W – rat ♀	86	Esophagus 70, Tongue 40, Forestomach 30
2,6-dimethylpiperidine	88	4.4	W – rat ♂		0
	88	4.4	W – rat ♀		0
2,2,6,6-tetramethylpiperidine	88	4.4	W – rat ♂		0
	88	4.4	W – rat ♀		0

Compound			Animal		Tumor sites
pipecolic acid	88	4.4	W – rat ♂		0
	88	4.4	W – rat ♀		0
iso-nipecotic acid	88	4.4	W – rat ♂		0
	88	4.4	W – rat ♀		0
α-phenyl-2-piperidineacetic acid, methyl ester	91	4.6	W – rat ♂		0
	91	4.6	W – rat ♀		0
3-chloropiperidine	17	0.5	W – rat ♂	40	Esophagus 95, Tongue 40, Forestomach 20
4-chloropiperidine	17	0.5	W – rat ♂	37	Esophagus 90, Tongue 60, Forestomach 30
3,4-dichloropiperidine	34	0.5	W – rat ♂	20	Esophagus 87, Trachea 40, Nasal 13
	17	0.36	W – rat ♂	24	Esophagus 85
3,4-dibromopiperidine	37	1.0	W – rat ♂	36	Esophagus 80, Trachea 33
4-tert-butylpiperidine	88	5.3	W – rat ♀	71	Esophagus 90, Tongue 60, Forestomach 20
4-phenylpiperidine	88	4.4	W – rat ♀	62	Liver 85, Esophagus 70
4-cyclohexylpiperidine	88	4.4	W – rat ♀		0
anabasine			*Oral – rat		Esophagus
			*S.C. – hamster		0
anatabine			*S.C. – rat		0
4-piperidone	90	2.7	W – rat ♀	45	Liver 95, Esophagus 95, Forestomach 40, Nasal 40, Tongue 15, Lung 10
	90	3.2	W – rat ♂	46	Nasal 93, Liver 20, Esophagus 20
	90	3.2	W – rat ♀	43	Liver 93, Nasal 93, Esophagus 13
4-hydroxypiperidine	88	3.2	W – rat ♂	48	Nasal 87
	88	3.2	W – rat ♀	43	Nasal 100, Liver 60

Table 7.5a (cont.)

	Dose (μmol/wk)	Total dose per animal† (mmol)	Treatment (Gavage, G; Drinking water, W)	Median time of death (weeks)	% of treated animals with tumours of each type
3-hydroxypiperidine	88	3.2	W – rat ♂	45	Esophagus 93, Nasal 87, Liver 27
	88	3.2	W – rat ♀	43	Esophagus 100, Nasal 86, Liver 43
3-methyl-4-piperidone	89	2.3	W – rat ♀	27	Esophagus 90, Nasal 60, Trachea 30, Tongue 20
Δ²-dehydropiperidine	89	2.3	W – rat ♀	30	Esophagus 90, Forestomach 50, Tongue 15
Δ³-dehydropiperidine	89	2.3	W – rat ♂	28	Liver 100
	89	2.2	W – rat ♀	29	Liver 95, Esophagus 80, Nasal 10
	36	0.9	W – rat ♀	41	Esophagus 95, Nasal 20
	14	0.36	W – rat ♀	77	Esophagus 80, Liver 20, Nasal 10
	5.7	0.4	W – rat ♀	88	Esophagus 100, Liver 10
	2.3	0.23	W – rat ♀	108	Esophagus 95, Liver 45
	0.9	0.09	W – rat ♀	117	Forestomach 50, Liver 35
3,4-epoxypiperidine	36	1.1	G – rat ♀	46	Liver 33, Nasal 25, Esophagus 17
guvacoline	87	4.4	W – rat ♂		0
	87	4.4	W – rat ♀		0
1,3-oxazolidine	34	1.7	W – rat ♀	53	Liver 100
	45	1.6	G – hamster ♂	43	Liver 100, Forestomach 10
2-methyl-1,3-oxazolidine	34	1.7	W – rat ♀	88	Liver 100
5-methyl-1,3-oxazolidine	34	1.7	W – rat ♀	85	Liver 100
	43	1.1	G – hamster ♂	35	Liver 100, Pancreas 15

Compound					
thiazolidine	35	3.5	W – rat ♀	54	0
tetrahydro-1,3-oxazine	34	1.7	W – rat ♀		Liver 100
morpholine	86	2.2	W – rat ♀	28	Liver 100, Esophagus 21
	34	1.7	W – rat ♂	57	Liver 100, Esophagus 35
	34	1.7	W – rat ♂	52	Liver 100, Esophagus 60, Nasal 20
	34	1.0	W – rat ♂	80	Liver 60
	34	1.4	W – rat ♀	52	Liver 96, Esophagus 54, Mouth 17
	23	1.1	W – rat ♀	64	Liver 100
	14	0.7	W – rat ♂	82	Liver 100, Esophagus 10
	14	0.7	W – rat ♀	79	Liver 92, Esophagus 13, Mouth 9
	5.5	0.55	W – rat ♀	97	Liver 96, Mouth 8
	5.5	0.27	W – rat ♀	97	Liver 58
	2.2	0.22	W – rat ♀	105	Liver 46
	1.0	0.1	W – rat ♀	108	Liver 31
	86	1.9	G – rat ♀	26	Liver 92, Esophagus 67, Thyroid 17
	86	2.6	I.U. – rat ♀	35	Nasal 100, Liver 58, Esophagus 17
	45	1.2	G – hamster ♂	35	Nasal 75, Trachea 30
2-methylmorpholine	35	0.9	W – rat ♀	27	Liver 100, Esophagus 50, Tongue 20
	42	1.3	G – hamster ♂	34	Nasal 85, Liver 60, Lung 55, Forestomach 40, Trachea 30
2,6-dimethylmorpholine (67% cis, 33% trans)	14	0.4	W – rat ♀	29	Esophagus 95, Nasal 75, Tongue 55
	35	1.0	W – rat ♂	30	Esophagus 87, Nasal 100, Forestomach 27, Trachea 20
	35	0.8	W – rat ♀	23	Esophagus 100, Forestomach 15
	35	1.0	W – rat ♀	28	Esophagus 100, Nasal 87, Trachea 13
cis	22	0.54	W – rat ♀	28	Esophagus 100, Forestomach 20, Nasal 10

Table 7.5a (cont.)

	Dose (μmol/wk)	Total dose per animal† (mmol)	Treatment (Gavage, G; Drinking water, W)	Median time of death (weeks)	% of treated animals with tumours of each type
trans	12	0.27	W – rat ♀	23	Esophagus 100
cis	9	0.3	W – rat ♀	45	Esophagus 85, Tongue 60, Nasal 30, Forestomach 15
trans	5	0.14	W – rat ♀	37	Esophagus 90, Nasal 55, Tongue 35, Forestomach 10
	140	1.7	G – rat ♂	14	Esophagus 100, Nasal 15
	140	1.7	G – rat ♀	26	Esophagus 100, Nasal 35, Lung 20, Forestomach 20
	86	1.5	I.U. – rat ♀	21	Esophagus 100
	42	1.3	G – hamster ♂	37	Liver 90, Pancreas 95, Lung 95, Nasal 60
	42	1.0	G – hamster ♀	29	Nasal 50, Liver 50, Pancreas 45, Lung 40
cis	42	1.0	G – hamster ♂	42	Lung 80, Liver 80, Pancreas 100, Kidney 20
trans	23	0.5	G – hamster ♂	80	Liver 35, Pancreas 20
3-methyl-2-phenylmorpholine	88	4.4	W – rat ♂		0
	88	4.4	W – rat ♀		0
2-hydroxymorpholine	23	1.1	W – rat ♂	108	0
	11	0.57	W – rat ♀	113	0
	170	8.5	G – rat ♀	101	Liver 33, Lung 33
thiomorpholine	125	5.8	W – rat ♂	56	Esophagus 40, Nasal 10, Trachea 10
	125	5.8	W – rat ♀	35	Esophagus 35, Tongue 12

3,5-dithiazine	37	1.8	G – rat ♀	0
thialdine	36	1.6	G – rat ♀	Esophagus 70, Tongue 35, Liver 30
piperazine	170	10.5	W – rat ♂	0
	170	10.5	W – rat ♀	0
	700	35	I.U. – rat ♀	Nasal 33, Bladder 17
4-methylpiperazine	1000	50	W – rat ♂	Nasal 100
	200	10	W – rat ♂	0
	200	10	W – rat ♀	0
3,5-dimethylpiperazine	75	2.0	W – rat ♀	Thymus 85, Nasal 25
4-acetyl-3,5-dimethylpiperazine	70	2.1	W – rat ♀	Esophagus 95, Tongue 40
4-benzoyl-3,5-dimethylpiperazine	70	3.5	W – rat ♀	Forestomach 35, Liver 30
	70	1.8	W – rat ♀	Thymus 90, Nasal 30
3,4,5-trimethylpiperazine	28	0.8	W – rat ♀	Nasal 90
	11	0.34	W – rat ♀	Nasal 65
	38	1.5	G – hamster ♂	Lung 90, Forestomach 70, Liver 20
hexamethyleneimine	86	2.6	W – rat ♀	Esophagus 90, Liver 80
	87	2.6	W – rat ♂	Esophagus 73, Liver 53, Nasal 47
heptamethyleneimine	140	2.1	W – rat ♂	Esophagus 80, Lung 60, Trachea 27
	70	0.9	W – rat ♂	Esophagus 70
	28	0.7	W – rat ♂	Esophagus 75
	70	0.9	W – rat ♀	Esophagus 35
	28	0.37	W – rat ♀	Esophagus 90, Nasal 10
	11	0.28	W – rat ♂	Esophagus 85
	4.5	0.23	W – rat ♂	Esophagus 70, Tongue 55, Nasal 15

Table 7.5a (cont.)

	Dose (μmol/wk)	Total dose per animal† (mmol)	Treatment (Gavage, G; Drinking water, W)	Median time of death (weeks)	% of treated animals with tumours of each type
	1.8	0.18	W – rat ♂	88	Esophagus 65, Tongue 50, Forestomach 15
	0.7	0.07	W – rat ♂	99	Tongue 30, Forestomach 25, Esophagus 15
	140	7	W – hamster	75	Forestomach 84, Trachea 64, Nasal 24
octamethyleneimine	130	4.3	W – rat ♀	43	Esophagus 80, Lung 50, Tongue 20
	32	1.2	W – rat ♂	48	Esophagus 70, Lung 30, Tongue 10
dodecamethyleneimine	225	11	G – rat ♀	92	Liver 40
N,N'-dinitroso-piperazine	70	3.5	W – rat ♂	58	Nasal 93, Liver 27, Esophagus 20
	70	3.5	W – rat ♀	51	Nasal 80, Liver 40, Esophagus 13
	70	2.6	I.U. – rat ♀	87	
2-methylpiperazine	70	2.3	W – rat ♂	36	Esophagus 93, Nasal 80
	70	2.3	W – rat ♀	33	Esophagus 93, Nasal 87
2,6-dimethylpiperazine	70	2.3	W – rat ♂	34	Nasal 93, Esophagus 87
	70	2.3	W – rat ♀	25	Esophagus 100, Nasal 93
	70	1.4	W – rat ♀	27	Esophagus 75
	14	0.3	W – rat ♀	41	Esophagus 42, Nasal 25
	41	1.4	G – hamster ♂	67	Forestomach 45, Lung 20, Liver 15

Compound				
2,5-dimethylpiperazine	70	3.5	W – rat ♂	53	Nasal 100, Esophagus 60
	70	3.5	W – rat ♀	46	Esophagus 93, Nasal 93
2,3,5,6-tetramethylpiperazine	70	3.5	W – rat ♂		0
	70	3.5	W – rat ♀		0
homopiperazine	70	2.2	W – rat ♀	36	Esophagus 100, Nasal 93
	70	2.2	W – rat ♂	35	Esophagus 100, Nasal 100
	70	2.0	W – rat ♀	39	Esophagus 95
	28	0.8	W – rat ♀	39	Esophagus 75
	11	0.34	W – rat ♀	59	Esophagus 80, Forestomach 20
	4.4	0.27	W – rat ♀	67	Esophagus 95, Nasal 35, Tongue 20, Forestomach 15
	1.8	0.21	W – rat ♀	99	Esophagus 85, Tongue 40
	4.4	0.13	W – rat ♀	75	Nasal 35, Esophagus 30, Forestomach 15
	0.7	0.08	W – rat ♀	126	Esophagus 15

* From the literature.
† Approximate body weight: ♂ Rat 400 g, ♀ Rat 250 g, ♂ Hamster 150 g, ♀ Hamster 120 g.

Table 7.5b. *Carcinogenic effects of nitroso-N-alkylureas and nitroso-N-alkylcarbamates in standard chronic experiments – comparative effects in rats and Syrian hamsters*

Compound	Dose (μmol/wk)	Total dose per animal† (mmol)	Treatment (Gavage, G; Drinking water, W)	Median time of death (weeks)	% of treated animals with tumours of each type
N-Nitroso-methylurea	100	1.0	I.V. – rat ♀	69	Uterus 85, Mammary 25, Lung 25
	41	0.8	G – rat ♂	35	Forestomach 100, Nervous system 50, Esophagus 17
	41	0.8	G – rat ♀	23	Forestomach 100, Nervous system 42, Thymus (lymphoma) 50
	20	0.4	G – rat ♂	46	Forestomach 100, Nervous system 58
	20	0.4	G – rat ♀	33	Forestomach 100, Nervous system 67
	10	0.2	I.U. – rat ♀	37	Bladder 92
	20	0.4	G – hamster ♂	18	Forestomach 100, Spleen 33
	20	0.4	G – hamster ♀	22	Forestomach 92, Spleen 83, Glandular stomach 17
	10	0.2	G – hamster ♂	31	Forestomach 92, Spleen 83, Glandular stomach 17
	10	0.2	G – hamster ♀	27	Forestomach 92, Spleen 92
ethylurea	41	1.1	G – rat ♂	30	Forestomach 90, Lung 45, Mesothelioma 20, Zymbal gland 15, Jejunum/Ileum 20, Duodenum 15, Colon 10, Nervous system 10

	41	0.8	G – rat ♀	25	Mammary 95, Forestomach 80, Zymbal gland 25, Lung 15, Tongue 15, Uterus 15
	21	0.4	G – rat ♂	45	Forestomach 90, Lung 65, Mesothelioma 35, Colon 35, Brain 30, Zymbal gland 30
	21	0.4	G – rat ♀	40	Forestomach 55, Mammary 45, Lung 25, Nervous system 20, Zymbal gland 15, Uterus 25, Bladder 20, Colon 10
	41	0.6	I.U. – rat ♀	26	Bladder 83, Mammary 10
	21	0.4	I.U. – rat ♀	27	Bladder 92
	21	0.4	G – hamster ♂	25	Spleen 90, Forestomach 85
	21	0.4	G – Hamster ♀	22	Spleen 95, Forestomach 80, Cervix 40, Uterus 10
	10	0.2	G – hamster ♂	56	Forestomach 80, Liver 50, Spleen 50, Colon 10
2-hydroxyethylurea	210	3.8	G – rat ♂	29	Forestomach 79, Tongue 58, Duodenum 53, Thyroid 47, Lung 32, Zymbal gland 32, Colon 16, Jejunum 16
	210	3.8	G – rat ♀	23	Forestomach 78, Tongue 39, Mammary 33, Duodenum 33, Thyroid 28, Zymbal gland 28, Jejunum 17
	105	2.6	G – rat ♂	50	Lung 65, Forestomach 50, Thyroid 25, Colon 20, Zymbal gland 20, Thymus 15, Glandular stomach 15, Bone 15, Duodenum 10, Tongue 10

Table 7.5b (cont.)

Compound	Dose (μmol/wk)	Total dose per animal† (mmol)	Treatment (Gavage, G; Drinking water, W)	Median time of death (weeks)	% of treated animals with tumours of each type
	42	1.3	G – rat ♂	49	Lung 55, Forestomach 25, Bone 15, Bladder 15, Colon 15, Thyroid 15, Duodenum 10, Bladder 15
	21	0.8	G – rat ♂	45	Lung 90, Forestomach 45, Colon 35, Ileum 20, Zymbal gland 25, Thyroid 20, Brain 15, Bone 10
	21	0.8	G – rat ♀	46	Uterus 35, Mammary 30, Thyroid 15, Thymus 15, Colon 10, Forestomach 10
	21	0.5	G – hamster ♂	49	Spleen 85, Forestomach 70, Pancreas 15
	21	0.5	G – hamster ♀	24	Spleen 80, Forestomach 20, Uterus 10
2-methoxyethylurea	44	1.2	G – rat ♂	30	Forestomach 92, Lung 58, Thyroid 25, Duodenum 17, Zymbal gland 25, Colon 17
	44	1.0	G – rat ♀	26	Forestomach 58, Mammary 58, Thymus 33, Lung 25, Colon 17, Duodenum 17, Zymbal gland 17
	22	0.65	G – rat ♂	38	Forestomach 92, Lung 58, Colon 50, Zymbal gland 25, Duodenum 25, Thyroid 17, Thymus 17

	22	0.65	G – rat ♀	33	Forestomach 75, Mammary 58, Uterus 42, Colon 25, Zymbal gland 25, Lung 17
2-fluoroethylurea	5	0.2	G – rat ♂	56	Forestomach 33, Lung 17
	5	0.13	G – rat ♀		0 – Toxic
	10	0.14	G – rat ♂		0 – Toxic
	10	0.13	G – rat ♀		0 – Toxic
2-phenylethylurea	41	0.9	G – rat ♂	31	Liver 92, Forestomach 92, Duodenum 33, Jejunum 17, Lung 17, Esophagus 17
	41	0.9	G – rat ♀	27	Forestomach 100, Liver 92
	21	0.6	G – hamster ♂	69	Spleen 33, Forestomach 42
	21	0.6	G – hamster ♀	71	Spleen 58, Forestomach 42
benzylurea			*Oral – rat		Forestomach
allylurea	40	1.2	G – rat ♂	35	Forestomach 67, Lung 50, Colon 42, Zymbal gland 42, Thymus 25, Tongue 25, Jejunum/Ileum 17
	40	1.2	G – rat ♀	32	Forestomach 58, Mammary gland 50, Thymus 33, Zymbal gland 33, Uterus 42, Tongue 17
n-butylurea	41	1.7	G – rat ♂	48	Forestomach 83, Lung 92, Colon 25, Mesothelioma 75
	41	1.7	G – rat ♀	41	Lung 75, Forestomach 92, Uterus 75, Mammary 50, Colon 25, Tongue 25
n-amylurea	40	2.0	G – rat ♂	62	Forestomach 100, Lung 63, Mesothelioma 50, Jejunum 25, Zymbal gland 25, Colon 12

Table 7.5b (cont.)

Compound	Dose (μmol/wk)	Total dose per animal† (mmol)	Treatment (Gavage, G; Drinking water, W)	Median time of death (weeks)	% of treated animals with tumours of each type
	40	2.0	G – rat ♀	54	Forestomach 75, Lung 58, Uterus 67, Mammary 25, Brain 17
n-hexylurea	40	2.0	G – rat ♂	60	Lung 100, Forestomach 100, Mesothelioma 78, Colon 33
	40	2.0	G – rat ♀	53	Uterus 100, Forestomach 75, Mammary gland 33
2-hydroxypropylurea	106	2.6	G – rat ♂	28	Thymus 70, Forestomach 35, Lung 15
	44	1.3	G – rat ♂	32	Thymus 65, Forestomach 15, Lung 15, Liver 10
	22	0.4	G – rat ♂	19	Thymus 95
	22	0.4	G – rat ♀	22	Thymus 95
	22	0.5	G – hamster ♀	36	Spleen 75, Forestomach 75
2-oxopropylurea	22	0.9	G – rat ♂	88	Kidney 50, Mesothelioma 25
	22	0.9	G – rat ♀	77	Kidney 67, Lung 17, Thyroid 17
	11	0.45	G – hamster ♂	59	Spleen 81
	11	0.45	G – hamster ♀	54	Spleen 100
3-hydroxypropylurea	44	1.3	G – rat ♂	82	Glandular stomach 33, Thyroid C-cell 42, Lung 25, Liver 17, Forestomach 17, Bladder 17
	44	1.3	G – rat ♀	70	Duodenum 27, Jejunum 18, Liver 36, Forestomach 36, Thyroid C-cell 36, Uterus 27, Kidney 18, Bladder 18

	22	0.9	G – rat ♂	75	Liver 25, Forestomach 25, Nervous system 17
	22	0.9	G – rat ♀	89	Uterus 33, Thyroid C-cell 42
methyl-N'-methylurea	41	1.2	G – rat ♂	81	Nervous system 67, Mesothelioma 50, Thyroid 33
	41	1.2	G – rat ♀	40	Brain 83, Lung 25
	21	0.65	G – rat ♂	83	Nervous system 17, Zymbal gland 17, Mesothelioma 17
	21	0.65	G – rat ♀	80	Nervous system 75, Thyroid 42, Kidney 25, Uterus 17
	41	1.2	W – rat ♂	86	Nervous system 33, Mesothelioma 42, Skin 25
	41	1.2	W – rat ♀	71	Nervous system 58, Kidney 50, Uterus 33, Thyroid 33
	21	0.6	W – rat ♂	95	Lung 25, Mesothelioma 25
	21	0.6	W – rat ♀	100	Thyroid 33
	41	1.2	I.U. – rat ♀	46	Nervous system 67, Kidney 33, Thyroid 33, Lung 17
	21	0.6	G – hamster ♂	36	Spleen 100
	21	0.6	G – hamster ♀	34	Spleen 92, Forestomach 17, Mammary 17
methyl-N'-ethylurea	41	1.2	G – rat ♂	28	Brain 75, Lung 17, Malignant lymphoma 17
	41	0.6	G – rat ♀	18	Thymus (lymphoma) 75, Nervous system 25, Forestomach 17
	21	0.6	G – rat ♂	48	Nervous system 58, Mesothelioma 25
	21	0.6	G – rat ♀	34	Nervous system 67
	41	1.2	W – rat ♂	43	Nervous system 80, Lung 10, Kidney 10
	41	1.2	W – rat ♀	36	Nervous system 92

Table 7.5b (*cont.*)

Compound	Dose (μmol/wk)	Total dose per animal† (mmol)	Treatment (Gavage, G; Drinking water, W)	Median time of death (weeks)	% of treated animals with tumours of each type
	21	0.6	W – rat ♂	71	Nervous system 67, Lung 33, Kidney 17, Skin 25
	21	0.6	W – rat ♀	53	Nervous system 50, Kidney 33
ethyl-*N*-ethylurea	41	1.2	G – rat ♂	33	Lung 58, Mesothelioma 42, Zymbal gland 42, Colon 33, Jejunum 17, Spleen 17
	41	0.9	G – rat ♀	24	Mammary 92, Lung 17, Uterus 17, Clitoris 17, Liver 17
	21	0.6	G – rat ♂	36	Nervous system 50, Lung 58, Mesothelioma 58, Colon 17, Forestomach 25, Duodenum 17, Bladder/Kidney pelvis 17
	21	0.6	G – rat ♀	29	Mammary 92
	41	1.2	W – rat ♂	49	Lung 83, Mesothelioma 75, Nervous system 50, Colon 42, Zymbal gland 25, Thyroid 17, Skin 58
	41	1.2	W – rat ♀	32	Mammary 92, Clitoris 42, Lung 25
	21	0.6	W – rat ♂	58	Mesothelioma 92, Nervous system 58, Lung 58, Liver 25, Thyroid 17, Colon 17
	21	0.6	W – rat ♀	47	Mammary 75, Uterus 42, Nervous system 25, Lung 33, Forestomach 17, Ileum/Jejunum 17

	41	0.9	I.U. – rat ♀	Mammary 100, Zymbal gland 25, Lung 33, Bladder 17
	21	0.5	G – hamster ♂	Spleen 92, Forestomach 42
	21	0.5	G – hamster ♀	Spleen 92, Forestomach 17
ethyl-N'-methylurea	41	1.2	G – rat ♂	Lung 67, Mesothelioma 42, Zymbal gland 42, Colon 25, Brain 17, Jejunum 17
	41	1.2	G – rat ♀	Mammary 83, Lung 17, Uterus 25, Colon 25, Zymbal gland 17
	21	0.6	G – rat ♂	Lung 58, Mesothelioma 50, Nervous system 25, Ileum/Jejunum 25, Zymbal gland 17
	21	0.6	G – rat ♀	Mammary 75, Uterus 67, Lung 25, Clitoris 25, Jejunum 25, Nervous system 25, Colon 17, Zymbal gland 17
	41	1.2	W – rat ♂	Lung 75, Mesothelioma 67, Nervous system 42, Colon 42, Ileum/Jejunum 25, Zymbal gland 25, Spleen 17
	41	1.2	W – rat ♀	Mammary 75, Uterus 25, Lung 17, Colon 17, Thyroid 17
	21	0.6	W – rat ♂	Mesothelioma 75, Lung 58, Thyroid C-cell 50, Nervous system 42, Zymbal gland 25
	21	0.6	W – rat ♀	Nervous system 67, Uterus 67, Mammary 33, Thyroid C-cell 33, Lung 25, Liver 17
ethyl-N'-2-hydroxyethylurea	45	1.4	G – rat ♂	Lung 100, Mesothelioma 70, Brain 45, Liver 30, Thyroid 30, Zymbal gland 25, Colon 20, Bladder 20

Table 7.5b (cont.)

Compound	Dose (μmol/wk)	Total dose per animal† (mmol)	Treatment (Gavage, G; Drinking water, W)	Median time of death (weeks)	% of treated animals with tumours of each type
	46	1.4	W – rat ♂	69	Mesothelioma 75, Lung 67, Skin 58, Nervous system 42, Bladder 25, Colon 17
	46	1.4	W – rat ♀	47	Nervous system 58, Mammary 42, Bladder 25, Uterus 50, Skin 42, Palate/Tongue 25, Colon 17
	23	0.7	W – rat ♂	77	Mesothelioma 73, Lung 45, Skin 50, Mouth 18, Nervous system 27
	23	0.7	W – rat ♀	68	Mammary 50, Uterus 50, Nervous system 33, Skin 33, Lung 25, Bladder 17
	46	1.4	I.U. – rat ♀	55	Bladder 67, Mammary 42, Brain 50, Lung 17
	22	0.7	G – hamster ♀	59	Cervix 35, Forestomach 30, Spleen 20
ethyl-2-oxopropylurea	44	1.3	G – rat ♂	39	Lung 100, Mesothelioma 75, Colon 58, Forestomach 58, Zymbal gland 58, Thyroid 42, Ileum/Jejunum 25, Liver 25, Nervous system 17, Skin 42
	44	1.3	G – rat ♀	29	Mammary 83, Lung 50, Uterus 33, Zymbal gland 17, Forestomach 25, Duodenum 17, Spleen 25

	22	0.65	G – rat ♂	Lung 100, Mesothelioma 75, Thyroid 25, Zymbal gland 33, Jejunum/Ileum 33, Colon 25, Nervous system 25, Spleen 25
	22	0.65	G – rat ♀	Mammary 100, Lung 42, Uterus 50, Spleen 17, Zymbal gland 17, Nervous system 17
	46	1.4	W – rat ♂	Lung 83, Colon 67, Mesothelioma 58, Zymbal gland 50, Nervous system 42, Skin 33, Jejunum/Ileum 17, Thyroid 17
	46	1.4	W – rat ♀	Mammary 67, Lung 50, Uterus 50, Lymphoma 25, Zymbal gland 17, Liver 17
	23	0.7	W – rat ♂	Mesothelioma 92, Nervous system 83, Lung 75, Skin 25, Colon 25
	23	0.7	W – rat ♀	Nervous system 42, Lung 33, Thyroid 33, Colon 25, Mammary 25, Uterus 25, Skin 17
	11	0.33	G – hamster ♂	Spleen 33, Nervous system 25, Liver 33, Colon 17, Pancreas 17, Harderian gland 50. Forestomach 42
	11	0.33	G – hamster ♀	Spleen 67, Nervous system 33, Forestomach 33, Harderian gland 17
dimethylaminoethyl-N'-ethylurea	44	1.6	G – rat ♂	Zymbal gland 17, Skin 33, Thyroid 17
	44	1.6	G – rat ♀	Mammary 25, Uterus 67, Lung 17, Tongue 25, Thyroid 17
2-hydroxyethyl-N'-ethylurea	45	1.4	G – rat ♂	Liver 55, Brain 50, Mesothelioma 50, Lung 45, Thyroid 35, Colon 15

Table 7.5b (cont.)

Compound	Dose (μmol/wk)	Total dose per animal† (mmol)	Treatment (Gavage, G; Drinking water, W)	Median time of death (weeks)	% of treated animals with tumours of each type
	46	1.4	W – rat ♂	66	Lung 58, Skin 58, Nervous system 50, Mesothelioma 50, Thyroid F-cell 25, Thyroid C-cell 33, Liver 33, Pancreas acinar cell 17, Colon 17, Preputial gland 17
	46	1.4	W – rat ♀	45	Mammary 63, Nervous system 12, Lung 12, Bladder 12, Colon 12, Kidney 12
	23	0.7	W – rat ♂	87	Lung 42, Mesothelioma 33, Liver 17, Nervous system 17, Thyroid 17, Skin 17
	23	0.7	W – rat ♀	68	Mammary 55, Uterus 36, Skin 18
	46	1.4	I.U. – rat ♀	65	Bladder 75, Uterus 67, Brain 17, Mammary 17, Thyroid 17
	22	0.5	G – hamster ♀	33	Spleen 85, Cervix 20, Forestomach 15, Trachea 15
2-hydroxyethyl-N'-2-chloroethylurea	41	1.3	G – rat ♂	78	Liver 100, Kidney 70
	20	0.85	G – rat ♂	105	Liver 90, Kidney 35, Lung 10
	20	0.85	G – rat ♀	94	Liver 63, Kidney 11
	20	0.65	W – rat ♂	89	Skin 58, Liver 33, Lung 17, Mesothelioma 17
	20	0.65	W – rat ♀	98	Uterus 50, Liver 25, Skin 17
2-chloroethyl-N'-2-hydroxyethylurea	8	0.1	G – rat ♂		0 – Toxic
	8	0.1	G – rat ♀		0 – Toxic

Compound				
2-chloroethyl-N'-2-hydroxypropylurea	8	0.1	G – rat ♂	0 – Toxic
	8	0.1	G – rat ♀	0 – Toxic
	8	0.17	W – rat ♂	62 Liver 18, Lung 45
	8	0.17	W – rat ♀	0 – Toxic
2-hydroxypropyl-N'-2-chloroethylurea	42	1.3	G – rat ♂	90 Liver 75, Lung 35, Thymus 15, Mesothelioma 10, Skin 10
	21	0.87	G – rat ♂	87 Liver 21, Kidney 16, Lung 11
	21	0.87	G – rat ♀	82 Liver 40, Uterus 25, Brain 10
	21	0.65	W – rat ♂	98 Skin 25
	21	0.65	W – rat ♀	103 Bladder 17, Thyroid 17
	42	1.3	I.U. – rat ♀	43 Bladder 50, Uterus 33
2-oxopropyl-N'-2-chloroethylurea	22	0.7	G – rat ♂	94 Nervous system 25, Mammary 17
	22	0.7	G – rat ♀	91 Liver 25
	22	0.66	W – rat ♂	100 Skin 42, Lung 17
	22	0.66	W – rat ♀	90 Uterus 25
	11	0.5	G – hamster ♂	63 Spleen 92, Harderian gland 17
	11	0.5	G – hamster ♀	63 Spleen 58
N,N',N'-trimethylurea	95	4.5	W – rat ♀	44 Brain 65, Forestomach 25
	38	1.9	W – rat ♂	90 —
	38	1.9	W – rat ♀	63 Nervous system 20
	95	3.0	W – rat ♀	33 Mammary gland 95, Forestomach 40, Uterus 30, Lung 20, Brain 10
N,N',N'-triethylurea	38	1.9	W – rat ♂	62 Nervous system 40
	38	1.9	W – rat ♀	43 Nervous system 7
	52	1.3	G – hamster ♂	30 Spleen 80, Forestomach 75, Pancreas 10
N-methyl-N',N'-diethylurea	94	3.1	W – rat ♀	32 Nervous system 90
	38	1.9	W – rat ♂	52 Nervous system 93
	38	1.9	W – rat ♀	45 Nervous system 60

Table 7.5b (*cont.*)

Compound	Dose (µmol/wk)	Total dose per animal† (mmol)	Treatment (Gavage, G; Drinking water, W)	Median time of death (weeks)	% of treated animals with tumours of each type
	50	1.3	G – hamster ♂	29	Forestomach 85, Spleen 70, Liver 20, Pancreas 10
N-ethyl-*N',N'*-dimethylurea	38	1.9	W – rat ♂	65	Nervous system 33
	38	1.9	W – rat ♀	44	Nervous system 20
N-methylurethane	11	0.2	G – rat ♀	65	Forestomach 100
	2.6	0.05	G – rat ♀	77	Forestomach 95
	10	0.02	I.U. – rat ♀	49	Lung 25
	5	0.01	I.U. – rat ♀	51	Lung 33, Bladder 17
N-methylcarbamic acid, phenyl ester		0.05	G – rat ♀	50	Forestomach 57
carbaryl (*N*-methylcarbamic acid, 1-naphthyl ester)		0.2	G – rat ♀	98	Forestomach 75
		1.3	G – rat ♂	58	Forestomach 75
baygon		0.5	G – rat ♀	70	Forestomach 80
carbofuran		0.5	G – rat ♀	60	Forestomach 54
landrin		0.5	G – rat ♀	92	Forestomach 100
bux-ten		0.5	G – rat ♀	65	Forestomach 69
methomyl		0.5	G – rat ♀	84	Forestomach 57
aldicarb		0.2	G – rat ♀	102	Forestomach 100
N-ethylurethane	11	0.2	G – rat ♀	80	Forestomach 100
	2.7	0.05	G – rat ♀	103	Forestomach 95
	11	0.15	I.U. – rat ♀	35	Bladder 83, Lung 17
	19	0.5	G – hamster ♀	37	Forestomach 75, Pancreas 15

Compound					
1,3-oxazolidone	59	2.3	G – rat ♂	79	Forestomach 80, Thymus 15
	41	0.3	I.U. – rat ♀	37	Bladder 20
	21	0.6	G – hamster ♀	67	Forestomach 25, Duodenum 20
5-methyl-1,3-oxazolidone	58	2.3	G – rat ♂	53	Forestomach 95
	21	0.6	G – hamster ♀	50	Forestomach 90
N-chloroethyl-N',N'-dimethylurea		0.2	G – rat ♂	87	Lung 69, Forestomach 38
N-chloroethyl-N',N'-diethylurea		0.2	G – rat ♂	75	Lung 23, Forestomach 15
N,N',N'-tris-chloroethylurea		0.05	G – rat ♂	91	Forestomach 40
N-methyl-N',N'-bis-chloroethylurea		0.1	G – rat ♀	68	Forestomach 57
N-methyl-N'-nitroguanidine	51	2.2	W – rat ♂	95	Glandular stomach 70
N-ethyl-N'-nitroguanidine			*W – rat		Glandular stomach
cimetidine	54	5.7	W – rat ♂	0	
	54	5.7	W – rat ♀	0	

* From the literature.
† Approximate body weight: ♂ Rat 400 g, ♀ Rat 250 g, ♂ Hamster 150 g, ♀ Hamster 120 g.

316 Structure–activity relations

piperidine were not notably different in carcinogenic activity, however (Wiessler and Schmähl, 1973).

The blockage of oxidation in 2,3,5,6-tetramethyl-1,4-dinitrosopiperazine (Fig. 7.9) is probably why it has failed to induce tumours after administration of doses equimolar with dinitrosopiperazine to rats for a long period (Lijinsky and Taylor, 1977a). Replacement of one or two α-hydrogens in dinitrosopiperazine with methyl groups, however, leads not to reduction in carcinogenic activity, but to an enhancement of carcinogenic activity. Dinitrosopiperazine induces tumours of the liver, esophagus and nasal mucosa in rats, and is less effective than an equimolar dose of nitrosopiperidine. In contrast, 2-methyldinitrosopiper-

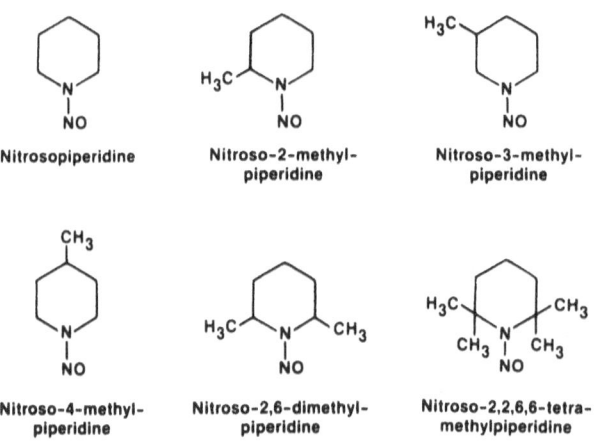

Fig. 7.8. Methylated nitrosopiperidines.

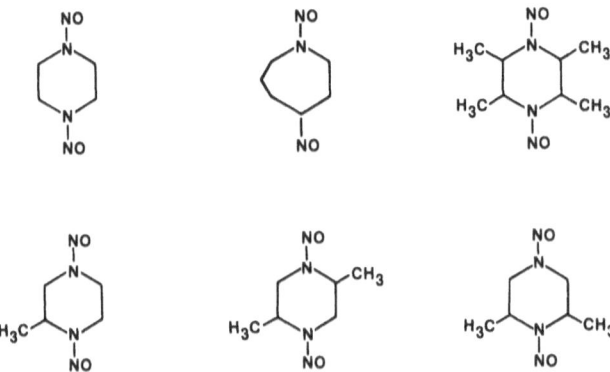

Fig. 7.9. Dinitrosopiperazine, dinitrosohomopiperazine, 2,3,5,6-tetramethyl-, 2-methyl-, 2,5-dimethyl-, and 2,6-dimethyl-dinitrosopiperazine.

azine and 2,6-dimethyldinitrosopiperazine are much more potent and induce tumours only of the esophagus and nasal mucosa, but not of the liver (Lijinsky and Taylor, 1975g). It seems likely that the rigidity imposed on the molecule by the substituents, with the axial hydrogens at the other end of the molecule orientated in the appropriate position, leads to more effective activation by the enzyme or enzymes, and therefore to an increase in carcinogenic effectiveness. Dinitroso-2,5-dimethylpiperazine, which is analogous in some ways to nitroso-2-methylpiperidine, is somewhat more potent than dinitrosopiperazine, but considerably less potent than either the 2-methyl- or 2,6-dimethyl- derivative; again only tumours of the esophagus and nasal mucosa are produced, but no liver tumours, although many of the treated rats lived as long as those in the dinitrosopiperazine-treated group (Lijinsky and Taylor, 1975g). Therefore, it seems that there is a compromise between the methyl substitution providing a more rigid molecule for enzymic activation and blockage of access to the enzyme by this molecule. The absence of liver tumours has some other explanation, perhaps related to the conformation effects of the methyl substituents. Few nitrosopiperazine derivatives induce liver tumours in rats.

Nitrosomethylphenidate (Fig. 7.10) is the nitroso derivative of a widely used drug and has a carboxymethyl ester substituent at the α-carbon. It was not carcinogenic following administration of a dose greater than that of nitrosopiperidine which induced 100% incidence of tumours after a short latent period (Lijinsky and Taylor, 1976c). The large bulk of the substituent (which contains a benzene ring) might partially explain the

Fig. 7.10. Nitrosophenmetrazine and nitrosomethylphenidate.

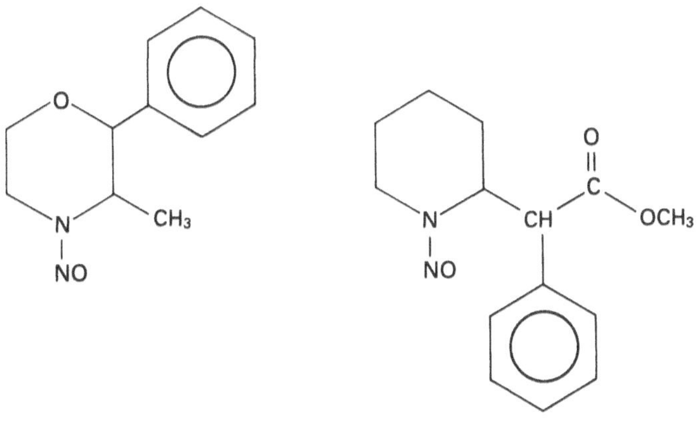

Nitrosophenmetrazine   **Nitrosomethylphenidate**

elimination of carcinogenic activity in this molecule, although 4-phenylnitrosopiperidine, which is also bulky, gave rise to tumours of the liver and esophagus in rats (Lijinsky *et al.*, 1981*d*). Only one α-substituted derivative of nitrosomorpholine has been tested for carcinogenic activity, nitroso-2-phenyl-3-methylmorpholine (nitrosophenmetrazine), and it was not carcinogenic at doses much higher than those of nitrosomorpholine that induced tumours in all treated rats (Lijinsky and Taylor, 1975*c*); it was mutagenic to bacteria, however.

Replacement of α-hydrogens in nitrosamines with groups other than methyl usually results in non-carcinogenic products, indeed, in compounds which are rather inactive. For example, cyano groups, as in nitrosimino-diacetonitrile, lead to loss of carcinogenic activity, although methyl-nitrosaminoacetonitrile is a quite potent carcinogen. Nitroso-2,6-dicyano-piperidine has not been tested for carcinogenic activity, but is non-mutagenic and quite unreactive. Nitrosiminodiacetic acid was not carcinogenic, even at high doses (Lijinsky *et al.*, 1973*b*) and nitroso-sarcosine (methylnitrosaminoacetic acid) was a weak carcinogen inducing tumours in the rat esophagus, but no other tumours. Nitrosoproline (Fig. 7.11) was not carcinogenic in a number of studies, in which doses as large as 0.1 mol have been given to rats (Lijinsky and Reuber, 1982*b*; Nixon

Fig. 7.11. Nitrosamino acids.

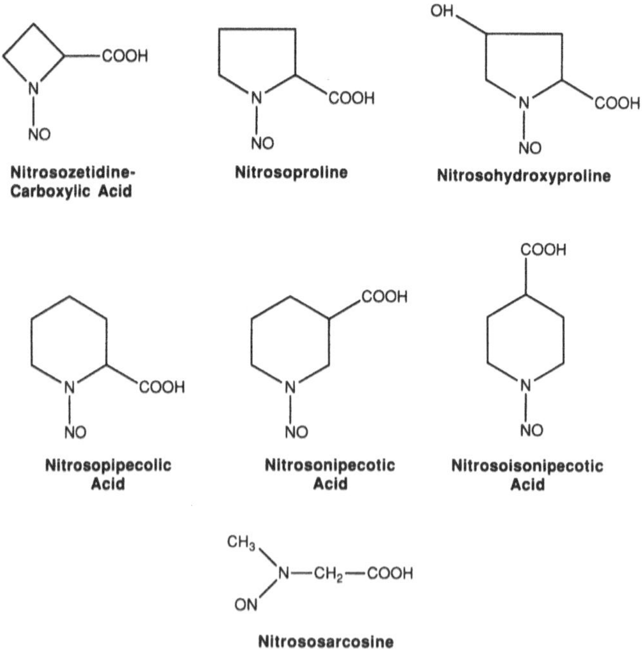

et al., 1976). Nitrosoproline is taken up by liver cells to only a very small extent and no metabolites could be detected (Chu and Magee, 1981). The question whether nitrosoproline fails to induce tumours because there is a strongly electronegative substituent on the α-carbon, or because it is a strong acid cannot be answered completely, although the Chu and Magee (1981) study suggests the latter. Oral administration of nitrosoproline ethyl ester also failed to yield tumours (Druckrey *et al.*, 1967). Nitrosohydroxyproline also seems to be a weak carcinogen or perhaps inactive after administration of quite high doses to rats (Lijinsky and Rueber, 1982*b*); in this experiment, unlike previous reports, there was a small incidence of hemangiosarcomas of the spleen, which is an uncommon tumour in those rats. Nitrosopipecolic acid was inactive also. It is difficult to decide whether the acidic properties of nitroso amino acids or the presence of the carboxyl group on the α-carbon is responsible for the lack of activity, although nitroso*iso*nipecotic acid and nitrosoguvacoline, having carboxy substituents not on the α-carbons, also failed to induce tumours in rats.

It seems clear that α-oxidation is an important step in activation of nitrosamines to carcinogenic forms, although this itself is not the sole criterion for carcinogenic activity. Among acyclic nitrosamines, α-oxidation produces one or more reactive intermediates the nature of which is not known for certain, although alkyldiazonium ions are popularly considered to be most likely. This is supported to some extent by examination of the activity of esters of the putative first oxidation product of nitrosamines, which are α-hydroxy nitrosamines. The esters of these relatively unstable compounds, such as α-acetoxydimethylnitrosamine, are stable until hydrolysed, when they become potent alkylating agents, mutagens and (presumably) carcinogens. However, α-acetoxynitrosamines do not induce the types of tumour induced by the parent nitrosamine (for example, liver tumours in rats in the case of α-acetoxydimethylnitrosamine), but instead they behave more like the directly acting alkylnitrosoureas (Wiessler and Schmähl, 1976). It is possible that conjugates unlike the acetate esters are formed from the α-hydroxynitrosamines, and that these have entirely different properties. Preparation of the unstable α-hydroxynitrosamines has been described by Mochizuki *et al.* (1980).

A more difficult question is posed in the case of the cyclic nitrosamines. These are potent bacterial mutagens when activated by liver enzymes and are assumed to be oxidised at the α-carbon atoms. The fate of the α-hydroxylated derivatives (which have been prepared and studied in some cases, nitrosopyrrolidine for example) is not known, or their possible role in carcinogenesis. However, interpolation of likely pathways of activation of cyclic nitrosamines is impaired because until now only in the case of nitrosopyrrolidine has it been possible to isolate in small quantities stable adducts to DNA in rat liver, for example, that have been derived from a cyclic nitrosamine. Reports many years ago suggested that such adducts were formed, but use of deuterium-labelled cyclic nitrosamines showed that the adducts did not contain deuterium and, therefore, could not be derived from the ring system of the cyclic nitrosamine (Lijinsky and Ross, 1969).

Another conundrum in the apparently simple relation of carcinogenesis by nitrosamines to α-oxidation, is the induction of bladder tumours in rats by nitrosodiphenylamine, which was a very important chemical used as a retarder of vulcanisation in the manufacture of rubber (Spiegelhalder et al., 1983). Nitrosodiphenylamine cannot be α-oxidised, and is not mutagenic to bacteria, probably for that reason (Andrews et al., 1978). Two other non-mutagenic nitrosamines which cannot be α-oxidised to form an alkylating agent are methylnitrosoaniline (Druckrey et al., 1967) and phenylnitrosobenzylamine (Lijinsky and Reuber, 1982b); both induce tumours of the esophagus in rats. These compounds differ from nitrosodiphenylamine, however, in that they can be α-oxidised, with formation of an α-hydroxy derivative which would break down to form a phenyldiazonium ion. This reactive product could conceivably be the proximate carcinogen, although evidence that it reacts with DNA so as to produce a mutagenic change is lacking. On the other hand, the α-oxidised product could itself be the intermediate which reacts with cellular macromolecules to begin neoplastic transformation, leading to formation of tumours. As discussed elsewhere, the relation between DNA alkylation and the appearance of tumours in the organ in which this occurs is not quantitative and not on as firm a footing as was once believed. The evidence relating mutagenesis with carcinogenesis is circumstantial, and it is not yet demonstrated that tumours are always – or even usually – associated with mutational changes in cells. This remains an interesting, plausible and provocative theory.

7.5   Organ specificity of carcinogenic *N*-nitroso compounds
7.5.1 *Esophageal carcinogens in rats*

Since about half of the nitrosamines examined induce tumours of the esophagus in rats, it is convenient to consider the role of structural changes, apart from substitution in the α-positions already described. The simplest esophageal carcinogen among nitrosamines is methylnitrosoethylamine, which only induces those tumours when administered at high dose rates to rats, in which species its main carcinogenic action is to induce liver tumours (Lijinsky and Reuber, 1981b). Substitution in the 2-carbon of the ethyl group of MNEA increases the tendency to produce tumours of the esophagus in rats; even replacement of hydrogens with deuterium has this effect, although the principal effect in this case remains the induction of liver tumours (Lijinsky et al., 1982e). Replacement of these three hydrogens with fluorine leads to a large increase in carcinogenic potency and to elimination of the property of inducing liver tumours; this compound, methylnitrosotrifluoroethylamine (Fig. 7.12) induced 100% incidence of tumours of the esophagus in rats, with a much reduced time-to-death-with-tumours (Lijinsky et al., 1982e). The strong electronegativity of the fluorine substituents increased the acidity of the α-methylene so that oxidation of that methylene was strongly inhibited (no trifluoroacetaldehyde was formed during in vitro metabolism – Farrelly and Stewart, 1982), although oxidation of the N-methyl group took place readily; methylnitrosotrifluoroethylamine was not activated to a bacterial mutagen by rat liver microsomes; whether it is converted to a methylating agent under any conditions in vivo is an unanswered question. Replacement of hydrogen in MNEA by a benzene ring, as in methylnitroso-2-phenylethylamine, also creates a powerful esophageal carcinogen, which, similarly, did not induce any liver tumours in rats. Its lower homologue, methylnitrosobenzylamine, which can be considered NDMA in which a benzene ring replaces a hydrogen atom, is a potent esophageal

Fig. 7.12. Methylnitrosoalkylamines with substituted alkyl groups.

$$ON-N<^{R_1}_{R_2}$$

Compound	R_1	R_2
METHYL-tert-BUTYL	CH_3	$C(CH_3)_3$
METHYL-NEOPENTYL	CH_3	$CH_2-C(CH_3)_3$
METHYL-BENZYL	CH_3	$CH_2-C_6H_5$
METHYL-PHENYLETHYL	CH_3	$CH_2-CH_2-C_6H_5$
PHENYL-BENZYL	C_6H_5	$CH_2-C_6H_5$
METHYL-TRIFLUOROETHYL	CH_3	CH_2-CF_3
METHYL-2-HYDROXYPROPYL	CH_3	$CH_2-CH(OH)-CH_3$
METHYL-2-OXOPROPYL	CH_3	$CH_2-CO-CH_3$
METHYL-2,3-DIHYDROXYPROPYL	CH_3	$CH_2-CH(OH)-CH_2OH$
METHYL-3-CARBOXYPROPYL	CH_3	$CH_2-CH_2-CH_2-COOH$
BENZYL-ACETOXYBENZYL	$CH_2C_6H_5$	$CH(O-CO-CH_3)-C_6H_5$

carcinogen in rats (Lijinsky et al., 1982e), which is oxidised in vitro (by rat liver preparations) to both formaldehyde and benzaldehyde (Farrelly and Stewart, 1982). Methylnitrosocyclohexylamine is also a potent esophageal carcinogen (Lijinsky et al., 1989a). The great increase in carcinogenic potency of MNEA following replacement of the terminal hydrogens by a benzene ring is not easy to explain, but might be due to some extent to changes in affinity for activating enzymes because of the bulk of the phenyl substituent. This possibility is fortified by the finding that methylnitroso-2,2-dimethylpropylamine (methylnitrosoneopentylamine), which can be considered nitrosodimethylamine in which a hydrogen atom is replaced by a tertiary butyl group, is an esophageal carcinogen in rats of considerable potency (Lijinsky et al., 1982e). In line with the tendency of methylnitrosoalkylamines to induce esophageal tumours in rats, whatever the alkyl group, even the nitrosamine containing the strongly basic N,N-dimethylaminoethyl group is a potent esophageal carcinogen, by several treatment regimens.

These powerful esophageal carcinogens, which do not induce liver tumours in rats, have in common a low tendency for oxidation of the α-carbon of the complex alkyl group, with the consequence of increased oxidation of the N-methyl. This, however, is incompatible with formation of a methylating agent from these compounds, and in contrast with the findings of Von Hofe et al. (1987) that the potent esophageal carcinogens methylnitroso-n-propylamine and its immediate higher homologues produce extensive *methylation* of DNA in the rat esophagus, which NDMA and MNEA do not. It is probable, therefore, that more than one characteristic of a nitrosamine lends itself to carcinogenic action in the rat esophagus, although this complicates our attempts to investigate the process. The presence of the methyl group as one of the two substituents on the amine nitrogen seems to be essential for induction of tumours of the esophagus by these compounds, since nitrosodibenzylamine, nitrosodicyclohexylamine (Druckrey et al., 1967) and nitrosobis-(trifluoroethyl)amine (Preussmann et al., 1981) were not carcinogenic although, surprisingly, nitrosodiphenylamine (Cardy et al., 1979) and nitrosophenylbenzylamine (Lijinsky and Reuber, 1982) are. A similar contrast is presented between the potent esophageal carcinogens methylnitrosoallylamine and methylnitrosaminopropionitrile (Wenke et al., 1984) and the corresponding symmetrical nitrosamines, which are very weak or inactive.

Nitrosarcosine is a weak esophageal carcinogen in rats (Druckrey et al., 1967) whereas its symmetrical counterpart, nitrosiminodiacetic acid, was entirely inactive (Lijinsky et al., 1973b). On the other hand, both methylnitroso-n-propylamine and nitrosodi-n-propylamine are potent esophageal carcinogens, and the symmetrical nitrosodipropylamine also induced liver tumours, especially when given by gavage (Lijinsky and Reuber, 1983a). Methylnitrosovinylamine also induced esophageal tumours (Druckrey et al., 1967); it has not been possible to prepare nitrosodivinylamine. Methylnitrosoamylamine (n-pentyl) is a potent esophageal carcinogen (Druckrey et al., 1967), whereas its symmetrical analogue, nitrosodi-n-amylamine induced liver tumours in rats, but was very much less potent; methylnitroso-n-butylamine shows a similar relation to nitrosodi-n-butylamine, although both of these induce tumours of the esophagus in rats (Lijinsky et al., 1983d; Druckrey et al., 1967), together with tumours of the lung, liver and bladder in the case of nitrosodi-n-butylamine. Methylnitroso-n-octylamine was less potent than the C-4 and C-5 homologues and induced tumours of the esophagus only when administered in drinking water, not by gavage, which instead gave rise to liver tumours and bladder tumours (Lijinsky et al., 1981c). Like many other symmetrical analogues, nitrosodi-n-octylamine failed to induce tumours of any kind following administration of quite large doses to rats (Lijinsky and Taylor, 1978e). Methylnitroso-n-heptylamine, like the octylamine derivative, induced esophageal tumours only when given to rats in drinking water, whereas methylnitroso-n-hexylamine induced 100 % incidence of esophageal tumours whether given in drinking water or by gavage (Lijinsky et al., 1983e). These differences suggest that there is some local effect of the nitrosamine on the rat esophagus, as well as a systemic one.

The common thread in esophageal carcinogenesis between the asymmetric nitrosamines is that they have an oxidisable N-methyl group and an alkyl substituent with an α-methylene resistant to oxidation. It is possible that the increased susceptibility to oxidation of the methyl group in these nitrosamines leads to increased formation of a reactive intermediate which has great affinity for an important receptor in the epithelial cells of the esophagus. How this might then lead to induction of tumours arising from those cells is quite unclear as, indeed, is the mechanism of action of any carcinogen.

A similar consistent pattern, but with uncertain meaning, is shown by the group of methylnitrosoalkylamines having oxygen substituents in their alkyl groups, and their corresponding symmetrical nitrosamine analogues. For example, methylnitroso-2-oxopropylamine is principally an esophageal carcinogen, although liver tumours also appear in female rats treated with high doses (Lijinsky et al., 1983d). The corresponding symmetrical nitrosamine, nitrosobis-(2-oxopropyl)amine (NBOPA) (which has been of great interest because it was one of the first compounds found to induce ductal carcinomas of the pancreas in Syrian hamsters (Pour et al., 1977), has not induced a single tumour of the esophagus in any of the multitude of experiments carried out with it. However, NBOPA is a potent carcinogen in rats, inducing liver and lung tumours in female rats, and tumours of the thyroid, lungs and bladder, but not liver tumours in male rats (Lijinsky et al., 1988f); the sex difference, as with MNOPA, is quite unusual in nitrosamine carcinogenesis. Methylnitroso-2-hydroxypropylamine also induced tumours of the esophagus in rats of both sexes, and was approximately equal in potency to the corresponding ketone (Lijinsky et al., 1983d). In this case the corresponding symmetrical nitrosamine, nitrosobis-(2-hydroxypropyl)amine (NBHPA) also induced a high incidence of tumours of the esophagus in rats, together with tumours of the thyroid, lungs, liver and nasal mucosa (Lijinsky et al., 1988e). Except for the thyroid tumours, this is a very similar response to that shown in rats to nitrosodi-n-propylamine (Reznik et al., 1975), which is similar in size. In spite of much seeking, however, there is no convincing evidence that the di-n-propyl compound acts as a carcinogen by conversion to the oxopropyl- or hydroxypropyl- derivative.

It is plausible that the two symmetrical oxygenated dipropylnitrosamines each has affinity for receptors in particular cells, and that this disposes them to give rise to tumours in particular 'target' organs. This probable relation to chemical structure is supported by the behaviour of (2-hydroxypropyl)-nitroso-(2-oxopropyl)amine (NHPOPA), which can be considered a hybrid of the two symmetrical oxygenated dipropylnitrosamines. This compound proved to be almost as potent as NBOPA and considerably more potent than NBHPA in inducing tumours in rats. In females it induced high incidences of tumours of both liver and esophagus, together with lung tumours; in males it induced few liver tumours (especially when given by gavage), but many tumours of the thyroid, lung, esophagus, nasal mucosa and, unusually, colon (Lijinsky et al., 1988f). It is not unreasonable to assume that the broad affinity of NHPOPA for so many organs in which tumours are induced has a chemical structural basis, related to the presence of both the alcohol and keto functions in the molecule.

Several other derivatives of nitrosodi-n-propylamine (Fig. 7.13) having oxygen substituents were esophageal carcinogens in rats, some very potent. Included was 2,3-dihydroxypropylnitroso-2'-hydroxypropylamine, a very hydrophilic compound, which nevertheless induced 100% incidence of tumours of the esophagus after administration of quite small doses (Lijinsky et al., 1984d); it was much more potent than NBHPA, which was more lipophilic. 2,3-Dihydroxypropylnitroso-2'-oxopropyl-

Organ specificity 325

Fig. 7.13. *N*-nitroso compounds with oxygen substituents and their main target organs in rats. NDELA, hydroxypropylnitrosoethanolamine, nitrosobis-(2-hydroxypropyl)amine, hydroxyethylnitrosourea, hydroxypropylnitrosourea, hydroxypropylnitrosooxopropylamine, hydroxyethylnitrosoethylurea, oxopropylnitrosourea, nitrosobis-(2-oxopropyl)amine, hydroxyethylnitrosochloroethylurea, hydroxypropylnitrosochloroethylurea.

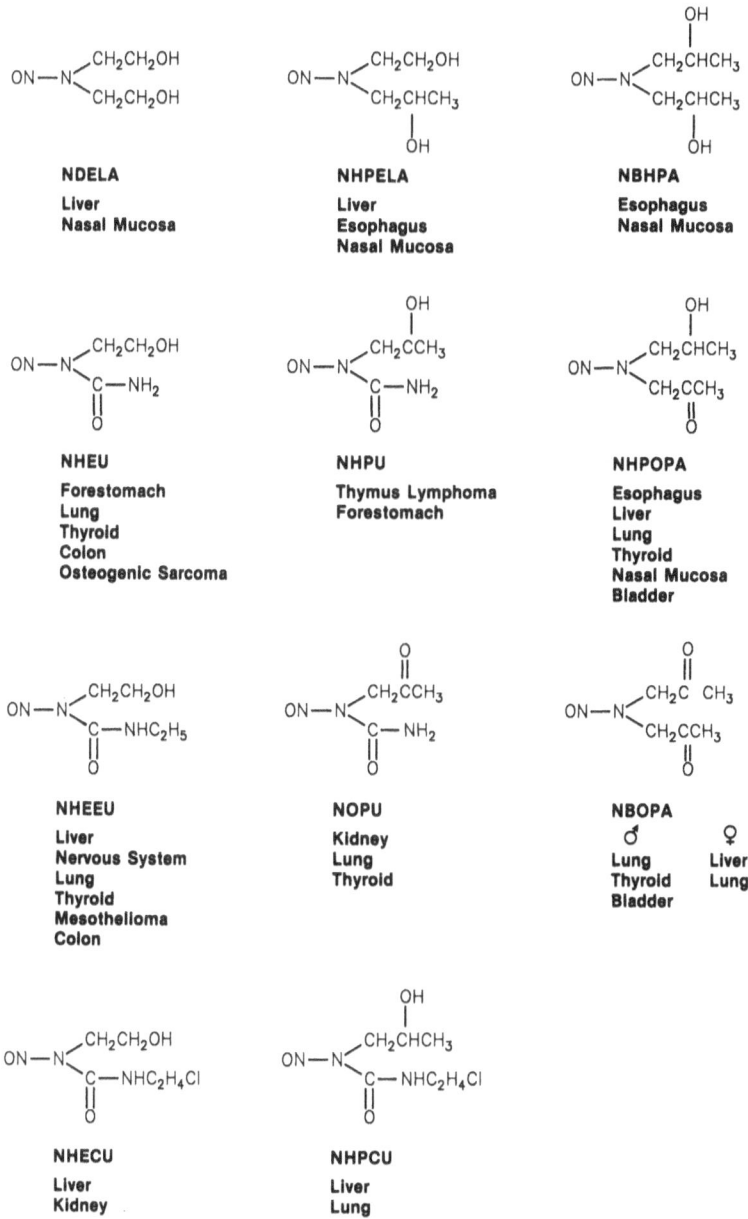

amine was even more effective in inducing esophageal tumours in rats than the trihydroxy compound, and did not give rise to any liver tumours in female rats (Lijinsky et al., 1984c), in spite of the presence of the 2-oxopropyl group. Compounds of this type were examined because of the resemblance of the dihydroxypropyl structure to glycerol and glyceraldehyde, which might underlie their very high carcinogenic potency. Methylnitroso-2,3-dihydroxypropylamine induced esophageal tumours in rats, but was weaker than MNHPA (Lijinsky and Reuber, 1984a), whereas allylnitroso-2,3-dihydroxypropylamine was more effective in inducing esophageal tumours than was allylnitroso-2-hydroxypropylamine (Lijinsky et al., 1984e), although the latter also induced liver tumours in high incidence. Perhaps these differences relate to the relative resistance of the allyl group to α-oxidation, compared with the methyl group; (nitrosodiallylamine failed to induce tumours of any kind in rats; Druckrey et al., 1967). Methylnitroso-2-hydroxyethylamine was mainly a liver carcinogen in rats when given by gavage, especially in females, but also induced tumours of the esophagus (Koepke et al., 1988a). Its symmetrical counterpart, nitrosodiethanolamine – an important environmental carcinogen – was considerably less potent, inducing liver tumours over a wide range of doses, but also low incidences of tumours of the esophagus at the high doses given (Lijinsky and Reuber, 1984b). In all of these cases, the presence of the N-methyl group seems to be important in giving the carcinogens a high affinity for the rat esophagus. However, the methyl group is not uniquely essential, since nitrosamines with structures not containing the N-methyl group are often esophageal carcinogens, sometimes very potent ones (nitrosodiethylamine, nitrosodi-n-propylamine).

Among the nitrosamines of other structures that are esophageal carcinogens in rats are a larger number of cyclic nitrosamines, with rings of different sizes and different shapes. The simplest cyclic nitrosamines, nitrosoazetidine and nitrosopyrrolidine have induced only liver tumours in rats, as also has nitroso-3-pyrroline (Garcia and Lijinsky, 1972). On the other hand, some derivatives of nitrosopyrrolidine have induced tumours of the esophagus, so the small ring size of nitrosopyrrolidine does not alone account for the absence of effect in the esophagus; neither nitroso-1,3-oxazolidine (Fig. 7.14), nor its methyl derivatives has induced tumours of the esophagus in rats. Nitroso-3,4-dichloropyrrolidine is enormously more potent than its parent, and induced only esophageal tumours, not liver tumours (Lijinsky et al., 1976d), while the tobacco-specific nitrosamine, nitrosonornicotine, induced esophageal tumours in high incidence when given to rats in drinking water (Hoffman et al., 1975), but

Organ specificity

Fig. 7.14. Nitrosomorpholine, 2-methyl- and 2,6-dimethyl-; nitroso-oxazolidine, 5-methyl- and 2-methyl-; nitrosotetrahydrooxazine.

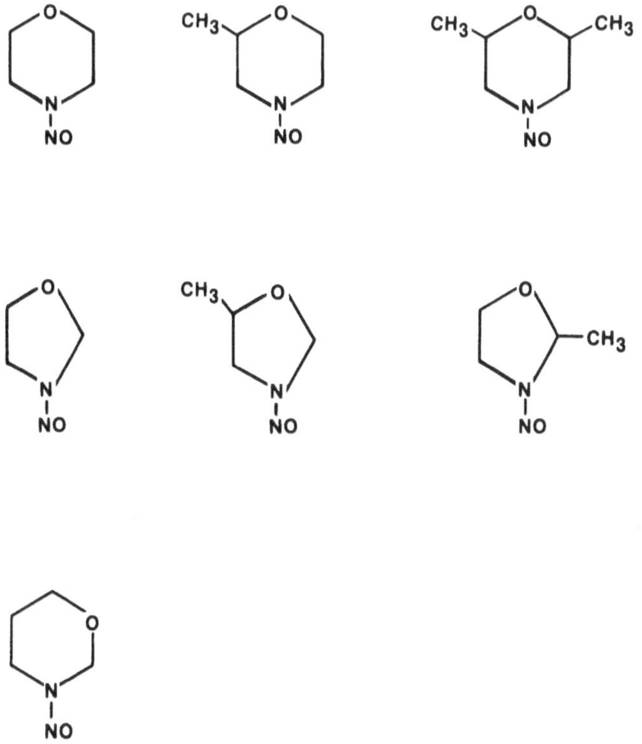

Fig. 7.15. Unsaturated nitrosopiperidines, Δ^2, Δ^3-, nitroso-3,4-epoxypiperidine.

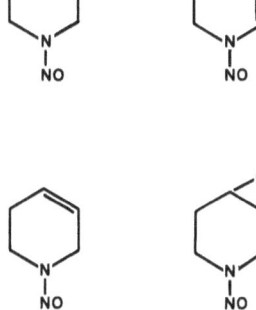

only tumours of the nasal mucosa when given subcutaneously. Again there seems to be a local effect of the carcinogen in the esophagus, but no influence of the large, basic pyridine ring, other than possibly to make nitrosonornicotine a relatively weak carcinogen.

Only a few derivatives of nitrosopyrrolidine have been tested for carcinogenic activity, whereas a large number of nitrosopiperidines (Fig. 7.15) have been examined, most of which have induced tumours of the esophagus in rats (Lijinsky, 1987). Nitrosopiperidine itself is principally an esophageal carcinogen in rats, although some liver tumours have also been produced in some experiments (and more frequently tumours of the nasal mucosa). Two unsaturated derivatives of nitrosopiperidine, nitroso-1,2,3,6-tetrahydropyridine and nitroso-1,2,3,4-tetrahydropyridine, induced a high incidence of esophageal tumours when given to rats in drinking water; they were of similar potency to nitrosopiperidine (Kupper et al., 1980). Nitroso-1,2,3,6-tetrahydropyridine also induced liver tumours in rats, especially when administered at high concentrations (Lijinsky et al., 1982a), a behaviour similar to that seen with nitrosodiethylamine (Lijinsky et al., 1981b); however, nitrosodiethylamine alkylates DNA *in vivo* extensively, while cyclic nitrosamines in general do not. It seems probable that toxicity to the liver at high doses is important in inducing tumours, especially the hemangiosarcomas which are common in these conditions, whereas lack of liver damage at low doses of these nitrosamines lets esophageal tumours prevail. It is of interest – and perplexing – that nitrosodiethanolamine shows the opposite behaviour, inducing high incidences of liver tumours at all doses, but esophageal tumours only at high doses in rats (Lijinsky and Reuber, 1984b; Lijinsky and Kovatch, 1985b). The same is true of nitrosomorpholine (Lijinsky et al., 1988b). These differences suggest that the pharmacokinetics of interaction of the nitrosamine with cellular receptors is the most important determinant of induction of tumours in one organ rather than another.

Substitution at various positions in the nitrosopiperidine ring does not often change the orientation towards inducing tumours of the esophagus in rats, although potency varies considerably and other tumours may also appear. Nitroso-3,5-dimethylpiperidine exists in two forms, *cis* and *trans*, of which the *trans* was the more potent carcinogen in rats. Both induced esophageal tumours, but the natural mixture (80% *cis*, 20% *trans*) was distinctly less potent than nitrosopiperidine, while the *trans* has similar potency to nitrosopiperidine (Lijinsky et al., 1982d). Nitroso-4-*tert*-butyl-piperidine (Fig. 7.3a) induced tumours of the esophagus in rats, as did the corresponding 4-phenyl-derivative (together with liver tumours); however, 4-cyclohexylnitrosopiperidine was completely inactive (Lijinsky et al.,

1981d). This unexpected result has not been explained by comparison of the metabolism of the 4-substituted nitrosopiperidines with the parent compound in rats that has been reported (Singer and MacIntosh, 1984). The conformational effects of these substituents on nitrosopiperidine were similar, but other effects must influence their carcinogenicity. Nitrosoanabasine, with a pyridyl substituent in the 2-position, was a weak esophageal carcinogen (Hoffmann et al., 1975).

Oxidation of nitrosopiperidine at various positions had different effects. Nitroso-4-hydroxypiperidine (Fig. 7.16) induced only tumours of the nasal mucosa and of the liver in female rats, but no tumours of the esophagus. On the other hand, nitroso-3-hydroxypiperidine induced a high incidence of esophageal tumours, as well as those of the nasal mucosa and liver in rats (Lijinsky and Taylor, 1975f). Nitroso-4-piperidone induced tumours of the liver and esophagus, as well as in the nasal mucosa and tongue. These three compounds were of similar potency to nitrosopiperidine. Nitroso-3-methyl-4-piperidone had increased potency compared with nitrosopiperidone and nitrosopiperidine, leading to earlier death of the treated rats, and induced a high incidence of tumours of the esophagus and nasal mucosa, but no liver tumours (Singer et al., 1984). These profound differences suggest strong steric effects in the process of carcinogenesis by these nitrosamines, perhaps by changing the affinity of the compounds for particular cellular receptors in an organ.

Halogen substitution in nitrosopiperidine has the effect of increasing

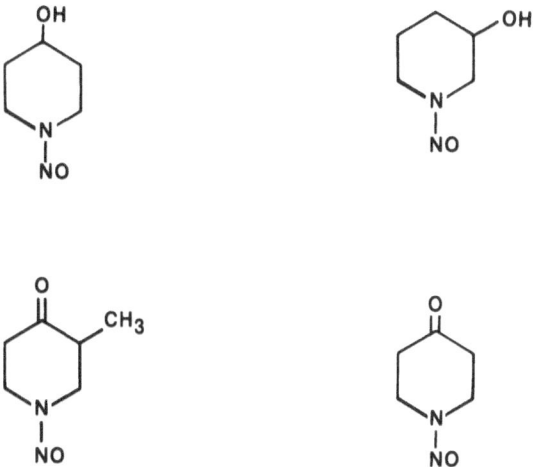

Fig. 7.16. 3-Hydroxy- and 4-hydroxy-nitrosopiperidine, 3-methylnitroso-4-piperidone, and nitroso-4-piperidone.

the potency of the compounds, and the main target organ remained the esophagus, as was the case with nitroso-3,4-dichloropyrrolidine. Both 3-chloro and 4-chloro-nitrosopiperidine induced tumours of the tongue and forestomach, in addition to the esophagus. In another strain of rat (Sprague-Dawley) nitroso-4-chloropiperidine induced hepatocellular or hemangioendothelial tumours in the liver of more than half of the rats (Lijinsky and Taylor, 1978a). Nitroso-3,4-dichloropiperidine was considerably more potent than either of the monochloro- compounds (which were similar to each other in potency – Lijinsky *et al.*, 1980e), while the 3,4-dibromo- derivative was less potent than the dichloro- compound (Lijinsky and Taylor, 1975b), and similar in potency to the monochloro- derivatives. The reasons for the particularly high potency of nitroso-3,4-dichloropiperidine are not clear, but one third of a millimol led to death of half of the treated rats with esophageal tumours in less than six months.

The cyclic higher homologue of nitrosopiperidine, nitrosohexamethyleneimine, induced esophageal tumours in rats, but also liver tumours in high incidence in several strains (Goodall *et al.*, 1968; Lijinsky and Taylor, 1979b; Lijinsky and Reuber, 1981a). The liver tumours included hemangiosarcomas, as well as hepatocellular carcinomas, the former being associated with extensive toxicity to liver cells; the lower homologue nitrosopyrrolidine is not very toxic to liver cells in rats, and induced only hepatocellular tumours (Lijinsky and Reuber, 1981a). The next two higher homologues, nitrosoheptamethyleneimine and nitrosooctamethyleneimine (Fig. 7.3), induced a high incidence of esophageal tumours in rats, the 8-carbon compound being distinctly less potent (Lijinsky *et al.*, 1969); in MRC-Wistar rats both compounds also induced squamous cell carcinomas of the lung when given in drinking water. In Sprague-Dawley rats nitrosoheptamethyleneimine again induced tumours of the esophagus and squamous cell lung tumours, which are accelerated by deliberate addition of an infectious bacterium (Schreiber *et al.*, 1972), but in Fischer (F344) rats, there was a high incidence of tumours of the esophagus, and some of the tongue, pharynx, etc., but almost no lung tumours (Lijinsky *et al.*, 1982b). The reason for the failure of the F344 rat lung to respond to nitrosoheptamethyleneimine might be due partially to the exceptional freedom of these rats from pathogens (at the Frederick Cancer Research Facility), although lung tumours in these same rats have been induced by many other *N*-nitroso compounds (Lijinsky, 1987), so other factors obviously are important. The only cyclic nitrosamine larger than C-8 examined is nitrosododecamethyleneimine (C-12), which induced liver tumours in two strains of rat (Lijinsky and Taylor, 1978e; Goodall and Lijinsky, 1984a); the ring system of this compound is quite convoluted, which might explain its relatively weak potency and its restricted carcinogenic action.

A group of cyclic nitrosamines which includes a number of potent esophageal carcinogens is the dinitrosopiperazines. The several mononitrosopiperazines that have been examined are non-carcinogenic (Love *et al.*, 1977) or weak carcinogens, which do not induce tumours of the esophagus in rats (Lijinsky and Reuber, 1982*b*; Singer *et al.*, 1981). That this lack of effect in the esophagus might be partly due to the fact that the mononitrosopiperazines are bases is supported to some extent by the induction of tumours of the esophagus by the *N*-acetyl- derivative of 3,5-dimethylnitrosopiperazine (Fig. 7.17); the unacetylated compound did not induce tumours of the esophagus, but thymic lymphomas (Singer *et al.*, 1981). The presence of the methyl groups in the 3 and 5 positions of 1-nitrosopiperazine increases the potency considerably, and suggests that these substituents create a structure, for example an appropriate conformation, that has strong affinity for receptors in the target cells of these carcinogens (Lyle *et al.*, 1976). This theme is recurrent in the evaluation of the relation of chemical structure to carcinogenic activity.

Dinitrosopiperazine itself induces tumours of the esophagus in rats (as well as liver tumours – Lijinsky and Taylor, 1975*g*; Garcia and Lijinsky, 1972) and is considerably less potent than nitrosopiperidine, which it resembles. Carcinogenic potency towards the rat esophagus is increased

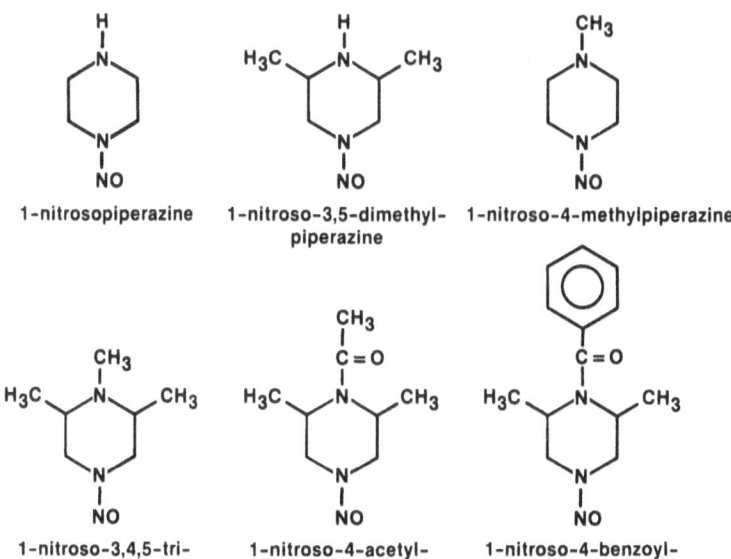

Fig. 7.17. Methylated mononitrosopiperazines and acylated derivatives.

considerably by methyl substitution, as in 2-methyl-, 2,6-dimethyl- and, to a lesser extent, in 2,5-dimethyldinitrosopiperazine (Lijinsky and Taylor, 1975g). The tetramethyl derivative, as discussed earlier, was inactive. Dinitroso-2,6-dimethylpiperazine was potent even at quite low concentrations (Lijinsky and Taylor, 1978d; Lijinsky et al., 1983c). The higher cyclic homologue of dinitrosopiperazine, 1,4-dinitrosohomopiperazine, was considerably more potent than dinitrosopiperazine and gave rise only to tumours of the esophagus and nasal mucosa (and occasionally of the tongue), but no liver tumours, over a large range of doses, as low as less than 0.1 millimole (Lijinsky et al., 1982a). It seems that molecular size plays an important role in the affinity of nitrosamines for the rat esophagus, as shown by the failure of the smallest (C-3 and C-4) and the largest (C-12) cyclic nitrosamines to induce esophageal tumours, while those neutral cyclic nitrosamines intermediate in size include very potent esophageal carcinogens.

This is emphasised by the findings with cyclic nitrosamines containing oxygen or sulphur in the ring. The 5-membered ring compounds, nitroso-1,3-oxazolidines, whether substituted or not, induced only liver tumours in rats (Lijinsky and Reuber, 1982c). Nitrosothiazolidine was, unexpectedly, devoid of carcinogenic activity (Lijinsky et al., 1988a). The higher cyclic homologues of nitrosooxazolidine, nitrosotetrahydrooxazine and nitrosomorpholine (Fig. 7.14), were mainly potent liver carcinogens in rats, although nitrosomorpholine (but not nitrosotetrahydrooxazine) at high dose rates also induced esophageal tumours (Garcia and Lijinsky, 1972; Lijinsky et al., 1988b). Nitroso-2-methylmorpholine also was a liver carcinogen in rats, considerably more potent than nitrosomorpholine (Lijinsky and Reuber, 1982c), and induced a high incidence of tumours of the esophagus. Nitroso-2,6-dimethylmorpholine was even more potent than the 2-methyl derivative (Lijinsky and Taylor, 1975c), especially the *trans* isomer (Lijinsky and Reuber, 1980c) and induced tumours of the esophagus, but no liver tumours. The effects of methyl substitution on carcinogenesis of nitrosomorpholine suggest, once again, that the particular structure containing the methyl groups is associated with a high affinity for some receptor or receptors in the rat esophagus, for which nitrosomorpholine itself has only a weak affinity, and that the reverse is true of the rat liver; in the hamster liver the methylated nitrosomorpholines induce liver tumours, whereas nitrosomorpholine itself does not, instead inducing tumours of the nasal mucosa (Lijinsky et al., 1984a).

Nitrosothiomorpholine (Fig. 7.18), in contrast with nitrosomorpholine (and with its homologue nitrosothiazolidine), is a potent inducer of esophageal tumours in rats (Garcia et al., 1970), but did not give rise to liver tumours. Another sulphur-containing nitrosamine, nitrosodithiazine, was not carcinogenic, although its trimethyl derivative, nitrosothialdine, induced tumours of both the liver and esophagus (Lijinsky et al., 1988a). These findings suggest that activation of those cyclic compounds containing two sulphur atoms in the ring is quite different from the carbocyclic analogues, in which substituents α- to the N-nitroso function eliminate carcinogenic activity. The possibility of generating from nitrosothialdine an analogue of activated nitrosodiethylamine has been discussed as an explanation of the unexpectedly strong carcinogenicity of this α-methylated cyclic nitrosamine (Lijinsky et al., 1988a).

The relation of chemical structure of nitrosamines to their induction of tumours of the esophagus in rats is not as simple as would be suggested by measurement of alkylation of DNA in the rat esophagus, even though this probably plays an important role in some cases (Von Hofe et al., 1987). The lack of evidence that cyclic nitrosamines, which are often potent esophageal carcinogens, form detectable quantities of DNA adducts in esophagus (Scanlan et al., 1980; Hecht et al., 1981) or even in liver (Farrelly and Hecker, 1984), suggest that the role of DNA

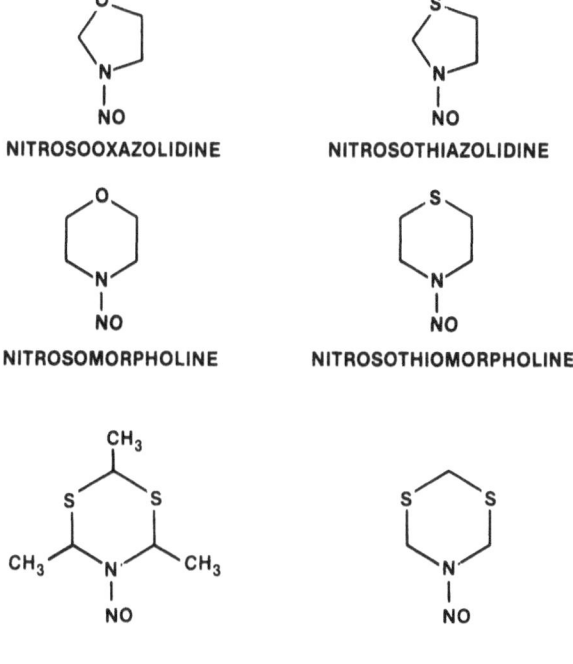

Fig. 7.18. Cyclic nitrosamines containing sulphur.

alkylation in esophageal carcinogenesis is unclear. Other facts make a simple explanation along those lines even more difficult. Included are the failure of the directly acting alkylating agents, alkylnitrosoureas, alkylnitrosocarbamates and alkylnitrosoguanidines to induce tumours in the rat esophagus more than very occasionally, even when given in drinking water, which bathes the esophagus. Also, and perhaps of even greater interest, is the failure of any of the very potent rat esophageal carcinogens to induce tumours of the esophagus in Syrian hamsters, or in guinea pigs. Several have induced tumours of the esophagus in mice with, however, lesser potency than in rats (Preussmann and Stewart, 1984; Goodall and Lijinsky, 1984b).

7.5.2 Liver carcinogens in rats and hamsters

The liver is the target of a large proportions of nitrosamines, and even of some alkylnitrosoureas, in rats. The liver is also the target of many carcinogenic nitrosamines in Syrian hamsters, although in several cases those that are liver carcinogens in rats have not induced tumours of the liver in hamsters (e.g. nitrosomorpholine – (Lijinsky et al., 1984a), and vice-versa (e.g. methylnitroso-n-propylamine, methylnitroso-n-butylamine and methylnitroso-n-amylamine – Lijinsky and Kovatch, 1988b).

Nitrosodimethylamine, the primordial carcinogenic nitrosamine, induces mainly hemangiosarcomas of the liver in rats when given in drinking water or in hamsters by gavage (Lijinsky and Reuber, 1984a; Lijinsky et al., 1987b). When given to rats by gavage or at low doses in drinking water (Lijinsky et al., 1987b; Peto et al., 1984) there is also a considerable incidence of hepatocellular neoplasms, and some cholangiocellular tumours. The liver cell damage produced by NDMA that plays a part in the formation of endothelial tumours in the liver must be different in character from that produced by the rat liver toxin carbon tetrachloride, administration of which concurrently with NDMA gave rise to hepatocellular tumours instead of angiosarcomas (Taylor et al., 1974). The less acutely toxic NDEA had a lesser tendency to induce hemangiosarcomas in the liver, and induced more hepatocellular neoplasms, in both rats and hamsters (Lijinsky et al., 1981b; 1987b). In connection with the induction of liver tumours in both rats and hamsters by these two nitrosamines, it is interesting that the order of their potency in rats and in hamsters is opposite; in hamsters NDMA is considerably more potent than is NDEA, while in rats NDEA is more potent than NDMA. However, there is not a large difference in the extent of liver DNA ethylation or methylation, respectively, by NDEA and NDMA between rat and hamster.

The 'hybrid' alkylating agent methylnitrosoethylamine also induces liver tumours in both rats and hamsters, with the tendency to resemble NDMA more than NDEA in both species (Lijinsky et al., 1987b). The tumours were commonly both hepatocellular and endothelial. Methylnitroso-n-propylamine, -n-butylamine and -n-amylamine induced liver tumours in hamsters, with more or less the same potency (Lijinsky and Kovatch, 1988b), but in rats only tumours of the esophagus. With methylnitroso-n-hexylamine, the capacity of methylnitroso-n-alkylamines to induce liver tumours in rats returns (Lijinsky et al., 1983e). The succeeding homologues, methylnitroso-n-heptylamine, -n-octylamine, -n-nonylamine and -n-undecylamine (Fig. 7.2) all induce liver tumours in rats, together with, in some cases, tumours of the esophagus, lung or bladder (Lijinsky et al., 1981c). Methylnitroso-n-decylamine, -n-dodecylamine and -n-tetradecylamine did not induce liver tumours, but bladder tumours only (Lijinsky et al., 1981c). Those of the same series, including the C-6, C-7 and C-8 compounds, that have been given to hamsters also induced liver tumours and, sometimes, lung tumours or bladder tumours (Lijinsky and Kovatch, 1988b). All of these compounds could be either methylating agents or could alkylate DNA in the liver with the larger alkyl group, or both. In conformity with this, all of the compounds gave rise to intermediates mutagenic to bacteria when activated by hamster liver enzymes, but many of them were non-mutagenic when activated by rat liver enzymes (Andrews and Lijinsky, 1984).

Among symmetrical dialkylnitrosamines, many induced liver tumours in rats, including nitrosodi-n-propylamine, nitrosodi-n-butylamine and nitrosodi-n-amylamine (Druckrey et al., 1967). These uniform responses of the rat liver led to Druckrey's conclusion that symmetrical dialkylnitrosamines induce liver tumours, whereas asymmetrical dialkylnitrosamines induce tumours of the esophagus. Neither of these claims is entirely true, although many nitrosamines of both classes fit into that mould. Several symmetrical substituted dialkylnitrosamines induce liver tumours, including NDELA (nitrosodiethanolamine), which also induces tumours of the esophagus, kidney and nasal mucosa (Lijinsky and Reuber, 1984b) in rats. In hamsters only tumours of the nasal mucosa appeared, and no liver tumours (Lijinsky et al., 1988e; Hoffmann et al., 1983), although the liver of the hamster is certainly not insusceptible to induction of tumours by nitrosamines. The ethyl and methyl ethers of nitrosodiethanolamine (Fig. 7.19) are also liver carcinogens in rats, and are more potent than the parent alcohol, possibly because the ethers are less easily excreted unchanged than the former (Lijinsky and Taylor, 1978c).

The diethylacetal of the symmetrical dialdehyde formed by oxidation of nitrosodiethanolamine (nitrosobis-diethoxyethylamine), on the other hand, was not carcinogenic at all (Lijinsky and Taylor, 1978c).

Nitrosobis-(2-hydroxypropyl)amine, a homologue of NDELA, was mainly an esophageal carcinogen in rats, and also induced tumours of the liver and thyroid when given by gavage (Lijinsky et al., 1988e), but not when administered in drinking water (Lijinsky et al., 1984d). In hamsters NBHPA induced tumours of the liver and of the pancreatic ducts (Lijinsky et al., 1984b). An oxidation product of NBHPA, nitrosobis-(2-oxopropyl)amine was a much more potent liver carcinogen in rats than the alcohol, but only in females; in males there were tumours of the thyroid follicles, lungs, kidney and bladder, but few of the liver, whether the compound was given by gavage or in drinking water (Lijinsky et al., 1988f). The females, on the other hand, did not develop tumours of the kidney or bladder. In hamsters NBOPA was also much more potent than the alcohol from which it was derived, and gave rise to tumours of the liver and pancreatic ducts (Lijinsky et al., 1984b). The ketoalcohol 2-hydroxypropylnitroso-2-oxopropylamine was intermediate in carcinogenic potency, inducing tumours of the liver and pancreatic ducts in hamsters, but in rats tumours of the liver, lungs, thyroid, esophagus and bladder in both males and females, and whether given by gavage or in drinking water (Lijinsky et al., 1984c, 1988f). These results suggest the existence in those organs of receptors with affinity for one or other alkyl group in the hybrid molecule.

It is clear from the broad spectrum of tumours induced by the β-oxidised nitrosodi-n-propylamines that these are not intermediates in the

Fig. 7.19. Diethylnitrosamine with symmetrical substituents.

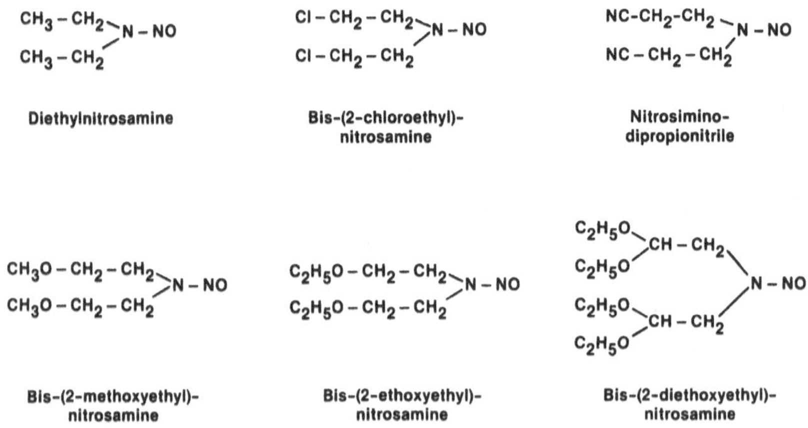

induction of liver tumours in rats by nitrosodi-n-propylamine, as was suggested by Krüger (1971). Nor does 4-hydroxybutylnitroso-n-butylamine, one of the oxidation products of nitrosodi-n-butylamine, appear to be involved in induction of tumours of the liver by NDBA in rats, since the hydroxy compound did not induce liver tumours.

It is probable that oxygen substitution in alkylnitrosamines provides a predilection for liver carcinogenesis, since many hydroxylated nitrosamines or their corresponding ketones were liver carcinogens in both rats and hamsters. Methylnitroso-2-hydroxyethylamine induced more liver tumours in female rats than in male rats (Koepke et al., 1988a), a sex difference which was displayed by some other oxygenated nitrosamines (Lijinsky et al., 1988f). For example, MNOPA induced liver tumours in 75% of female rats, but in only 10% of male rats (Lijinsky et al., 1983d). MNHPA induced liver tumours in only a small number of rats of either sex (Lijinsky and Reuber, 1984a), while both the alcohol and the ketone were potent liver carcinogens in hamsters (Lijinsky and Kovatch, 1985a). An additional hydroxyl group, as in methylnitroso-2,3-dihydroxypropylamine, reduced carcinogenic potency in rats and hamsters but increased the effect in the rat liver (Lijinsky and Reuber, 1984a); only nasal tumours appeared in hamsters (Lijinsky and Kovatch, 1985a). Other methylnitrosoalkylamines with substituted alkyl chains, including the tobacco-specific NNK (Hoffmann et al., 1984b; Rivenson et al., 1988), induced liver tumours in rats and hamsters.

All nitrosamines that are derivatives of ethanolamine seem to have an ability to induce liver tumours in rats, but much less so in hamsters (Lijinsky et al., 1985c), including hydroxyethylnitroso-2-hydroxypropylamine,-2-oxopropylamine, -2,3-dihydroxypropylamine and -allylamine, and they differed considerably in potency (Lijinsky et al., 1984d), the dihydroxypropyl derivative being particularly weak. These results suggest that the nitrosoethanolamine function provides an affinity of the compound for receptors in the liver of rats or hamsters.

Nitrosamines containing a 2-hydroxypropyl group or a 2-oxopropyl group do not consistently provide an orientation towards the liver of rats. While allylnitroso-2-hydroxypropylamine and allylnitroso-2-oxopropylamine induced a high incidence of liver tumours in rats (Lijinsky et al., 1984e), but not in hamsters (Lijinsky et al., 1985c), neither 2-hydroxypropylnitroso-2,3-dihydroxypropylamine nor 2-oxopropylnitroso-2,3-dihydroxypropylamine induced liver tumours in rats or in hamsters, although both of the latter were potent carcinogens (Lijinsky et al., 1984c,d, 1985c).

The failure of some of these compounds, seemingly with appropriate structures, to induce liver tumours is surprising, since the liver of rats and hamsters is obviously susceptible to induction of tumours by nitrosamines. In several experiments one or two animals of a group developed liver tumours, even when the response of other organs to a particular nitrosamine was overwhelming.

Among cyclic nitrosamines a large number induced liver tumours in rats and often in hamsters, also. However, a much smaller number of cyclic nitrosamines has been tested in hamsters. As in the acyclic series, there is no consistent difference in relative carcinogenic potency of cyclic nitrosamines between rats and hamsters, even when liver tumours are induced in both species.

The simplest cyclic nitrosamine is nitrosoazetidine, first prepared as long ago as 1898 (Marckwald and Droste-Huelshorff, 1898), which induced liver tumours in rats with reasonable potency, and virtually no other tumours (Lijinsky et al. 1967; Lijinsky and Taylor, 1977b). In both mice and in Syrian hamsters the effectiveness in inducing liver tumours (and lung tumours in mice) was much lower than in rats; indeed, in the first study of the compound in hamsters, it was reported noncarcinogenic (Lijinsky et al., 1970). The next homologue nitrosopyrrolidine, to which there is broad human exposure (Lijinsky and Epstein, 1970), was a carcinogen for the rat liver of considerable potency (Lijinsky and Reuber, 1981a); Preussmann et al., 1977), but in mice and hamsters induced only lung tumours (Greenblatt and Lijinsky, 1972b). Nitroso-3-hydroxypyrrolidine also induced liver tumours in rats, although it was somewhat less potent than the parent compound (Eisenbrand et al., 1980), while the unsaturated derivative nitroso-3-pyrroline was similar in potency to nitrosopyrrolidine in inducing liver tumours in rats (Garcia and Lijinsky, 1972); neither the hydroxy derivative nor the unsaturated compound has been tested in hamsters.

Ascending the homologous series, the next cyclic nitrosamine is nitrosopiperidine, which induced some liver tumours in rats and mice, but was mainly an esophageal carcinogen (Garcia and Lijinsky, 1972), and in hamsters induced tumours of the trachea and nasal mucosa, but no liver tumours (Haas et al., 1973). The three methyl derivatives of nitrosopiperidine all induced tumours of the esophagus in rats, but the least potent among them, 2-methylnitrosopiperidine, also gave rise to liver tumours (Lijinsky and Taylor, 1975a). Hydroxylation of nitrosopiperidine increased the tendency to induce liver tumours in rats, as was also the case with the 4-ketone (nitroso-4-piperidone), and these derivatives also gave rise to tumours of the nasal mucosa (Lijinsky and Taylor, 1975f); the 3-methyl derivative of nitroso-4-piperidone was a more potent carcinogen (for the esophagus) in rats, but failed to induce liver tumours (Singer et al., 1984), a combination of properties common to cyclic nitrosamines made more rigid by alkyl substitution in the ring (Lijinsky and Taylor, 1975c). On the other hand, 4-phenylnitrosopiperidine induced liver tumours as well as tumours of the esophagus in rats, whereas the very similar 4-*tert*-butyl-nitrosopiperidine induced only tumours of the esophagus (Lijinsky et al., 1981d), and 4-cyclohexylnitrosopiperidine was inactive.

One of the two unsaturated derivatives of nitrosopiperidine, nitroso-Δ^3-dehydropiperidine, was a much more potent liver carcinogen than nitrosopiperidine, whereas its isomer nitroso-Δ^2-dehydropiperidine was an equally potent carcinogen in rats, but induced only tumours of the esophagus (Kupper et al., 1980). It was possible that the unsaturated nitrosopiperidines owed their activity to formation of the epoxides, but nitroso-3,4-epoxypiperidine was a less potent carcinogen than nitroso-Δ^3-dehydropiperidine, although it did give rise to liver tumours in rats; this makes it unlikely that carcinogenesis by the unsaturated compounds is mediated through formation of the epoxides.

Nitrosohexamethyleneimine, the next higher homologue to nitrosopiperidine, was a potent liver carcinogen in rats, giving rise mainly to hemangiosarcomas, in addition to tumours of the esophagus (Goodall et al., 1968; Lijinsky and Taylor, 1979b; Lijinsky and Reuber, 1981a). Once again, in hamsters this cyclic nitrosamine did not induce liver tumours, but tumours of the trachea (Althoff et al., 1973), and in mice liver and lung tumours (Strickland et al., 1988). Neither nitrosoheptamethyleneimine nor nitrosooctamethyleneimine induced liver tumours in rats, but nitrosododecamethyleneimine gave rise to liver tumours following administration of large doses to several strains of rat (Lijinsky and Taylor, 1978e; Goodall and Lijinsky, 1984a), and also in mice (Goodall et al., 1973).

The cyclic nitrosamines having atoms other than carbon in the ring seem to differ considerably from the nitrosoazetidine series in their ability to induce liver tumours in rats. For example, in contrast with nitrosopiperidine, the six-membered cyclic nitrosamine nitrosomorpholine induces mainly liver tumours in rats, and few esophageal tumours except at very high doses (Lijinsky et al., 1988b). In hamsters, on the other hand, nitrosomorpholine induced only tumours of the nasal mucosa, but no liver tumours (Lijinsky et al., 1984a). Nitroso-2-methylmorpholine (having the methyl group β-to the N-nitroso function) gave rise to many tumours of the esophagus, as well as liver tumours in rats, and was more potent than nitrosomorpholine (Lijinsky and Reuber, 1982c). Contrasting with nitrosomorpholine in hamsters, nitroso-2-methylmorpholine induced a high incidence of tumours of the liver and nasal mucosa (Lijinsky et al., 1984a), and nitroso-2,6-dimethylmorpholine induced in hamsters tumours of the liver and pancreas ducts in high incidence (Lijinsky et al., 1982f), but not in rats. These differences between rats and hamsters in response to nitrosomorpholines can be ascribed to the tendency of β-oxidation to predominate as a metabolic pathway in hamsters, and α-oxidation to predominate in rats (Gingell et al., 1976; Underwood and Lijinsky, 1982), leading to ring opening in the case of β-oxidation. In both rats and hamsters the carcinogenic properties of 2-methylnitrosomorpholine were a combination of those of nitrosomorpholine and 2,6-dimethylnitrosomorpholine, a reflection of possession of the chemical attributes of both. In guinea pigs nitrosodimethylmorpholine induced liver hemangiosarcomas in high incidence (Cardy et al., 1980).

The lower homologue of nitrosomorpholine, nitroso-1,3-oxazolidine was almost identical to nitrosomorpholine in liver carcinogenicity in rats, but the 2-methyl and 5-methyl derivatives were considerably less effective, while inducing also a high incidence of tumours in the liver; these were predominantly hepatocellular carcinomas in the case of the two methyl derivatives, but both hepatocellular and hemangioendothelial tumours, usually in the same animal, in the case of nitrosooxazolidine itself (Lijinsky and Reuber, 1982c). Unlike nitrosomorpholine, in hamsters nitrosooxazolidine induced liver tumours in almost all of the animals, most of which had both hepatocellular and cholangiocellular neoplasms, but 5-methylnitrosooxazolidine was more potent than the parent compound, inducing many hemangiosarcomas of the liver, as well as hepatocellular and cholangiocellular neoplasms; there were also a small number of animals with tumours of the pancreas duct (Lijinsky et al., 1984a), reminiscent of the action of nitroso-2,6-dimethylmorpholine. An isomer of nitrosomorpholine, also having a six-membered ring, is nitrosotetrahydro-1,3-oxazine, which has not been tested in hamsters, but in rats is very similar in potency and effectiveness to nitrosomorpholine, inducing hemangiosarcomas and hepatocellular carcinomas in the liver of almost all treated rats. These oxygen-containing cyclic nitrosamines had a similar affinity for rat liver, but the hamster liver was highly selective, only responding by tumour formation to particular chemical structures. These differences indicate that the mechanisms by which they induce liver tumours were quite different in the two species.

The replacement of the oxygen atom in the ring of nitrosomorpholine by sulphur changes the target organ in rats from the liver to the esophagus, whereas changing the oxygen atom in nitrosooxazolidine for sulphur, as in nitrosothiazolidine, changes the compound from a liver carcinogen in rats to a non-carcinogen. The reasons for these sharp changes in activity are not known, but might be related to the difference in oxidation state of the neighbouring carbon atoms from oxygen to sulphur; the resulting product of ring opening might be quite different in reactivity (Lijinsky et al., 1988a).

In the comparative study of sulphur-containing nitrosamines just referred to, the six-membered cyclic nitrosamines containing sulphur at positions 3 and 5 (instead of methylene groups) have unexpected carcinogenic activities. For example, nitrosothialdine (Fig. 7.18) induces in rats tumours of the liver and esophagus (like NDEA), possibly due to opening of the ring at the sulphur atoms forming a product related to that formed by activation of nitrosodiethylamine. Although a parallel process would convert nitrosodithiazine into a product analogous to that formed by activation of nitrosodimethylamine, nitrosodithiazine, which lacks the α-methyl substituents of nitrosothialdine was without carcinogenic activity. It is obvious that the presence of α-methyl substituents in nitrosothialdine does not inhibit carcinogenic activity, as does analogous methyl substitution in nitrosopiperidine or nitrosopyrrolidine; this means also that α-oxidation of nitrosothialdine is not an important step in activation to carcinogenic products.

Among derivatives of nitrosopiperazine, another six-membered cyclic nitrosamine, only dinitrosopiperazine and nitroso-4-carbethoxypiperazine induced liver tumours in rats (Druckrey et al., 1967); all of the others induced tumours of the esophagus, which dinitrosopiperazine also induced. Dinitrosopiperazine also induced liver tumours in DBA mice, and has not been tested in hamsters. However, nitroso-2,6-dimethylpiperazine induced tumours of the liver in both hamsters and guinea pigs (Cardy and Lijinsky, 1980), but not in rats. Nitrosoindoline produced liver tumours when given by gavage to rats (Urban and Danz, 1976). Nitroso-2-methoxy-2,6-dimethylmorpholine induced tumours of the liver and a variety of other organs (including pancreas) in hamsters, although, unlike the hydroxy compound, it cannot easily form the acyclic isomer (Pour et al., 1981).

The differences between cyclic nitrosamines in their induction of liver tumours seems to depend on the activity of the products of α-oxidation in rats and β-oxidation in hamsters, although the nature of the intermediates is not known. Studies of mutagenesis following activation of these compounds by rat and hamster liver microsomes (Chapter 6) have shed no light on the matter.

Among directly acting N-nitroso compounds few induced liver tumours in rats or mice, and only a single one of those examined in hamsters, the natural product streptozotocin, a glycosidic derivative of nitrosomethylurea (Berman et al., 1973). The failure of the directly acting alkylating agents, which alkylate DNA in the liver of animals even when given by mouth (Lijinsky, 1988b), to induce liver tumours more than occasionally is one of the most perplexing problems in carcinogenesis; particularly when the pattern of alkylation of DNA they produce is little different from that produced by nitrosamines and azoxyalkanes which require metabolic activation and do induce liver tumours.

The finding that some of the directly acting compounds induce liver tumours in rats, however, suggest that there is no inherent failing of alkylnitrosoureas, for example, to evoke a carcinogenic response in the rat liver. The alkylnitrosoureas inducing liver tumours in rats include the cyclic compound, nitroso-3,6-dihydrouracil (Bulay et al., 1979). The only other monoalkylnitrosourea that induced a high incidence of liver tumours was 2-phenylethylnitrosourea, which gave rise to almost 100% incidence of hemangiosarcomas, and was a potent bacterial mutagen, in spite of the large alkyl group which is presumed to be inserted in the DNA to produce the mutations (and the tumours).

The other liver carcinogens among the nitrosoureas were dialkylnitrosoureas, but only those with a particular type of structure. For example, while neither ethylnitrosourea nor hydroxyethylnitrosourea induced liver tumours in rats (although they induced a broad spectrum of tumours of other types – Lijinsky and Reuber, 1983b; Lijinsky et al., 1985e), both ethylnitrosohydroxyethylurea and hydroxyethylnitrosoethylurea induced liver tumours by gavage. The latter was more effective, giving rise to liver tumours in more than half of the animals, and most were hepatocellular carcinomas, whereas the liver tumours in six rats given the ethylnitroso-compound were all adenomas (Lijinsky et al., 1985e). When administered to rats in drinking water, however, both isomers induced a similar variety of tumours as by gavage treatment, but only the hydroxyethylnitroso compound gave rise to a few liver tumours. The reason for the discrepancy between the two methods of treatment might be that the bolus dose by gavage treatment has an effect in the liver not produced by the lower local doses of the compounds delivered in drinking water. This is the opposite of the effects of compounds requiring metabolic activation, such as NDMA and azoxymethane, which induce many liver tumours in rats when given in drinking water, but few or no liver tumours when given by gavage (Lijinsky et al., 1987b).

Hydroxyethylnitrosochloroethylurea and its homologue hydroxypropylnitrosochloroethylurea both induced primarily liver tumours in rats, more so when given by gavage than in drinking water (Lijinsky et al.,

1986a); the hydroxyethyl compound was more potent than the hydroxypropyl compound, and the former also gave rise to a high incidence of tubular cell neoplasms of the kidney. Although the dialkylnitrosoureas are similar in bacterial mutagenic potency to the corresponding monoalkylnitrosourea, they were considerably weaker carcinogens than the latter, and induced high incidences of liver tumours in rats, which the monoalkylnitrosoureas did not. The reason for the differences is not clear, but it seems that the presence of the alkyl group (particularly 2-chloroethyl-) on the N'-amino group creates a carcinogen with particular affinity for some molecular receptor in the rat liver, which then leads to transformation of liver cells. This receptor is presumably absent from the hamster liver, since the directly acting nitrosoalkylureas do not induce tumours in the hamster liver, although they all induce tumours in the hamster spleen (Lijinsky and Kovatch, 1989c).

7.5.3 Compounds inducing lung tumours

Like the liver, lung is a common target for induction of tumours by N-nitroso compounds in a number of species in which many compounds have been tested; some of the agents inducing tumours of the lung have been discussed in relation to their chemical structure in a review (Lijinsky, 1983a). As in the case of liver tumours it is common for one of these compounds to give rise to lung tumours in one species, but other types of tumour in other species, with a nevertheless consistent pattern. This leads to the conclusion that particular chemical structural characteristics are associated with induction of lung tumours in a particular species, and this is the case also with N-nitroso compounds having other organs as targets. In addition there are many types of cell in the lung, in several of which tumours develop in response to treatment with N-nitroso carcinogens. Some of these differences have been elaborated by Reznik-Schüller, in particular the association of some types of lung tumour with particular nitrosamine structures (Reznik-Schüller and Hague, 1981; Reznik-Schüller and Lijinsky, 1979).

The simplest nitrosamine, nitrosodimethylamine, has induced lung tumours in rats and mice, but rarely in other species, including hamsters. The tumours in the lung were alveolar/bronchiolar adenomas and carcinomas, which were the most common lung tumours induced in rats or mice by N-nitroso compounds in general. Nitrosodimethylamine did not induce many lung tumours in rats when administered in drinking water, liver tumours being by far the most common (Peto et al., 1984; Lijinsky and Reuber, 1984a). However, when rats are given NDMA by gavage, tumours of the lung (and kidney) become more common than those in liver (Lijinsky et al., 1987b; Lijinsky and Kovatch, 1989a) in both sexes. This was true of the very similar methylnitrosoethylamine, which induced liver tumours when administered in drinking water to rats (Lijinsky et al., 1980a), but also lung tumours when it was given by gavage (Lijinsky et al., 1987b). It is possible that the relatively large bolus doses of the two compounds evade 'first pass' destruction in the liver and reach the lung in sufficient amounts to be converted there to reactive carcinogenic metabolites. Lung and not liver of the rat is the principal target of NDMA when given by intravesicular injection (Lijinsky and Kovatch, 1989a), which perhaps circumvents the principal metabolising organ, the liver. However, MNEA administered intravesicularly gives rise mainly to liver tumours (Lijinsky, 1990d).

While nitrosodiethylamine induced lung tumours by several routes of administration to mice, in rats it did not give rise to lung tumours. This contrasts with the effects of MNEA and NDMA, and suggests that methylation of DNA (as O^6-MeG, for example) in the rat lung is more effective in inducing tumours there than is ethylation, which is found at much lower levels than methylation. No lung tumours were produced by NDEA in rats whether given in drinking water (Lijinsky et al., 1981b), or by gavage (Lijinsky et al., 1987b) or by injection (Svoboda and Higginson, 1968). On the other hand, NDEA induced lung tumours in hamsters, whether given by mouth or by injection (Dontenwill and Mohr, 1961; Dontenwill et al., 1962), but rarely in any of the other 40 species in which this compound has been tested (Bogovski and Bogovski, 1981; Preussmann and Stewart, 1984).

Nitrosodi-*n*-propylamine induces lung tumours in both rats and hamsters, as does nitrosodi-*n*-butylamine; the latter also induces lung tumours in mice, rabbits, Chinese hamsters and European hamsters. Nitrosodi-*iso*butylamine, a considerably weaker carcinogen than NDBA, gave rise to lung tumours in rats (Lijinsky and Taylor, 1979b) and in hamsters (Althoff et al., 1975). Nitrosodi-*n*-amylamine, continuing the pattern, also gave rise to lung tumours in rats (Druckrey et al., 1967) and in mice (Zabeszhinskii and Bolanski, 1977). Whether or not these large molecules undergo α-oxidation in the lung to form the proximate carcinogen, assumed to be an agent alkylating DNA or other cellular macromolecules, is not known. The heterogeneity of the lung makes it difficult to perform meaningful metabolic studies which can be related with precision to induction of tumours, since the cells giving rise to the tumours comprise only a fraction of the lung tissue.

Considering the series of methylnitroso-*n*-alkylamines, the first two, NDMA and MNEA both induce lung tumours in rats by gavage, but not in hamsters. Methylnitroso-*n*-propylamine induced lung tumours in hamsters, but not in rats, a pattern shown also by methylnitroso-*n*-butylamine and methylnitroso-*n*-amylamine (Lijinsky and Kovatch, 1988*b*). It is possible that early death of the rats from esophageal tumours prevented development of lung tumours. The higher homologues, methylnitroso-*n*-hexylamine, -*n*-heptylamine and -*n*-octylamine all induced lung tumours in both rats and hamsters, but with decreasing potency as the chain length increased; it is possible that the decreased effectiveness could be related to the increase in liposolubility and decrease in water solubility as the molecular weight increased. The methylnitroso-*n*-alkylamines from C-9 to C-14 all induced lung tumours (Lijinsky *et al.*, 1981*c*) in rats (again with decreasing potency as the chain length increased), and only C-12 has been tested in hamsters, producing some lung tumours (Althoff and Lijinsky, 1977). The methylnitroso-*n*-alkylamines with long carbon chains might undergo extensive metabolism before forming proximate carcinogens for the lung, which might be structurally similar or even identical from all the members of the series (c.f. Singer *et al.*, 1981).

Among the simplest oxygen-substituted nitrosamines are the methyl ethers of α-hydroxynitrosamines, the latter being the putative primary products of activation of nitrosamines and precursors of the alkyl-diazonium ions believed to be the alkylating agents for DNA which initiate carcinogenesis. The isomeric methoxymethylnitrosoethylamine (Fig. 7.20) and 1-methoxyethylnitrosomethylamine are very different in carcinogenic activity, the latter giving rise in rats to tumours of the lung

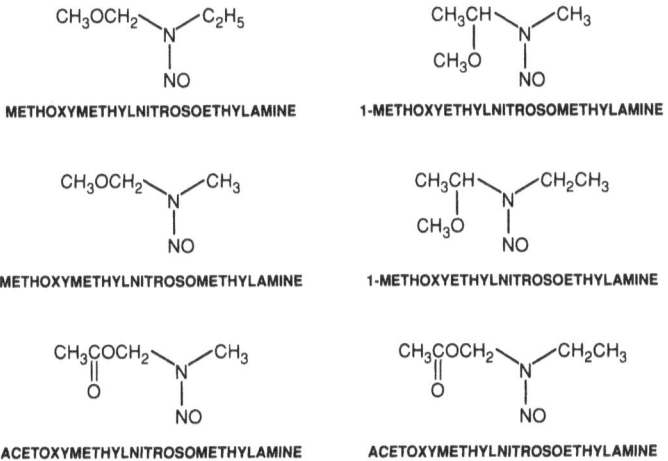

Fig. 7.20. Ethers and esters of α-hydroxynitrosamines.

and esophagus, while the former gives rise only to liver tumours. Methoxymethylnitrosomethylamine was a lung carcinogen, as was 1-methoxyethylnitrosoethylamine, which also induced liver tumours (Wiessler and Schmähl, 1976). This suggests a sharp difference between methylating and ethylating agents in their affinity for these three organs of the rat. The acetate ester of hydroxymethylnitrosoethylamine, which can give rise after hydrolysis to an ethylating agent, induced lung tumours in rats, but no liver tumours (Maekalia and Ogiu, 1982). The same was true of the acetate of hydroxymethylnitrosomethylamine, which is reminiscent of the failure of most other directly acting alkylating N-nitroso compounds (alkylnitrosoureas, etc.) to induce liver tumours in rats; perhaps metabolism is important for carcinogenesis in the liver.

Several methylnitrosoalkylamines with oxygen or other substituents in the alkyl chain are lung carcinogens in rats, for example methylnitroso-3-hydroxypropylamine (Koepke et al., 1988a), methylnitroso-2-hydroxypropylamine (but not its oxidation product, methylnitroso-2-oxopropylamine) and methylnitroso-2,3-dihydroxypropylamine (Lijinsky and Reuber, 1984a); only methylnitroso-2-hydroxypropylamine, of the three propyl derivatives, induced lung tumours in hamsters (Lijinsky and Kovatch, 1985a). Since all of these compounds could have in common formation of a methylating agent, it seems that other characteristics of these molecules are the determinants of their induction of lung tumours. Similarly n-propylnitroso-2-hydroxypropylamine induced lung tumours

Fig. 7.21. 4-(Methylnitrosamino)-1-(3-pyridyl)-butanone and its alcohol.

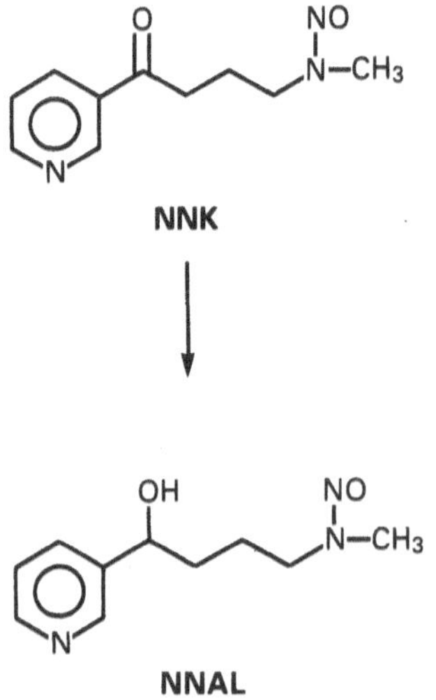

NNK

NNAL

in rats and mice (Reznik *et al.*, 1975), whereas the corresponding 2-oxopropyl compound produced lung tumours in mice (Dickhaus *et al.*, 1978), but not in rats.

Two important tobacco-specific methylnitrosamines (Fig. 7.21) are what are called NNK (methylnitrosamino-3-pyridylbutanone) and NNA1 (methylnitrosamino-3-pyridylbutanol, the corresponding secondary alcohol), which are lung carcinogens in mice (Castonguay *et al.*, 1983*a*) and in rats (Hecht *et al.*, 1980; Rivenson *et al.*, 1988); NNK is also a lung carcinogen in hamsters, a species in which NNA1 has not yet been tested. NNK has been the subject of considerable investigation to determine the mechanism of its carcinogenic action, notably by Belinsky *et al.* (1987) on the relation between dose and the extent of DNA methylation in lung and liver of rats; the assumption is that methylation of DNA is the interaction of importance in carcinogenesis by this nitrosamine. However, it is possible that further metabolism of the other alkyl group is involved, including elimination of the pyridyl moiety, or alkylation of DNA by that large group (Hecht *et al.*, 1987). This issue is unlikely to be resolved easily. NNK might be the most important carcinogenic nitrosamine to which people are exposed, since it appears to be the most potent carcinogen in tobacco and tobacco smoke (Hoffmann *et al.*, 1979).

The presence of oxygen (as hydroxyl or carbonyl) in an alkyl chain of a nitrosamine seems to dispose it to have the lung as a target for carcinogenesis. A large number of nitrosamines of this ilk have been examined, and most of them have induced tumours in the lung, as well as in other organs, particularly of the rat. For example nitrosobis-(2-hydroxypropyl)amine induced lung tumours in mice, in several strains of rat, in hamsters and in rabbits, but not in guinea pigs (Rao and Reddy, 1977); the diacetate ester of NBHPA also induced tumours of the lung (and other organs) in mice and hamsters, probably undergoing ready hydrolysis to the alcohol *in vivo* (Green *et al.*, 1980; Pour *et al.*, 1976). Nitrosobis-(2-oxopropyl)amine (NBOPA) is a more potent carcinogen than NBHPA and also gave rise to lung tumours, in addition to tumours of other types, in mice, rats, and hamsters, but again not in guinea pigs (Rao and Pour, 1978). In guinea pigs the only response to these two nitrosamines, NBHPA and NBOPA, was formation of liver tumours, and the liver seems to be the only organ of guinea pigs that develops neoplasms in response to nitrosamine treatment, an important example of carcinogenic species specificity which will be discussed elsewhere. Nitroso-2-hydroxypropyl-2-oxopropylamine, which can be considered a hybrid of NBHPA and NBOPA, induced tumours of the lung in both rats and hamsters, the only species in which it has been examined (Lijinsky *et al.*,

1984b; 1988f; Pour et al., 1979). All three compounds give rise to methylating agents *in vivo* and lead to methylation of DNA in liver, lung and other organs of rats and hamsters (Lijinsky et al., 1988e; Lijinsky, 1988b), and might partly explain their similar carcinogenic effects which, however, also show some differences.

Nitrosamines containing a 2-hydroxypropyl group or a 2-oxopropyl group together with allyl, hydroxyethyl or 2,3-dihydroxypropyl (Fig. 7.22) did not give rise to lung tumours in rats or hamsters, although several were potent carcinogens for other organs. This suggests that the properties of the entire nitrosamine molecule, rather than the properties of one alkyl group, were determinants of the induction of neoplasms in a particular organ. Interaction with some receptor molecule, dictated by the structure of the nitrosamine, is presumably followed by formation of an electrophile which reacts to begin the carcinogenic process in certain cells.

Takahashi et al. (1982) examined a number of acetates of 1-hydroxydi-n-butylnitrosamine and derivatives in the pathway of its oxidation to a bladder carcinogen, including acetoxybutylnitroso-3-carboxypropylamine and the corresponding lactone, all of which produced lung tumours in rats, in addition to local tumours at the site of injection. The appearance of lung tumours in animals treated with the remaining approximately 100 acyclic nitrosamines that have been tested was sporadic and showed no particular pattern related to chemical structure (Preussmann and Stewart, 1984).

Fig. 7.22. Asymmetric nitrosamines with oxygen substituents.

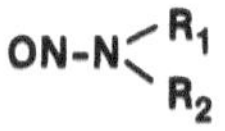

$$ON-N\begin{matrix}R_1\\R_2\end{matrix}$$

R_1	R_2
CH_2-CH_2OH	$CH_2-CH(OH)-CH_3$
CH_2-CH_2OH	$CH_2-CO-CH_3$
CH_2-CH_2OH	$CH_2-CH(OH)-CH_2OH$
CH_2-CH_2OH	$CH_2-CH=CH_2$
$CH_2-CH_2=CH_2$	$CH_2-CH(OH)-CH_3$
$CH_2-CH=CH_2$	$CH_2-CH(OH)-CH_2OH$
$CH_2-CH=CH_2$	$CH_2-CO-CH_3$
$CH_2-CO-CH_3$	$CH_2-CH(OH)-CH_3$
$CH_2-CO-CH_3$	$CH_2-CH(OH)-CH_2OH$
$CH_2-CH(OH)-CH_3$	$CH_2-CH(OH)-CH_2OH$

In spite of the large differences between cyclic nitrosamines and acyclic nitrosamines in their capacity to alkylate nucleic acids *in vivo*, the former in most instances failing to produce detectable levels of DNA adducts (Lijinsky *et al.*, 1973c; Lijinsky, 1976; Farrelly and Hecker, 1984), many cyclic nitrosamines were equally effective as acyclic nitrosamines in inducing lung tumours. The simplest cyclic nitrosamine, nitrosoazetidine, induced tumours of the lung in both mice and rats (Lijinsky *et al.*, 1967), but not in hamsters, whereas nitrosopyrrolidine, the next higher homologue, induced lung tumours in mice and hamsters, but not in rats (Greenblatt and Lijinsky, 1972; Dontenwill, 1968). Its 2-pyridyl derivative, nitrosonornicotine, was a lung carcinogen in mice (Boyland *et al.*, 1964a), but not in rats or hamsters. The nitrosonornicotine derivatives hydroxylated at positions 3 or 4 of the pyrrolidine ring, and the N-oxide, possible metabolic intermediates from nitrosonornicotine, also gave rise to lung tumours in mice (Hoffmann *et al.*, 1982).

Ascending the homologous series of cyclic nitrosamines, nitrosopiperidine was a lung carcinogen in mice and hamsters, but not in rats (Greenblatt and Lijinsky, 1972; Haas *et al.*, 1973). None of the 25 derivatives of nitrosopiperidine that have been tested in rats was a lung carcinogen, and few of them have been studied in other species. One conclusion is that this particular six-membered cyclic nitrosamine, nitrosopiperidine, lacks affinity for the receptors in lung cells which are important in producing neoplasia.

Nitrosomorpholine induced lung tumours in mice (Bannasch and Müller, 1964), but not in rats or hamsters. In contrast, nitroso-2-methylmorpholine and nitroso-2,6-dimethylmorpholine induced high incidences of lung tumours in hamsters (Reznik *et al.*, 1978; Lijinsky *et al.*, 1984b), but not in rats or guinea pigs; they have not been tested in mice. Nitroso-3,4,5-trimethylpiperazine also induced lung tumours in hamsters, but not in rats (Lijinsky *et al.*, 1983f), as did also dinitroso-2,6-dimethylpiperazine, while dinitrosopiperazine itself induced lung tumours in mice, but also not in rats (Druckrey *et al.*, 1967; Garcia and Lijinsky, 1972). These results confirm the pattern that six-membered cyclic nitrosamines do not have an affinity for the lung of rats, but are carcinogenic for hamster lung; the mechanisms underlying these differences are not understood.

The larger cyclic nitrosamines, nitrosohexamethyleneimine, nitrosoheptamethyleneimine and nitrosooctamethyleneimine have very different effects in the rat lung. Nitrosohexamethyleneimine induced lung tumours in mice (Althoff *et al.*, 1972; Strickland *et al.*, 1988; Goodall and Lijinsky, 1984b), but not in rats or hamsters, while nitrosoheptamethyleneimine and nitrosooctamethyleneimine induced squamous cell carcinomas of the lung in rats when administered in drinking water (Lijinsky et al., 1969). Nitrosoheptamethyleneimine induced few lung tumours in another rat strain (Lijinsky et al., 1982b), but was a lung carcinogen in European hamsters (Reznik et al., 1978a), although not in Syrian hamsters (Lijinsky et al., 1970a). The specificity with which only cyclic nitrosamines of particular structures induce lung tumours in one species or another suggests that specific receptor molecules in particular cells of the lung play an important role in interacting with the active carcinogens.

Many directly acting N-nitroso compounds, particularly alkynitrosoureas, give rise to lung tumours in rats and in mice, but in hamsters only when injected into the trachea or lungs (Yarita and Nettesheim, 1978), but not systemically. This is another example of the sharp difference between rats and hamsters in response to carcinogenic N-nitroso compounds that require no metabolic activation, and one of the most perplexing problems in chemical carcinogenesis.

Lung tumours have been seen in mice treated with the simplest alkylnitrosoureas, methyl- and ethyl-nitrosourea, but in some strains these have been the normally spontaneous lung adenomas accelerated and increased in number by the treatment. In other strains the lung tumours were adenocarcinomas or squamous cell carcinomas.

Methyl- and ethyl-nitrosourethanes induced high incidences of lung adenocarcinomas when injected I.V. into rats (Druckrey et al., 1967) or when painted on the skin of Swiss mice (Lijinsky and Reuber, 1988); ethylnitrosourethane gave 100% incidence of lung tumours in the painted mice.

Lung tumours were often induced by painting some nitrosoalkylureas on the skin of (Swiss) mice, and usually were in inverse proportion to the skin tumours produced (Lijinsky and Reuber, 1988). Alkylnitrosoureas yielding a high incidence of lung adenocarcinomas following skin painting of mice were ethylnitrosourea, n-butylnitrosourea, n-amylnitrosourea, n-hexylnitrosourea and phenylnitrosourea; nitrosooxazolidone also produced an elevated incidence of lung adenocarcinomas. Methylnitrosobis-(2-chloroethyl)urea (Fig. 5.1) and nitrosotris-(2-chloroethyl)urea induced high incidences of squamous cell carcinomas of the lung following skin painting of mice, and few skin tumours (Lijinsky and Reuber, 1988); squamous lung tumours are rarely seen in these mice. In contrast chloroethylnitrosodimethylurea and chloroethylnitrosodiethylurea painted on mouse skin did not produce lung tumours, even benign ones that appear at low incidence in controls.

In rats many alkylnitrosoureas administered orally gave rise to lung tumours, mainly alveolar-bronchiolar neoplasms (Lijinsky and Kovatch, 1989b), and these were usually more numerous in males than in females (as also were tumours of the intestinal tract: Lijinsky, 1988a). Methylnitrosourea did not induce lung tumours in rats, whereas ENU did (Lijinsky and Kovatch, 1989b). Hydroxyethylnitrosourea induced lung tumours in male rats at several doses, but not in females, whereas its methyl ether, methoxyethylnitrosourea induced lung tumours in both sexes, but much lower incidences in females (Lijinsky and Kovatch, 1988a). Lung tumours appeared about equally in male and female rats treated with n-butylnitrosourea and n-amylnitrosourea, but the next homologue, n-hexylnitrosourea induced lung tumours in 100% of male rats, but none in female rats (Lijinsky and Kovatch, 1989b). Allylnitrosourea showed the same pattern.

The specificity shown by alkylnitrosoureas of particular structures for male rather than female rats suggest an influence of sex hormonal status on susceptibility of the lung to some alkylnitrosoureas, but not to others. The effect is opposite to that shown by the female rat liver, which is more susceptible to carcinogenesis by some nitrosamines than is male rat liver (Lijinsky et al., 1988f). 2-Hydroxypropylnitrosourea and 3-hydroxypropylnitrosourea induced very few lung tumours (in either sex) in rats, as was also true of dimethylnitrosourea and methylnitrosoethylurea. On the other hand, lung tumours were common in rats given diethylnitrosourea or ethylnitrosomethylurea, especially in males (Lijinsky et al., 1989c); the results were similar when the dialkylnitrosoureas were given by gavage or in drinking water (since they are relatively stable in water) (Lijinsky et al.,

Fig. 7.23. Chloroethylnitrosoureas.

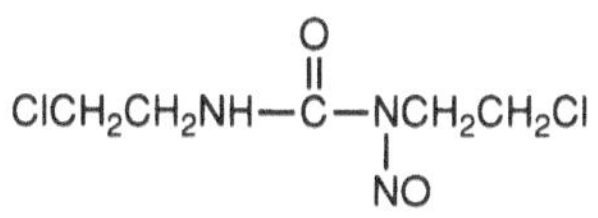

$$ClCH_2CH_2NH-\underset{\underset{NO}{|}}{\overset{\overset{O}{\|}}{C}}-NCH_2CH_2Cl$$

1-Nitroso-bis-(chloroethyl)-
urea (BCNU)

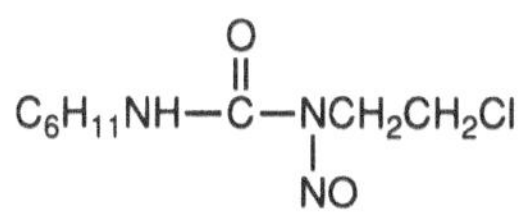

$$C_6H_{11}NH-\underset{\underset{NO}{|}}{\overset{\overset{O}{\|}}{C}}-NCH_2CH_2Cl$$

1-Nitroso-1-chloroethyl-
3-cyclohexylurea (CCNU)

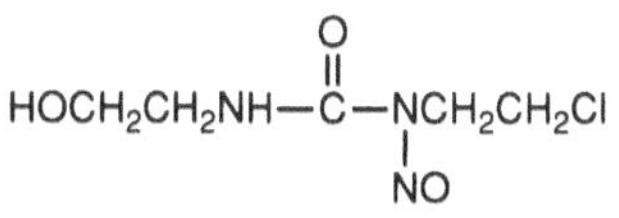

$$HOCH_2CH_2NH-\underset{\underset{NO}{|}}{\overset{\overset{O}{\|}}{C}}-NCH_2CH_2Cl$$

1-Nitroso-1-chloroethyl-
3-hydroxyethylurea

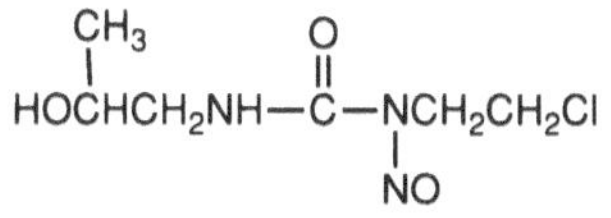

$$HO\underset{\underset{}{|}}{\overset{\overset{CH_3}{|}}{C}}HCH_2NH-\underset{\underset{NO}{|}}{\overset{\overset{O}{\|}}{C}}-NCH_2CH_2Cl$$

1-nitroso-1-chloroethyl-
3-(2-hydroxypropyl)-urea

1987a). There seems to be a consistently greater affinity of ethylnitrosoureas for the rat lung than of methylnitrosoureas.

Male rats were susceptible to lung carcinogenesis by ethylnitrosohydroxyethylurea, ethylnitroso-2-oxopropylurea and hydroxyethylnitrosoethylurea, given by gavage or in drinking water, and there were many fewer females in which lung tumours developed, in some cases none. In rats chloroethylnitrosodimethylurea and chloroethylnitrosodiethylurea gave rise to lung tumours when administered by gavage, whereas methylnitrosobis-(2-chloroethyl)urea and nitrosotris-(2-chloroethyl)urea did not (Lijinsky and Taylor, 1979a), completely opposite to the effects of the latter two compounds in mice following skin painting (Lijinsky and Reuber, 1988). None of the alkylnitrosocarbamates that have been given orally has induced lung tumours in rats, although they are directly acting alkylating agents.

The dialkylnitrosoureas that are cancer-therapeutic agents HECNU, CCNU and BCNU (Fig. 7.23) have induced lung tumours in rats when given I.V. (Eisenbrand, 1984), as did the glycoside chlorozotocin, MNNG and methylnitrosourethane (Druckrey et al., 1967).

7.5.4 Compounds having other target organs
7.5.4.1 Nervous system

Among N-nitroso compounds only alkylnitrosoureas induce tumours of the nervous system of rats and often in mice, but not in hamsters or in guinea pigs. Few alkylnitrosoureas have been tested in other species, and MNU has given rise to tumours of the nervous system in rabbits and dogs, but not in monkeys (Adamson et al., 1977); neither has ENU (Rice et al., 1977). This negative result raises the question whether the human brain would be responsive to the carcinogenic effects of alkylnitrosoureas.

As shown by Ivankovic a number of alkylnitrosoureas induced tumours of the nervous system in the offspring of pregnant rats treated with these agents during the period of organogenesis of the fetuses (Ivankovic, 1979); the alkylnitrosoureas contained n-alkyl chains from methyl to n-octyl, and they differed in effectiveness, the larger molecules being less potent. In adults the number of compounds inducing nervous system tumours was much more limited. MNU and dimethylnitrosourea were potent nervous system carcinogens (Druckrey et al., 1967; Lijinsky et al., 1989b), as were methylnitrosoethylurea (Fig. 7.24), trimethylnitrosourea (Fig. 7.5) and methylnitrosodiethylurea (Lijinsky et al., 1980b) in rats. None of these carcinogens gave rise to tumours of the nervous system in hamsters, although all were potent carcinogens in this species, inducing hemangiosarcomas of the spleen (Lijinsky et al., 1985b; Lijinsky and Kovatch, 1989c). ENU also gave rise to tumours of the nervous system in rats, but it was less effective than MNU; ENU also induced a broad range of other tumours, whereas the methylnitrosoureas induced in rats tumours of the nervous system almost exclusively (Lijinsky et al., 1989b). Alkylnitrosoureas with larger alkyl chains, which induced tumours of the nervous system in rats when administered transplacentally (Ivankovic et al., 1981), gave rise to few nervous system tumours when given to adults orally, although they induced a host of other tumours (Lijinsky and Kovatch, 1989b). It is also notable that 2-oxopropylnitrosourea, which behaves in many ways as if it is converted to a common product with MNU, as predicted by Leung and Archer (1984), induced kidney mesenchymal tumours and thyroid follicular cell tumours in rats, but not induce tumours of the nervous system (Lijinsky et al., 1990).

The reason for the predilection of methylnitrosoureas for the rat nervous system, but not in other species, is probably an affinity of these compounds for some receptor in supporting cells of the nervous system. These have a lesser affinity for ethylnitrosoureas and for alkylnitrosoureas with longer alkyl chains, which are less effective carcinogens for the rat nervous system. To narrow the specificity even more, methylnitrosocarbamate esters, such as methylnitrosourethane, which lack the second amine nitrogen of the urea, do not give rise to tumours of the nervous system in rats, although they are potent carcinogens for the forestomach of rats (or hamsters) (Lijinsky and Reuber, 1982a; Lijinsky and Schmähl, 1978; Lijinsky et al., 1985b).

Fig. 7.24. Nitrosodialkylureas.

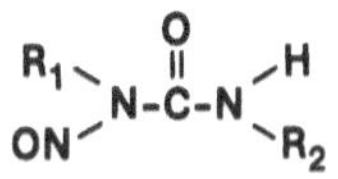

$$R_1 \diagdown \underset{ON \diagup}{N} - \overset{\overset{O}{\|}}{C} - N \underset{\diagdown R_2}{\diagup H}$$

R_1 = Methyl
Ethyl
2-Hydroxyethyl
2-Chloroethyl
2-Oxopropyl

R_2 = Methyl
Ethyl
2-Hydroxyethyl
2-Chloroethyl
2-Hydroxypropyl
2-Oxopropyl

Alkylnitrosoureas with substituents in the alkyl chain, such as hydroxyl groups, phenyl or halogen, were often potent carcinogens in rats, but did not induce tumours in the nervous system. For example, hydroxyethylnitrosourea produced a very large spectrum of tumours (Lijinsky and Reuber, 1983b), but they did not include tumours of the nervous system. However, hydroxyethylnitrosoethylurea, with an ethyl group on the 3-amino nitrogen, gave rise to nervous system tumours in half of the treated rats; in parallel, ethylnitrosohydroxyethylurea, its isomer, also gave rise to tumours of the nervous system in half of the treated rats, as opposed to a 10% incidence in rats treated with ENU (Lijinsky et al., 1985e). These results suggest that the receptors in the nervous system of rats for these alkylnitrosoureas recognise the entire molecule, not simply the alkyl group next to the N-nitroso function which is responsible for the alkylation of DNA produced by these compounds (Lijinsky, 1988b; Ludeke et al., 1991a).

Ethylnitrosomethylurea given to rats in drinking water is more effective in inducing tumours of the nervous system than is ethylnitrosourea, or than ethylnitrosomethylurea given by gavage. This dependency on dose rate, as well as on the chemical structure of the nitrosourea suggests that pharmacokinetics play an equally important role with alkylation of DNA in tumour induction. The ability of alkylnitrosoureas to cross the blood–brain barrier is important, but is an incomplete explanation of their carcinogenic action in the brain and other parts of the nervous system. The refractoriness of hamsters, guinea pigs and other species to induction of tumours of the nervous system by alkylnitrosoureas illustrates the importance of a match between chemical structure and reactivity of the carcinogen with the presence of certain components in the cells of a particular species for tumourigenesis to ensue. If we assume that particular cells in mammals, especially closely related rodents, have the same or similar functions developed through evolution, it follows that the differences between those responding to a carcinogen and those not responding, are quite subtle. This makes prediction of response from one species to another difficult.

7.5.4.2 Gastrointestinal tract

The anterior of the gastrointestinal tract is the esophagus, which has already been discussed extensively as the most common target in rats of carcinogenic nitrosamines. In most rodents the esophagus merges into the forestomach, which is also lined by squamous epithelium; in guinea pigs and many other mammals the esophagus changes abruptly into the glandular stomach. Very few N-nitroso compounds induce tumours in the glandular stomach of rats, mice or hamsters, the foremost being methyl- and ethyl-nitrosonitroguanidine (Sugimura and Fujimura, 1967), methylnitrosourea administered in drinking water to rats (Hirota et al., 1987), 3-hydroxypropylnitrosourea (Lijinsky and Kovatch, 1989b) and a small number of unrelated nitrosamines, including nitrosododecamethyleneimine (Goodall et al., 1973). There was no indication that species with only a glandular stomach, such as the guinea pig or rabbit, are therefore susceptible to induction of tumours by N-nitroso compounds in that organ.

In contrast to the relative resistance of the glandular stomach to carcinogenesis by N-nitroso compounds, the non-glandular or forestomach of rats, mice and hamsters is one of the most common sites of tumour induction by these carcinogens. In fact, tumours of the forestomach are the only neoplasms induced in both rats and hamsters by nitrosamines and nitrosamides (Tables 7.1, 7.2). It is perhaps not surprising that many N-nitroso compounds administered orally (by gavage, in drinking water or in food) give rise to tumours of the forestomach, since the compound remains in contact with the epithelium of the stomach for some time. What is more interesting is the reason so many N-nitroso compounds given orally to rats or hamsters do not induce tumours of the forestomach, although some of them resemble very closely in structure those that do induce forestomach tumours.

As with N-nitroso compounds that are esophageal carcinogens in rats, the induction of tumours of the forestomach is not entirely a local effect, since S.C., I.P. or I.V. administration of nitrosamines or nitrosamides in many cases gives rise to tumours of the forestomach. However, the incidence is usually lower than following oral administration, indicating that in the latter a local effect of the carcinogen is a contributory factor. Skin painting of a number of alkylnitrosoureas on mice gave rise to tumours of the forestomach, usually in low incidences (Lijinsky and Reuber, 1988), whereas skin painting of alkylnitrosocarbamates on mice did not, although alkylnitrosocarbamates are potent inducers of forestomach tumours when given by mouth (Lijinsky and Reuber, 1982a; Lijinsky and Reuber, 1983b; Lijinsky et al., 1985b), and this is usually the only tumour that is produced. A series of methylnitrosocarbamate esters that are derivatives of commonly used insecticides (such as carbaryl, baygon and methomyl) gave rise only to tumours of the forestomach when given to rats by gavage (Lijinsky and Schmähl, 1978).

The action of N-nitroso compounds in the stomach can be compared with the effects of the same compounds in the esophagus; this comparison is restricted to rats, since hamsters do not develop tumours of the esophagus apparently, although forestomach tumours are often produced in this species. Alkylnitrosoureas usually induce forestomach tumours in rats, but dialkylnitrosoureas usually do not. In contrast, no alkylnitrosourea has induced esophageal tumours in rats, a finding alluded to previously. Methylnitrosourethane in one study in rats gave rise to a few tumours of the esophagus (Lijinsky and Reuber, 1982a). It seems highly unlikely that alkylnitrosoureas given by mouth would not alkylate DNA in cells of the esophagus when a series of methylnitrosoalkylamines does, and the latter give rise to high incidences of esophageal tumours (Von Hofe et al., 1987).

NDMA did not induce tumours of the forestomach and NDEA induced very few in rats or hamsters. Nitrosodi-n-propylamine induced forestomach tumours when given in drinking water, but not by gavage, while nitrosodi-n-butylamine did produce forestomach tumours by gavage. Forestomach tumours in small numbers of rats were produced by methylnitroso-n-propylamine in drinking water, but not in hamsters by gavage, and they were common in rats given methylnitroso-n-butylamine treatment by gavage or in drinking water, and in hamsters (Lijinsky and Kovatch, 1988b). Tumours of the forestomach were more common in hamsters (sometimes close to 100%) than in rats treated with the higher homologous methylnitroso-n-alkylamines; in fact, forestomach tumours were essentially absent from rats treated with the C-6, C-7 and C-8 compounds, which induced high incidences of forestomach tumours in hamsters. The hamster forestomach, therefore, appears to be more susceptible than the rat forestomach to methylnitroso-n-alkylamines, which are presumed to induce tumours through methylation of DNA.

Among a large group of nitrosamines with oxygen-containing alkyl groups only 2-oxopropylnitroso-2,3,-dihydroxypropylamine and the closely related 2-hydroxypropylnitroso-2,3-dihydroxypropylamine induced a high incidence of forestomach tumours in both rats and hamsters.

While many cyclic nitrosamines induced tumours of the esophagus in rats, few gave rise to tumours of the forestomach. On the other hand, a number of cyclic nitrosamines gave rise to high incidences of forestomach tumours in hamsters; these were 2-methylnitrosomorpholine, nitroso-3,4,5-trimethylpiperazine, 2,6-dimethyldinitrosopiperazine and nitrosoheptamethyleneimine. There are some resemblances between these compounds, but other nitrosamines resembling them even more closely did not give rise to forestomach tumours.

Most of the directly acting alkylnitrosoureas induced high incidences of forestomach tumours in rats (Lijinsky and Kovatch, 1989b), and in hamsters (Lijinsky and Kovatch, 1989c); incidences often approach 100%

in both sexes. Alkylnitrosoureas with substituents in the alkyl chain were equally effective. Most dialkylnitrosoureas had no effect in the forestomach, although they are directly acting alkylating agents equally active with the corresponding alkylnitrosoureas. Possibly the greater stability of the dialkylnitrosoureas enables a larger proportion of the compound inserted into the stomach to be absorbed into the circulation and avoid breakdown in the stomach. The nitrosotrialkylureas, which require metabolic activation gave rise to few forestomach tumours in rats. In contrast, monoalkyl-, dialkyl- and trialkyl-nitrosoureas were almost equally effective in hamsters in producing a high incidence of tumours of the forestomach.

The conclusions are that the forestomach is not merely a passive organ in which tumours appear, induced by the local presence of the orally administered *N*-nitroso compound. To the contrary, the forestomach appears to be highly selective, because a minor proportion of the *N*-nitroso compounds that have been tested by the oral route of administration in rats and hamsters have given rise to forestomach tumours, and some of these compounds have been structurally related. Therefore, tumours of the forestomach in rodents are as good an indicator of a probable carcinogenic effect in humans as any other target organ.

In the organs of the gastrointestinal tract below the stomach relatively few *N*-nitroso compounds have given rise to tumours, and almost invariably more in male rats than in females. Of the nitrosamines, only NHPOPA and NBOPA have been reported to induce colon tumours in rats (Lijinsky et al., 1988f; Pour, 1978). However, many alkylnitrosoureas have induced tumours in the colon, ileum, jejunum or duodenum of rats (but not in hamsters). Much of this information has been reviewed (Lijinsky, 1988a).

MNU and MNNG have produced tumours in the colon when they were introduced in solution into the colon of rats through the anus (Watanabe et al., 1979), but ENU produced colon tumours following oral administration to rats (Lijinsky et al., 1985e). HENU induced tumours of the duodenum, jejunum and colon, more in males than in females, and its methyl ether had essentially the same effects, except that there were no tumours of the jejunum (Lijinsky and Kovatch, 1988a). Allylnitrosourea and *n*-amylnitrosourea induced tumours of the colon and jejunum in male rats, but *n*-hexyl- and *n*-butyl-nitrosourea only induced colon tumours, while phenylethylnitrosourea produced tumours in the duodenum and jejunum, but not in the colon. There seems to be a relation between chemical structure and the induction of tumours of the lower gastrointestinal tract, so that it is not simply contact of these directly acting alkylating agents with the epithelium in the various segments of the tract that produces the tumours. It is perplexing that some alkylnitrosoureas produce tumours of the colon, but not in the other parts of the gastrointestinal tract through which they must pass to reach the colon, and vice-versa for the compounds that induce tumours of the duodenum, but not of the colon in which they can reside for some time.

When given orally, MNU, methylnitrosoethylurea and dimethylnitrosourea did not induce tumours in the lower gastrointestinal tract, whereas ethylnitrosomethylurea and diethylnitrosourea induced tumours of the colon and jejunum (but not duodenum) and in higher incidences than ENU. Ethylnitrosohydroxyethylurea induced colon tumours in rats (as did its isomer hydroxyethylnitrosoethylurea), whereas ethylnitroso-2-oxopropylurea (which is very similar as an ethylating agent) induced tumours of the colon and the jejunum. These results point to the affinity of alkylnitrosoureas for certain types of cell, in which they induce tumours if their action is sufficiently great and persistent. It is of considerable importance that the characteristics of the tumour formed, even if originating from the same stem cells, often are related to the chemistry of the particular carcinogen, for example three N-nitroso compounds and a hydrazine that induce colon carcinoma in rats (Amberger, 1986).

7.5.4.3 Tumours of the bladder

The number of N-nitroso compounds which induce tumours of the bladder is small, but the relation of chemical structure to the ability to induce bladder tumours is particularly striking among nitrosamines. In contrast with other target organs for nitrosamine carcinogenesis, there seems to be little qualitative difference between species in response of the bladder epithelium to carcinogenesis. As the most studied example, nitrosodi-n-butylamine, first reported by Druckrey *et al.* (1967) to produce bladder tumours when administered to rats, has given rise to tumours of the bladder in many species, including mice, rats, hamsters (Syrian, Chinese, European), guinea pigs and rabbits. This suggests that the mechanism of induction of bladder tumours by this nitrosamine might be very similar in all species, and the bladder epithelium might be almost passive, and simply responding to the presence of a proximate carcinogen in the urine.

Quite early in the study of bladder carcinogenesis by nitrosodi-n-butylamine the product of omega oxidation of one of the n-butyl groups was tested. This compound, 4-hydroxybutylnitroso-n-butylamine, proved to be a bladder carcinogen in rats, somewhat more potent than NDBA; it induced only bladder tumours and not tumours of the liver and esophagus additionally produced by NDBA (Druckrey et al., 1967). The 4-hydroxydibutylnitrosamine, furthermore, induced bladder tumours in dogs as well as in mice, rats and hamsters. Further work by Okada's group showed that carcinogenesis in the bladder by NDBA was further mediated by oxidation of the 4-hydroxy derivative to the corresponding carboxylic acid, butylnitroso-3-carboxypropylamine, which gave rise to bladder tumours in rats whether given by mouth or by intravesicular instillation (Hashimoto et al., 1972, 1974). The role of the n-butyl group in this nitrosamino acid was not crucial for the formation of bladder tumours, since the corresponding methylnitroso-3-carboxypropylamine was also a bladder carcinogen in rats when given by mouth (Lijinsky et al., 1983d) or by intravesicular instillation (Thomas et al., 1988).

It has already been mentioned that several methylnitroso-n-alkylamines having even numbers of carbon atoms in the alkyl chain induced bladder tumours when given orally to rats or hamsters (but not in guinea pigs – Cardy and Lijinsky, 1980). The first of these examined was methylnitroso-n-dodecylamine (Lijinsky and Taylor, 1975e, 1978b) and prompted the suggestion by Okada et al. (1976b) that omega oxidation of the 12-carbon chain followed by successive β-oxidations with elimination of acetate, would eventually result in methylnitroso-3-carboxypropylamine, an analogue of the aforementioned butylnitroso-3-carboxypropylamine. Methylnitroso-3-carboxypropylamine was found in the urine of rats fed methylnitroso-n-dodecylamine. Further study of the metabolism of methylnitroso-n-alkylamines showed that the urine of rats fed those with even number of carbons in the chain contained a series of methylnitrosamino acids differing in length by 2 carbons, and in addition small amounts of two methylnitrosoalkylamines, methylnitroso-2-hydroxypropylamine and methylnitroso-2-oxopropylamine (Singer et al., 1981). The concentrations of these two nitrosamines in the urine of rats fed methylnitroso-n-alkylamines having odd numbers of carbons in the chain was very small by comparison with the even-numbered compounds.

This supported Okada's suggestion strongly, and extended to the possibility that an additional β-oxidation of methylnitroso-3-carboxypropylamine occurred, followed by spontaneous decarboxylation of the β-keto acid to form MNOPA. This compound seemed a good candidate for the proximate bladder carcinogen formed from the methylnitrosoalkylamines. However, on oral administration to rats MNOPA gave rise only to tumours of the esophagus (Lijinsky et al., 1983d) and liver; in a gavage study one bladder tumour was observed. In a recent experiment MNOPA administered by intravesicular instillation to rats gave rise to a high incidence of bladder tumours; the corresponding alcohol and methylnitroso-3-carboxypropylamine were considerably less effective by the same route, but also gave rise to bladder tumours (the induction time was greater); there were tumours of other organs, indicating that the compounds were absorbed into the circulation through the bladder wall (Thomas et al., 1988). It appeared that failure of MNOPA to induce bladder tumours after oral administration to rats was due to the death of animals from esophageal tumours prior to the appearance of tumours in the bladder.

It was notable that methylnitroso-n-hexylamine induced tumours of the liver and esophagus, but no bladder tumours by oral administration to rats, while in hamsters, in which there was no early death from tumours of the esophagus, this nitrosamine did induce bladder tumours (Lijinsky and Kovatch, 1988b), as well as liver and lung tumours. It seems from this latter study that a similar pathway of activation of methylnitrosoalkylamines is responsible for their induction of bladder tumours in a variety of species, probably including humans, who are exposed to some of the compounds (in toiletries – Hecht et al., 1982a) that have induced bladder tumours in experimental animals. The similar long-chain methylnitrosododecylamine has induced bladder tumours in hamsters (Althoff and Lijinsky, 1977), as well as rats. Apparently methylnitroso-n-butylamine did not undergo oxidation to methylnitroso-3-carboxypropylamine, since neither in rats (Lijinsky et al., 1980d) nor in hamsters (Lijinsky and Kovatch, 1988b) did it give rise to any tumours of the bladder, even when the dose was sufficiently small for rats to survive beyond six months and hamsters beyond a year. In addition, intravesicular instillation of MNBA in rats did not lead to the appearance of bladder tumours; all of the animals died with tumours of the esophagus (Lijinsky, 1990d).

Apart from the methylnitroso-n-alkylamines few nitrosamines were bladder carcinogens. Nitrosodiphenylamine given to rats in food at high concentrations resulted in a large incidence of bladder tumours (Cardy et al., 1979). This was a surprising result, since nitrosodiphenylamine is not an alkylating agent, nor would it be expected to be, so its mechanism of action is even more obscure than that of other carcinogenic N-nitroso compounds. Nitrosodiphenylamine did not induce bladder tumours in mice. The similar aromatic nitrosamine, phenylnitrosobenzylamine is also unexpectedly carcinogenic (being a non-mutagen), but induced tumours of the esophagus in rats (Lijinsky and Reuber, 1982b).

Nitrosobis-(2-oxopropyl)amine induced bladder tumours in male rats when administered by gavage, but not in females, and not when given in drinking water (Lijinsky et al., 1988f); there were also few bladder tumours induced in male rats that had been feminised by castration or by implantation of estradiol (Lijinsky et al., 1988g). A close analogy with methylnitroso-2-oxopropylamine is apparent. Nitrosobis-(2-hydroxypropyl)amine did not induce bladder tumours in rats, male or female, whereas 2-hydroxypropylnitroso-2-oxopropylamine induced a small, but significant, number of bladder tumours when given by gavage to male rats. It seems that the 2-oxopropylnitrosamine structure is related to induction of bladder tumours among this group of nitrosamines. Nitroso-2,6-dimethylmorpholine, although converted metabolically *in vivo* to NHPOPA (Gingell et al., 1976), did not induce bladder tumours in rats. None of this group of compounds gave rise to bladder tumours in hamsters.

Nitrosobis-(2-oxopropyl)amine is converted *in vivo* into one or more intermediates which methylate DNA in the liver and other organs (Lijinsky et al., 1988e; Lawson et al., 1981a), yet the mechanism by which it induces tumours in the bladder of rats is not known. It is possible, but unlikely, that it is converted into MNOPA, which induces tumours in the bladder. However, this reaction could occur equally well in the hamster, and the hamster is quite susceptible to induction of bladder tumours produced presumably by the same mechanism as in the rat from methylnitroso-*n*-alkylamines, yet NBOPA does not induce bladder tumours in hamsters by oral administration. However, NBOPA given by intravesicular instillation has induced bladder tumours in rats. Several other 2-oxopropylnitrosamine derivatives, including 2-oxopropylnitrosoethanolamine and 2-oxopropylnitroso-2,3-dihydroxypropylamine given orally have not induced bladder tumours in rats (Lijinsky et al., 1984c). None of the large number of cyclic nitrosamines so far examined has given rise to tumours of the bladder in any species.

Several alkylnitrosoureas have induced bladder tumours in rats, but by no means all, even though they are directly acting alkylating agents and can presumably reach the bladder following administration by mouth or by injection, as those that induce bladder tumours must. Indeed, methylnitrosourea injected into the bladder in small single doses has induced bladder tumours readily (Hicks and Wakefield, 1972), but no bladder tumours resulted when it was given by mouth. Occasional bladder tumours were seen in rats given ethylnitrosourea orally however (Lijinsky and Kovatch, 1989b), certainly in numbers large enough to be related to the treatment. Intravesicular administration of repeated small doses (1 mg) of ethylnitrosourea to rats has given rise to bladder tumours, together with a few mammary tumours.

The fact that ethyl- and methyl-nitrosoalkylureas are very similar chemically and are of similar stability, suggests that the induction of bladder tumours by the orally administered ethyl compounds reflects a particular affinity of these compounds for some receptor in the rat bladder, which is not responsive to methylnitrosoalkylureas. This is a constant theme to attempt to explain the much broader carcinogenic effect in many organs by ethylnitroso compounds, compared with their methyl analogues. Hydroxyethylnitrosourea gave rise to a small incidence of bladder tumours in rats, but its methyl ether did not (Lijinsky and Kovatch, 1988a). Only 3-hydroxypropylnitrosourea of the many other monoalkylnitrosoureas examined in rats (Lijinsky and Kovatch, 1989b) induced bladder tumours. This is surprising since bladder tumours were induced in rats by oral administration of a number of dialkylnitrosoureas, and many of these resembled the corresponding monoalkylnitrosoureas in mutagenic activity, for example. It might be that the greater stability of the dialkylnitrosoureas allowed them to reach the bladder epithelium after oral administration, whereas the alkylnitrosoureas did not. This might explain the failure of orally administered methylnitrosourea to induce bladder tumours, whereas it was a very potent bladder carcinogen when administered intravesically.

Ethylnitroso-hydroxyethylurea administered orally by gavage or in drinking water to rats gave rise to a small, but consistent, incidence of bladder tumours (Lijinsky et al., 1985e, 1990), whereas neither ethylnitrosomethylurea nor diethylnitrosourea gave rise to any tumours in the bladder, although both could be expected to reach the bladder epithelium as easily as the ethylhydroxyethyl compound. This suggests that the presence of the hydroxyethyl group increases the affinity of the molecule for some target in the bladder epithelium, which was not a receptor for the ethylnitrosoalkylureas lacking the hydroxyethyl at the 3-position. In contrast, orally administered hydroxyethylnitrosoethylurea and ethylnitroso-2-oxopropylurea did not induce bladder tumours in rats, while 2-hydroxypropylnitrosochloroethylurea induced some bladder tumours, underlining the importance of a particular chemical structure for formation of bladder tumours, in addition to the assumed need for alkylation of DNA in the bladder epithelium. Hydroxyethylnitrosoethylurea, ethylnitrosohydroxyethylurea and 2-hydroxypropylnitrosochloroethylurea (but not dimethylnitrosourea or diethylnitrosourea) induced bladder tumours in high incidence in rats by intravesicular administration; tumours of several other organs were also induced by this route. The alkylnitrosocarbamates, ethylnitrosourethane and methylnitrosourethane also gave rise to bladder tumours by intravesicular administration. The dependence of the type of tumour induced on particular chemical structures is not unique either to the urinary bladder, or to carcinogenesis by N-nitroso compounds. No alkylnitrosourea is known to induce bladder tumours in hamsters by oral administration (Lijinsky and Kovatch, 1989c).

7.5.4.4 Kidney

The kidney is a complex organ and several types of cell are the target for certain N-nitroso compounds. Most N-nitroso compounds, however, have no discernible effect in the kidney, although those containing halogenated alkyl groups frequently produce glomerulosclerosis, for example the haloalkylnitrosoureas used in cancer therapy. Tumours of the kidney are very rare in untreated rats, for example less than 1% in Fischer rats (Solleveld et al., 1984), so that even small incidences in treated rats can be considered related to the treatment.

The induction of kidney tumours in rats by nitroso compounds seems often related to the rate at which the dose is delivered. It has been known since the beginning of studies of nitrosamine carcinogenesis that nitrosodimethylamine (NDMA) administered to rats in drinking water or food (i.e. at low dose rates) gives rise almost solely to tumours of the liver (Peto et al., 1984), whereas single large (acutely toxic) doses give rise to kidney tumours, but not to liver tumours, perhaps a consequence of the extensive liver necrosis in those animals (Magee and Barnes, 1956; Hard and Butler, 1971). It is difficult to understand why at least some of the liver cells damaged by the large NDMA dose, but not killed, fail to develop into tumours, since it is known that there is extensive methylation of DNA in liver cells under these circumstances (Magee and Farber, 1962). Treatment of mice with large doses of NDMA gave rise to kidney tumours, but treatment of hamsters did not. The rat kidney tumours were mesenchymal, not tubular cell renal neoplasms, which is surprising since the proportion of mesenchymal tissue in the kidney of rats is small.

Lower doses of NDMA than those close to the LD_{50}, delivered by injection, also give rise to kidney tumours, often accompanied by tumours of the liver and lung; multiple treatments with these lower gavage doses or other injection procedures are needed to induce tumours. The same is true of the isomeric azoxymethane, also a potent methylating agent *in vivo*, producing more extensive alkylation than the nitrosamine (Lijinsky, 1985, 1988b). A comparison of the effects of the two compounds has been reported (Lijinsky et al., 1987b), and shows that by gavage azoxymethane induces in rats tumours of lung and kidney, but not of the liver, whereas liver tumours and kidney tumours are induced by drinking water treatment (Lijinsky et al., 1985d). NDMA gave rise to tumours of the liver, but not of the lung or kidney, when given to rats in drinking water, but tumours of kidney, liver and lung when given by gavage.

Nitrosodiethylamine gives rise to occasional kidney tumours, as well as the much more common tumours of the liver and esophagus, whether the treatment of the rats is by gavage or in drinking water, but these are tubular cell neoplasms (Lijinsky et al., 1987b). Methylnitrosoethylamine, which is a hybrid of NDMA and NDEA, and a methylating agent as well as an ethylating agent, has not induced tumours of the kidney, of any type, whether administered by gavage or in drinking water. This lack of effect of MNEA in the kidney suggests that its methylating or ethylating properties (and both are quantitatively similar in the DNA of liver, kidney and other organs of rats to those produced by NDMA or NDEA – Lijinsky, 1988b) are not as important in giving rise to tumours as are other properties of these molecules.

N-Nitroso compounds have not been associated with induction of kidney tumours in hamsters, although methylation of kidney DNA in hamsters by NDMA, MNEA or azoxymethane is similar to that in rats by these compounds (Lijinsky, 1988b). The failure of the hamster kidney to respond to carcinogenesis by these powerful alkylating agents is a surprise. The kidney of the hamster cannot be considered unresponsive to carcinogenic influence, since several sex hormones (such as estradiol, estriol and diethylstilbestrol) have induced kidney tumours on chronic administration to hamsters (Li et al., 1983).

In addition to NDMA several other N-nitroso compounds have induced mesenchymal tumours in the rat kidney. All were methylating agents, presumably forming a methyldiazonium ion as the proximate alkylating agent. The other compounds were dimethylnitrosourea and methylnitrosoethylurea, both directly acting, and 2-oxopropylnitrosourea, which has been proposed to form a methylating agent in the manner described by Leung and Archer (1984). Although many other N-nitroso compounds that have been examined are methylating agents *in vivo* (Von Hofe et al., 1987), none has given rise to mesenchymal tumours in the rat kidney, which seems to be a property of compounds of very specific structure. This suggests, in turn, that binding to some particular receptor molecule is important in producing tumours in mesenchymal cells of the rat kidney, but not in mesenchymal cells in other organs, or in other species.

Tubular cell adenomas and carcinomas of the kidney are induced in rats and mice, but again not in hamsters, by a variety of N-nitroso compounds, including NDEA. One of the most effective kidney carcinogens is 2-hydroxyethylnitrosoethylamine (Hiasa et al., 1979). Nitrosodiethanolamine also gave rise to kidney tumours in rats, but was much less effective than hydroxyethylnitrosoethylamine. Few of the compounds inducing kidney tumours had the kidney as the main target and, of course, few have been studied over a wide range of doses and with different methods of administration, as has been done with NDMA. More complex nitrosamines have induced kidney tumours (including transitional cell neoplasms of the kidney pelvis closely related to the bladder tumours induced by the same compounds) in small incidences. Among these agents are propylnitrosamines containing oxygen on the β-carbon, such as NBOPA, NHPOPA and others with similar structure (Lijinsky et al., 1988f). These can, as previously discussed, form a methylating agent *in vivo* – presumably a methyldiazonium ion – yet they did not give rise to mesenchymal tumours in the kidney of rats, instead forming tubular cell neoplasms. These results also point to the relation of specific chemical structures of nitrosamines with induction of kidney tumours in rats.

Kidney tumours were very rarely a consequence of treating rats with cyclic nitrosamines. The simplest cyclic nitrosamine, nitrosoazetidine, induced a small number, as did nitrosomorpholine and 2,6-dimethylnitrosomorpholine, but no other compounds, as far as is known.

Although 2-hydroxyethylnitrosourea (HENU) induced tumours in a great variety of organs in rats, these did not include the kidney. This is surprising since hydroxyethylnitrosoethylamine and nitrosodiethanolamine, which give rise to the same alkylating intermediate, a hydroxyethyldiazonium ion, induce in rats tumours of the kidney, among other organs. Other alkylnitrosoureas did give rise to kidney tumours, including hydroxyethylnitroso-2-chloroethylurea (HENCU), which induced a high incidence of kidney tumours in rats (Lijinsky et al., 1986a). How the structural change represented by the presence of a chloroethyl substituent on the second nitrogen atom could change hydroxyethylnitrosourea from a carcinogen not affecting the kidney (or liver), to one having the kidney (and liver) as a major target, is quite unknown. However, it does suggest that the entire molecule, not simply the part giving rise to the alkylating intermediate, is involved in the induction of kidney tumours (and liver tumours) in rats by HENCU. Also 2-hydroxypropylnitroso-2-chloroethylurea gave rise to a small number of kidney tumours in rats, but it was much weaker than its lower homologue (Lijinsky et al., 1986a).

The cyclic nitrosourea, nitroso-5,6-dihydrouracil (Fig. 7.25) also induced kidney tumours in rats (together with liver tumours) when given by mouth (Mirvish and Garcia, 1973), as did nitrosohydantoin (Bulay et al., 1979). Methylnitrosourethane and ethylnitrosourethane, on the other hand, gave rise to kidney tumours only when administered by I.V. injection to rats, not when given orally (Druckrey et al., 1967).

It is not known what characteristics of N-nitroso compounds unite those that induce kidney tumours, particularly those giving rise to tubular cell neoplasms. The activity of HENCU in the kidney suggests that this particular structure has affinity for susceptible cells in the rat kidney, which apparently HENU did not have, although it is a broadly acting carcinogen in rats, as is also an analogous compound without effect in the kidney, hydroxyethylnitrosoethylurea; the important role of the chloroethyl group is hard to elucidate. Also, the reason for the failure of N-nitroso compounds to induce kidney tumours in hamsters is not understood, since alkylation of kidney DNA occurs.

Several N-nitroso compounds give rise to tumours in the kidney pelvis, in addition to those that induce tubular cell neoplasms. Sometimes tumours of both types are present in the kidneys of the same animal. The kidney pelvis tumours are neoplasms of the transitional epithelium common in the urinary tract, and are occasionally present in the urethra also. In the main those compounds that induce neoplasms in the kidney pelvis also give rise to tumours in the urinary bladder. So nitrosobis-(2-oxopropyl)amine and compounds structurally and metabolically associated with it induce kidney pelvis neoplasms (Lijinsky et al., 1988f). The series of metabolic intermediates from the bladder carcinogens that are methylnitroso-n-alkylamines with even numbers of carbons, including methylnitroso-3-carboxypropylamine, MNHPA and MNOPA, when administered intravesically also induce tumours of the kidney pelvis (Thomas et al., 1988). Occasionally these compounds have also induced tumours of the kidney pelvis when administered by mouth.

The induction of kidney pelvis tumours by both nitrosamines and alkylnitrosoureas follows the same pattern of structural relationship shown by compounds that induce tumours of the urinary bladder, except that the methylnitrosoalkylamines (Lijinsky et al., 1981c) did not induce tumours in the kidney pelvis. In general there was a lesser frequency of kidney pelvis tumours in rats than of bladder tumours induced by the same carcinogen. This suggests that at least part of the process of bladder

Fig. 7.25. Nitrosated cyclic ureas.

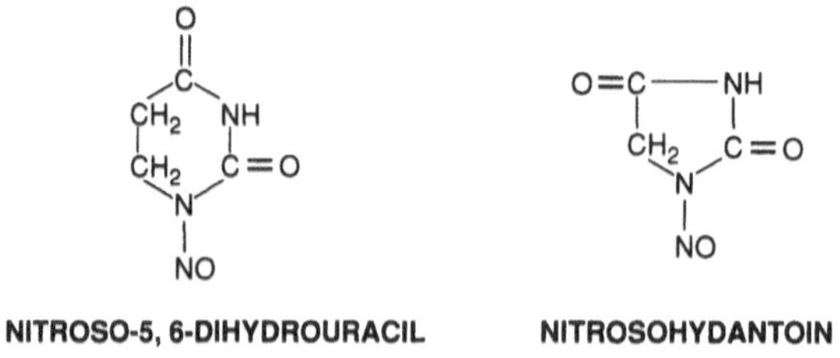

NITROSO-5, 6-DIHYDROURACIL   NITROSOHYDANTOIN

tumour development is dependent on the persistence of exposure of the bladder mucosa to the compound in the urine, as opposed to the transient exposure in the kidney and urethra, or that activation of the compound takes place more extensively in the bladder mucosa than in the epithelium of the kidney pelvis and urethra, or both.

7.5.4.5 Thyroid

Like the rat kidney, the thyroid gland of the rat is a target of few carcinogenic N-nitroso compounds, but they seem to constitute a group of structurally related compounds. C-Cell adenomas or carcinomas of the thyroid are not uncommon in untreated rats (F344), the incidence in males at two years being 9%, and in males allowed to live out their lifespan 27%; the corresponding incidences in females were, respectively, 8.5% and 20% (Solleveld et al., 1984). However, the incidence of follicular cell tumours of the thyroid in rats was very much smaller, in males less than 2% at two years, 8% in lifetime observation and in females less than 1% at two years, less than 8% lifetime (Solleveld et al., 1984).

Few of the N-nitroso compounds examined give rise to increases in the incidence of C-cell neoplasms of the thyroid, appearing before the two years at which the spontaneous tumour incidence becomes considerable. The compounds that have this effect are a few alkylnitrosoureas or dialkylnitrosoureas, including 3-hydroxypropylnitrosourea (Lijinsky and Kovatch, 1989b), ethylnitrosomethylurea (Lijinsky et al., 1989a) and hydroxyethylnitrosoethylurea (Lijinsky et al., 1985e). There is no good explanation why these particular alkylnitrosoureas and not others have C-cells of the thyroid gland as their target in rats.

The number of N-nitroso compounds which give rise to follicular cell neoplasms in the rat thyroid is much larger. Incidences in some cases have approached 100%, implying that these particular cells of the thyroid are prime targets of these compounds. The simplest nitrosamine, NDMA, induced a small number of follicular cell tumours in rat thyroid when given by gavage or intravesically to females, but not in male rats. None of the higher homologues of NDMA gave rise to thyroid tumours. The three metabolically related nitrosodipropylamine derivatives with 2-oxygen substituents NBOPA, NBHPA and NHPOPA, all induced high incidences of follicular cell thyroid tumours in male rats, whether given in drinking water or by gavage, but few or none in female rats (Lijinsky et al., 1988e, f). This is reminiscent of a number of types of tumour in rats that are more frequently induced in male rats than in female rats; others are tumours of the lung, colon and intestine. It is reasonable to assume that induction of these tumours is dependent on the sex hormone status of the rats, as has been shown in the case of NBOPA (Lijinsky et al., 1988g). The oxygen at the 2-position in the n-propyl group seems to be necessary for access of these nitrosamines to the follicular cells of the thyroid in rats, but apart from these three nitrosamines, others containing a 2-oxopropyl- or 2-hydroxypropyl- group with some other substituent on the opposite side of the nitrogen were without effect on the thyroid. Cyclic nitrosamines did not induce thyroid tumours, even nitroso-2,6-dimethylmorpholine which is metabolised by β-oxidation to NHPOPA (Gingell et al., 1976).

Several alkylnitrosoureas gave rise to follicular cell tumours of the thyroid in rats, occasionally accompanied by C-cell tumours. One was 2-hydroxyethylnitrosourea which induced thyroid tumours in male and female rats at a variety of doses (Lijinsky and Reuber, 1983b; Lijinsky and Kovatch, 1988a). Its methyl ether also induced thyroid follicular cell tumours, but only in males. Nitrosodimethylurea induced thyroid tumours but not methylnitrosoethylurea, nitrosodiethylurea or ethylnitrosomethylurea. Ethylnitrosohydroxyethylurea gave rise to thyroid tumours only by gavage in males (Lijinsky et al., 1985e), whereas ethylnitroso-2-oxopropylurea and hydroxyethylnitrosoethylurea induced follicular cell tumours of the thyroid mainly in male rats by gavage or in drinking water (Lijinsky et al., 1990; Lijinsky 1989). It seemed that compounds containing a β-oxygenated alkyl group were favoured for inducing follicular cell tumours in the rat thyroid, suggesting the existence in those cells of receptors for which these N-nitroso compounds have a particular affinity. Nitrosodimethylurea and NDMA were the only compounds lacking a β-oxygen in the alkyl chain that gave rise to thyroid tumours. No thyroid tumours have been reported in hamsters or mice treated with N-nitroso compounds.

7.5.4.6 Thymus (lymphoma)

Few N-nitroso compounds give rise to leukemia in experimental animals, and mononuclear cell leukemia is a common spontaneous neoplasm in Fischer 344 rats which are often used for carcinogenesis studies. The incidence and progression of these leukemias in male rats do not appear to be affected by treatment with N-nitroso compounds, even those which are powerful directly acting alkylating agents. However, a small number of N-nitroso compounds, most of which are structurally related, do induce lymphoma of the thymus in rats and in mice, but not, to the present, in hamsters. This rat lymphoma is a highly malignant neoplasm, frequently leading to early death of the animals (within three to four months following the beginning of treatment – Lijinsky et al.,

1986a). The lymphoma is transplantable and has been partially characterised (Konishi et al., 1988).

As in the case of thyroid follicular cell tumours, the N-nitroso compounds inducing thymic lymphomas included directly acting alkylnitrosoureas usually with a β-oxygen substituent in the alkyl chain, but also n-propylnitrosourea (Suzuki et al., 1984). Nitrosamines with an oxygen substituent in the β-position, such as NBOPA, NBHPA and NHPOPA did not give rise to tumours of the thymus, whereas the two nitrosamines that did were bases, derivatives of mononitrosopiperazine. These two compounds, 1-nitroso-3,5-dimethylpiperazine (Fig. 7.17) and 1-nitroso-3,4,5-trimethylpiperazine are closely related, and were effective in the rat thymus only at high concentrations in drinking water (Singer et al., 1981). At lower concentrations nitrosotrimethylpiperazine induced only tumours of the nasal mucosa (Lijinsky et al., 1983f), which was the common tumour induced by 1-nitroso-4-methylpiperazine at much higher concentrations in water (Lijinsky and Reuber, 1982b); at lower concentrations the monomethyl compound was ineffective. None of the dinitrosopiperazines induced tumours in the rat thymus, even those with methyl substituents in the 3 and 5 positions. The capacity to induce these tumours seems, therefore, to reside with mononitrosopiperazines having methyl groups in the 3 and 5 positions, again showing the chemical specificity required for interaction with receptors in stem cells of the thymus, thereby giving rise to lymphoma. Apparently none of the other nitrosamines tested has this property.

Presumably the nitrosamines that give rise to thymic lymphomas need to be activated by enzymes in the target cells, since the nitrosamines are biologically inert without metabolic activation. On the other hand, several alkylnitrosoureas requiring no metabolic activation are potent inducers of thymic lymphomas in rats. The most potent of these is 2-hydroxypropylnitrosourea (but not 3-hydroxypropylnitrosourea), which induces those tumours within 12 to 14 weeks following the beginning of treatment with as little as three milligrams per week; all of the rats were dead at week 20 (Lijinsky et al., 1986a). Other alkylnitrosoureas inducing thymic lymphomas in rats, but less potent than the 2-hydroxypropyl compound, were 2-hydroxyethylurea and its methyl ether, 2-methoxyethylnitrosourea (Lijinsky and Kovatch, 1988a), 2-hydroxypropylnitrosochloroethylurea, methylnitrosourea and methylnitrosoethylurea, the last two not containing a β-oxygen substituent, but being directly acting methylating agents. Allylnitrosourea also gave rise to tumours of the thymus (Lijinsky and Kovatch, 1989b), and it is conceivable that it could undergo metabolism to 2-hydroxypropylnitrosourea, which would account for its activity towards the thymus; *n*-propylnitrosourea has induced thymic lymphomas by oral administration (Ogiu *et al.*, 1975). One closely related cyclic alkylnitrosocarbamate, nitrosooxazolidone, also induced some thymic lymphomas when given to rats by gavage, possibly because, like hydroxyethylnitrosourea, it can form a hydroxyethylating agent (and is a potent bacterial mutagen).

The failure of the many other alkylnitrosoureas to induce tumours in the rat thymus is unexplained, since it can be assumed that they all can reach the thymus and alkylate DNA in cells of that organ. Another anomaly is that methylnitrosourea induced thymic lymphoma in rats, but not ethylnitrosourea, whereas in mice ethylnitrosourea induced thymic lymphoma (Swenson *et al.*, 1979), but not methylnitrosourea, although dimethylnitrosourea did (Hiraki, 1971). Here again it must be assumed that specific chemical structural characteristics are responsible for the affinity of those compounds which induce thymic lymphomas for those target cells in the rat thymus. This, independently of the alkylation and mutation of DNA (although probably requiring it), propels the affected cells to neoplasia. The importance of the β-oxygenated alkyl group might be its resemblance to the active portion of the normal substrate of the thymic receptor, perhaps a steroid hormone. This might be a fruitful area for further investigation, particularly since none of the compounds affected the thymus in the hamster.

7.5.4.7 Mammary gland and uterus

Tumours of the mammary gland and uterus that are associated with treatment with nitrosamines have not been reported. Tumours of both organs are occasionally seen in untreated rats and mice (but not mammary tumours in hamsters or guinea pigs), so that a sizable incidence is needed in carcinogen-treated animals in order to associate them with the treatment. Fibroadenomas are present in a high proportion of female rats of several strains given no treatment when they approach two years of age, but these are benign tumours. The incidence of carcinomas of the mammary gland in untreated control F344 rats, reported by Solleveld *et al.* (1984) was 0.3% and 2.1% at two years in males and females respectively, and 1.5% and 11% in rats allowed to live their normal lifespan. The corresponding number at two years and lifespan for tumours of the uterus were about 1% for carcinomas and adenomas, but approached 20% in the case of endometrial stromal polyps and sarcomas.

A series of alkylnitrosoureas give rise to carcinomas of the mammary gland and to tumours of the uterus in rats and in mice, but relatively few compounds have been examined in mice. Mammary tumours have rarely been seen in hamsters treated with alkylnitrosoureas, but several of this class of N-nitroso compound have induced tumours of the uterus and cervix in hamsters. The mode of administration seems to have a large effect on the type of tumours induced by alkylnitrosoureas. For example, single or multiple I.V. doses of methylnitrosourea have induced mammary carcinomas and tumours of the uterus in rats, the latter at relatively greater incidences; a single large I.P. dose of methylnitrosourea delivered to rats at about seven weeks of age induces a high incidence of mammary carcinomas (Gullino et al., 1975), a model much studied by investigators working on the molecular biology of carcinogenesis (Zarbl et al., 1985). Single intraportal administration of methylnitrosourea or ethylnitrosourea to female rats 10 to 12 weeks old failed to give rise to uterine or mammary tumours (Lijinsky et al., 1972c). Multiple small oral doses of methylnitrosourea over 20 weeks beginning at eight weeks of age also failed to give rise to mammary carcinomas, instead inducing a high incidence of tumours of the nervous system, already discussed (Lijinsky and Kovatch, 1989b). Analogously, other methylnitrosoureas, including methylnitrosodiethylurea (Fig. 3.1), dimethylnitrosourea, methylnitrosoethylurea and trimethylnitrosourea did not induce mammary carcinomas in rats, while inducing nervous system tumours in high incidence (Lijinsky and Taylor, 1975d; Lijinsky et al., 1980b) and some tumours of the uterus.

There seems to be a perhaps subtle difference in susceptibility between the uterus and the mammary gland of the rat in response to treatment with carcinogenic methylnitrosoureas. On the other hand, ethylnitrosourea and higher homologues readily induce mammary carcinomas and tumours of the uterus in rats given the compounds in multiple small doses by mouth. Diethylnitrosourea gives rise primarily to mammary carcinomas when administered to rats intravesically in multiple doses. This suggests that there are receptors in the mammary gland for ethyl- and higher alkylnitrosoureas, which are unaffected by methylnitrosoureas. Such specificity, as has been mentioned earlier, is not unusual in carcinogenesis by N-nitroso compounds. The great interest in this specificity among alkylnitrosoureas is that they are directly acting compounds, which act independently of activating enzymes that are so important in modulating the biological effect of nitrosamines.

In addition to ethylnitrosourea, the analogous ethylnitrosodimethylurea, ethylnitrosomethylurea (Fig. 3.1), triethylnitrosourea, diethylnitrosourea, ethylnitrosohydroxyethylurea, hydroxyethylnitrosoethylurea and ethylnitroso-2-oxopropylurea all induce carcinomas of the mammary gland and of the uterus in female rats, with similar potency. Hydroxyethylnitrosochloroethylurea induced tumours of the uterus, but not in the mammary gland. The relative potency of some of the higher homologues of ethylnitrosourea, such as butyl-, amyl-, hexyl- and allyl-nitrosourea, which also give rise to mammary tumours and to tumours of the uterus, is somewhat lower; rats treated with larger doses of these took longer to die with tumours than rats given ENU. Methoxyethylnitrosourea induced more mammary tumours in rats than hydroxyethylnitrosourea, suggesting that methylation of the hydroxyl group made its action resemble that of ethylnitrosourea. These results indicate that the alkylnitrosoureas with larger alkyl groups have different interaction in the uterus and mammary gland of rats from methylnitrosoureas, possibly because different receptors are involved.

It is notable that alkylnitrosoureas, which so commonly give rise to mammary carcinomas in rats and mice, together with tumours of the uterus in those species, have uniformly failed to induce mammary tumours in hamsters. On the other hand, several dialkylnitrosoureas have induced tumours of the uterus or cervix in hamsters, again showing the distinct preference of *N*-nitroso compounds for particular organs in a given species (Lijinsky *et al.*, 1985*b*; Lijinsky and Kovatch, 1989*c*).

7.5.4.8 Pancreas

Like the esophagus in rats, the pancreas of the hamster shows a remarkable response to carcinogenic *N*-nitroso compounds of particular structure sharply different from the response of the pancreas in other species, in which the exocrine pancreas rarely develops tumours. The *N*-nitroso compounds which induced tumours of the pancreas ducts in hamsters were mainly nitrosamines having a 2-oxopropyl- or 2-hydroxypropyl- group as one or both alkyl substituents. The first discovered of these was nitrosobis-(2-hydroxypropyl)amine (Krüger *et al.*, 1974), closely followed by its oxidation product NBOPA (Pour *et al.*, 1977). Then followed nitroso-2,6-dimethylmorpholine (Mohr *et al.*, 1977), which is converted by oxidation at the β-carbon into the cyclic form of NHPOPA (Gingell *et al.*, 1976), to which both NBHPA and NBOPA can be also converted, by oxidation and reduction, respectively. This is shown by examination of the metabolites of these compounds excreted in the urine of a number of species (Underwood and Lijinsky, 1982).

More recently a number of *N*-nitroso compounds of similar structure have been tested in hamsters and some, but not all, have induced significant numbers of pancreas duct tumours. Although there have been small incidences of pancreas duct tumours in hamsters treated with alkylnitrosoureas, whether containing a 2-oxopropyl- or 2-hydroxypropyl- group or not, none of these tumour incidences (Lijinsky and Kovatch, 1989c) was sufficiently elevated above the occasional appearance of this tumour in an untreated control hamster to ensure that they were related to the treatment, although it is likely that they were.

The most potent of the pancreatic carcinogens among nitrosamines was NBOPA which, under most conditions, also gave rise to liver tumours in the treated hamsters, as did the other nitrosamines causing pancreas tumours; the liver tumours often arose from bile duct cells, also true of NBOPA-induced liver tumours in rats (Lijinsky et al., 1988f). Nitroso-2-hydroxypropyl-2-oxopropylamine was similar in potency in inducing pancreas tumours to NBOPA, but not more potent, which is necessary if the former is to be considered a more proximate carcinogenic product, as was claimed by Pour et al. (1979); however, it is definitely more potent than NBHPA (Lijinsky et al., 1984b).

Among nitrosamines containing only one oxopropyl or hydroxypropyl group, several gave rise to tumours of the pancreas ducts in hamsters, but others did not. For example, 2-oxopropylnitroso-2,3-dihydroxypropylamine (Fig. 7.26) and the analogous 2-hydroxypropylnitroso- compound

Fig. 7.26. Nitrosamines containing oxopropyl- and hydroxypropyl-groups.

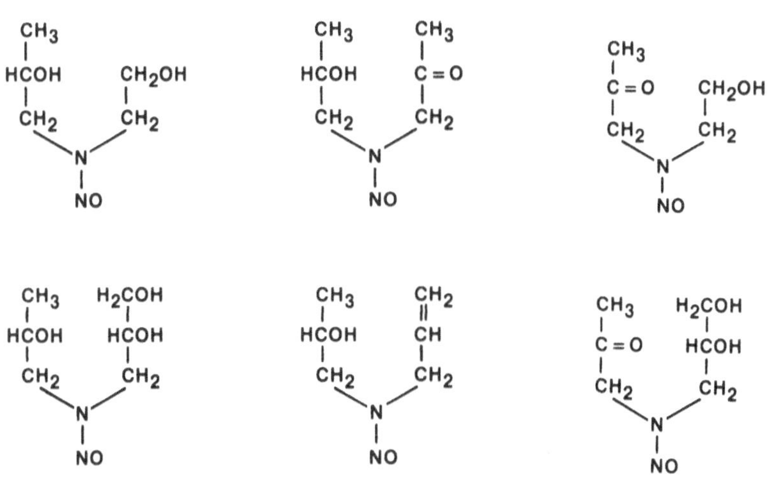

gave rise to a small number of pancreas tumours, as did 2-hydroxypropylnitrosoethanolamine; 2-oxopropylnitrosoethanolamine was a much more potent carcinogen than the latter, and induced a high incidence of pancreas tumours (Lijinsky *et al.*, 1985c). Methylnitroso-2-hydroxypropylamine induced pancreas tumours in almost half of female hamsters (Lijinsky and Kovatch, 1985a). The analogous methylnitroso-2-oxopropylamine did not induce pancreas tumours in that study, although it was reported in another laboratory to be a potent inducer of pancreas duct tumours in hamsters, following a different route of administration (Pour *et al.*, 1980); these differences in effect are probably dependent on the pharmacokinetics of the compound delivered at different dose rates.

Neither allylnitroso-2-hydroxypropylamine nor allylnitroso-2-oxopropylamine gave rise to pancreas tumours in hamsters, although they were quite potent carcinogens in this species. The failure of these two nitrosamines to induce pancreas tumours in hamsters might be due to the resistance of the allyl group to oxidation, the compounds therefore failing to form a 2-oxopropyl- or 2-hydroxypropyl-diazonium ion. Arguing against this, however, is the failure of either 2-hydroxypropylnitrosourea or 2-oxopropylnitrosourea to induce pancreas tumours in hamsters, although both can give rise to the corresponding alkyldiazonium ion directly, without metabolic activation; like all alkylnitrosoureas tested, the two β-oxidised propylnitrosoureas induced mainly hemangiosarcomas of the spleen in hamsters (Lijinsky and Kovatch, 1989c).

These results, which are quite clear-cut, suggest once again that formation of the expected alkyldiazonium ion from an *N*-nitroso compound is not the most important step in organ-specific carcinogenesis by these compounds. No better examples exist for a complementary or alternative set of steps in carcinogenesis than the induction of pancreas duct tumours in hamsters by certain *N*-nitroso compounds, but not in rats, and induction of tumours of the esophagus in rats by many nitrosamines, but not in hamsters. Very recently Rivenson *et al.* (1988) reported the first example of an *N*-nitroso compound inducing tumours of the pancreas ducts in rats, 4-(methylnitrosamino)-3-pyridylbutanol (NNAl). The importance of this is in showing that the rat pancreas is not innately resistant to the induction of tumours by carcinogens of appropriate structure.

7.5.4.9 Skin

There has been no evidence that any nitrosamine gives rise to tumours of the skin, whether applied systemically or directly to the skin. This negative information derives from several studies of skin painting of nitrosamines on mice, which are well known to be susceptible to topical application of carcinogens such as polynuclear aromatic compounds, which induce skin tumours, and suggests that the enzymes necessary to activate nitrosamines are lacking in the skin.

On the other hand, a number of directly acting N-nitroso compounds, particularly alkylnitrosoureas, have readily induced tumours of the skin by multiple treatments in mice (Graffi et al., 1967), the species in which most such experiments have been performed. However, methylnitrosourea was not a good initiator of skin tumours in mice (Waynforth and Magee, 1975), nor was MNNG, in the two-stage system developed by Berenblum and Shubik (1947). Perhaps equally interesting are the findings that a number of alkylnitrosoureas administered systemically to rats have given rise to skin tumours, showing that they reach the skin in activated form; dialkylnitrosoureas seem to be particularly effective in this regard (Lijinsky and Kovatch, 1989b; Lijinsky et al., 1989b, 1990), although not all compounds of this structure are equally effective.

A series of alkylnitrosoureas beginning with the simplest, methylnitrosourea, was painted on mouse skin in acetone or methanol solution for 40 or 50 weeks, following which tumours developed over a period. Swiss mice and CD-1 mice were markedly less susceptible to MNU skin carcinogenesis than Balb/c or sencar mice, but Swiss mice were nevertheless used in most of these studies. Methyl- and ethyl-nitrosourea were similar in carcinogenic effectiveness, the ethyl compound being a little less active (Lijinsky and Winter, 1981). In contrast, ethylnitrosonitroguanidine was considerably more effective in inducing tumours on mouse skin than MNNG (Lijinsky, 1982). This perplexing result suggests that formation of a methyl- or ethyl-diazonium ion from these compounds was not the principal determinant of formation of skin tumours in mice, but that other properties of the nitrosourea and nitrosonitroguanidine, respectively, are major factors. This is further supported by the failure of methylnitrosourethane and ethylnitrosourethane to induce skin tumours when painted on mice, although both are potent mutagens and readily induce tumours of the forestomach when given orally (Lijinsky and Reuber, 1982a). The analogous non-volatile nitrosocarbamate ester nitrosocarbaryl, a very potent bacterial mutagen, induced forestomach tumours in rats (Lijinsky and Taylor, 1976b), but was weakly carcinogenic by mouse skin painting (Lijinsky and Winter, 1981), suggesting something lacking in the alkylnitrosocarbamate structure needed for transformation of cells in the mouse skin, and apparently present in the alkylnitrosourea structure. In contrast, the cyclic alkylnitrosocarbamates, nitrosooxazolidone and nitroso-5-methyloxazolidone, were quite effective carcinogens

Table 7.6. Carcinogenic response to N-nitroso compounds painted on mouse skin – 20 female Swiss mice, 8 weeks old, 25 µl of solution, 2 × week

	Mutagenicity Salmonella TA 1535 (revertants/µmol)	Total dose (µmol/mouse)	Median week of death	Week of 1st tumour on skin	Average latent period (weeks)	No. of mice with skin tumours	No. of mice with other tumours
N-Nitroso-							
methylurea	4000	80	42	23	34	18	Lung 1
ethylurea	100	80	60	28	50	12	Lung 9
n-propylurea	200	100	79	65	73	5	Mammary 5, Uterus 6, Lung 3
isopropylurea	90	100	87	65	80	3	Lung 1
allylurea	560	100	64	—	—	0	Mammary 15, Uterus 3
n-butylurea	140	100	73	60	75	8	Mammary 4, Uterus 4, Ovary 4, Lung 11
isobutylurea	730	100	94	—	—	0	Glandular stomach 3
sec-butylurea	390	100	97	—	—	0	Glandular stomach 4
n-amylurea	530	100	81	40	63	11	Mammary 3, Uterus 3, Ovary 5
n-hexylurea	1200	100	81	55	66	11	Lung 9, Mammary 2
n-undecylurea	320	100	101	48	60	3	Uterus 2
n-tridecylurea	5	100	98	45	73	2	Uterus 4
cyclohexylurea	52	100	98	70	78	2	Glandular stomach 2, Mammary 2
phenylurea	660	100	102	70	—	1	Lung 9, Uterus 3

Compound							Tumors
benzylurea	11	100	98	—	—	0	Glandular stomach 3, Mammary 3, Uterus 3
2-phenylethylurea	4600	100	54	—	—	0	Oropharynx 7
dimethylurea	73	100	113	—	—	0	Lung 5, Lymphoma 3, Schwannoma 3
diethylurea	34	100	100	—	—	0	Lung 6, Lymphoma 2, Schwannoma 3
2-fluoroethylurea	2700	20	70	40	54	15	Uterus 6, Glandular stomach 2
2-hydroxyethylurea	2400	90	78	42	48	15	Mammary 3, Uterus 5, Ovary 4
2-hydroxypropylurea	3100	80	58	31	45	18	Glandular stomach 3, Uterus 3, Ovary 2
3-hydroxypropylurea	220	90	73	35	50	17	
chloroethyl-diethylurea	NT	50	30	—	—	0	Thymus lymphoma 9
methylbis-(chloroethyl)-urea	NT	80	45	46	52	5	Lung squamous cell carcinoma 9
tris(chloroethyl)urea	NT	70	37	—	66	2	Lung squamous cell carcinoma 12
methylurethane	4500	120	61	55	—	1	Lung 3
carbaryl	33000	100	100	60	72	6	Lung 4
ethylurethane	1800	160	84	—	—	0	Lung 19
oxazolidone	9300	100	76	50	54	11	Lung 9
5-methyloxazolidone	5200	100	67	43	47	16	Lung 7
methylnitroguanidine	12000	12	109	60	82	4	
ethylnitroguanidine	5800	12	76	35	60	14	
cimetidine	3600	110	102	—	—	1	
Control (solvent)	—	—	110	—	—	0	Lung 5, Uterus 1

NT = not tested.

by mouse skin painting (Lijinsky and Reuber, 1983b), as well as potent bacterial mutagens. They induced stomach tumours in rats by oral administration of quite low doses.

Among the alkylnitrosoureas there were quite large differences in potency as mouse skin carcinogens, depending to some extent on the size and intimate structure of the alkyl group (Table 7.6, which includes bacterial mutagenicity). For example, undecyl- and tridecyl-nitrosourea were very weak skin carcinogens, almost to the point of calling them inactive (Lijinsky and Winter, 1981). They are virtually insoluble in water, which might be an important factor in diminishing their carcinogenic activity. This might also be the case with 2-phenylethylnitrosourea (Fig. 7.27), which failed to induce tumours on mouse skin, although it is a potent bacterial mutagen and quite carcinogenic by oral administration to rats, in which it induces liver tumours (Lijinsky and Kovatch, 1989b). Phenylnitrosourea, cyclohexylnitrosourea and *iso*propylnitrosourea are particularly unstable alkylnitrosoureas, possibly for steric reasons, and did not induce tumours in mouse skin; nor did *sec*-butylnitrosourea. *Iso*butylnitrosourea was equally ineffective as a mouse skin carcinogen, although it is neither sterically hindered nor particularly unstable; there seems to be considerable specificity in structural requirements for induction of skin tumours in mice by this type of carcinogen.

This is borne out further by the weak carcinogenicity of *n*-propylnitrosourea (Fig. 7.28), in contrast to the potent effect of ethylnitrosourea; *n*-butylnitrosourea was somewhat more potent, while *n*-amyl- and *n*-hexylnitrosourea were much more potent than the *n*-

Fig. 7.27. Phenyl-, phenylethyl- and substituted alkylnitrosoureas.

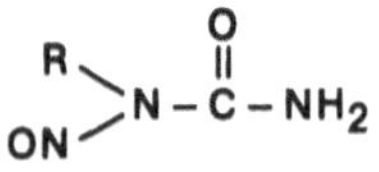

$$\begin{array}{c} R \\ ON \end{array} \!\!\! > \!\! N - \overset{\overset{\displaystyle O}{\|}}{C} - NH_2$$

R	
PHENYL	$C_6H_5\cdot$
CYCLOHEXYL	$C_6H_{11}\cdot$
BENZYL	$C_6H_5CH_2\cdot$
PHENYLETHYL	$C_6H_5CH_2CH_2\cdot$
FLUOROETHYL	$FCH_2CH_2\cdot$
HYDROXYETHYL	$HOCH_2CH_2\cdot$
2-HYDROXYPROPYL	$CH_3\underset{\underset{\displaystyle OH}{\|}}{CH}CH_2\cdot$
3-HYDROXYPROPYL	$HOCH_2CH_2CH_2\cdot$

propyl compound. The complete homologous series was not examined, so that the compounds between C-6 and C-11 are of unknown relative potency.

The alkylnitrosoureas with substituents in the alkyl group were usually potent skin carcinogens. For example, fluoroethylnitrosourea produced a high incidence of skin tumours in mice after application of quite small doses. At higher doses this compound was very toxic, leading to severe ulceration of the skin and early death of the mice, long before tumours would be expected to appear. Hydroxyethylnitrosourea was also a very potent skin carcinogen, equivalent to ethylnitrosourea, and its homologue, 2-hydroxypropylnitrosourea (Fig. 7.27), had very similar potency; its isomer, 3-hydroxypropylnitrosourea, was also similar in potency for mouse skin to the isopropanol derivative (Lijinsky and Reuber, 1988), although the former was a weak mutagen and also a weak carcinogen when given systemically to rats (Lijinsky and Kovatch, 1989b). As previously mentioned, the cyclic analogues of the hydroxyalkylnitrosoureas, nitroso-1,3-oxazolidone and nitroso-5-methyloxazolidone were potent skin carcinogens in mice, although they are alkylnitrosocarbamates, the acyclic members of which group were inactive or very weakly carcinogenic in mouse skin. The nitrosooxazolidones are potent mutagens and hydroxyalkylate DNA *in vivo* as effectively as do the corresponding hydroxyalkylnitrosoureas (Lijinsky, 1988b). It must be

Fig. 7.28. *n*-Alkylnitrosoureas.

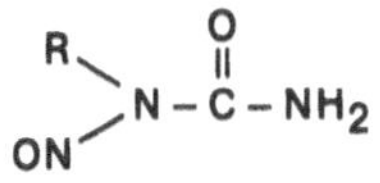

$$\underset{ON}{\overset{R}{>}}N-\overset{O}{\underset{\|}{C}}-NH_2$$

R	
METHYL	CH_3-
ETHYL	CH_3CH_2-
n-PROPYL	$CH_3CH_2CH_2-$
iso-PROPYL	CH_3CHCH_3
n-BUTYL	$CH_3CH_2CH_2CH_2-$
iso-BUTYL	CH_3CHCH_2-
	$\quad\;\; CH_3$
sec-BUTYL	$CH_3CH_2CHCH_3$
n-AMYL	$CH_3[CH_2]_3CH_2-$
n-HEXYL	$CH_3[CH_2]_4CH_2-$
n-UNDECYL	$CH_3[CH_2]_9CH_2-$
n-TRIDECYL	$CH_3[CH_2]_{11}CH_2-$

asked, therefore, whether it is the acyclic nitrosocarbamate structure itself that is not conducive to induction of tumours in mouse skin.

The number of dialkyl- and trialkyl-nitrosoureas examined by mouse skin painting is not large, but so far no dialkylnitrosourea has produced skin tumours in mice by painting, although they have given rise to tumours of internal organs, such as lung, when painted on mouse skin. Yet dialkylnitrosoureas are directly acting compounds, equivalent in systemic carcinogenic effectiveness to the corresponding monoalkylnitrosourea in many cases. The presence of the second alkyl group seems in some way to be inimical to formation of tumours in the skin of mice, as if some key receptor in skin cells were inaccessible to these compounds. Trialkylnitrosoureas, which require metabolic activation, would not be expected to give rise to skin tumours by painting on mice – and they do not.

While monoalkylnitrosoureas given systemically to rats give rise to few tumours of the skin, these are quite common as a result of administering several of the dialkylnitrosoureas chronically by mouth, especially in males given them in drinking water (Lijinsky *et al.*, 1990). It might be, of course, that the dialkylnitrosoureas are sufficiently stable to reach the cells of the skin, whereas the monoalkylnitrosoureas are too unstable for this. However, the skin receives a quite good blood flow, and it is known that distribution of an oral dose of an N-nitroso compound is rapid. Nevertheless, differences in pharmacokinetics seems to be the best explanation for these perplexing differences.

7.5.4.10 Nasal mucosa

Less attention has been given to tumours of the nasal mucosa, which are commonly induced by nitrosamines in rats and hamsters but, like tumours of the esophagus, not by the directly acting alkylnitrosamides. One reason for this lack of attention is that the nasal structures of rodents are different from those of humans, and there is no human tumour analogous to those rodent tumours. On the other hand, since the nasal mucosa is a target for nitrosamines, induction of those tumours is a valid index of carcinogenicity of a chemical. Indeed, attention has been paid to the possible risks to humans of exposure to formaldehyde, based solely on the results of exposure of rats to inhaled formaldehyde, which gave rise to tumours of the nasal mucosa; no other tumours have been induced in experimental animals by formaldehyde (IARC, 1982*b*). On the other hand, nasal tumours in humans have been associated with woodworking and furniture making, and with eating of smoked fish by Chinese people (Fong and Chan, 1976; Huang *et al.*, 1981). In another analogy with esophageal tumours, tumours of the nasal mucosa in untreated or vehicle-treated rats are so rare (Solleveld et al., 1984) as to be non-existent.

Like all nitrosamine-induced tumours, those of the nasal mucosa are not a response to treatment with all compounds of that class, but are related specifically to particular chemical structures. Investigators (Tjälve and Castonguay, 1983) have shown that there is a particularly high level of nitrosamine-metabolising enzymes in cells of the nasal mucosa of rats and hamsters. This could be the source of the common induction of nasal mucosal tumours by nitrosamines, but not by alkylnitrosoureas, namely that the metabolism is itself part of the process of carcinogenesis in these cells; in the absence of interaction with an activating enzyme, carcinogenesis is not initiated.

A small number of the nitrosamines examined have induced only tumours of the nasal mucosa, although most have induced other types of tumour as well. Those that have induced mainly nasal tumours include nitrosodi-*iso*propylamine, nitrosodi-*iso*butylamine and 4-methyl-1-nitrosopiperazine in rats, requiring relatively high doses (for a nitrosamine) to induce the tumours; lower doses failed to induce any tumours. Nitrosodiallylamine (Althoff et al., 1977a) and nitrosodiethanolamine also induced almost exclusively tumours of the nasal mucosa in hamsters (Hilfrich et al., 1977; Pour and Wallcave, 1981), whereas the former was inactive in rats (Druckrey et al., 1967). Since these compounds can give rise, through metabolism, to aldehydes – and several aldehydes have induced tumours of the nasal mucosa in rats and hamsters – it is possible that formation of the aldehydes is the key to their carcinogenic action in those cells. Nitrosodimethylamine, although rapidly metabolised in the liver, has induced tumours of the nasal mucosa in rats when given by gavage and in special circumstances, such as by inhalation exposure (Druckrey et al., 1967) or when given together with ethanol (Griciute et al., 1981); formaldehyde is a major product of metabolism of NDMA. However, nitrosomorpholine also induced mainly nasal tumours in hamsters (Lijinsky et al., 1984a).

One-third of the nitrosamines tested have induced tumours of the nasal mucosa in rats, and two-thirds of the smaller number of nitrosamines examined in hamsters have induced those tumours. The nasal mucosa, then, must be considered one of the most susceptible organs to nitrosamine carcinogenesis. Its very active metabolism of these compounds must be a factor. The very frequency of these tumours, across all types of nitrosamine structure, makes it more difficult to discern a pattern related to specific chemical properties and reactions.

All of the nitrosamines of low molecular weight give rise to tumours of the nasal mucosa in rats, mice and hamsters, although the incidences vary considerably with the mode of application and the dose. In many studies the nasal cavity was not examined microscopically, so it is likely that tumours in that organ were missed, because they were small when tumours of other sites caused early death of the animals. This might be particularly true of studies in the mouse, in which relatively few nitrosamines have been tested. Tumours of the nasal mucosa were commonly induced by symmetrical and asymmetrical nitrosamines, by those with oxygen or halogen substituents and those without, by those with saturated and unsaturated carbon chains, by cyclic and acyclic nitrosamines, and by volatile and non-volatile nitrosamines – but not by alkylnitrosoureas or alkylnitrosocarbamates. This pattern in rats is strikingly similar to the pattern of N-nitroso compounds which induce tumours of the esophagus. With few exceptions those nitrosamines which induce tumours of the nasal mucosa in rats also induce tumours of the esophagus, and vice-versa. Among those that induce nasal tumours, but no tumours in the rat esophagus are nitrosodi-*iso*butylamine, 1-nitroso-4-methylpiperazine, 1-nitroso-3,5-dimethylpiperazine and 1-nitroso-3,4,5-trimethylpiperazine; the last did not induce nasal tumours in hamsters. Many of the same compounds gave rise to tumours of the nasal mucosa in hamsters, but none induced tumours in the hamster esophagus. Therefore, whatever parallelism exists between the esophagus and nasal mucosa in susceptibility to (and activation of) nitrosamines in the rat, does not exist in the hamsters; the hamster esophagus apparently lacks whatever it is that makes the rat esophagus, or the hamster nasal mucosa, susceptible. These sharp differences between species are the most noticeable and difficult to understand, but are likely, if resolved, to provide great insight into mechanisms by which chemicals induce tumours.

Only two types of nitrosamine seem not to induce tumours of the nasal mucosa in rats or hamsters. They are high molecular weight nitrosamines, such as methylnitrosononylamine and larger (methylnitrosooctylamine induced some nasal tumours in hamsters, but not in rats) and cyclic nitrosamines having five atoms or less in the ring. Included among the latter are nitrosopyrrolidine, nitrosooxazolidine and their derivatives. These do not induce tumours of the esophagus in rats either, again suggesting that the structural criteria for induction of tumours in these two organs of the rat are similar. The failure of the directly acting N-nitroso compounds to induce tumours in the nasal mucosa indicates that alkylation of nucleic acids in those cells by N-nitroso compounds, which can be assumed to occur, is not the determinant of tumour induction.

7.5.4.11 Other organs

(1) *Mesotheliomas.* Among other organs in which N-nitroso compounds induce tumours which are related to the treatment is the tunica vaginalis of the rat, where mesotheliomas develop in response to several alkylnitrosoureas. Mesotheliomas are common in rats treated with ethylnitrosoureas, whether there is a substituent on the second nitrogen or not. The atypical compound is methylnitrosourea, which did not induce mesotheliomas, whereas dimethylnitrosourea and methylnitrosoethylurea gave rise to low incidences of this tumour (Lijinsky et al., 1989b). Mesotheliomas did not appear in hamsters in response to alkylnitrosoureas, and they were not induced at all by nitrosamines. Therefore, the cells of origin were unable to activate nitrosamines, or the important target molecules were receptive only to alkylnitrosoureas, not to nitrosamines; alkylnitrosocarbamates also failed to give rise to mesotheliomas, although they are similar as alkylating agents to the analogous alkylnitrosourea.

The primary sex organs, ovary and testis seem quite refractory to the carcinogenic action of N-nitroso compounds, even the directly acting ones. Except for an isolated report (Pour, 1983), unconfirmed from other quarters, N-nitroso compounds have not given rise to tumours of the prostate gland, which remains the only type of common human tumour for which there is no satisfactory animal model.

(2) *Trachea.* Tumours of the trachea have an affinity with lung tumours, and are induced by some N-nitroso compounds in both rats and hamsters, although in high incidences only by a few compounds. For example, methylnitroso-*n*-hexylamine gives rise to some tumours of the trachea in rats, and its homologue methylnitroso-*n*-heptylamine induces even more, but only when given by gavage (Lijinsky et al., 1983e), not when administered in drinking water. Neither induces tracheal tumours in hamsters (Lijinsky and Kovatch, 1988b), nor do any of the higher homologues, from C-8 to C-14, induce tracheal tumours in either species. Some tracheal tumours were seen in rats treated with methylnitroso-2-oxopropylamine and in hamsters by the tobacco-specific nitrosamines NNK and nitrosonornicotine (Hoffmann et al., 1981; Hilfrich et al., 1977). Allylnitroso-2,3-dihydroxypropylamine, the cyclic nitrosamines 3,4-dichloro- and 3,4-dibromo-nitrosopiperidine (Lijinsky and Taylor, 1975b), nitroso-3-methyl-

4-piperidone (Singer *et al.*, 1984) (but not nitroso-4-piperidone) induced tracheal tumours in rats, nitrosomorpholine and nitroso-2-methylmorpholine in hamsters (Lijinsky *et al.*, 1984a), but not in rats. Nitrosoheptamethyleneimine has given rise to tracheal tumours in Wistar rats (Garcia and Lijinsky, 1972), but not in Fischer rats. There is no striking pattern in these results which facilitates relating the induction of tracheal tumours to chemical structural characteristics of nitrosamines. Neither has any alkylnitrosourea given rise by oral administration to tumours of the trachea in hamsters or rats, although they were induced by direct injection of methylnitrosourea in the trachea (Herrold, 1970; Harris *et al.*, 1973; Yarita and Nettesheim, 1978).

(3) *Tongue.* Tumours of the tongue are induced by a variety of *N*-nitroso compounds, both nitrosamines and nitrosamides, in rats. Tongue tumours are occasionally seen in untreated rats, although the incidence is low. The tongue was not an organ routinely examined, so it cannot be assumed that small tumours of the tongue never appeared in the carcinogenesis studies. Nevertheless, there are compounds in this series which have given rise to high incidences of tongue tumours in rats (but apparently not in hamsters). Neither NDMA nor NDEA gave rise to tongue tumours in rats, nor did methylnitrosoethylamine. However, some of the higher homologues did, beginning with methyl-nitroso-*n*-butylamine and continuing with methylnitroso-*n*-hexyl-amine (Lijinsky *et al.*, 1983e), methylnitrosobenzylamine and methylnitrosophenylethylamine (Lijinsky *et al.*, 1982e); larger molecules did not induce tongue tumours. 2-Oxopropylnitroso-2,3-dihydroxypropylamine gave a high incidence of tongue tumours, as did also 2-hydroxypropylnitroso-2,3-dihydroxy-propylamine (Lijinsky *et al.*, 1984c, d). These latter results suggest that there are receptors in the tongue for nitrosamines with that particular polyhydroxylated propylamine structure.

(4) *Spleen.* While lymphosarcomas of the spleen are occasionally produced by oncogenic viruses in animals, the spleen is not a common site of action of chemical carcinogens; lymphosarcomas of the spleen arise in rats spontaneously in occasional animals (Solleveld *et al.*, 1984). Hemangiosarcomas of the spleen in rats, which are equally rare spontaneously, have been induced only by nitrosohydroxyproline and by phenylnitrosobenzylamine; none of the nitrosamines of analogous structure, such as methyl-

nitrosophenylamine or methylnitrosobenzylamine or nitrosoproline, induced tumours of the spleen in rats. Neither did any alkylnitrosourea. In contrast, several N-nitroso compounds examined have induced hemangiosarcomas of the spleen in Syrian hamsters. All but one, methylnitrosophenylamine (methylnitrosoaniline) have been alkylnitrosoureas. Methylnitrosoaniline induced tumours in the esophagus of rats, but not in the spleen, although the analogous phenylnitrosobenzylamine did. Every alkylnitrosourea studied – 14 in all – induced hemangiosarcomas of the spleen in hamsters, regardless of the structure of the alkylnitrosourea (Lijinsky and Kovatch, 1989c). Mono- and dialkylnitrosoureas, which are directly acting, were equally effective with trialkylnitrosoureas, which require enzymic activation. Apart from locally formed forestomach tumours which appeared following administration of the alkylnitrosoureas by gavage, hemangiosarcomas of the spleen were almost the only tumours induced in hamsters (some induced tumours of the cervix also). This suggests a particular affinity of alkylnitrosoureas for some receptor in endothelial cells of the hamster spleen. This is not the case in endothelial cells in other organs of the hamsters, nor in the spleen of the rat, mouse or guinea pig, none of which develop hemangiosarcomas as a result of treatment with alkylnitrosoureas. Nor do directly acting alkylnitrosocarbamates induce hemangiosarcomas of the spleen in hamsters, although they differ from alkylnitrosoureas only in lacking the second amino nitrogen, again indicating great specificity by the alkylnitrosoureas.

It is noteworthy that splenectomy in hamsters eliminated hemangiosarcoma induction by hydroxyethylnitrosourea, but there was no compensating increase in incidence of any other tumour (Lijinsky *et al.*, 1988c).

(5) *Zymbal gland.* This is near the ear of rats, and has no equivalent in humans, but is of interest because some carcinogens produce tumours in it, and these tumours are an index of the carcinogenic effect of the agent. Tumours of the Zymbal gland are seldom seen in untreated rats (Solleveld *et al.*, 1984), and can be considered related to the treatment whenever they appear in several animals of a group. No tumours of the Zymbal gland have been seen in rats treated with nitrosamines, other than those partly activated by conversion into α-acetoxy derivatives, such as methylnitrosoacetoxymethylamine (Berman *et al.*, 1979). Several alkylnitroso-

ureas induced tumours of the Zymbal gland in rats, another example in the spectrum of rat tumours induced by alkylnitrosoureas, but not by nitrosamines.

The alkylnitrosoureas which gave rise to Zymbal gland tumours in rats seemed to fit a definite pattern. They were induced by ethylnitrosourea, allylnitrosourea and amylnitrosourea, but not by butylnitrosourea or hexylnitrosourea (Lijinsky and Kovatch, 1989b), both of which had even numbers of carbon atoms in the chain. In addition to ENU, Zymbal gland tumours were also induced in rats by diethylnitrosourea, ethylnitrosomethylurea, ethylnitrosohydroxyethylurea and ethylnitroso-2-oxopropylurea, but not by MNU, dimethylnitrosourea or methylnitrosoethylurea (Lijinsky, 1989). The propensity to affect the Zymbal gland seems to lie with ethylating nitroso compounds rather than with methylating nitroso compounds, as is the case with induction of mammary tumours, tumours of the uterus and mesotheliomas. Curiously, the methylating agent azoxymethane, which so resembles NDMA in many ways, induced tumours of the Zymbal gland when given by gavage to rats (Lijinsky and Kovatch, 1989a), which no nitrosamine did. The reason for the sharp difference between methylnitrosoureas and ethylnitrosoureas in the response of so many organs of the rat is intriguing and potentially revealing. Again, alkylnitrosocarbamates did not induce tumours of the Zymbal gland.

7.6 Dose–response studies

Apart from the academic interest in mechanisms of induction of tumours, most experiments with N-nitroso compounds have been uninformative about human cancer and methods for its prevention, which should be one of the primary goals of cancer research, the other being treatment. However, there is one type of experiment in animals which can be very helpful in understanding cancer in humans. Those are dose–response studies with N-nitroso compounds to which humans are – or likely to be – exposed, and there are a number of these as discussed in a previous chapter. Dose–response studies have been carried out with additional compounds because of their intrinsic interest or because they can shed light on the results of studies with the more relevant compounds.

The first dose–response study with an N-nitroso compound was that of NDEA conducted by Druckrey, Preussmann and their colleagues (Druckrey *et al.*, 1963), feeding various concentrations of the compound in drinking water to groups of rats for their lifetime. Almost all of the rats died of tumours induced by the treatment; rats within each dose group died in a very linear manner, mostly with liver tumours, but many also with tumours of the esophagus. The average time of mortality was plotted graphically and led Druckrey to propose a formula relating time of death (T) to dose (D), namely $DT^n = $ constant; n was 2.6 in that study.

Since that time dose–response studies with a number of nitrosamines have been undertaken, for a variety of reasons. Human exposure is common to some of them (NDMA, nitrosopyrrolidine and nitrosodiethanolamine), probable to others (nitrosopiperidine, nitrosomorpholine and NDEA). Studies of other compounds (nitroso-1,2,3,6-tetrahydropyridine and dinitrosohomopiperazine) were undertaken for comparison with nitrosamines of similar carcinogenic properties. The range of doses used in these studies was large and this distinguishes them from many other studies in which a small range of doses was employed, and from which, therefore, much more limited conclusions can be drawn.

The most comprehensive of these studies was that carried out by Peto, Grasso and their colleagues (Peto et al., 1984), involving more than 5,000 rats, mice and hamsters. The largest group studied was rats given NDMA. The experiment with NDEA also included groups of rats treatment of which began at different ages. The analysis of the results was complex, and included consideration of competing risks posed by the appearance of different tumours at various times. An important conclusion was that a no-effect level could not with confidence be predicted for nitrosamines, even in experiments of this size; doses of 0.1 mg/kg body weight per day gave rise to significant incidences of tumours during the two- to three-year lifetimes of the animals (Peto et al., 1984). The studies of nitrosopyrrolidine and nitrosopiperidine were on a much smaller scale than those of NDMA and NDEA, and complemented those previously conducted by Preussmann et al. (1977), Habs et al. (1980), and Eisenbrand et al. (1980).

A dose–response study with nitrosomorpholine in rats involved a large number of animals, almost 700 female rats (Lijinsky et al., 1988b). In that study liver tumours were commonly induced at all doses, which ranged from 70 parts per billion in drinking water (a total dose of 0.7 mg per animal in two years of treatment) to 100 ppm (a total dose of 250 mg per animal in six months of treatment). The increased incidence of liver tumours with increasing dose is shown in Fig. 7.29. At the highest dose rates, tumours of the esophagus and oral mucosa were also seen. There was an elevated incidence of liver tumours compared with controls even at the lowest dose rate, which cannot, therefore, be considered a no-effect level, although exposure of humans to doses as large or larger occur among rubber workers (Fajen et al., 1979; Spiegelhalder and Preussmann,

1982) and in users of certain kinds of smokeless tobacco (Hoffmann et al., 1984a). A dose–response study with NDEA in the same strain (Fischer 344) of rats (Lijinsky et al., 1981b), showed that at the lowest doses (450 parts per billion in drinking water) and increasing in steps of 2.5, tumours of the esophagus were induced and were the cause of death in many cases, while at the highest doses (113 ppm for 20 weeks) liver tumours were common in addition to those in the esophagus. Again the lowest dose was not a no-effect level, nor could one be calculated from the results. The reason for the different relative frequencies of tumours of the liver and esophagus induced by nitrosomorpholine and NDEA is not known or understood, since both are mutagenic, but only NDEA has formed stable adducts with liver DNA *in vivo*. Both nitrosamines show similar hepatotoxicity to rats, if that is a modulating factor in liver carcinogenesis (see Chapter 5).

In the studies at Frederick Cancer Research Facility time-to-death with tumours is a good index of relative carcinogenic effectiveness of the different doses of the carcinogenic nitrosamines, at least at the higher doses, and a plot of average time of death in each group against total dose of nitrosamine administered lies along a straight line (Fig. 7.30), from which a tangent can be calculated. These tangents or slopes are not the same for different carcinogens, nor, in different strains of animal are they necessarily the same for a given nitrosamine (Druckrey et al. (1963) calculate 2.6 for NDEA in BD-rats, Peto et al. (1984) calculate 2.3 in MRC/Wistar rats). At low doses there is no significant life-shortening, since the induced tumours do not develop quickly enough to cause early

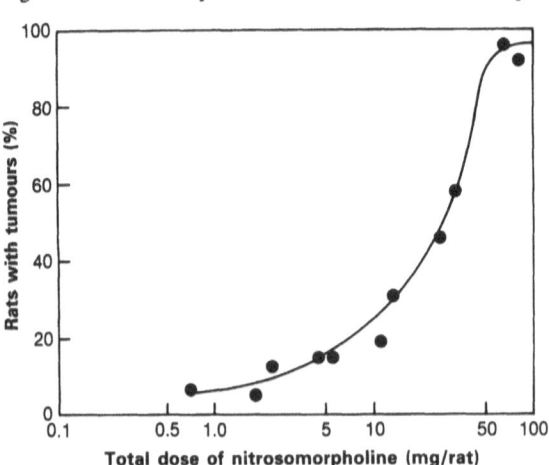

Fig. 7.29. Tumour response of rats related to nitrosomorpholine dose.

Dose–response studies

death of the animals, so at these low dose rates frequency of tumours (or tumour incidence) must be used as an index of carcinogenic effectiveness of the treatment.

It is apparent in the studies of Preussmann using nitrosopiperidine (Eisenbrand *et al.*, 1980), that the intervals between the doses (5 ×) was too large, so that an apparent no-effect level was reached in too few steps, and an extrapolation using the relatively small numbers of animals was not useful. Again there were differences in tumourigenic effect at different dose levels, and only nitrosopyrrolidine seemed to be a 'pure' inducer of one type of tumour in rats, hepatocellular carcinomas. It is interesting that in the NDEA study a change in the type of tumour which is the principle cause of death, from liver tumours at high doses to esophageal tumours at

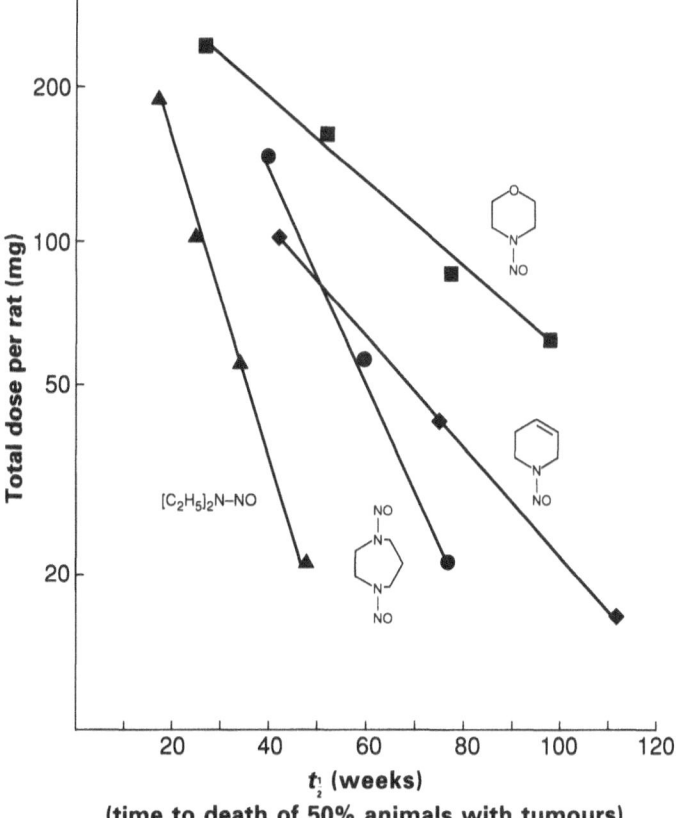

Fig. 7.30. Dose–response of rats to nitrosodiethylamine, dinitrosohomopiperazine, nitroso-1,2,3,6-tetrahydropyridine and nitrosomorpholine.

lower doses does not affect the linearity of average time of death versus dose.

In dose–response studies with two other nitrosamines, nitroso-1,2,3,6-tetrahydropyridine (NTHP) and dinitrosohomopiperazine (DNHP), the most common tumours were in the rat esophagus as with NDEA. Like NDEA the unsaturated cyclic nitrosamine induced liver tumours at the highest dose levels (in addition to tumours of the esophagus), but only esophageal tumours at lower dose rates; nevertheless, the median time of death was related linearly with dose administered across the range where the tumour pattern changed (Lijinsky et al., 1982a). Dinitrosohomopiperazine induced only tumours of the esophagus in rats across the entire range of doses. The difference between the effects produced by these two cyclic nitrosamines is probably related to liver toxicity produced by NTHP, but not by DNHP; the reason for such differences in toxicity is not known. The resemblance between the two cyclic nitrosamines and NDEA is striking, because the latter is an alkylating agent *in vivo* and its alkylation of DNA is believed to be the mechanism by which it induces tumours and perhaps toxicity. In contrast, all of the cyclic nitrosamines examined, except nitrosopyrrolidine, fail to give rise to stable, identifiable, or even detectable adducts with DNA *in vivo*. The cyclic nitrosoheptamethyleneimine was the subject of a dose–response study in rats, in which both males and females were used. The principal tumours induced were in the esophagus and, again, the median time of death was linearly related to dose (Lijinsky et al., 1982b).

Nitrosodiethanolamine is another carcinogenic nitrosamine that interacts with DNA *in vivo* to only a small extent and, unlike the other compounds, is not detectably mutagenic. In dose–response studies it induces liver tumours in rats at all doses, but tumours of the esophagus and nasal mucosa at higher doses (Lijinsky and Reuber, 1984b; Lijinsky and Kovatch, 1985b), as does nitrosomorpholine. Again the response, as measured by the median time of death with tumours in a group, was linearly related to dose at higher dose rates, but at lower dose rates tumours did not decrease survival below that of control rats. It was notable that even the lowest dose administered induced a significant incidence of liver tumours in female rats. This concentration was 28 ppm in drinking water, which is a lower concentration of this nitrosamine than is found in some sources of human exposure. Nitrosodiethanolamine was for many years considered a very weak carcinogen, an opinion that now must be reconsidered and revised. Alkylation of DNA by nitrosodiethanolamine has been measured at much higher doses (Farrelly et al.,

Table 7.7. Tumour mortality by nitrosamines related to dose

Compound	Slope (tangent)	Rat sex	Lowest dose (mmol)	Target organ(s)*	Duration of treatment (weeks)	Median week of death	Mode (Gavage, G; Drinking Water, W)
N-nitroso-							
dimethylamine	1.4	♀	0.22	L	33	91	W
diethylamine	2.4	♀	0.2	L, E	30	45	W
methylethylamine	2.6	♂	0.2	L	30	105	W
methyl-oxopropylamine	1.1	♂	0.5	E	30	42	G
methyl-n-hexylamine	0.8	♂	1.2	L, E	33	63	G
methylphenylethylamine	3.6	♂	0.2	E	33	33	W
diethanolamine	4.5	♂	4.8	L	100	108	W
	3.4	♀	4.8	L	100	113	W
Δ^3-dehydropiperidine	1.8	♀	0.4	L, E	70	88	W
heptamethyleneimine	1.1	♂	0.23	E	50	61	W
morpholine	1.2	♀	0.7	L	50	79	W
trimethylpiperazine	2.0	♀	0.34	Nose	30	62	W
Dinitroso-							
dimethylpiperazine	4.0	♀	0.3	E	21	41	W
homopiperazine	2.3	♀	0.34	E	30	67	W

*Target organ: E = Esophagus; L = Liver.

1987), but must be vanishingly small at low doses, yet a tumourigenic effect is easily seen.

The acyclic nitrosamine, methylnitroso-2-phenylethylamine was administered to rats at a wide range of doses, down to three ppm in drinking water, which was not a no-effect dose even after only 30 weeks' administration. This compound would be expected to be a methylating agent, as shown by Kleihues for many methylnitrosoalkylamines that induce tumours in the rat esophagus (Von Hofe et al., 1987). The same linear relation between average time of death and dose was shown at higher doses (Lijinsky et al., 1982c). The 'tangents' for the dose responses in these studies are in Table 7.7, and vary from 0.8 to 4.5.

Many other dose–response studies with a variety of N-nitroso compounds have been carried out, but on a more limited scale (a small range of doses), which makes them less useful in illuminating the dose–response relationship in animals or in extrapolation of risk to humans. These limited studies include nitrosomorpholine in hamsters (Ketkar et al., 1983), nitrosodiethanolamine in rats (Preussmann et al., 1982) and in hamsters (Pour and Wallcave, 1981; Hoffmann et al., 1983), methylnitroso-n-hexylamine, 2-hydroxypropylnitroso-2,3-dihydroxypropylamine, nitroso-3,4,5-trimethylpiperazine, dinitroso-2,6-dimethylpiperazine, hydroxyethylnitrosourea and 2-hydroxypropylnitrosourea (Table 7.7).

The results of these dose–response studies, whether over a large range of doses or abbreviated, show that whatever the target organ and whatever the nature of the carcinogenic N-nitroso compounds, higher dose rates and higher doses lead to increased incidences of tumours and/or decreased survival of animals with the tumours. In turn this suggests that there is a pool of affected cells created by each dose, from which emerge after a long interval some few cells that are partially transformed (although they are not individually identifiable as such), and one or more of these progress to tumours. The larger the dose of carcinogen, the larger the pool from which cells destined to become tumours are formed, and therefore the larger the number of such preoplastic cells; they also appear earlier in animals receiving the higher doses. The progressiveness of this process argues to some extent against the concept that mutations occurring sporadically in cells are responsible for the dose-related appearance of tumours of particular types in a non-random manner. Furthermore, the failure of treatment of rats with the powerful alkylating mutagenic N-nitroso compounds to augment or accelerate the many spontaneous tumours of the pituitary, adrenal medulla, testis, thyroid, and pancreas in rats (Lijinsky – unpublished), suggests that progenitor cells of these tumours are unresponsive to the mutagenic effects of the carcinogens, and these tumours arise by a different mechanism.

If the N-nitroso compound is acting through induction of particular mutations, then continued treatment with the carcinogen, or toxicity induced by high doses, must play an important role in progression of the transformed cells, since high doses lead to more malignant tumours, measured by more invasion and increasing metastatic potential. An explanation is needed also for the similarity in pattern of dose-related appearance of tumours between carcinogenic N-nitroso compounds that are powerful alkylating agents and potent mutagens (NDEA, NDMA), those that are mutagenic but do not readily form DNA adducts (cyclic nitrosamines), and those that are not mutagens (nitrosodiethanolamine). There is at the moment no common biochemical thread connecting these carcinogens with their similar carcinogenic effects, although clues exist in the association of some chemical structural characteristics with certain types of tumour in particular species. Unravelling the mystery might depend on investigating these clues.

7.7 Transplacental carcinogenesis by N-nitroso compounds

The possibility that some human cancers are caused through *in utero* exposure of a fetus to carcinogens acquired by the pregnant mother has inspired an interest in the transplacental action of environmental carcinogens. A fetus, because of its small size and the high proportion of rapidly dividing cells, is likely to be susceptible to considerably smaller doses of a carcinogen than is the adult. Furthermore, the fetus is exposed to reactive metabolites of carcinogens formed in the mother's tissues (mainly the liver), but itself having a low capacity for detoxication of foreign chemicals.

The ubiquity of N-nitroso compounds, and their potent carcinogenic activity as a class, makes their possible transplacental action a matter of concern, and has impelled some examination of transplacental carcinogenesis by selected N-nitroso compounds, but still only a small number. It began in the early 1960s with NDEA in mice (Mohr and Althoff, 1965), Syrian hamsters (Mohr *et al.*, 1965) and later in rats by Pielsticker *et al.* (1967), and by Ivankovic and Druckrey (Ivankovic *et al.*, 1966) with ethylnitrosourea in rats. The alkylnitrosourea induced tumours of the nervous system in the offspring, as it did also in adult rats. This model has since received a great deal of attention, including the painstaking elucidation of the optimal stage in embryogenesis for induction of tumours, by determining the incidence of nervous system tumours that arise eventually in rats born following administration of ENU on successive days of gestation (Ivankovic and Druckrey, 1968). Prior to day 11 administration of ENU did not lead to subsequent development of tumours. Following day 12 there was an increasing incidence of tumours of the nervous system in the offspring until day 19 or 20, just before birth, when it levelled off. Few tumours of other types were seen in the offspring, indicating a particular susceptibility of the developing nervous system to the direct acting ENU. The correlation of DNA ethylation in those cells with tumour induction has been a preoccupation of several investigators, including M. F. Rajewsky (Goth and Rajewsky, 1974), and this is discussed elsewhere (Chapter 4); little attention has been given to ethylation in other organs and tissues, which did not lead to tumour induction. ENU also induced tumours of the nervous system transplacentally in hamsters (Mennel and Zülch, 1972), but involved only peripheral nerves, not brain or spinal cord. Nervous system tumours were not induced by alkylnitrosoureas in adult hamsters (Lickhachev et al., 1983).

In addition to the existence of an optimal development stage of the embryo for transplacentally induced nervous system tumours, there was an increasing response at a given stage (day 15 of gestation) to increasing doses of the carcinogenic ENU, which manifested itself as an earlier onset of the tumour, or earlier death of the offspring from tumours. Many alkylnitrosoureas were shown by Ivankovic to induce nervous system tumours when rats were exposed to them transplacentally, although with much lesser effectiveness as the alkyl group increased in length beyond propyl (Ivankovic et al., 1981). It is probable that they reach the target cells in the embryos unchanged and react there directly with receptor molecules, presumably DNA. Methylnitrosourea also was less effective than ENU (Alexandrov, 1969; Druckrey, 1973).

Most N-nitroso compounds, particularly those of small molecular weight, are readily diffusible. However, the nitrosamines which reach the embryo following transplacental exposure probably cannot be activated, since embryos are deficient in detoxifying enzymes. The carcinogenic action of transplacentally administered nitrosamines is, therefore, expected to be different from that of alkylnitrosoureas, and dependent on the products of metabolism of the nitrosamine in the mother. Presumably the activated forms of the nitrosamines are sufficiently stable to circulate in the blood and cross the placenta. The fact that nitrosamines can act transplacentally suggests that stable transport forms of nitrosamines can also induce tumours in distant organs of adults, although as yet there has been no convincing demonstration of the identity of such a transport molecule.

Table 7.8. *Tumours induced transplacentally by nitrosamines*

Compound	Dose (mg/kg)	Species	Tumours in offspring	(Incidence %)
N-nitroso-				
dimethylamine	12.5	Hamster	Trachea, Forestomach	16
	40	Rat	Kidney	4
diethylamine	120	Mouse	Esophagus, Liver	30
	20	Hamster	Trachea	40
	120	Rat	Kidney	6
di-*n*-propylamine	100	Hamster	Trachea, Forestomach	19
di-*n*-butylamine	30	Hamster	Trachea	6
piperidine	100	Hamster	—	0
hexamethyleneimine	20	Hamster	Larynx	2
methylurethane	40	Rat	Lung, Nervous system	18

Both nitrosodiethylamine and nitrosodimethylamine have been tested transplacentally in rats and in hamsters. Although metabolic activation is believed to give rise to the same proximate carcinogenic alkylating agent, an alkyldiazonium ion, as the corresponding alkylnitrosourea forms directly (see Chapter 4), the tumour response to the nitrosamine is quite different from that to the alkylnitrosourea; this is also true in adults. When given to pregnant female rats in the last stages of pregnancy, both NDMA and NDEA give rise to kidney tumours in the offspring when they mature. The kidney tumours induced by NDMA are mesenchymal, the same as are induced by a single large dose of NDMA in adult rats, accompanied by (or following) extensive liver necrosis (Hard and Butler, 1970). The kidney tumours induced transplacentally by NDEA were of tubular cell origin, not mesenchymal. It can be assumed that both nitrosamines are activated in the maternal liver (and perhaps other organs) to reactive metabolites transported across the placenta, and it is surprising that tumours are induced only in the kidney of the offspring, not in other organs. Adult rats given pulsed doses of NDMA or NDEA respond with a variety of tumours, in the liver, lung, nasal mucosa and esophagus, as well as in the kidney. These results suggest that response to an agent that alkylates DNA, as the reactive intermediates are assumed to do, is not the sole parameter of tumour induction in an organ, and might not even be the most important.

Most of the other studies of the transplacental action of nitrosamines have been carried out by Althoff, Mohr and their associates in Syrian Hamsters; a few studies in mice have also been reported. A summary of the results of these studies is in Table 7.8. The most common site for transplacental tumourigenesis in hamsters is the lung and other parts of the respiratory tract, and frequently also the liver. Most interesting in these results are the tumours that are induced in adult hamsters by these nitrosamines, which do not appear in the offspring. For example, nitrosodibutylamine induced bladder tumours in adults, but not in the offspring treated transplacentally, which developed a few tumours of the trachea and larynx (Althoff et al., 1976). Several nitroso derivatives of 2-hydroxypropylamine induced pancreas tumours in adult hamsters, but not in the offspring treated transplacentally, although there was a small incidence of tumours of other organs in the offspring (Althoff et al., 1977b). Neither NDMA nor NDEA given transplacentally to hamsters induced kidney tumours in the offspring (in contrast with rats), or liver tumours, which are the prime tumour induced by them in adult hamsters. These results suggest that the fetal organs have a different response to the activated nitrosamine from those of the same organs in the adult hamsters, although organs such as the liver, bladder and gastrointestinal tract of the fetus are responsive to nitrosamines with other structures, so they are not innately resistant. The transplacental route of treatment with nitrosamines and its results, unfortunately, make the resolution of the mechanisms of carcinogenesis by N-nitroso compounds more complicated, not less.

In a recent study, Pour (1986a) described the transplacental effect of a single dose of 10 mg/kg body weight of nitrosobis-(2-oxopropyl)amine in Syrian Hamsters given to pregnant females on days 8, 10, 12 and 14 of gestation in succession, a total of 40 mg/kg. The incidence of pancreas duct adenomas in the mothers was high (seven of nine animals) and four of nine had cholangiomas; there were a few tumours of lung and kidney, but none of the thyroid gland. In 27 male and 24 female offspring there were very few with tumours at week 46, the termination of the experiment. One male in three of the litters had a tumour of the pancreas duct, but none of the females. One female in four of the litters had a cholangioma, but none of the males, and one female in five of the litters had thyroid adenomas, which neither the male offspring nor the mothers had. It is possible that this weak transplacental carcinogenic effect is due to the killing of the animals at an arbitrary 46 weeks, although it is equally possible that no more tumours would have developed later. Pregnant rats given NBOPA on days 14, 18 and 20 of gestation gave birth to offspring with ovarian or testicular tumours, but few tumours of the types induced in adult rats (Pour, 1986b). The weakness of the transplacental effect of NBOPA, however, contrasts with the strong effect of alkylnitrosoureas given transplacentally to rats and with the high sensitivity of newborns to very small doses of nitrosamines like NBOPA. The results suggest that relatively little of the administered NBOPA or its metabolites formed in the mother crosses the placenta. However, recently H. Schuller has shown in hamsters that doses of 100–200 mg of the tobacco-specific NNK to pregnant hamsters gives rise to considerable incidences of tumours of the lung, pancreas and other organs in the offspring.

The similar limited responses of transplacentally exposed fetuses to other nitrosamines given to pregnant females suggest two explanations. First, that nitrosamines do not readily cross the placenta, while alkylnitrosoureas and probably alkylnitrosocarbamates (Tanaka, 1973) do. Second, lack of metabolising enzymes in the fetus might prevent activation of the nitrosamine to proximate carcinogenic forms; a corrollary is that transport forms of activated nitrosamines produced by the mother do not usually cross the placenta either.

Although it has been shown that methylnitrosourethane induces some tumours of the nervous system of rats when applied transplacentally, as well as some tumours of kidney and lung (Tanaka, 1973), the incidence was much lower than with alkylnitrosoureas. Methylnitrosourethane has induced only tumours of the forestomach in adult rats; tumours of the nervous system, kidney and lung have not been reported (Lijinsky and Reuber, 1982a; Preussmann and Stewart, 1984). The transplacental effect of methylnitrosourethane is obviously not universal, since administration of a mixture of the N-methylcarbamate ester carbaryl (a common insecticide and readily nitrosated) together with sodium nitrite to pregnant female rats on successive days up to shortly before delivery, failed to produce any effect in the offspring, either of teratogenicity or carcinogenicity (Lijinsky and Taylor, 1977d). This contrasts sharply with the many tumours induced transplacentally by feeding ethylurea together with sodium nitrite to rats (Osske et al., 1972). The negative result with carbaryl and nitrite suggests that ingestion by female humans of this and the many other insecticides that are N-methylcarbamate esters poses little transplacental carcinogenic risk through in vivo formation of the corresponding N-nitroso derivatives.

Other nitrosamines that have been tested transplacentally include NDMA and NDEA in rats and a number of cyclic nitrosamines and those having carbonyl or hydroxy groups in the chain (which are frequently pancreas duct carcinogens) in hamsters. A report by Althoff and Grandjean (1979) described the transplacental carcinogenic effects in hamsters of 10 nitrosamines. The nitrosamines dissolved in olive oil were injected s.c. into the pregnant females (which might not be the optimal route of administration). They include nitroso-dimethylamine, -di-n-

propylamine, -methylpropylamine, -propyloxopropylamine, -propylhydroxypropylamine, -piperidine and -hexamethyleneimine. All of them showed only a small effect, that is a low incidence of tumours, in the offspring, usually less than in the mothers, except those that induced no tumours at all in the mothers. With few exceptions the incidences of tumours in the offspring were below 10%, often less than 5%; these results were only bordering on significance in many cases. Some tumours that might have been expected, such as bladder tumours with nitrosodi-n-butylamine and nitrosobutyl-4-hydroxybutylamine, were not seen in the offspring. No pancreas duct tumours were induced in the offspring of hamsters given nitrosobis-(2-hydroxypropyl)amine, which induces those tumours in adult hamsters. There were low incidences of tumours in the respiratory tract, including trachea and larynx, but few in the lungs and nasal mucosa of offspring treated with nitrosamines which induced high incidences of these same tumours in adults. The 'latent periods' were also much longer in the offspring than in adults. These results suggest that fetuses of hamsters (as in rats) are not particularly sensitive to nitrosamines administered to the mothers. This could be because the fetuses are incapable of activating nitrosamines that reach them through the placenta, and because active metabolites formed in the mother also have a low probability or capacity for crossing the placenta.

7.8 Response of different species to N-nitroso compounds

What sets N-nitroso compounds apart from other classes of carcinogens is their broadly different effects in different species, particularly the variety of organs and tissues in which they cause tumours to appear. It is rare among them that an N-nitroso compound carcinogenic to one species will fail to elicit tumours of some type in another species, even a quite unrelated one, having a lifespan of a year or more. The time for development of tumours produced by treatment with an N-nitroso compound is unrelated to the lifespan of the species, and is roughly proportional to the dose, the larger the dose the earlier tumours appear, and vice-versa. This is likely to be true for all carcinogens, but is usually difficult to establish, because so many species are refractory to one carcinogen or another – but not to N-nitroso compounds.

It is worth noting which of the many species that have been treated with N-nitroso compounds have responded with various types of tumour, and in which particular organs. These results probably provide the best clues that can lead us to an understanding of the mechanisms by which N-nitroso compounds induce cancer – and perhaps to an understanding of carcinogenesis in general.

Table 7.9. *Species responding to carcinogenic N-nitroso compounds*

Animal	Treatment	Tumours induced
Mammals		
Mouse	NDMA, NDEA	Liver, Esophagus
Rat	NDMA, NDEA	Liver, Esophagus
Syrian Hamster	NDMA, NDEA	Liver, Nasal cavity
Chinese Hamster	NDMA, NDEA	Lung, Esophagus
European Hamster	NDMA, NDEA	Lung, Nasal cavity
Gerbil	NDEA	Liver, Nasal cavity
Guinea pig	NDMA, NDEA	Liver
Rabbit	NDMA, NDEA	Liver
Dog	NDEA	Liver
Pig	NDEA	Liver
Cat	NDEA	Liver
Hedgehog	NDEA	Liver, Lung
Monkey	NDMA, NDEA	Liver
Fox	NDMA	Liver
Opossum	ENU	Brain, Kidney, Liver
Mink	NDMA	Liver
Bushbaby	NDEA	Liver
Shrew	NBHPA	Lung
Mastomys	NDMA	Liver
Birds		
Duck	NDMA	Liver
Chicken	NDEA	Liver, Kidney
Parakeet	NDEA	Liver
Reptiles		
Python	NDEA	Liver, Kidney
Amphibians		
Frog	NDMA, NDEA	Liver, Kidney
Newt	NDMA	Liver
Xenopus	NDMA	Liver, Kidney
Fish		
Guppy	NDMA	Liver
Zebra Fish (*Brachybanio rerio*)	NDMA	Liver, Esophagus
Medaka (*Oryzias latides*)	NDMA	Liver
Rivulus ocellatus	NDEA	Pancreas
Poecilla reticulata	NDMA, NDEA	Liver
Trout	NDMA, NDEA	Liver
Platyfish (*Lebistes*)	MNU	Sarcoma
Mollusc		
Unio pictorum	NDMA, NDEA	Liver

Relatively few *N*-nitroso compounds have been tested in more than two species, a large proportion in only one. Only three compounds, nitrosodiethylamine, nitrosodimethylamine and methylnitrosourea have been examined in more than a handful of species, and that because of the

Table 7.10. *Tumours induced in different species by some N-nitroso compounds*

Compound	Species	Route	Tumours
N-nitroso-			
dimethylamine	Mouse	Oral	Liver, Lung, Kidney
	Rat	Oral	Liver, Lung, Kidney, Nasal cavity
	Hamster	Oral	Liver, Nasal
	Guinea pig	Oral	Liver
	Rabbit	Oral	Liver
	Frog	Immersion	Liver
	Duck	Oral	Liver
	Trout	Oral	Liver
methylurea	Mouse	I.P.	Lung, Kidney, Lymphoma
	Rat	Oral	Nervous system, Glandular stomach, Mammary gland, Kidney
	Hamster	Oral	Spleen, Forestomach
	Guinea pig	Oral	Stomach, Pancreas
	Rabbit	I.V.	Nervous system, Intestine
	Pig	Stomach	Stomach
	Dog	I.V.	Nervous system, Lung
	Monkey	Oral	Esophagus, Pharynx
ethylurea	Mouse	I.P.	Liver, Lung, Nervous system, Mammary gland
	Rat	Oral	Mammary gland, Lung, Intestine, Nervous system
	Hamster	Oral	Spleen, Forestomach
	Opossum	Oral	Liver, Nervous system, Kidney
	Gerbil	I.P.	Melanoma
	Monkey	I.V.	Ovary, Uterus, Osteosarcoma
di-n-butylamine	Mouse	Oral	Esophagus, Bladder
	Rat	Oral	Liver, Bladder, Lung
	Hamster	Oral	Bladder, Lung
	Guinea pig	Oral	Liver, Bladder
	Rabbit	S.C.	Bladder, Lung
methyl-n-	Rat	Oral	Bladder
dodecylamine	Hamster	Oral	Bladder
	European hamster	S.C.	Bladder, Lung
	Guinea pig	Oral	Liver
methylnitroguanidine	Mouse	Oral	Stomach, Liver
	Rat	Oral	Stomach, Intestine
	Hamster	Oral	Stomach, Duodenum
	Guinea pig	I.R.	Colon

Table 7.10. (cont.)

Compound	Species	Route	Tumours
	Rabbit	I.V.	Trachea
	Mastomys	Oral	Lymphoma, Duodenum
	Dog	Oral	Stomach, Duodenum
bis-(2-hydroxypropyl)-amine	Mouse	Oral	Liver, Lung
	Rat	Oral	Lung, Liver, Thyroid, Esophagus
	Hamster	Oral	Pancreas, Lung, Liver, Nasal cavity
	Guinea pig	S.C.	Liver
	Rabbit	Oral	Liver, Lung
pyrrolidine	Mouse	Oral	Lung
	Rat	Oral	Liver
	Hamster	S.C.	Lung
morpholine	Mouse	Oral	Liver, Lung
	Rat	Oral	Liver, Esophagus, Nasal cavity
	Hamster	Oral	Nasal cavity, Trachea
	Fish	Immersion	Liver
2,6-dimethylmorpholine	Mouse	Oral	Forestomach, Liver
	Rat	Oral	Esophagus, Nasal cavity
	Hamster	Oral	Pancreas, Liver, Lung, Nasal cavity
	Guinea pig	Oral	Liver

enthusiasm of a few biologists for exploring the comparative biology of these alkylating carcinogens. More than 40 species have responded to the carcinogenic action of N-nitroso compounds, most to NDEA (Bogovski and Bogovski, 1981) and a fuller description of many of the studies is given by Schmähl et al. (1978). In most species the main target was the liver, although a variety of other tumours were seen occasionally in one species or more. The commonness of liver tumours induced in these species (Table 7.9), which include several members of some genera, suggests the existence of similar mechanisms of activation and of carcinogenic action of NDEA in liver parenchymal cells of many of them; probably there are similar receptors in liver cells of all or most species, for which NDEA has affinity, but the nature of such receptors is not known.

If we look at the differences in the types of tumour between the species affected by a number of other N-nitroso compounds, the explanation might lie in the existence of activating enzymes (which might be among the receptors in question) in a particular organ of some species, but not in

Table 7.11. *Carcinogenesis by nitrosodiethylamine in several species*

Species	Total dose (mg/kg) (approximate)	Approximate lifespan (years)	Tumours and approximate time of appearance (weeks)	
Mouse	1000	2	Liver	50
Rat	700	3.5	Liver	35
Syrian hamster	600	4	Liver	50
Chinese Hamster	900	?	Esophagus, Lung	60
European Hamster	400	8	Nasal cavity	25
Gerbil	400	8	Nasal cavity	50
Guinea pig	800	8	Liver	50
Rabbit	1700	10	Liver	60
Hedgehog	800	10	Liver	40
Pig	1200	20	Liver	70
Cat	900	20	Liver	80
Dog	400	30	Liver	60
Monkey	> 1000	30	Liver	60
Chicken	1000	25	Liver	80
Parakeet	2500	50	Liver	37
Snake (python)	600	100	Liver, Kidney	95

others. This is not the entire explanation of the interspecies differences, however, since the directly acting alkylnitrosoureas show quite as large differences (Table 7.10), but are thought to need no enzymic activation. One of the most striking differences between species is the response of rats and hamsters to alkylnitrosoureas. In hamsters mainly hemangiosarcomas of the spleen appear, but in rats a great variety of tumours (not including hemangiosarcomas of the spleen, however). Table 7.3 compares the effects in rats of some dialkylnitrosamines and the corresponding alkylnitrosourea.

Guinea pigs seem to respond with liver tumours, or, in some cases (e.g. nitrosoheptamethyleneimine – Cardy and Lijinsky, 1980) not at all. Large mammalian species have seldom been studied, because of the difficulty and expense, and most of the results we have are from rodents. Which organs of humans respond to N-nitroso compounds to which we are exposed is not known, and cannot be speculated about with our present lack of knowledge of mechanisms of carcinogenesis.

The range of species susceptible to carcinogenic N-nitroso compounds comprises many families and includes invertebrates, so as to suggest that a common susceptibility to carcinogens is an ancient evolutionary character. We do not know whether ancient species developed cancer in the absence of exposure to the many synthetic chemical carcinogens we have produced in the past century or so, and the common habits (smoking) associated with cancer in man.

An evaluation of the response of many species to similar doses of an N-nitroso compound such as NDEA (mainly from the work of D. Schmähl) shows that tumours develop after similar times, which are not related to the lifespan of the species (Table 7.11). This suggests that tumourigenesis occurs with similar dynamics, depending on the multiple biological and biochemical effects of the carcinogenic N-nitroso compound in the cells of the target organ. Programming of the cells for short or long survival seems not to be important in the response of the animal. Ageing of cells is not a factor, since the tissues of old rats are less susceptible to the action of carcinogens than those of young adults, and tumours appear later in the former in response to the same dose (Lijinsky and Kovatch, 1986).

8

Conclusions – The importance of *N*-nitroso compounds as environmental carcinogens and as experimental models for investigating cancer

The growing interest in *N*-nitroso compounds is due to their unique properties as cancer-causing agents. That people may be exposed to them in industrialised countries adds another area of interest, because in those countries cancer is a dread disease and there are inadequate explanations for the high incidence of many human cancers. In addition, production of cancer in experimental animals by *N*-nitroso compounds provides the best models – and sometimes the only models – of many human cancers. The experimental animal models afford the opportunity to study mechanisms by which *N*-nitroso compounds are formed and induce their many biological effects. The carcinogenicity of NDEA in a great variety of species has led to the conclusion that the time-to-tumour is a function of dose, independent of the animal's lifespan. Models using nitrosamines have demonstrated the influence of hormones on the genesis of particular cancers (Lijinsky *et al.*, 1988g) and have shown that old animals are less susceptible than young adults to induction of tumours (Lijinsky and Kovatch, 1986). The models also allow study of the reversal of those effects – several of the most effective cancer therapeutic agents are *N*-nitroso compounds. Exploration of the action of *N*-nitroso compounds in animal models and other biological systems might be the most informative about carcinogenesis in general, as this book has attempted to show.

8.1 Environmental carcinogenesis

One indication of the role of nitrosamines in human cancer is the occurrence of cancer of the mouth associated with the use of snuff or chewing tobacco. Chewing tobacco contains several carcinogenic nitrosamines, and no other identified carcinogens. Some of these compounds induce tumours in the tongue and oral cavity, as well as the esophagus, in rats and other experimental animals, but they are not usually regarded as locally acting carcinogens. Nevertheless, there seems to be a firm link between the use of 'smokeless' tobacco and oral cancer in many parts of the World, including Scandinavia, the U.S.A. and India, where it is often used with lime and betel nut. This link between carcinogenic nitrosamines in tobacco and oral cancer was not sufficient to persuade a jury of causality in the case of a young snuff-user in Oklahoma who died in his teens of cancer of the tongue, but the circumstantial evidence is as convincing as any in the area of occupational cancer.

These observations place humans among the more than 40 species in which N-nitroso compounds are carcinogenic, and makes it overwhelmingly likely that testing N-nitroso compounds in experimental animals provides indications of effects in humans exposed to them. That being so, there is no reason to assume that man is less sensitive than rodents, for example, to the carcinogenic effects of N-nitroso compounds. Therefore, the low concentrations of nitrosamines to which some populations are exposed are likely to be important in increasing the risk of cancer, because the exposures can be almost continuous and can last a lifetime.

There are no other specific human cancers which have yet been attributed to exposure to N-nitroso compounds in a way similar to cancers associated with occupational exposure of well defined populations to aromatic amines, polynuclear hydrocarbons, vinyl chloride and some other carcinogens. While continuing to seek additional epidemiologic evidence (which is difficult to gather because of almost universal exposure to some level of N-nitroso compounds), we must bear in mind that there is no species that has been found resistant to the action of N-nitroso compounds. The few compounds that have been tested at low doses in experimental animals have given rise to statistically significant incidences of tumours at dose rates or concentrations even lower (e.g. parts per 10^9) than those to which some humans have been exposed. Nitrosamines are among the most potent carcinogens yet discovered, as little as one to two milligrams of, for example, nitrosomorpholine in a lifetime having induced tumours in rats.

In some occupational settings, such as tyre and rubber factories, and in factories in which pesticides are formulated using secondary amines, the exposure of workers to N-nitroso compounds can be quite substantial; there have been indications of increased incidences of cancer in workers in such factories. Another heavy exposure of workers to nitrosamines occurs in machine shops in which synthetic cutting oils containing nitroso-diethanolamine are used; no reports of increased cancer incidences associated with these exposures have been made, but it is probable that they have been in use for too short a time.

The most important human exposure to carcinogenic nitrosamines is probably in tobacco smoke, which is responsible for nearly one million cases of cancer a year worldwide. There are carcinogens of other types in tobacco smoke, particularly polycyclic hydrocarbons, but they are less potent and are present in smaller concentrations than nitrosamines. It is probable that the carcinogenic nitrosamines and the other carcinogens in tobacco smoke act together in concert to evoke the large carcinogenic effect of the smoking habit, because the total dose of nitrosamines to a smoker of one pack of cigarettes (20) has been estimated as approximately 20 μg (0.3 μg/kg body weight). Cancers that have been related to cigarette smoking are lung, bladder, pancreas, esophagus, oral cavity, larynx and kidney. It is very likely that the carcinogenic nitrosamines in tobacco smoke add their effects to those of other carcinogens encountered in daily life in the formation and progression of those cancers.

Much less is known about the importance of carcinogen exposures other than tobacco, but cancers are plentiful in people of Western societies who do not smoke. Japanese smokers and people in non-industrialised countries have a relatively low incidence of lung cancer, although smoking is common in many of them, suggesting that in Western industrialised countries factors other than the carcinogens in tobacco smoke are important in generating cancer of the lung and other organs.

It is difficult to assess exposure of people to carcinogenic nitrosamines in sources other than tobacco, because the concentrations in food, air, rubber and other materials are so variable. Averages are not very useful, because people differ so much in their environments and habits. Variations are even larger in exposure to N-nitroso compounds formed endogenously from ingested amines which react with nitrite in the stomach; the nitrite can come from food and from nitrite in saliva formed by bacterial reduction of nitrate. Because nitrosation is second-order in nitrite, the differences in yield due to differences in concentration of nitrite are magnified. People differ from one another in their intake of nitrite-cured meat, which can range from a hundred grams or more a day to zero in people who do not eat any, such as this author. Many people have infections in the bladder, harbouring bacteria that can reduce urinary nitrate to nitrite, which can then form N-nitroso compounds with amines present; such N-nitroso compounds are easily absorbed through the bladder wall to exert their carcinogenic effects systemically in appropriate target organs. Although methods have been developed for assessing the capacity for endogenous formation of N-nitroso compounds in humans, these have lately been criticised for inadequacy, and the results may reflect the situation only within an order of magnitude or more.

8.2 Models for carcinogenesis

Apart from their probable importance in increasing the risk of a variety of cancers in humans – and they have as a class induced almost every human tumour in some experimental animal – N-nitroso compounds are important as a focus for study of mechanisms of carcinogenesis. A majority of those studied are alkylating agents, and their activation and interactions with cellular macromolecules have been widely used to support the current hypothesis of carcinogenesis mechanisms through mutagenic interactions with cellular DNA. However, some experimental data do not support the hypothesis; for example, the DNA alkylation by these reactive molecules often occurs equally in organs in which tumours arise and in those in which tumours do not arise (Fong et al., 1990; Lijinsky, 1988b, 1991). The specificity with which N-nitroso compounds of particular structure induce tumours only in particular cells of certain organs in a given species, not in other organs, and not necessarily in those same cells in a different species, indicate that other interactions of the N-nitroso compound are of great importance.

There are myriads of publications on DNA alkylation and mutagenesis. Little attention, however, is given to other chemical attributes of N-nitroso compounds, such as pharmacokinetics, differential metabolism and chemical reactions other than DNA alkylation, particularly as affected by physiological factors such as hormones. The result is that we seem little closer to understanding how the simplest nitrosamine, nitrosodimethylamine, induces its particular spectrum of tumours, and no others, in rats, mice or hamsters, than we were when Magee, Hultin and Farber first reported methylation of proteins and nucleic acids, almost 30 years ago. That the mechanism of action of this very simple molecule has proved so elusive does not reflect well on the intellectual prowess of modern cancer research, with all the tools of biochemistry and molecular biology at its command. On the contrary, it suggests that the present narrow focus on the technologically driven *in vitro* techniques is a misplaced emphasis, and that a greater regard for the reactions and interactions that occur in cells of an organ within an animal body, in an approach free of dogma, is more likely to produce the answer. As J. M. Barnes suggested to me more than 20 years ago, nitrosamines provide such wonderful models of carcinogenesis, that resolving how they can produce cancer might well provide a general and comprehensive explanation of the phenomenon.

The organ- and species-specific carcinogenic effects of N-nitroso compounds show that there is a strong relationship between chemical structure and the tumourigenic response of animals. There is a different relationship between their chemical structure and mutagenic activity in a number of systems; the latter probably depends on the tendency of the compounds to form particular alkylating agents. For example, the directly acting (i.e. without metabolism) α-acetoxynitrosamines and the analogous alkylnitrosoureas yield the same alkylating intermediate (probably an alkyldiazonium ion), and have comparable mutagenic activities. Yet they have very different carcinogenic effects in animals, including the induction always of local sarcomas by the acetoxynitrosamines, but not often by alkylnitrosoureas (also directly acting) when given by S.C. injection. The induction of sarcomas is common by S.C. injection of locally acting carcinogens, including polynuclear compounds, and production of sarcomas is the usual means of demonstrating that 'oncogenes' activated by alkylating and other carcinogens are the significant targets and mediators of carcinogens.

The induction of liver tumours in rats by nitrosamines appears to be largely independent of the size or shape of the molecule, since such a diversity of structures have the liver as target, including some directly acting alkylnitrosoureas. On the other hand, there are restrictions on the size, shape and chemical reactivity of N-nitroso compounds that are compatible with induction of tumours of the esophagus in rats. Directly acting compounds, such as alkylnitrosoureas, are excluded from this category, as are nitrosodimethylamine and cyclic nitrosamines with fewer than six atoms in the ring. Nitrosamines with more than eight carbons are also excluded, although methylnitroso-n-octylamine is at the borderline. Thus, nitrosooctamethyleneimine induces esophageal tumours, but nitrosododecamethyleneimine does not; neither do nitrosodi-n-amylamine or methylnitroso-n-nonylamine. Nitrosomorpholine induces some esophageal tumours in rats, but nitrosooxazolidine does not (Fig. 8.1). There is some anomaly between the weak esophageal carcinogenicity of methylnitrosoethylamine, the very potent esophageal carcinogenicity of nitrosodiethylamine and the failure of nitrosopyrrolidine to induce esophageal tumours, although it is the cyclic analogue of nitrosodiethylamine. There are subtleties of chemistry among cyclic nitrosamines that are presently unfathomable, but among the remainder of N-nitroso compounds the affinity for the rat esophagus bespeaks varying degrees of interaction with a cellular receptor of certain dimensions, perhaps analogous to the manner in which molecules of defined shape have particular odours because they fit within certain olfactory receptors.

Interaction with the 'receptor' in this model assumes a critical role in the process of neoplastic transformation.

For carcinogenesis in some organs interaction with enzymes seems to be critical because of the failure of directly acting N-nitroso compounds (alkylnitrosoureas, alkyl-nitrosocarbamates) to induce tumours of the esophagus or nasal mucosa, whereas a large proportion of nitrosamines, which must be metabolised, induce tumours in the rat esophagus and in the nasal mucosa of rats and hamsters.

The current popularity of research into the role of oncogenes and changes in their expression in the process of cancer formation has defined the gap between environmental chemistry and biomolecular events in toxicology. The number of oncogenes identified and sequenced is quite

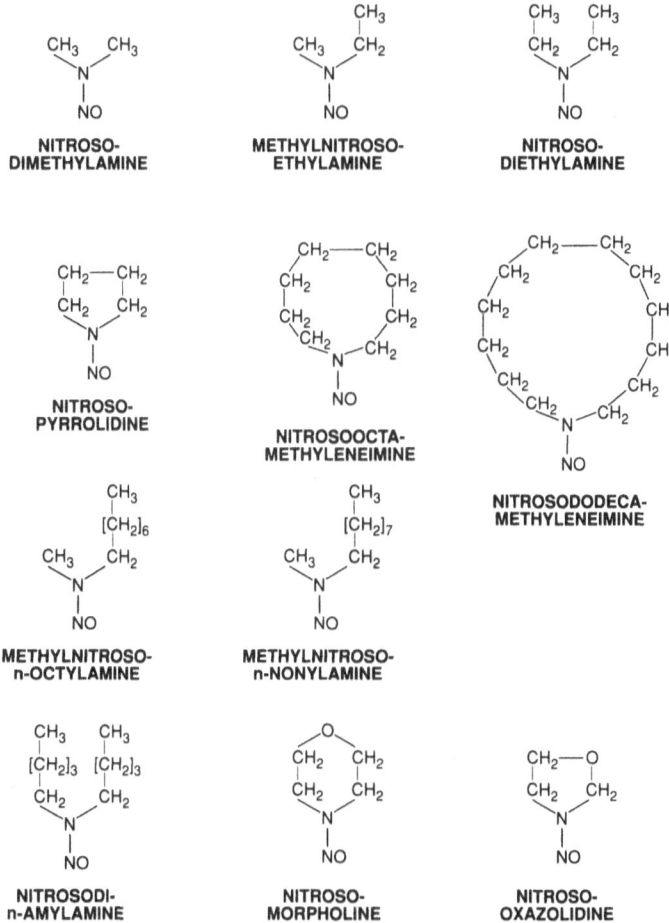

Fig. 8.1. Large and small carcinogenic nitrosamine molecules.

large and growing, yet the link to events that precipitate cancer in humans is tenuous. Considering bladder cancer, a proportion of those found in people have displayed an activated oncogene, but not all of them. Surely if the oncogene activation were causative, all of the cancers should have it. Even so, there is no link to factors that gave rise to the tumours, which may be caused by chemical exposure to aromatic amines (a proven link) or to certain nitrosamines.

The achievement of Barbacid and his group was considerable in identifying an oncogene that is activated by exposure of female rats at puberty to a single dose of methylnitrosourea and locating the altered gene on a chromosome in cells of the mammary gland, correlating with induction of tumours in that organ. However, it is prudent to ask questions before accepting that finding as a prototype of chemical induction of tumours. For example, did this directly acting alkylating agent also activate the same or different genes in other target organs of the rat (brain and kidney) in which tumours appear following chronic administration of MNU? Repeated small doses of MNU have not given rise to mammary tumours in rats, while repeated small doses of ENU and other ethylnitrosoureas have induced high incidences of mammary cancer in rats (but not in hamsters). This suggests that much more is involved in induction of mammary cancer in rats than activation of a single gene by alkylation of DNA.

Alkylation of DNA by MNU probably occurs in many cells of most organs (Fong *et al.*, 1990), but tumours arise only in a few organs. Additional organs with similar levels of methylation are susceptible to tumour induction by other methylating agents, such as nitrosodimethylamine, methylnitroso-*n*-butylamine, etc. What other actions, or effects, by the *N*-nitroso compound are needed for tumours to develop? Are these effects capable of causing tumours to develop in the absence of alkylation of DNA, which might explain the appearance of 'spontaneous' tumours and the action of 'non-genotoxic' carcinogens? What of other *N*-nitroso compounds that are structurally different but induce the same tumours (as MNU) in rats, for example ethylnitrosourea (ENU) (see Chapter 7)? Is the same oncogene activated? At the same locus? ENU, like most ethylnitroso compounds, is more potent (or effective) than MNU and other methylating nitroso compounds in rats, notwithstanding that the extent of alkylation by ethylating agents is two orders of magnitude smaller than by an equal dose of the corresponding methylating agent. The range of tumours induced in rats by a set of directly acting ethylnitrosoureas is much greater than the rather limited range of tumours induced by corresponding methylnitrosoureas (Fig. 8.2). Few, if any, comparative studies of oncogene activation have been done with the same

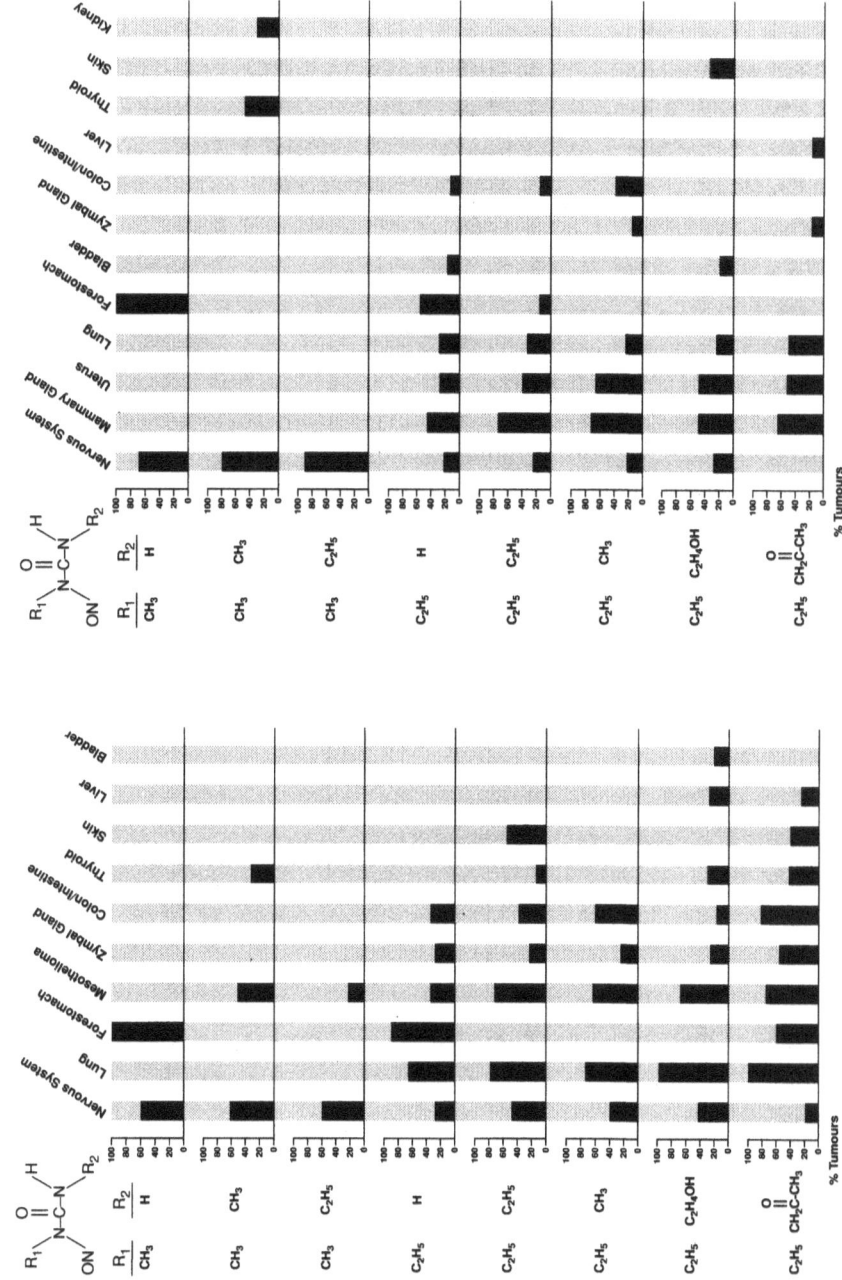

Fig. 8.2. Tumours in male (left) and female (right) rats treated with methyl- and ethyl-nitrosoureas.

carcinogen in different organs, or with several carcinogens in a single organ in which they all induce tumours. This is a crucial requirement to prove that cancer causation by activation of certain oncogenes is more than an attractive hypothesis. There is need for a fusion of the knowledge and experience of chemists and pathologists with those of molecular biologists, but this has not yet happened.

Alkylnitrosocarbamates (Fig 8.3) induce tumours in very few organs of rats (only the forestomach when given orally) although they are mutagenic alkylating agents equivalent to the corresponding alkylnitrosoureas, which induce tumours in many rat organs. Why is the second amino nitrogen so important in alkylnitrosourea carcinogenesis? Why in hamsters do alkylnitrosoureas, whatever their structure, all induce mainly hemangiosarcomas of the spleen, but alkylnitrosocarbamates do not? Are there special receptors in endothelial cells of the hamster spleen for which alkylnitrosoureas have affinity, but which are absent from endothelial cells in the aorta, liver, kidney, and other organs of the hamster, or the rat, in which alkylnitrosoureas do not induce hemangiosarcomas? What could be the nature of so specific a receptor? When the spleen is removed from the hamster the alkylnitrosourea does not induce tumours of other organs instead.

The particular chemical structure of the N-nitroso compound must be the principal determinant of organ-specific carcinogenesis. Chemical carcinogenesis is a process driven by the reactivity of the chemical, and it is certain that the clues to the problem of organ specificity – especially in the case of carcinogens that are directly acting, such as alkylnitrosoureas – lie in the interaction of the carcinogen with critical molecules in certain cells, some of which go on, after an interval, to self-perpetuating cancer cells. It is difficult to consider, on the basis of evidence summarised here, that this critical molecule has to be DNA, which is adducted similarly whether tumours arise or not. It might seem a diversion from the current vogue of looking only at the changes found in the sodium salt of DNA, but it seems likely that more attention to the remaining wondrous

Fig. 8.3. Alkylnitrosourea and alkylnitrosocarbamate structures.

Alkylnitrosourea   Alkylnitrosocarbamate–Ester

structure of cells, and the changes produced in it by carcinogens, will lead to an understanding of the genesis of cancer and its progression. The important effect might be interference by carcinogenic N-nitroso compounds with the structure and maintenance of the plasma membrane, through which pass signals controlling cell behaviour. Tumours might be another differentiated state, rather than mutagenically altered cells. I believe that further exploration of the chemistry and toxicology of N-nitroso compounds will lead us to the goal of understanding their action as carcinogens, and of carcinogenesis in general.

References

Adamson, R. H., Krolakowski, F. J., Correa, P., Siebra, S. M. & Dalgard, D. W. (1977). Carcinogenicity of 1-Methyl-1-Nitrosourea in non-human primates. *J. Natl. Cancer Inst.*, **59**: 415–22.
Adelberg, E. A., Mandel, M. & Chang, G. C. C. (1965). Optimal conditions for mutagenesis by N-Methyl-N'-Nitro-N-Nitrosoguanidine in *Escherichia coli* K12. *Biochem. Biophys. Res. Comm.*, **18**: 788–95.
Aidjanov, M. M. & Sharmanov, T. S. (1982). Nitrosamines and their precursors in some Kazakh foodstuffs. *IARC Scientific Publication* No. 41: 267–76.
Alexandrov, V. A. (1969). Transplacental blastomogenic action of N-nitrosomethylurea on rat offspring. *Vopr. Onkol.*, **15**: 55–61.
Althoff, J., Cardesa, A. & Mohr, U. (1973). Carcinogenic effect of N-nitrosohexamethyleneimine in Syrian Golden Hamsters. *J. Natl. Cancer Inst.*, **50**: 323–9.
Althoff, J., Eagen, M. & Grandjean, C. (1975). Carcinogenic effect of 2,2'-dimethyldipropylnitrosamine in Syrian hamsters. *J. Natl. Cancer Inst.*, **55**: 1209–11.
Althoff, J., Grandjean, C. & Gold, B. (1977a). Diallylnitrosamine: A potent respiratory carcinogen in Syrian Golden hamsters. *J. Natl. Cancer Inst.*, **59**: 1569–71.
Althoff, J., Grandjean, C. & Pour, P. (1977b). Transplacental effect of nitrosamines in Syrian hamsters. IV. Metabolites of dipropyl- and dibutylnitrosamine. *Z. Krebsforsch.*, **90**: 119–26.
Althoff, J. & Grandjean, C. (1979). *In vivo* studies in Syrian Golden Hamsters: A transplacental bioassay of ten nitrosamines. *National Cancer Institute Monograph* No. 51, 251–5.
Althoff, J. & Lijinsky, W. (1977). Urinary bladder neoplasms in Syrian hamsters after administration of N-nitroso-N-methyl-N-dodecylamine. *Z. Krebsforsch.*, **90**: 227–31.
Althoff, J., Pour, P., Grandjean, C. & Eagen, M. (1976). Transplacental effects of nitrosamines in Syrian hamsters. I. Dibutylnitrosamine and nitrosohexamethyleneimine. *Z. Krebsforsch.* **86**: 69–75.
Althoff, J., Pour, P., Grandjean, C. & Marsh, S. (1977c). Transplacental effects of nitrosamines in Syrian hamsters. *Z. Krebsforsch.*, **90**: 79–86.
Amberger, H. (1986). Different autochthonous models of colorectal cancer in the rat. *J. Cancer Res. Clin. Oncol.*, **111**: 157–9.
Ames, B. N., Durston, W. E., Yamasaki, E. & Lee, F. D. (1973). Carcinogens

are mutagens: A simple test system combining liver homogenates for activation and bacteria for detection. *Proc. Natl. Acad. Sci. U.S.A.*, **70**: 2281–85.
Andrews, A. W., Fornwald, J. A. & Lijinsky, W. (1980). Nitrosation and mutagenicity of some amine drugs. *Toxicol. Appl. Pharmacol.*, **52**, 237–44.
Andrews, A. W. & Lijinsky, W. (1980). Mutagenicity of 45 nitrosamines in *Salmonella typhimurium*. *Teratogenesis, Carcinogenesis and Mutagenesis*, **1**: 295–303.
Andrews, A. W. & Lijinsky, W. (1984). N-nitrosamine mutagenicity using the *Salmonella*/mammalian-microsome mutagenicity assay. In: *Genotoxicology of N-Nitroso Compounds* (Rao, T. K. *et al.*, ed.), Plenum Publishing Corp., New York, pp. 13–43.
Andrews, A. W., Lijinsky, W. & Snyder, S. W. (1984). Mutagenicity of amine drugs and their products of nitrosation. *Mutation Res.*, **135**, 105–8.
Andrews, A. W., Thibault, L. H. & Lijinsky, W. (1978). The relationship between mutagenicity and carcinogenicity of some nitrosamines. *Mutation Res.*, **51**: 319–26.
Archer, M. C. & Labuc, G. E. (1982). On the mode of action of N-nitrosomethylbenzylamine, an esophageal carcinogen in the rat. *Banbury Report* No. 12, Cold Spring Harbor, 87–101.
Autrup, H. & Stoner, G. D. (1982). Metabolism of N-nitrosamines by cultured human and rat esophagus. *Cancer Res.* **42**: 1307–11.
Bachmann, W. E. & Deno, N. C. (1951). The nitrosation of hexamethylenetetramine and related compounds. *J. Amer. Chem. Soc.* **73**: 2777–9.
Bannasch, P. & Massner, B. (1976). Histogenesis und Cytogenese von Cholangiofibromen und Cholangiocarcinomen bei Nitrosomorpholin-vergiften Ratten. *Z. Krebsforsch.*, **87**: 239–55.
Bannasch, P. & Muller, H. A. (1964). Lichtmikroskopische Untersuchungen uber die Wirkung von N-Nitrosomorpholin auf die Leber von Ratte und Maus. *Arzneimittel-Forsch.*, **14**: 805–14.
Barnes, J. M. & Magee, P. N. (1954). Toxic properties of dimethylnitrosamine. *Brit. J. Ind. Med.*, **11**: 167–74.
Barrett, J. C., Hesterberg, T. W. & Thomassen, D. G. (1984). Use of cell transformation systems for carcinogenicity testing and mechanistic studies of carcinogenicity. *Pharmacol. Rev.*, **36**: 53S–70S.
Bartsch, H., Malaveille, C. & Montesano, R. (1975). *In vitro* metabolism and microsome-mediated mutagenicity of dialkylnitrosamines in rat, hamster and mouse tissues. *Cancer Res.*, **35**: 644–51.
Bartsch, H., Malaveille, C. & Montesano, R. (1976). The predictive value of tissue-mediated mutagenicity assays to assess the carcinogenic risk of chemicals. In: *Screening Tests in Chemical Carcinogenesis*, IARC Scientific Publication No. 12: 467–86.
Belinsky, S. A., Devereux, T. R., Maronpot, R. R., Stoner, G. D. & Anderson, M. W. (1989). Relationship between the formation of promutagenic adducts and the activation of the K-ras protooncogene in lung tumors from A/J mice treated with nitrosamines. *Cancer Res.*, **49**: 5305–11.
Belinsky, S. A., White, C. M., Devereux, T. R., Swenberg, J. A. & Anderson, M. W. (1987). Cell selective alkylation of DNA in rat lung following low dose exposure to the tobacco specific carcinogen 4-(N-methyl-N-nitrosamino)-1-(3-pyridyl)-1-butanone. *Cancer Res.*, **47**: 1143–8.
Bellander, T. & Osterdahl, B. G. (1983). Determination of N-mononitrosopiperazine and N,N'-dinitrosopiperazine in human urine, gastric juice and blood. *J. Chromatogr.*, **278**: 71–80.

Berenblum, I. & Shubik, P. (1947). A new quantitative approach to the study of the stages of chemical carcinogenesis in the mouse's skin. *Brit. J. Cancer*, **1**: 983–91.

Berman, L. D., Hayes, J. A. & Sibay, T. M. (1973). Effect of streptozotocin in the Chinese Hamster (*Cricetulus griseus*). *J. Natl. Cancer Inst.*, **51**: 1287–94.

Berman, J. J., Rice, J. M., Wenk, M. C. & Roller, P. P. (1979). Dependence of tumor spectrum on route of administration in Sprague-Dawley rats as a result of single or multiple injections of methyl(acetoxymethyl)nitrosamine. *J. Natl. Cancer Inst.*, **63**: 93–100.

Bigger, C. A. H., Tomaszewski, J. E., Dipple, A. & Lake, R. S. (1980). Limitations of metabolic activation systems used with *in vitro* tests of carcinogens. *Science*, **209**: 503–5.

Blattmann, L. & Preussmann, R. (1974). Biotransformation of carcinogenic dialkylnitrosamines. Additional urinary metabolites of Di-*n*-butyl- and Di-*n*-pentylnitrosamine. *Z. Krebsforsch.*, **81**: 75–8.

Blevins, R. D., Lijinsky, W. & Regan, J. D. (1977). Nitrosated methylcarbamate insecticides: Effect on the DNA of human cells. *Mutation Res.*, **44**: 1–7.

Bogovski, P. & Bogovski, S. (1981). Animal species in which *N*-Nitroso compounds induce cancer. *Internatl. J. Cancer*, **27**: 471–4.

Boveri, T. (1914). *The Origin of Malignant Tumors*, Jena, G. Fischer-Verlag.

Boyland, E., Roe, F. J. C., & Gorrod, J. W. (1964a). Induction of pulmonary tumours by nitrosonornicotine, a possible constituent of tobacco smoke. *Nature*, **202**: 1126.

Boyland, E., Roe, F. J. C., Gorrod, J. W. & Mitchley, B. V. C. (1964b). The carcinogenicity of nitrosoanabasine, a possible constituent of tobacco smoke. *Brit. J. Cancer*, **18**: 265–70.

Bronaugh, R. L., Congdon, E. R. & Scheuplein, R. J. (1981). The effect of cosmetic vehicles on the penetration of *N*-nitrosodiethanolamine through excised human skin. *J. Invest. Dermatol.*, **76**: 94–6.

Brookes, P. & Lawley, P. D. (1960). The reaction of mustard gas with nucleic acids *in vitro* and *in vivo*. *Biochem. J.*, **77**: 478–84.

Buckley, N. (1987). A regioselective mechanism for mutagenesis and oncogenesis caused by alkylnitrosourea sequence-specific DNA alkylation. *J. Amer. Chem. Soc.*, **109**: 7918–20.

Bulay, S., Mirvish, S. S., Garcia, H., Pelfrene, A. F., Gold, B. & Eagen, M. (1979). Carcinogenicity test of six nitrosamines and a nitrosocyanamide administered orally to rats. *J. Natl. Cancer Inst.*, **62**: 1523–8.

Camus, A., Bertram, B., Krüger, F. W., Malaveille, C. & Bartsch, H. (1976). Mutagenicity of β-oxidized *N*,*N*-di-*n*-propylnitrosoamine derivatives in *S. typhimurium* mediated by rat and hamster tissues. *Z. Krebsforsch.*, **86**: 293–302.

Cardy, R. H. & Lijinsky, W. (1980). Comparison of the carcinogenic effects of five nitrosamines in Guinea pigs. *Cancer Res.*, **40**: 1879–84.

Cardy, R. H., Lijinsky, W. & Hildebrandt, P. (1979). Neoplastic and non-neoplastic urinary bladder lesions induced in Fischer 344 rats and B6C3F1 hybrid mice by *N*-nitrosodiphenylamine. *Ecotoxicology & Environ. Safety*, **3**: 29–35.

Castonguay, A., Lin, D., Stoner, G. D., Radok, P., Furuya, K., Hecht, S. S., Schut, H. A. J. & Klaunig, J. E. (1983a). Comparative carcinogenicity in A/J mice and metabolism by cultured mouse peripheral lung of *N*-nitrosonornicotine, 4-(methylnitrosamino)-1-(3-pyridyl)-1-butanone and their analogues. *Cancer Res.*, **43**: 1223–9.

Castonguay, A., Stoner, G. B., Schut, H. A. J. & Hecht, S. S. (1983b).

Metabolism of tobacco-specific N-nitrosamines by cultured human tissues. *Proc. Natl. Acad. Sci. U.S.A.*, **80**: 6694–7.

Challis, B. C. (1973). Rapid nitrosation of phenols and its implications for health hazards from dietary nitrites. *Nature*, **244**: 466.

Challis, B. C. & Bartlett, C. D. (1975). Possible cocarcinogenic effects of coffee constituents. *Nature*, **254**: 532–3.

Challis, B. C. & Challis, J. A. (1982). N-Nitrosamines and N-nitrosimines. In: Patai, S., ed., *The Chemistry of Amino, Nitroso and Nitro Compounds and Their Derivatives*, Wiley, New York, pp. 1151–223.

Challis, B. C., Hopkins, A. R., Milligan, J. R., Mitchell, R. C. & Massey, R. C. (1984). Nitrosation of peptides. *IARC Scientific Publication* No. 57: 61–70.

Challis, B. C. & Kyrtopoulos, S. A. (1979). Nitrosation of amines by the two-phase interaction of amines in solution with gaseous oxides of nitrogen. *J. Chem. Soc. Perkin Trans. I*, 299–304.

Challis, B. C., Milligan, J. R. & Mitchell, R. C. (1984). Synthesis and stability of N-nitrosopeptides. *J. Chem. Soc. Chem. Commun.*, 1050–1.

Challis, B. C. & Osborne, M. R. (1973). The chemistry of nitroso-compounds. Part VI. Direct and indirect transnitrosation reactions of N-nitrosodiphenylamine. *J. Chem. Soc., Perkin II*, 1526–33.

Chow, Y. L. (1973). Nitrosamine photochemistry: Reactions of aminium radicals. *Accounts Chemical Research*, **6**: 354–60.

Chow, Y. L. & Colon, C. J. (1968). Nuclear magnetic resonance studies on the configuration and conformation of heterocyclic nitrosamines. *Can. J. Chem.*, **46**: 2827–33.

Chu, C. & Magee, P. M. (1981). Metabolic fate of nitrosoproline in the rat. *Cancer Res.*, **41**: 3653–7.

Chung, F.-L. & Hecht, S. S. (1983). Formation of cyclic $1,N^2$-adducts by reaction of deoxyguanosine with 2-acetoxy-N-nitrosopyrrolidine, 4-(carbethoxynitrosamino)butanol, or crotonaldehyde. *Cancer Res.*, **43**: 1230–5.

Chung, F.-L. & Hecht, S. S. (1985). Cyclic $1,N^2$-glyoxal-deoxyguanosine adducts. *Carcinogenesis*, **6**: 1671–3.

Chung, F.-L., Wang, M. & Hecht, S. S. (1989a). Detection of exocyclic guanine adducts in hydrolysates of hepatic DNA of rats treated with N-nitrosopyrrolidine and in calf thumus DNA reacted with α-acetoxy-N-nitrosopyrrolidine. *Cancer Res.*, **49**: 2034–41.

Chung, F.-L., Young, R. & Hecht, S. S. (1989b). Detection of cyclic $1,N^2$-propanodeoxyguanosine adducts in DNA of rats treated with N-nitrosopyrrolidine and mice treated with crotonaldehyde. *Carcinogenesis*, **10**: 1291–7.

Clive, D. (1983). A report of the U.S. Environmental Protection Agency Gene-Tox Program. *Mutation Res.*, **115**: 225–51.

Clive, D., Flamm, W. G., Machesko, M. R. & Bernheim, M. J. (1972). A mutational assay system using the thymidine kinase locus in mouse lymphoma cells. *Mutat. Res.*, **16**: 77–87.

Coulston, F. & Dunne, J. F. (1980). *The Potential Carcinogenicity of Nitrosatable Drugs*. Ablex Publishing Corp., Norwood, New Jersey.

Craddock, V. M. (1984). Repair and replication of DNA containing O^6-methylguanine in fetal and adult animal tissues in relation to their susceptibilities to cancer induction by N-nitroso-N-alkylureas. *IARC Scientific Publications* No. 57: 571–4.

von Dahn, H., Loewe, L. & Bunton, C. A. (1960). Über Die oxydation von Ascorbinsäure durch Saltpetrige Säure. Teil VI. Übersicht und Diskussion der Ergebnisse. *Helv. Chim. Acta*, **43**: 320–33.

Devik, O. G. (1967). Formation of nitrosamines by Maillard reaction. *Acta Chem. Scand.*, **21**: 2302–3.

Dickhaus, S., Reznik, G., Green, V. & Ketkar, M. (1978). The carcinogenic effect of β-oxidized dipropylnitrosamine in mice. II. 2-hydroxypropyl-*n*-propylnitrosamine and 2-oxopropyl-*n*-propylnitrosamine. *Z. Krebsforsch.*, **91**: 189–93.

DiPaolo, J. A. & Donovan, P. (1967). Properties of Syrian hamster cells transformed in the presence of carcinogenetic hydrocarbons. *Exp. Cell Res.*, **48**: 361–77.

DiPaolo, J. A., Nelson, R. L., Donovan, P. J. & Evans, C. H. (1973). Host mediated *in vivo – in vitro* assay for chemical carcinogenesis. *Arch. Pathol.*, **95**: 380–5.

Djerassi, C., Lund, E., Bunnenberg, E. & Sjöberg, B. (1961). Optical rotatory dispersion studies. XLVII. The nitroso chromophore. *J. Amer. Chem. Soc.*, **83**: 2307–12.

Dontenwill, W. (1968). Experimental studies on the organotropic effect of nitrosamines in the respiratory tract. *Food Cosmet. Toxicol.*, **6**, 571.

Dontenwill, W. & Mohr, U. (1961). Carcinome des Respirationstractus nach Behandlung von Goldhamstern mit Diäthylnitrosamin. *Z. Krebsforsch.*, **64**: 305–12.

Dontenwill, W., Mohr, U. & Zagel, M. (1962). Über die unterschiedliche Lungencarcinogene Wirkung des Diäthylnitrosamin bei Hamster und Ratten. *Z. Krebsforsch.*, **64**: 499–502.

Douglass, M. L., Kabacoff, B. L., Anderson, G. A. & Cheng, M. C. (1978). The chemistry of nitrosamine formation, inhibition and destruction. *J. Soc. Cosmetic Chemists*, **29**: 581–606.

Driver, H. E., White, I. N. H. & Butler, W. H. (1987). Dose–response relationships in chemical carcinogenesis: Renal mesenchymal tumours induced in the rat by single dose dimethylnitrosamine. *Br. J. Exp. Path.*, **68**: 133–43.

Druckrey, H. (1973). Chemical structure and action in transplacental carcinogenesis and teratogenesis. *IARC Scientific Publication* No. 4: 45–58.

Druckrey, H. (1979). Quantitative aspects in chemical carcinogenesis. *J. Envir. Path. & Tox.*, **3**: 60–78.

Druckrey, H. & Preussmann, R. (1962). Zur Entstehung carcinogener Nitrosamine am Belspiel des Tabakrauchs. *Naturwissenschaften*, **49**: 498–9.

Druckrey, H., Preussman, R., Ivankovic, S. & Schmähl, D. (1967). Organotrope carcinogene Wirkungen bei 65 verschiedenen *N*-Nitroso-Verbindungen an BD-Ratten. *Z. Krebsforsch.*, **69**: 103–201.

Druckrey, H., Schildbach, A., Schmähl, D., Preussman, R. & Ivankovic, S. (1963). Quantitative Analyse der carcinogen Wirkung von Diäthylnitrosamin. *Arzneimittel-Forschung*, **13**: 841–51.

Dubsky, J. V. & Spritzmann, M. (1916). Die Salzbildung der Nitroso-, Nitro- und Phënyliminodiessigsäure. *J. Prakt. Chem.*, **96**: 105–11.

Duden, P. & Scharff, M. (1895). Uber die Konstitution des Hexamethylentetramins. *Liebigs Annalen*, **288**: 218–21.

Dutton, A. H. & Heath, D. F. (1956). Demethylation of dimethylnitrosamine in rats and mice. *Nature*, **178**: 644.

Eisenbrand, G. (1984). Anticancer nitrosoureas: investigations on antineoplastic, toxic and neoplastic activities. *IARC Scientific Publication* No. 57: 695–708.

Eisenbrand, G., Denkel, E. & Pool, B. (1984). Alcoholdehydrogenase as an activating enzyme for *N*-nitrosodiethanolamine (NDELA). *In vitro* activation of NDELA to a potent mutagen in *Salmonella typhimurium*. *J. Cancer Res. Clin. Oncol.*, **108**: 76–80.

Eisenbrand, G., Fiebig, H. & Zeller, W. J. (1976). Some new congeners of the anticancer agent 1,3-bis (2-chloroethyl)-1-nitrosourea (BCNU). Synthesis of bifunctional analogues and water soluble derivatives and preliminary evaluation of their chemotherapeutic potential. *Z. Krebsforsch.*, **86**: 279–86.
Eisenbrand, G., Habs, M., Schmähl, D. & Preussmann, R. (1980). Carcinogenicity of *N*-nitroso-3-hydroxypyrrolidine and dose–response study with *N*-nitrosopiperidine in rats. *IARC Scientific Publication* No. 31: 657–66.
Eiter, K., Hebenbrock, K. & Kabbe, H. (1972). Neuoffenkettige und cyclische α-Nitrosaminoalkyather. *Justus Liebigs Ann. Chem.*, **765**: 55–77.
Elespuru, R. K. (1976). Mutagenicity of nitrosocarbaryl and other methylating nitrosamides is related to uptake in *Haemophilus influenzae*. *Environmental Mutagenesis*, **1**: 249–57.
Elespuru, R. K. (1984). Induction of bacteriophage lambda by *N*-nitroso compounds. *In: Genotoxicology of* N-*Nitroso Compounds*, 91–118.
Elespuru, R. K. (1991). Discrimination of mutagenic intermediates derived from alkylating agents by mutational patterns generated in *Escherichia coli*. *Carcinogenesis*, **18**: 1161–7.
Elespuru, R. K. & Lijinsky, W. (1973). The formation of carcinogenic *N*-nitroso compounds from nitrite and some types of agricultural chemicals. *Food Cosmet. Toxicol.*, **11**: 807–17.
Elespuru, R. K. & Lijinsky, W. (1976). Mutagenicity of cyclic nitrosamines in *E. coli* following activation with rat liver microsomes. *Cancer Res.*, **36**: 4099–101.
Elespuru, R. K., Lijinsky, W. & Setlow, J. K. (1974). Nitrosocarbaryl as a potent mutagen of environmental significance. *Nature*, **247**: 386–7.
Elespuru, R. K. & Yarmolinsky, M. (1979). A colorimetric assay of lysogenic induction designed for screening potential carcinogenic and carcinostatic agents. *Environmental Mutagenesis*, **1**: 65–78.
Emmett, G. C., Michejda, C. J., Sansone, E. B. & Keefer, L. K. (1980). Limitations of photodegradation in the decontamination and disposal of chemical carcinogens. *In:* D. B. Walters, ed., *Safe Handling of Chemical Carcinogens, Mutagens, Teratogens and Highly Toxic Substances*, Ann Arbor Science, pp. 535–53.
Emmons, W. D. (1954). Peroxytrifluoroacetic acid. I. The oxidation of nitrosamines to nitramines. *J. Amer. Chem. Soc.*, **76**: 3468–70.
Ender, F. & Ceh, L. (1968). Occurrence of nitrosamines in foodstuffs for human and animal consumption. *Food Cosmet. Toxicol.*, **6**: 569–71.
Ender, F., Havre, G., Helgebostad, A., Koppang, W., Madsen, R. & Ceh, L. (1964). Isolation and identification of a hepatotoxic factor in herring meal produced from sodium nitrite preserved herring. *Naturwissenschaften*, **51**: 637–8.
Fahmy, M. & Fahmy, O. (1984). Genetic activity of the carcinogen *N*-Nitrosodiethanolamine (NDELA) on unstable mutant alleles of the white locus in *Drosophila melanogaster*. *Teratogen. Carcinogen. Mutagen.*, **4**: 437–67.
Fajen, J. M., Carson, G. A., Rounbehler, D. P., Fan, T. Y., Vita, R., Goff, U. E., Wolf, M. H., Edwards, G. S., Fine, D. H., Reinhold, V. & Biemann, K. (1979). *N*-Nitrosamines in the rubber and tire industry. *Science* **205**: 1262–4.
Fan, T. Y., Goff, U., Song, L. & Fine, D. H. (1977*a*). *N*-nitrosodiethanolamine in cosmetics, lotions and shampoos. *Food Cosmet. Toxic.*, **15**: 423–30.
Fan, T. Y., Morrison, J., Rounbehler, D. P., Ross, R., Fine, D. H., Miles, W. & Sen, N. P. (1977*b*). *N*-nitrosodiethanolamine in synthetic cutting fluids: A part-per-hundred impurity. *Science*, **196**: 70–71.
Farrelly, J. G. & Hecker, L. I. (1984). The relationship between metabolism

and mutagenicity of two cyclic nitrosamines. *In: Genotoxicology of N-Nitroso Compounds*, (ed: T. K. Rao, W. Lijinsky & J. L. Epler), Plenum, New York, 167–87.

Farrelly, J. G., Saavedra, J. E., Kupper, R. J. & Stewart, M. L. (1987a). The metabolism of *N*-nitrosobis(2-oxopropyl)amine by microsomes and hepatocytes from Fischer 344 rats. *Carcinogenesis*, **8**: 1095–9.

Farrelly, J. G. & Stewart, M. L. (1982). The metabolism of a series of methylalkylnitrosamines. *Carcinogenesis*, **3**: 1299–302.

Farrelly, J. G., Stewart, M. L., Farnsworth, D. W. & Saavedra, J. E. (1988). Metabolism of the Z and E isomers of *N*-nitroso-*N*-methyl-(2-oxopropyl)amine by rat hepatocytes. *Cancer Res.*, **48**: 3347–9.

Farrelly, J. G., Stewart, M. L. & Lijinsky, W. (1984). The metabolism of nitrosodi-*n*-propylamine, nitrosodiallylamine and nitrosodiethanolamine. *Carcinogenesis*, **5**: 1015–19.

Farrelly, J. G., Stewart, M. L., Saavedra, J. E. & Lijinsky, W. (1982). Relationship between carcinogenicity and *in vitro* metabolism of nitrosomethylethylamine, nitrosomethyl-*n*-butylamine, and nitrosomethyl-(2-phenylethyl)amine labeled with deuterium in the methyl and α-methylene positions. *Cancer Res.*, **42**: 2105–9.

Farrelly, J. G., Thomas, B. J. & Lijinsky, W. (1987b). Metabolism and cellular interactions of *N*-nitrosodiethanolamine. *In: N-Nitroso Compounds: Relevance to Human Cancer*, eds. H. Bartsch, I. O'Neill & R. Schulte-Hermann. *IARC Scientific Publication* No. 84: 87–90.

Fazio, T., & Havery, D. C. (1982). Volatile *N*-nitrosamines in direct flame dried processed foods. *IARC Scientific Publication* No. 41: 277–86.

Fazio, T., Howard, J. W. & White, R. (1972). Multidetection method for analysis of volatile nitrosamines in food. *IARC Scientific Publication* No. 3: 16–30.

Fiala, E. S. (1977). Investigations into the metabolism and mode of action of the colon carcinogens 1,2-dimethylhydrazine and azoxymethane. *Cancer*, **40**: 2436–45.

Fiddler, W. (1973). Occurrence and determination of *N*-nitroso compounds. *J. Food Sci.*, **38**: 1084–90.

Fine, D. H., Ross, R., Rounbehler, D. P., Silverglied, A. & Song, L. (1977). Formation *in vivo* of volatile *N*-nitrosamines in man after ingestion of cooked bacon and spinach. *Nature*, **265**: 753–5.

Fine, D. H., Rounbehler, D. P., Sawicki, E., Krost, K. & DeMarrais, G. A. (1976). *N*-Nitroso compounds in the ambient community air of Baltimore, MD. *Analyt. Letters*, **9**: 595–604.

Fine, D. H. & Rufeh, F. (1974). Description of the thermal energy analyser for *N*-nitroso compounds. *IARC Scientific Publication* No. 9: 40–4.

Fine, D. H., Rufeh, F. & Lieb, D. (1974). Group analysis of volatile and non-volatile *N*-nitroso compounds. *Nature*, **249**: 309–10.

Fong, L. Y. Y. & Chan, W. C. (1976). Methods for limiting the content of dimethylnitrosamine in Chinese marine salt fish. *Food Cosmet. Toxicol.*, **14**: 95–8.

Fong, L. Y. Y., Jensen, D. E. & Magee, P. N. (1990). DNA methyl-adduct dosimetry and O^6-alkylguanine-DNA alkyltransferase activity determinations in rat mammary carcinogenesis by procarbazine and *N*-methylnitrosourea. *Carcinogenesis*, **11**: 411–17.

Fong, L. Y. Y., Sivak, A. & Newberne, P. M. (1978). Zinc-deficiency and methylbenzylnitrosamine-induced esophageal cancer in rats. *J. Natl. Cancer Inst.*, **61**: 145–50.

Foulds, L. (1969). *Neoplastic Development*, Academic Press, London.

Frank, N. & Wiessler, M. (1983). Alkyldiazohydroxides are stable intermediates in the degradation of N-nitroso-(acetoxyalkyl)-alkylamines in rat serum. *Carcinogenesis*, **4**: 751–3.

Frantz, C. N. & Malling, H. V. (1975). Factors affecting metabolism and mutagenicity of dimethylnitrosamine and diethylnitrosamine. *Cancer Res.*, **35**: 2307–14.

Fraser, R, R., Grindley, T. B. & Passannanti, S. (1975). Stereoselectivity in the reactions of nitrosopiperidine carbanions. Steric vs. stereoelectronic controls. *Can. J. Chem.*, **53**: 2473–80.

Freund, H. A. (1937). Clinical manifestations and studies in parenchymatous hepatitis. *Ann. Int. Med.*, **10**: 1144–55.

Fridman, A. L., Mukhametshin, F. M. & Novikov, S. S. (1971). Advances in the chemistry of aliphatic nitrosamines. *Russian Chemical Rev.*, **40**: 34–50.

Fussgaenger, R. D. & Ditschunheit, H. (1980). Lethal exitus of a patient with N-nitrosodimethylamine poisoning, 2.5 years following the first ingestion and signs of intoxication. *Oncology*, **37**: 273–7.

Gaffield, W., Keefer, L. & Lijinsky, W. (1972). Chiroptical properties of nitrosamino acids and their relationhip to the nitrosamine sector rule. *Tetrahedron Lett.*, **9**: 779–82.

Gaffield, W., Lundin, R. E. & Keefer, L. K. (1981). Chiroptical properties of N-nitrosopyrrolidines and N-nitrosamino acids. Implications for the nitrosamine sector rule. *Tetrahedron*, **37**: 1861–9.

Gangolli, S. D., Shilling, W. H. & Lloyd, A. G. (1974). A method for the destruction of nitrosamines in solution. *Food Cosmet. Toxicol.*, **12**: 168–72.

Garcia, H., Keefer, L., Lijinsky, W. & Wenyon, C. E. M. (1970). Carcinogenicity of nitrosothiomorphine and 1-nitrosopiperazine in rats. *Z. Krebsforsch.*, **74**: 179–84.

Garcia, H. & Lijinsky, W. (1972). Tumorigenicity of five cyclic nitrosamines in MRC rats. *Z. Krebsforsch.*, **77**: 257–61.

Garcia, H. & Lijinsky, W. (1973). Studies on the tumorigenic effect in feeding of nitrosamino acids and of low concentrations of amines and nitrite to rats. *Z. Krebsforsch.*, **79**: 141–4.

Garro, A. J., Sfitz, H. K. & Lieber, C. S. (1981) Enhancement of dimethylnitrosamine metabolism and activation to a mutagen following chronic ethanol consumption. *Cancer Res.*, **41**: 120–4.

Geuther, A. (1863). Ueber die Einwirkung von salpetrigsaurem Kali auf salzsaures Diäthylamin. *Liebigs Annalen*, **128**: 151–6.

Gichner, T. & Veleminsky, J. (1986). Organic solvents inhibit the mutagenicity of promutagens dimethylnitrosamine and methylbutylnitrosamine in a higher plant. *Arabidopsis thaliana. Mutagenesis*, **1**: 107–9.

Gingell, R., Brunk, G., Nagel, D. & Pour, P. (1979). Metabolism of three radiolabeled pancreatic carcinogenic nitrosamines in hamsters and rats. *Cancer Res.*, **39**: 4579–83.

Gingell, R., Wallcave, L., Nagel, D., Kupper, R. & Pour, P. (1976). Common metabolites of N-nitroso-2,6-dimethylmorpholine and N-nitroso-bis-(2-oxopropyl)amine in the Syrian hamster. *Cancer Lett.*, **2**: 47–52.

Goff, E. V. & Fine, D. H. (1979). Analysis of voltatile N-nitrosamines in alcoholic beverages. *Food Cosmet. Toxicol.*, **17**: 569–73.

Gold, B. & Hines, L. (1984). Evidence for the penetration of the nuclear envelope by N-nitrosomethylhydroxymethylamine. *IARC Scientific Publication* No. 57: 453–8.

Gold, B., Salmasi, S., Linder, W. & Althoff, J. (1981). Biological and

chemical studies involving methyl-*t*-butylnitrosamine, a non-carcinogenic nitrosamine. *Carcinogenesis*, **2**: 529–32.
Goodall, C. M. & Kennedy, T. H. (1976). Carcinogenicity of dimethylnitramine in NZR rats and NZO mice. *Cancer Lett.*, **1**: 295–8.
Goodall, C. M. & Lijinsky, W. (1976). Oncogenicity tests of *p*-nitroso-*N*,*N*-dimethylaniline and *p*-nitroso-*N*,*N*-diethylaniline in NZR rats and NZO mice. *Pathology*, **8**: 143–9.
Goodall, C. M. & Lijinsky, W. (1984*a*). Carcinogenicity of nitrosododecamethyleneimine in NZR/Gd inbred rats. *Carcinogenesis*, **5**: 537–40.
Goodall, C. M. & Lijinsky, W. (1984*b*). Strain and sex differences in *N*-nitrosohexamethyleneimine carcinogenesis in NZB, NZC, NZO and NZY mice. *J. Natl. Cancer Inst.*, **73**: 1215–18.
Goodall, C. M. & Lijinsky, W. (1985). Carcinogenesis by *N*-nitrosohexamethyleneimine in NZO inbred mice. *Toxicology*, **33**: 251–9.
Goodall, C. M., Lijinsky, W., Keefer, L. & D'Ath, E. F. (1973). Oncogenic activity of *N*-nitrosododecamethyleneimine in liver, glandular stomach and other tissue of NZO/B1 mice. *Internat. J. Cancer*, **11**: 369–76.
Goodall, C. M., Lijinsky, W., Tomatis, L. & Wenyon, C. E. M. (1970). Toxicity and oncogenicity of nitrosomethylaniline and nitrosomethylcyclohexylamine. *Toxicol. Appl. Pharmacol.*, **17**: 426–32.
Goth, R. & Rajewsky, M. J. (1974). Molecular and cellular mechanisms associated with pulse-carcinogenesis in the rat nervous system by ethylnitrosourea: Ethylation of nucleic acids and elimination rates of ethylated bases from the DNA of different tissues. *Z. Krebsforsch.*, **82**: 37–64.
Gough, T. A., McPhail, M. F., Webb, K. S., Wood, B. J. & Coleman, R. F. (1977). An examination of some foods for the presence of volatile nitrosamines. *J. Sci. Food Agric.*, **28**: 345–51.
Graffi, A., Hoffmann, F. & Schütt, M. (1967). *N*-Methyl-*N*-nitrosourea as a strong topical carcinogen when painted on the skin of rodents. *Nature*, **214**: 611.
Green, L. C., Wagner, D. A., Ruiz de Luzuriaga, K., Rand, W., Istean, N., Young, V. R. & Tannenbaum, S. R. (1981). Nitrate biosynthesis in man. *Proc. Natl. Acad. Sci. U.S.A*, **78**: 7764–8.
Green, U. & Althoff, J. (1982). Carcinogenicity of vinylethylnitrosamine in Syrian Golden hamsters. *J. Cancer Res., Clin., Oncol.*, **102**: 227–33.
Green, U., Konishi, Y., Ketkar, M. B. & Althoff, J. (1980). Complete study of the carcinogenic effect of BHP and BAP on NMR1 mice. *Cancer Lett.*, **9**: 257–61.
Greenblatt, M., Kommineni, V., Conrad, E., Wallcave, L. & Lijinsky, W. (1972). *In vivo* conversion of phenmetrazine into its *N*-nitroso derivative. *Nature New Biology*, **236**: 25–6.
Greenblatt, M. & Lijinsky, W. (1972*a*). Failure to induce tumors in Swiss mice after concurrent administration of amino acids and sodium nitrite. *J. Natl. Cancer Inst.*, **48**: 1389–92.
Greenblatt, M. & Lijinsky, W. (1972*b*). Nitrosamine studies: Neoplasms of liver and genital mesothelium in nitrosopyrrolidine treated MRC rats. *J. Natl. Cancer Inst.*, **48**: 1687–96.
Greenblatt, M. & Lijinsky, W. (1974). Carcinogenesis and chronic toxicity study of nitrilotriacetic acid (NTA) in Swiss mice. *J. Natl. Cancer Inst.*, **52**: 1123–6.
Greenblatt, M., Mirvish, S. S. & So, B. T. (1971). Nitrosamine studies: Induction of lung adenomas by concurrent administration of sodium nitrite and secondary amines in Swiss mice. *J. Natl, Cancer Inst.*, **46**: 1029–34.

Griciute, L., Castegnaro, M. & Bereziat, J. C. (1981). Influence of ethyl alcohol on carcinogenesis with N-nitrosodimethylamine. *Cancer Lett.*, **13**: 345–52.

Gullino, P. M., Pettigrew, H. M. & Grantham, F. H. (1975). N-Nitrosomethylurea as mammary gland carcinogen in rats. *J. Natl. Cancer Inst.*, **54**: 401–14.

Guttenplan, J. B. (1980). Enhanced mutagenic activities of N-nitroso compounds in weakly acidic media. *Carinogenesis*, **1**: 439–44.

Guttenplan, J. B. (1984). Effects of pH and structure on the mutagenic activity of N-nitroso compounds. In: *Genotoxicology of* N-*Nitroso Compounds*, ed. T. K. Rao, W. Lijinsky & J. L. Epler, Plenum Press, New York, pp. 59–90.

Guttenplant, J. B. (1989). An important role for cytosol in the microsomal metabolism of N-nitrosodimethylamine to a mutagen: evidence for two different mutagenic metabolites. *Cancer Lett.*, **47**: 63–7.

Güttner, J., Schmidt, A. & Jungstand, W. (1971). Unterschiedliche Onkogenität von N-Methyl-N-nitroso-beta-D-glucosylamin, N-Methyl-N-nitroso-beta-D-galactosylamin und entsprechenden Zuckeralkoholen bei Ratten. *Z. Krebsforsch.*, **75**: 296–300.

Haas, H., Mohr, U. & Krüger, F. W. (1973). Comparative studies with different doses of N-nitrosomorpholine, N-nitrosopiperidine, N-nitrosomethylurea and dimethylnitrosamine in Syrian Golden hamsters. *J. Natl. Cancer Inst.*, **51**: 1295–301.

Habs, M., Habs, H. & Schmähl, D. (1980). Effect of the intermittent administration of N-nitrosopyrrolidine (NPYR) on the tumor incidence in Sprague-Dawley rats. *Int. J. Cancer*, **26**: 47–51.

Hadidian, Z., Frederickson, T. N., Weisburger, E. K., Weisburger, J. H., Glass, R. M. & Mantel, N. (1968). Tests for chemical carcinogens. Report on the activity of derivatives of aromatic amines, nitrosamines, quinolines, nitroalkanes, amides, epoxides, aziridines, and purine antimetabolites. *J. Natl. Cancer Inst.*, **41**: 985–1036.

Hanna, C. & Schueler, F. W. (1952). The reaction of disubstituted nitrosamines with lithium aluminum hydride. *J. Amer. Chem. Soc.*, **74**: 3693–4.

Hard, G. C. & Butler, W. H. (1970). Cellular analysis of renal neoplasia: Light microscopic study of the development of interstitial lesions induced in the rat kidney by a single carcinogenic dose of dimethylnitrosamine. *Cancer Res.*, **30**: 2806–15.

Hard, G. C. & Butler, W. H. (1971). Morphogenesis of epithelial neoplasms induced in the rat kidney by dimethylnitrosamine. *Cancer Res.*, **31**: 1496–505.

Harington, J. S., Nunn, J. R. & Irwig, L. (1973). Dimethylnitrosamine in the human vaginal vault. *Nature*, **241**: 491–50.

Harris, C. C., Autrup, H., Stoner, G. D., McDowell, E. M., Trump, B. F. & Schafer, P. W. (1977). Metabolism of acyclic and cyclic N-nitrosamines in cultured human bronchi. *J. Natl. Cancer Inst.*, **59**: 1401–6.

Harris, C. C., Autrup, H., Stoner, G. D., Trump, B. F., Hillman, E., Schafer, P. W. & Jeffrey, A. M. (1979). Metabolism of benzo(a)pyrene, N-nitrosodimethylamine, and N-nitrosopyrrolidine and identification of the major carcinogen–DNA adducts formed in cultured human esophagus. *Cancer Res.*, **39**: 4401–4.

Harris, C. C., Kaufman, D. G., Sporn, M. B., Smith, J. M. & Saffiotti, U. (1973). Ultrastructural effects of N-methyl-N-nitrosourea on the tracheobronchial epithelium of the Syrian Golden Hamster. *Int. J. Cancer*, **12**: 259–83.

Hashimoto, Y., Suzuki, K. & Okada, M. (1972). Induction of urinary bladder tumors in ACI/N rats by butyl (3-carboxypropyl) nitrosamine, a major urinary metabolite of butyl(4-hydroxybutyl)-nitrosamine. *GANN*, **63**: 637–8.

Hashimoto, Y., Suzuki, K. & Okada, M. (1974). Induction of urinary bladder tumors by intravesicular instillation of butyl-(4-hydroxybutyl)nitrosamine and its principal urinary metabolite, butyl-(3-carboxypropyl)nitrosamine in rats. *GANN*, **65**: 69–71.

Hawksworth, G. M. & Hill, M. J. (1974). The in vivo formation of N-nitrosamines in the rat bladder and their subsequent absorption. *Brit. J. Cancer*, **29**: 353–8.

He, C. & Farrelly, J. G. (1989). Microsomal metabolism of the Z and E isomers of N-nitroso-N-methyl-N-n-pentylamine. *Cancer Lett.*, **45**: 189–94.

Heath, D. F. (1961). Mechanism of the hepatotoxic action of dialkynitrosamines. *Nature*, **192**: 170.

Heath, D. F. & Dutton, A. (1958). The detection of metabolic products from dimethylnitrosamine in rats and mice. *Biochem. J.*, **70**: 619–26.

Heath, D. F. & Jarvis, J. A. E. (1955). The polarographic determination of dimethylnitrosamine in animal tissue. *Analyst*, **80**: 613–16.

Hecht, S. S. (1984). N-Nitroso-2-hydroxymorpholine, a mutagenic metabolite of N-nitrosodiethanolamine. *Carcinogenesis*, **5**: 1745–47.

Hecht, S. S., Chen, S. B., McCoy, G. D. & Hoffman, D. (1979). 2-Hydroxylation of N-nitrosopyrrolidine and N'-nitrosonornicotine by human liver microsomes. *Cancer Lett.*, **8**: 35–41.

Hecht, S. S., Chen, C. B., Ohmori, T. & Hoffmann, D. (1980). Comparative carcinogenicity in F344 rats of the tobacco specific nitrosamines, N'-nitrosonornicotine and 4-(N-methyl-N-nitrosamino)-1-(3-pyridyl)-1-butanone. *Cancer Res.*, **40**, 298–302.

Hecht, S. S. & Kozarich, J. W. (1973). Mechanism of the base-induced decomposition of N-nitroso-N-methylurea. *J. Org. Chem.*, **38**: 1821–4.

Hecht, S. S., Lijinsky, W., Kovatch, R. M., Chung, F.-L. & Saavedra, J. E. (1989). Comparative tumorigenicity of N-nitroso-2-hydroxymorpholine, N-nitrosodiethanolamine, and N-nitrosomorpholine in A/J mice and F344 rats. *Carcinogenesis*, **10**: 1475–7.

Hecht, S. S., Lin, D., Castonguay, A. & Rivenson, A. (1987). Effects of α-Deuterium substitution on the tumorigenicity of 4-(methylnitrosamino)-(3-Pyridyl)-1-Butanone in F344 rats. *Carcinogenesis*, **8**: 291–4.

Hecht, S. S., McCoy, G. D., Chen, C. B. & Hoffmann, D. (1981). The metabolism of cyclic nitrosamines. In: *N-Nitroso Compounds, ACS Symposium Series* **174**, 49–75.

Hecht, S. S., Morrison, J. B. & Wenninger, J. A. (1982a). N-nitroso-N-methyldodecylamine and N-nitroso-N-methyltetradecylamine in hair-care products. *Food Chem. Toxicol.*, **20**: 165–70.

Hecht, S. S., Reiss, B., Lin, D. & Williams, G. M. (1982b). Metabolism of N-nitrosonornicotine by cultured rat esophagus. *Carcinogenesis*, **3**: 453–6.

Hecht, S. S., Trushin, N., Castonguay, A. & Rivenson, A. (1986). Comparative carcinogenicity and DNA methylation in F344 rats by 4-(methylnitrosamino)-1-(3-pyridyl)-1-butanone and N-nitrosodimethylamine. *Cancer Res.*, **46**: 498–502.

Hecht, S. S., Young, R., Rivenson, A. & Hoffmann, D. (1982). On the metabolic activation of N-nitrosomorpholine and N'-nitrosonornicotine: Effects of deuterium substitution. *IARC Scientific publication* No. 41: 499–507.

Hedler, L. & Marquardt, P. (1974). Determination of volatile N-nitroso compounds, and particularly of nitrosodimethylamine, nitrosodiethylamine and nitrosodipropylamine, in soya bean oil: Effect of oil storage period on recovery rates. *IARC Scientific Publication* No. 9: 183–91.

Hein, G. E. (1963). Reactions of tertiary amines with nitrous acid. *J. Chem. Education*, **40**: 181–4.
Hemminki, K. (1982). Dimethylnitrosamine adducts excreted in rat urine. *Chem. Biol. Interactions*, **39**: 139–48.
Herr, R. R., Jahnke, H. K. & Arogoudelis, A. D. (1967). The structure of streptozotocin. *J. Amer. Chem. Soc.*, **89**: 4808–9.
Herrold, K. M. (1970). Upper respiratory tract tumors induced in Syrian hamsters by N-methyl-N-nitrosourea. *Int. J. Cancer*, **6**: 217–48.
Herron, D. C. & Shank, R. C. (1979). Quantitative high pressure liquid chromatographic analysis of methylated purines in DNA of rats treated with chemical carcinogens. *Analyt. Biochem.*, **100**: 58–63.
Herron, D. C. & Shank, R. C. (1980). Methylated purines in human liver DNA after probable dimethylnitrosamine poisoning. *Cancer Res.*, **40**: 3116–17.
Heyns, K. & Koch, H. (1971). Zur Frage der Entstehung von Nitrosaminen bei der Reaktion von Monosacchariden mit Aminosäuren (Maillard-Reaktion). *Z. Lebensmitt.-Untersuch.*, **145**: 576–84.
Heyns, K. & Röper, H. (1974). Gas chromatographic trace analysis of volatile nitrosamines in various types of wheat flour after application of different nitrogen fertilisers to the wheat. *IARC Scientific Publication* No. 9: 166–72.
Hiasa, Y., Ohshima, M., Iwata, C. & Tanikake, T. (1979). Histopathological studies on renal tubular cell tumors in rats treated with N-ethyl-N-hydroxyethylnitrosamine. *GANN*, **70**: 817–20.
Hicks, R. M. & Wakefield, J. S. (1972). Rapid induction of bladder cancer in rats with N-methyl-N-nitrosourea. I. Histology. *Chem. Biol. Interact.*, **5**: 139–52.
Hilfrich, J., Hecht, S. S. & Hoffmann, D. (1977). A study of tobacco carcinogenesis. XV. Effects of N'-nitrosonornicotine and N'-nitrosoanabasine in Syrian Golden Hamsters. *Cancer Lett.*, **2**: 169–75.
Hilfrich, J., Schmeltz, J. & Hoffmann, D. (1977). Effects of N-nitrosodiethanolamine and 1,1-diethanolhydrazine in Syrian Golden Hamsters. *Cancer Lett.*, **4**: 55–60.
Hiraki, S. (1971). Induction of malignant lymphomas by N,N'-dimethylnitrosourea in adult mice. *GANN*, **62**: 135–7.
Hirota, N., Aonuma, T., Yamada, S., Kawai, T., Saito, K. & Yokoyama, T. (1987). Selective induction of glandular stomach carcinomas in F344 rats by N-methyl-N-nitrosourea. *GANN*, **78**: 634–8.
Ho, T., San Sebastian, J. R. & Hsie, A. W. (1984). Mutagenic activity of nitrosamines in mammalian cells: Study with the CHO/HGPRT and human leukocyte SCE assays. *In: Genotoxicology of* N-*Nitroso Compounds*, T. K. Rao et al., eds., Plenum, pp. 129–47.
Hodgson, R., Swann, P. F., Clothier, R. & Balls, M. (1980). The persistence in *Xenopus laevis* DNA of O-6-methylguanine produced by exposure to N-methyl-N-nitrosourea. *Eur. J. Cancer*, **16**: 481–6.
Hoffman, D., Adams, J. D., Brunnemann, K. D. & Hecht, S. S. (1979). Assessment of tobacco-specific N-nitrosamines in tobacco products. *Cancer Res.*, **39**, 2505–9.
Hoffmann, D., Brunnemann, K. D., Adams, J. D. & Hecht, S. S. (1984a). Formation and analysis of N-nitrosamines in tobacco products and their endogenous formation in consumers. *IARC Scientific Publication* No. 57: 743–62.
Hoffmann, D., Brunnemann, K. D., Adams, J. D., Rivenson, A. & Hecht, S. S. (1982). N—nitrosamines in tobacco carcinogenesis. *Banbury Report* No. 12, 211–25, Cold Spring Harbor, New York.

Hoffmann, D., Castonguay, A., Rivenson, A. & Hecht, S. S. (1981).
Comparative carcinogenicity and metabolism of 4-(methylnitrosamino)-1-(3-pyridyl)-1-butanone and N'-nitrosonornicotine in Syrian Golden Hamster. *Cancer Res.*, **41**: 2386–93.

Hoffmann, D. & Hecht, S. S. (1985). Nicotine-derived N-nitrosamines and tobacco-related cancers: Current status and future directions. *Cancer Res.*, **45**: 935–42.

Hoffmann, D., Raineri, R., Hecht, S. S., Maronpot, R. & Wynder, E. L. (1975). A study of tobacco carcinogenesis. XIV. Effects of N^1-nitrosonornicotine and N^1-nitrosoanabasine in rats. *J. Natl. Cancer Inst.*, **55**: 977–81.

Hoffmann, D., Rivenson, A., Adams, J. D., Juchatz, A., Vinchkoski, N. & Hecht, S. S. (1983). Effect of route of administration and dose on the carcinogenicity of N-nitrosodiethanolamine in the Syrian Golden Hamster. *Cancer Res.*, **43**: 2521–4.

Hoffmann, D., Rivenson, A., Amin, S. & Hecht, S. S. (1984*b*). Dose–response study of the carcinogenicity of tobacco-specific nitrosamines in F344 rats. *J. Cancer Res. Clin Oncol.*, **108**: 81–6.

Hotchkiss, J. H., Libbey, L. M., Barbour, J. F. & Scanlan, R. A. (1980). Combination of a GC-TEA and a GC-MS-data system for the μg/kg estimation and confirmation of volatile N-nitrosamines in foods. *IARC Scientific Publication* No. 31: 361–76.

Hotchkiss, J. H., Scanlan, R. A., Lijinsky, W. & Andrews, A. W. (1979). Mutagenicity of nitrosamines formed from nitrosation of spermidine. *Mutat. Res.*, **68**: 195–9.

Hsie, A. W., Machanoff, R., Couch, D. B. & Holland, J. M. (1978). Mutagenicity of dimethylnitrosamine and ethylmethane-sulfonate as determined by the host-mediated CHO/HGPRT assay. *Mutat. Res.*, **51**: 77–84.

Huang, D. P., Ho, J. H. C., Webb, K. S., Wood, B. J. & Gough, T. A. (1981). Volatile nitrosamines in salt-preserved fish before and after cooking. *Food Cosmet. Toxicol.*, **19**: 167–71.

Huberman, E. & Sachs, L. (1974). Cell-mediated mutagenesis of mammalian cells with chemical carcinogens. *Internat. J. Cancer*, **13**: 326–33.

Huisgen, R. & Reimlinger, H. (1956). Nitroso-acyl-amine und Diazo-ester X. Die Isomerisierung der Nitroso-acyl-alkylamine zu Diazo-estern und ihre Kinetik. *Liebigs Ann. Chem.*, **599**: 162–82.

IARC Monograph (1982*a*). *The Rubber Industry*. No. 28

IARC Monograph (1982*b*). *Some Industrial Chemicals and Dyestuffs*. No. 29: 345–89.

Issenberg, P., Conrad, E. E., Nielsen, J. W., Klein, D. A. & Miller, S. E. (1984). Determination of N-nitrosobis(2-hydroxypropyl)amine in environmental samples. *IARC Scientific Publication* No. 57: 43–50.

Ivankovic, S. (1978). Carcinogene Wirkung von N-Benzyl-N-Nitrosoharnstoff (BzNH) an BD-Ratten. *Z. Krebsforsch.*, **91**: 63–7.

Ivankovic, S. (1979). Teratogenic and carcinogenic effects of some chemicals during prenatal life in rats, Syrian Golden Hamsters, and minipigs. *In: Perinatal Carcinogenesis, National Cancer Institute Monograph* No. 51, 103–15.

Ivankovic, S. & Druckrey, H. (1968). Transplacentare Erzeugung Maligner Tumoren Des Nervensystems. I. Äthylnitroso-Harnstoff an BD IX-Ratten. *Z. Krebsforsch.*, **71**: 320–60.

Ivankovic, S., Druckrey, H. & Preussmann, R. (1966). Erzeugung Neurogener Tumoren bei den Nachkommen nach einmaliger Injektion von Athylnitrosoharnstoff an schwangere Ratten. *Naturwiss.*, **53**: 410.

Ivankovic, S., Klimpel, F., Wiessler, M. & Preussmann, R. (1981). Carcinogenicity in BD-IX rats of 7 homologues of N-nitroso-N-n-alkylureas in different stages of postnatal development. *Arch. Geschwulstforsch.*, **51**: 187–203.

Iverson, O. H. (1980). Tumorigenicity of N-nitroso-diethyl, -dimethyl and -diphenyl-amines in skin painting experiments. *Eur. J. Cancer*, **16**: 695–8.

Iversen, P. E. (1971). Organic electrosyntheses. III. Reduction of N-nitrosamines. *Acta Chem. Scand.*, **25**: 2337–44.

Iwoaka, W. & Tannenbaum, S. R. (1976). Photohydrolytic detection of N-nitroso compounds in high-performance liquid chromatography. *IARC Scientific Publication* No. 14: 51–6.

Janzowski, C., Eisenbrand, G. & Preussmann, R. (1978). Occurrence and determination of N-nitroso-3-hydroxypyrrolidine in cured meat products. *J. Chromatog.*, **150**: 216–20.

Jensen, D. E. (1983). Denitrosation as a determinant of nitrosocimetidine in vivo activity. *Cancer Res.*, **43**: 5258–67.

Ji, C., Mirvish, S. S., Nickols, J., Ishizaki, H., Lee, N. J. & Yang, C. S. (1989). Formation of hydroxy derivatives, aldehydes, and nitrite from N-nitrosomethyl-n-amylamine by rat liver microsomes and by purified cytochrome P-450. *Cancer Res.*, **49**: 5299–304.

Jones, A. R., Lijinsky, W. & Singer, G. M. (1974). Steric effects in the nitrosation of piperidines. *Cancer Res.*, **34**: 1079–81.

Jones, C. A., Marlino, P. J., Lijinsky, W. & Huberman, E. (1981). The relationship between the carcinogenicity and mutagenicity of nitrosamines in a hepatocyte mediated mutagenicity assay. *Carcinogenesis*, **2**: 1075–7.

Kamm, J. J., Dashman, T. Conney, A. H. & Burns, J. J. (1973). Protective effect of ascorbic acid on hepatotoxicity caused by nitrite plus aminopyrine. *Proc. Natl. Acad. Sci. U.S.A.*, **70**: 747–9.

Karabatsos, G. J. & Taller, R. A. (1964). Structural studies by nuclear magnetic rsonance. IX. Configurations and conformations of N-nitrosamines. *J. Amer. Chem. Soc.*, **86**: 4373–8.

Kawabata, T., Matsui, M., Ishibashi, T., Hamano, M. & Ino, M. (1982). Formation of N-nitroso compounds during cooking of Japanese food. *IARC Scientific Publication* No. 41: 287–97.

Kawabata, T., Uibu, J., Ohshima, H., Matsui, M., Hamano, M. & Tokiwa, H. (1980). Occurrence, formation and precursors of N-nitroso compounds in the Japanese diet. *IARC Scientific Publication* No. 31: 481–92.

Keefer, L. K., Anjo, T., Wade, D., Wang, T. & Yang, C. S. (1987). Concurrent generation of methylamine and nitrite during denitrosation of N-nitrosodimethylamine by rat liver microsomes. *Cancer Res.*, **47**: 447–52.

Keefer, L. K. & Fodor, C. H. (1970). Facile hydrogen isotope exchange as evidence for an α-nitrosamino carbanion. *J. Am. Chem. Soc.*, **92**: 5747–8.

Keefer, L., Lijinsky, W. & Garcia, H. (1973). Deuterium isotope effect on the carcinogenicity of dimethylnitrosamine in rat liver. *J. Natl. Cancer Inst.*, **51**: 299–302.

Keefer, L. K. & Roller, P. P. (1973). N-Nitrosation by nitrite ion in neutral and basic medium. *Science*, **181**: 1245–7.

Ketkar, M. B., Althoff, J. & Lijinsky, W. (1981). The carcinogenic effect of nitrosomethyldodecylamine in European Hamsters. *Cancer Lett.*, **13**: 165–8.

Ketkar, M. B., Holste, J., Preussmann, R. & Althoff, J. (1983). Carcinogenic effect of nitrosomorpholine administered in the drinking water to Syrian Golden Hamsters. *Cancer Lett.*, **17**: 333–8.

Khoda, K., Hakuru, A., Ninmomiaya, S. & Kawazoe, Y. (1986). Spontaneous and enzymatic activations of carcinogenic N-butyl-N-dimethyl-N-nitrosourea. *Chem. Pharm. Bull.*, **34**: 5056–62.

Kleihues, P., Hodgson, R. M., Veit, C., Schweinsberg, F. & Wiessler, M. (1983). DNA modification and repair *in vivo*: towards a biochemical basis of organ-specific carcinogenesis by methylating agents. In: *Organ and Species Specificity in Chemical Carcinogenesis*, ed. R. Langenbach, S. Nesnow & J. M. Rice, Plenum Press, New York, pp. 509–29.

Kleihues, P. & Wiestler, O. D. (1984). Involvement of thiols in gastric cancer induced by N-methyl-N'-nitro-N-nitrosoguanidine: Biochemical and autoradiographic studies. *IARC Scientific Publication* No. 57: 603–8.

Koepke, S. R., Creasia, D. R., Knutsen, G. L. & Michejda, C. J. (1988a). Carcinogenicity of hydroxyalkylnitrosamines in F344 rats: Contrasting behavior of β- and γ-hydroxylated nitrosamines. *Cancer Res.*, **48**: 1533–6.

Koepke, S. R., Kroeger-Koepke, M. B., Bosan, W., Thomas, B. J., Alvord, W. G. & Michejda, C. J. (1988b). Alkylation of DNA in rats by N-nitrosomethyl-(2-hydroxyethyl)amine: Dose response and persistence of the alkylated lesion *in vivo*. *Cancer Res.*, **48**: 1537–42.

Koepke, S. R., Kroeger-Koepke, M. B. & Michejda, C. J. (1991). The mechanism of DNA binding by the esophageal carcinogen, N-nitroso-N-methylaniline. *IARC Scientific Publication* No. 105, 346–50.

Koepke, S. R., Kupper, R. & Michejda, C. J. (1979). Unusually facile solvolysis of primary tosylates: A case for participation by the N-nitroso group. *J. Org. Chem.*, **44**: 2718–22.

Kokinnakis, D. M., Hollenberg, P. F. & Scarpelli, D. G. (1984). Metabolism of the *cis* and *trans* isomers of N-nitroso-2,6-dimethylmorpholine and their deuterated analogs by liver microsomes of rat and hamster. *Carcinogenesis*, **5**: 1009–14.

Kokinnakis, D. M., Hollenberg, P. F. & Scarpelli, D. G. (1985). Major urinary metabolites in hamsters and rats treated with N-nitroso-(2-hydroxypropyl)(2-oxopropyl)amine. *Cancer Res.*, **45**: 3586–92.

Kokinnakis, D. M., Hollenberg, P. F. & Scarpelli, D. G. (1986). The role of hepatic sulfotransferases in the activation of β-hydroxynitrosamines to potential mutagenic agents. *Proc. Am. Assoc. Cancer Res.*, **27**: 119.

Kokinnakis, D. M., Scarpelli, D. G., Rao, M. S. & Hollenberg, P. F. (1983). Metabolism of pancreatic carcinogens N-nitroso-2,6-dimethylmorpholine and N-nitrosobis(2-oxopropyl)amine by microsomes and cytosol of hamster pancreas and liver. *Cancer Res.*, **43**: 5761–7.

Kolar, G. F. & Carubelli, R. (1979). Urinary metabolite of 1-(2,4,6-trichlorophenyl)-3,3-dimethyltriazene with an intact diazoamino structure. *Cancer Lett.*, **7**: 209–14.

Kondo, H., Ikeda, T., Yoshimura, Y., Hoshimura, H. & Konishi, T. (1978). Carcinogenic effect of N-nitrosobis(2-hydroxypropyl)amine in rabbits. *Cancer Lett.*, **5**: 339–43.

Konishi, M., Ward, J. M., Reynolds, C. W. & Lijinsky, W. (1988). Thymic T-cell lymphoma with the CD8+ (OX-8), CD4+ (W3/25) phenotype, induced in F344/NCr rats by nitroso-2-hydroxypropylurea. *THYMUS*, **12**: 225–37.

Koppang, N. (1964). An outbreak of toxic liver injury in ruminants. *Nord. Vet.-Med.*, **16**: 305–22.

Koppang, N. & Helgebostad, A. (1987). Vascular changes and liver tumors induced in mink by high levels of nitrite in feed. *IARC Scientific Publication* No. 84: 256–60.

Kraft, P. L., Skipper, P. L., Charnley, G. & Tannenbaum, S. R. (1981). Urinary excretion of dimethylnitrosamine: a quantitative relationship between dose and urinary excretion. *Carcinogenesis*, **2**, 609–13.

Kroeger-Koepke, M. B., Koepke, S. R., McClusky, G. A., Magee, P. N. & Michejda, C. J. (1981) α-Hydroxylation pathway in the *in vitro* metabolism of carcinogenic nitrosamines: *N*-nitrosodimethylamine and *N*-nitroso-*N*-methylaniline. *Proc. Nat. Acad. Sci. U.S.A.*, **78**: 6489–93.
Kroeger-Koepke, M. B. & Michejda, C. J. (1979). Evidence for several demethylase enzymes in the oxidation of dimethylnitrosamine and phenylmethyl-nitrosamine by rat liver fractions. *Cancer Res.*, **39**: 1587–91.
Kroeger-Koepke, M. B., Reuber, M. D., Iype, P. T., Lijinsky, W. & Michejda, C. J. (1983). The effect of substituents in the aromatic ring on carcinogenicity of *N*-nitrosomethylaniline in F344 rats. *Carcinogenesis*, **4**: 157–60.
Krüger, F. W. (1971). Metabolism of nitrosamines *in vivo*. I. Evidence for β-oxidation of aliphatic dialkylnitrosamines. Simultaneous formation of 7-methylguanine and 7-propyl or 7-butylguanine after application of dipropyl or dibutylnitrosamine. *Z. Krebsforsch.*, **76**: 145–54.
Krüger, F. W. & Bertram, B. (1973). Metabolism of carcinogenic nitrosamines. *Z. Krebsforsch.*, **80**: 189–96.
Krüger, F. W., Pour, P. & Althoff, J. (1974). Induction of pancreas tumors by diisopropanolnitrosamine. *Naturwiss.*, **61**: 328.
Kupper, R., Reuber, M. D., Blackwell, B.-N., Lijinsky, W., Koepke, S. R. & Michejda, C. J. (1980). Carcinogenicity of the isomeric *N*-nitroso-Δ^3- and *N*-nitroso-Δ^2-piperidines in rats and the *in vivo* isomerization of the Δ^3- to the Δ^2-isomer. *Carcinogenesis*, **1**: 753–7.
Labuc, G. F. & Archer, M. C. (1982). Esophageal and hepatic microsomal metabolism of *N*-nitrosomethylbenzylamine and *N*-nitrosodimethylamine in the rat. *Cancer Res.*, **42**: 3181–6.
Lai, D. Y. & Arcos, J. S. (1980). Dialkylnitrosamine bioactivation and carcinogenesis. *Life Sci.*, **27**: 2149–65.
Lai, D. Y., Arcos, J. S. & Argus, M. F. (1980). Interaction of the tobacco-specific nitrosamines, methylethylnitrosamine and *N*-nitrosonornicotine with DNA and guanosine. *Res. Commun. Chem. Pathol. Pharmacol.*, **28**: 87–103.
Langenbach, R. (1986). Mutagenic activity and structure–activity relationships of short-chain dialkyl *N*-nitrosamines in a hamster hepatocyte V79 cell-mediated system. *Mutat. Res.*, **163**, 303–11.
Langenbach, R., Gingell, R., Kuszynski, C., Walker, B., Nagel, D. & Pour, P. (1980). Mutagenic activities of oxidized derivatives of *N*-nitrosodipropylamine in the liver cell-mediated and *Salmonella typhimurium* assays, *Cancer Res.*, **40**: 3463–7.
Lawson, T. A., Gingell, R., Nagel, D., Hines, L. A. & Ross, A. E. (1981*a*). Methylation of hamster DNA by the carcinogen nitrosobis-(2-oxopropyl)amine. *Cancer Lett.*, **11**: 251–5.
Lawson, T. A., Helgeson, A. S., Grandjean, C. J., Wallcave, L. & Nagel, D. (1981*b*). The formation of *N*-nitrosomethyl(2-oxopropyl)amine from *N*-nitrosobis-(2-oxopropyl)amine *in vivo*. *Carcinogenesis*, **2**: 845–9.
Lawson, T. & Nagel, D. (1988). The production and repair of DNA damage by *N*-nitrosobis(2-oxopropyl)-amine and azaserine in hamster and rat pancreas acinar and duct cells. *Carcinogenesis*, **9**: 1007–10.
Lee, K. Y. & Lijinsky, W. (1966). Alkylation of rat liver RNA by cyclic *N*-nitrosamines *in vivo*. *J. Nat. Cancer Inst.*, **37**: 401–7.
Lee, M., Ishizaki, H., Brady, J. F. & Yang, C. S. (1989). Substrate specificity and alkyl group selectivity in the metabolism of *N*-nitrosodialkylamines. *Cancer Res.*, **49**: 1470–4.
Lethco, E. J., Wallace, W. C. & Brouner, E. (1982). The fate of

nitrosodiethanolamine after oral and topical administration to rats. *Food Chem. Toxicol.*, **20**: 401–6.

Leung, K. H. & Archer, M. C. (1984). Studies on the metabolic activation of β-keto nitrosamines: Mechanisms of DNA methylation by N-(2-oxopropyl)-N-nitrosourea and N-nitroso-N-acetoxymethyl-N-2-oxopropylamine. *Chem. Biol. Interact.*, **48**: 169–79.

Leung, K. H., Park, K. K. & Archer, M. C. (1980). Methylation of DNA by N-nitroso-2-oxopropylpropylamine: formation of O^6- and 7-methylguanine and studies on the methylation mechanism. *Toxicol. Appl. Pharmacol.*, **53**: 29–34.

Li, J. J., Li, S. A., Klicka, J. K., Parsons, J. A. & Lam, L. K. T. (1983). Relative carcinogenic activity of various synthetic and natural estrogens in the Syrian Hamster kidney. *Cancer Res.*, **43**: 5200–4.

Libbey, L. M., Scanlan, R. A. & Barbour, J. F. (1980). N-nitrosodimethylamine in dried dairy products. *Food Cosmet. Toxicol.*, **18**: 459–61.

Liberato, D., Saavedra, J. E., Farnsworth, D. & Lijinsky, W. (1989). Thermospray liquid chromatography/mass spectrometry studies on mechanisms of nucleic acid alkylation by some deuterated carcinogens. *Chem. Res. in Toxicol.*, **2**: 307–11.

Lickhachev, A. J., Ohshima, H., Anisimov, V. N., Ovsyannikov, A. I., Revskoy, S. Y., Keefer, L. K. & Reist, E. J. (1983a). Carcinogenesis and aging. II. Modifying effect of aging in metabolism of methyl (acetoxymethyl) nitrosamine and its interaction with DNA of various tissues in rats. *Carcinogenesis*, **4**: 967–74.

Lickhachev, A. F., Ivanov, M. N., Bresil, H., Planche-Martel, G., Montesano, R. & Margison, G. P. (1983b). Carcinogenicity of single doses of N-nitroso-N-methylurea and N-nitroso-N-ethylurea in Syrian Golden Hamsters and the persistence of alkylated purines in the DNA of various tissues. *Cancer Res.*, **43**: 829–33.

Lijinsky, W. (1971). Hearings on regulation of food additives and medicated animal feeds, intergovernmental relations subcommittee of the comittee on Government Operations, Chairman, L. H. Fountain, March 16. U.S. Government Printing Office, Stock No. 5270–1144.

Lijinsky, W. (1974). Reaction of drugs with nitrous acid as a source of carcinogenic nitrosamines. *Cancer Res.*, **34**: 255–8.

Lijinsky, W. (1976a). Interaction with nucleic acids of carcinogenic and mutagenic N-nitroso compunds. *Progress in Nucleic Acid Research*, **17**: 247–69.

Lijinsky, W. (1976b). Health problems associated with nitrites and nitrosamines. *Ambio*, **5**: 67–72.

Lijinsky, W. (1982). Comparison of the carcinogenic effectiveness in mouse skin of methyl- and ethyl-nitrosourea, nitrosourethane and nitrosonitroguanidine and the effect of deuterium labeling. *Carcinogenesis*, **3**: 1289–91.

Lijinsky, W. (1983a). Chemistry and metabolism of N-nitroso compounds related to respiratory tract carcinogenesis. *In: comparative Respiratory Tract Carcinogenesis*, ed. H. M. Reznik-Schuller, Vol. II, CRC Press, 95–107.

Lijinsky, W. (1983b). Species specificity in nitrosamine carcinogenesis. *In: Organ and Species Specificity in Chemical Carcinogenesis*, ed. R. Langenbach, S. Nesnow & J. M. Rice. Plenum Publishing Corp., 63–75.

Lijinsky, W. (1984a). Induction of tumors of the nasal cavity in rats by concurrent feeding of thiram and sodium nitrite. *J. Toxicol Env. Health*, **13**: 609–14.

Lijinsky, W. (1984b). Induction of tumors in rats by feeding nitrosatable amines together with sodium nitrite. *Fd Chem. Toxicol.*, **22**: 715–20.

Lijinsky, W. (1985). The metabolism and cellular interactions of some aliphatic nitrogenous carcinogens. *Cancer Lett.*, **26**: 33–42.
Lijinsky, W. (1986a). Deuterium isotope effects in carcinogenesis by *N*-nitroso compounds and related carcinogens. *J. Cancer Res. Clin. Oncol.*, **112**: 229–39.
Lijinsky, W. (1986b). Significance of *N*-nitroso compounds as environmental carcinogens. *Environmental Carcinogenesis Reviews*, C4: 1–45.
Lijinsky, W. (1987). Structure–activity relations in carcinogenesis by *N*-nitroso compounds. *Cancer and Metastasis Reviews*, **6**: 301–56.
Lijinsky, W. (1988a). Intestinal cancer induced by *N*-nitroso compounds. *Toxicologic Pathology*, **16**: 198–204.
Lijinsky, W. (1988b). Nucleic acid alkylation by *N*-nitroso compounds related to organ-specific carcinogenesis. In: *Chemical Carcinogens, Activation Mechanisms, Structural & Electronic Factors, and Reactivity*, ed. P. Politzer & L. Roberts, Elsevier Science Publishers, pp. 242–63.
Lijinsky, W. (1989). A view of the relation between carcinogenesis and mutagenesis. *Environmental and Molecular Mutagenesis* **14**: (Supplement 16): 78–84.
Lijinsky, W. (1990a). Occupational and environmental exposure to *N*-nitroso compounds. In: *Environmental and Occupational Cancer: Scientific Update*, pp. 189–207, Princeton University Press.
Lijinsky, W. (1990b). Structure–activity relations in reactions of drugs with nitrite. In: *The Significance of* N-*nitrosation of Drugs*, ed. G. Eisenbrand, G. Bozler, & H. Von Nicolai, G. Fischer-Verlag, Stuttgart, pp. 93–107.
Lijinsky, W. (1990c). Non-genotoxic environmental carcinogens. *Environmental Carcinogenesis Revs.*, C8, 45–87.
Lijinsky, W. (1990d). Tumors outside the bladder induced by intravesicular administration of *N*-nitroso compounds. *15th International Cancer Congress, Poster Abstracts*, p. 76.
Lijinsky, W. (1991). Alkylation of DNA related to organ-specific carcinogenesis by *N*-nitroso compounds. In: *Relevance to Human Cancer of* N-*nitroso Compounds, Tobacco and Mycotoxins. IARC Scientific Publication* No. 105: 305–10.
Lijinsky, W. & Andrews, A. W. (1979). The mutagenicity of nitrosamides in *Salmonella typhimurium. Mutation Res.*, **68**: 1–8.
Lijinsky, W. & Andrews, A. W. (1983). The superiority of hamster liver microsomal fraction for activating nitrosamines to mutagens in *Salmonella typhimurium. Mutation Res.*, **111**: 135–44.
Lijinsky, W., Andrews, A. W., Elespuru, R. K. & Farrelly, J. G. (1985a). Lack of genetic and *in vitro* metabolic activity of potently carcinogenic azoxyalkanes. *Mutation Res.*, **157**: 23–7.
Lijinsky, W., Christie, W. H. & Rainey, W. T. (1973a). Mass spectra of *N*-nitroso compounds. *Oak Ridge National Laboratory, Technical Report* TM-4359.
Lijinsky, W., Conrad, E. & Van de Bogart, R. (1972a). Carcinogenic nitrosamines formed by drug/nitrite interactions. *Nature*, **239**: 165–7.
Lijinsky, W., Conrad, E. & Van de Bogart, R. (1972b). Formation of carcinogenic nitrosamines by interaction of drugs with nitrite. In: N-*Nitroso compounds, Analysis and Formation, IARC Scientific Publication* No. 3: 130–3.
Lijinsky, W. & Elespuru, R. K. (1976). Mutagenicity and carcinogenicity of *N*-nitroso derivatives of carbamate insecticides. In: *Environmental* N-*Nitroso Compounds, Analysis and Formation, IARC Scientific Publication* No. 14: 425–8.
Lijinsky, W., Elespuru, R. K. & Andrews, A. W. (1987a). Relative mutagenic

and prophage-inducing effects of mono- and di-alkylnitrosoureas. *Mutation Res.*, **178**: 157–65.

Lijinsky, W. & Epstein, S. S. (1970). Nitrosamines as environmental carcinogens. *Nature*, **225**: 21–3.

Lijinsky, W., Ferrero, A., Montesano, R. & Wenyon, C. E. M. (1970a). Tumorigenicity of cyclic nitrosamines in Syrian Golden Hamsters. *Z. Krebsforsch.*, **74**: 185–9.

Lijinsky, W., Garcia, H., Keefer, L. & Loo, J. (1972c). Carcinogenesis and alkylation of rat liver nucleic acids by nitrosomethylurea and nitrosoethylurea administered by intraportal injection. *Cancer Res.*, **32**: 893–7.

Lijinsky, W. & Greenblatt, M. (1972). Carcinogen dimethylnitrosamine produced *in vivo* from sodium nitrite and aminopyrine. *Nature New Biology*, **236**: 177–8.

Lijinsky, W., Greenblatt, M. & Kommineni, C. (1973b). Feeding studies of nitrilotriacetic acid and derivatives in rats. *J. Natl. Cancer Inst.*, **50**: 1061–3.

Lijinsky, W., Keefer, L., Conrad, E. & Van de Bogart, R. (1972d). The nitrosation of tertiary amines and some biologic implications. *J. Natl. Cancer Inst.*, **49**: 1239–49.

Lijinsky, W., Keefer, L. & Loo, J. (1970b). The preparation and properties of some nitrosamino acids. *Tetrahedron*, **26**: 5137–53.

Lijinsky, W., Keefer, L, Loo, J. & Ross, A. E. (1973c). Studies of alkylation of nucleic acids in rats by cyclic nitrosamines. *Cancer Res.*, **33**: 1634–41.

Lijinsky, W., Keefer, L. K., Saavedra, J. E., Hansen, T. H., Kovatch, R. M., Fiddler, W. E. & Miller, A. T. (1988a). Carcinogenesis in rats by cyclic *N*-nitrosamines containing sulfur. *Fd. Chem. Toxicol.*, **26**: 3–7.

Lijinsky, W., Knutsen, G. M. & Kovatch, R. M. (1985b). Carcinogenic effect of nitrosoalkylureas and nitrosoalkylcarbamates in Syrian hamsters. *Cancer Res.*, **45**: 542–45.

Lijinsky, W., Knutsen, G. M. & Kovatch, R. M. (1985c). Comparative carcinogenesis by hydroxylated nitrosopropylamines in Syrian hamsters. *J. Natl. Cancer Inst.*, **74**: 923–6.

Lijinsky, W., Knutsen, G. M. & Reuber, M. D. (1983f). Carcinogenicity of methylated nitrosopiperazines in rats and hamsters. *Carcinogenesis*, **4**: 1165–7.

Lijinsky, W. & Kovatch, R. M. (1985a). Carcinogenesis by oxygenated nitrosomethylpropylamines in Syrian hamsters. *J. Cancer Res. Clin. Oncol.*, **109**: 1–4.

Lijinsky, W. & Kovatch, R. M. (1985b). Induction of liver tumors in rats by nitrosodiethanolamine at low doses. *Carcinogenesis*, **6**: 1679–81.

Lijinsky, W. & Kovatch, R. M. (1986). The effect of age on susceptibility of rats to carcinogenesis by two nitrosamines. *GANN*, **77**: 1222–6.

Lijinsky, W. & Kovatch, R. M. (1988a). Carcinogenesis by nitrosohydroxyethylurea and nitrosomethoxyethylurea in F344 rats. *GANN*, **79**: 181–6.

Lijinsky, W. & Kovatch, R. M. (1988b). Comparative carcinogenesis by nitrosomethylalkylamines in Syrian Golden Hamsters. *Cancer Res.*, **48**: 6648–52.

Lijinsky, W. & Kovatch, R. M. (1989a). Carcinogenesis by nitrosamines and azoxyalkanes by different routes of administration to rats. *Biomed. Env. Science*, **2**: 154–9.

Lijinsky, W. & Kovatch, R. M. (1989b). Similar carcinogenic actions of nitrosoalkylureas of varying structure given to rats by gavage. *Toxicol. Ind. Health*, **5**: 925–35.

Lijinsky, W. & Kovatch, R. M. (1989c). The uniform carcinogenic action of nitrosoalkylureas in Syrian Hamsters. *Biomed. Env. Science*, **2**: 167–73.

Lijinsky, W., Kovatch, R. M. & Knutsen, G. L. (1984a). Carcinogenesis by nitrosomorpholines, nitrosooxazolidines and nitrosoazetidine given by gavage to Syrian Golden Hamsters. *Carcinogenesis*, **5**: 875–8.
Lijinsky, W., Kovatch, R. M. & Riggs, C. W. (1983a). Altered incidences of hepatic and hemopoietic neoplasms in F344 rats fed sodium nitrite. *Carcinogenesis*, **4**: 1189–91.
Lijinsky, W., Kovatch, R. M. & Riggs, C. W. (1987b). Carcinogenesis by nitrosodialkylamines and azoxyalkanes given by gavage to rats and hamsters. *Cancer Res.*, **47**: 3968–72.
Lijinsky, W., Kovatch, R. M., Riggs, C. W. & Walters. P. T. (1988b). A dose–response carcinogenesis study of nitrosomorpholine in F-344 rats. *Cancer Res.*, **48**: 2089–95.
Lijinsky, W., Kovatch, R. M. & Singer, S. S. (1986a). Carcinogenesis in F344 rats by nitrosohydroxyalkylchloroethylureas. *J. Cancer Res. Clin. Oncol.*, **112**: 221–8.
Lijinsky, W., Kovatch, R. M. & Thomas, B. J. (1988c). Carcinogenesis by nitroso-2-hydroxyethylurea in splenectomized hamsters. *Cancer Lett.*, **41**: 199–202.
Lijinsky, W., Lee, K. Y., Tomatis, L. & Butler, W. H. (1967). Nitrosoazetidine – A potent carcinogen of low toxicity. *Naturwiss.*, **54**: 518.
Lijinsky, W., Loo, J. & Ross, A. (1968). Mechanism of alkylation of nucleic acids by nitrosodimethylamine. *Nature*, **218**: 1174–5.
Lijinsky, W., Losikoff, A. M. & Sansone, E. P. (1980a). The penetration of rat skin by some nitrosamines of environmental importance. In: N-Nitroso Compounds: Analysis, Formation and Occurrence., IARC Scientific Publication No. 31: 705–13.
Lijinsky, W., Losikoff, A. M. & Sansone, E. P. (1981a). Penetration of rat skin by *N*-nitrosodiethanolamine and *N*-nitrosomorpholine. *J. Natl. Cancer Inst.*, **66**: 125–7.
Lijinsky, W., Milner, J. A., Kovatch, R. M. & Thomas, B. J. (1989a). Lack of effect of selenium on induction of tumors of esophagus and bladder in rats by two nitrosamines. *Toxicology & Industrial Health*, **51**: 63–72.
Lijinsky, W. & Reuber, M. D. (1980a). Carcinogenicity in rats of nitrosomethylethylamines labeled with deuterium in several positions. *Cancer Res.*, **40**: 19–21.
Lijinsky, W. & Reuber, M. D. (1980b). Carcinogenicity of deuterium-labeled *N*-nitroso-*N*-methylcyclohexylamine in rats. *J. Natl. Cancer Inst.*, **64**: 1535–6.
Lijinsky, W. & Reuber, M. D. (1980c). Comparison of carcinogenesis by two isomers of nitroso-2,6-dimethylmorpholine. *Carcinogenesis*, **1**: 501–3.
Lijinsky, W. & Reuber, M. D. (1980d). Tumors induced in Fischer 344 rats by the feeding of disulfiram together with sodium nitrite. *Fd. Cosmet. Toxicol.*, **18**: 85–7.
Lijinsky, W. & Reuber, M. D. (1981a). Carcinogenic effect of nitrosopyrrolidine, nitrosopiperidine and nitrosohexamethyleneimine in Fischer rats. *Cancer Lett.*, **12**: 99–103.
Lijinsky, W. & Reuber, M. D. (1981b). Comparative carcinogenesis by some aliphatic nitrosamines in Fischer rats. *Cancer Lett.*, **14**: 297–302.
Lijinsky, W. & Reuber, M. D. (1981c). Comparative carcinogenicity of two isomers of dimethylnitrosomorpholine in guinea pigs. *Cancer Lett.*, **14**: 7–11.
Lijinsky, W. & Reuber, M. D. (1982a). Studies of a deuterium isotope effect in carcinogenesis by *N*-nitroso-*N*-alkylurethanes in rats. *Cancer Lett.*, **16**: 273–9.
Lijinsky, W. & Reuber, M. D. (1982b). Transnitrosation by nitrosamines *in*

vivo. In: N-*Nitroso Compounds: Occurrence and Biological Effects. IARC Scientific Publication* No. 41: 625–31.
Lijinsky, W. M. & Reuber, M. D. (1982c). Comparative carcinogenesis by nitrosomorpholines, nitrosooxazolidines and nitrosotetrahydroxazine in rats. *Carcinogenesis*, **3**: 911–15.
Lijinsky, W. & Reuber, M. D. (1983a). Carcinogenesis in Fischer rats by nitrosodipropylamine, nitrosodibutylamine and nitrosobis(2-oxopropyl)-amine given by gavage. *Cancer Lett.*, **19**: 207–13.
Lijinsky, W. & Reuber, M. D. (1983b). Carcinogenicity of hydroxylated alkylnitrosoureas and of nitrosooxazolidones by mouse skin painting and by gavage in rats. *Cancer Res.*, **43**: 214–21.
Lijinsky, W. & Reuber, M. D. (1984a). Carcinogenesis in rats by nitrosodimethylamine and other nitrosomethylalkylamines at low doses. *Cancer Lett.*, **22**: 83–8.
Lijinsky, W. & Reuber, M. D. (1984b). Dose response study with N-nitrosodiethanolamine in F344 rats. *Fd. Chem. Toxicol.*, **22**: 23–6.
Lijinsky, W. & Reuber, M. D. (1984c). Comparison of nitrosocimetidine with nitrosomethylnitroguanidine in chronic feeding tests in rats. *Cancer Res.*, **44**: 447–9.
Lijinsky, W. & Reuber, M. D. (1988). Neoplasms of the skin and other organs observed in Swiss mice treated with nitrosoalkylureas. *J. Cancer Res. Clin. Oncol.*, **114**: 245–9.
Lijinsky, W., Reuber, M. D. & Blackwell, B.-N. (1980b). Carcinogenicity of nitrosotrialkylureas in Fischer rats. *J. Natl. Cancer Inst.*, **65**: 451–3.
Lijinsky, W., Reuber, M. D. & Blackwell, B.-N. (1980c). Liver tumors induced in rats by chronic oral administration of the common antihistaminic methapyrilene hydrochloride. *Science*, **209**: 817–19.
Lijinsky, W., Reuber, M. D., Davies, T. S. & Riggs, C. W. (1982a). Dose–response studies with nitroso-1,2,3,6-tetrahydropyridine and dinitrosohomopiperazine. *Ecotoxicol. Envir. Safety*, **6**: 513–27.
Lijinsky, W., Reuber, M. D., Davies, T. S. & Riggs, C. W. (1982b). Dose–response studies with nitrosoheptamethyleneimine and its alpha deuterium labeled derivative in F344 rats. *J. Natl. Cancer Inst.*, **69**: 1127–33.
Lijinsky, W., Reuber, M. D., Davies, T. S., Saavedra, J. E. & Riggs, C. W. (1982c). Dose–response studies in carcinogenesis by nitrosomethyl-2-phenylethylamine in rats and the effect of deuterium. *Fd. Cosmet. Toxicol.*, **20**: 393–9.
Lijinsky, W., Reuber, M. D. & Manning, W. B. (1980c). Potent carcinogenicity of nitrosodiethanolamine in rats. *Nature*, **288**: 309–10.
Lijinsky, W., Reuber, M. D. & Reznik-Schüller, H. M. (1982f). Contrasting carcinogenic effects of nitroso-2,6-dimethylmorpholine given by gavage to F344 rats and Syrian golden hamsters. *Cancer Lett.*, **16**: 281–6.
Lijinsky, W., Reuber, M. D. & Riggs, C. W. (1981b). Dose–response studies in rats with nitrosodiethylamine. *Cancer Res.*, **41**: 4997–5003.
Lijinsky, W., Reuber, M. D. & Riggs, C. W. (1983c). Carcinogenesis by combinations of N-nitroso compounds in rats. *Fd. Chem. Toxicol.*, **21**: 601–5.
Lijinsky, W., Reuber, M. D., Saavedra, J. E. & Blackwell, B.-N. (1980d). Effect of deuterium on the carcinogenicity of nitroso-methyl-*n*-butylamine. *Carcinogenesis*, **1**: 157–60.
Lijinsky, W., Reuber, M. D., Saavedra, J. E. & Singer, G. M. (1983d). Carcinogenesis in F344 rats by nitrosomethyl-*n*-propylamine derivatives. *J. Natl. Cancer Inst.*, **70**: 959–63.
Lijinsky, W., Reuber, M. D. & Singer, G. M. (1983e). Induction of tumors of

the esophagus in rats by nitrosomethylalkylamines. *J. Cancer Res. Clin. Oncol.*, **106**: 171–5.
Lijinsky, W., Reuber, M. D., Singer, G. M. (1982d). Carcinogenesis by isomers of 3,5-dimethylnitrosopiperidine. *J. Natl. Cancer Inst.*, **68**: 989–91.
Lijinsky, W. & Ross, A. (1969). Alkylation of rat liver nucleic acids not related to carcinogenesis by N-nitrosamines. *J. Natl. Cancer Inst.*, **42**: 1095–100.
Lijinsky, W., Saavedra, J. E., Knutsen, G. M. & Kovatch, R. M. (1984b). Comparison of the carcinogenic effectiveness of nitrosobis(2-hydroxypropyl)amine, nitrosobis(2-oxopropyl)amine, nitrosohydroxypropyloxopropylamine and nitroso-2,6-dimethylmorpholine in Syrian hamsters. *J. Natl. Cancer Inst.*, **72**: 685–8.
Lijinsky, W., Saavedra, J. E. & Kovatch, R. M. (1988e). Carcinogenesis and nucleic acid alkylation by some oxygenated nitrosamines in rats and hamsters. *Chemico.-Biol. Interactions*, **66**: 37–47.
Lijinsky, W., Saavedra, J. E. & Kovatch, R. M. (1988f). Carcinogenesis by nitrosobis-(2-oxopropyl)-amine labeled with deuterium and by nitroso-2-hydroxypropyl-2-oxopropylamine in rats and hamsters. *Cancer Lett.*, **42**: 37–41.
Lijinsky, W., Saavedra, J. E. & Kovatch, R. M. (1989b). Carcinogenesis in rats by nitrosodialkylureas containing methyl and ethyl groups given by gavage and in drinking water. *J. Toxicol. Env. Health*, **28**: 27–38.
Lijinsky, W., Saavedra, J. E. & Kovatch, R. M. (1990). Carcinogenesis in rats by nitrosodialkylureas containing oxygenated alkyl groups. *In Vivo*, **4**: 1–6.
Lijinsky, W., Saavedra, J. E. & Kovatch, R. M. (1991). Carcinogenesis in rats by substituted dialkylnitrosamines given by gavage. *In Vivo*, **5**: 85–90.
Lijinsky, W., Saavedra, J. & Reuber, M. D. (1981c). Induction of carcinogenesis in Fischer rats by methylalkylnitrosamines. *Cancer Res.*, **41**: 1288–92.
Lijinsky, W., Saavedra, J. E. & Reuber, M. D. (1984c). Carcinogenesis in F344 rats by nitrosobis-(2-oxopropyl)amine and related compounds administered in drinking water. *J. Cancer Res. Clin. Oncol.*, **107**: 178–82.
Lijinsky, W., Saavedra, J.E. & Reuber, M. D. (1984d). Carcinogenesis in rats by some hydroxylated acyclic nitrosamines. *Carcinogenesis*, **5**: 167–70.
Lijinsky, W., Saavedra, J. E. & Reuber, M. D. (1985d). Organ-specific carcinogenesis in rats by methyl- and ethyl-azoxyalkanes. *Cancer Res.*, **45**: 76–9.
Lijinsky, W., Saavedra, J. E., Reuber, M. D. & Blackwell, B.-N. (1980e). Carcinogenicity of 3-chloro-, 4-chloro- and 3,4-dichloronitroso- piperidine in Fischer rats. *Cancer Res.*, **40**: 3325–7.
Lijinsky, W., Saavedra, J. E., Reuber, M. D. & Blackwell, B.-N. (1980f). The effect of deuterium labeling on the carcinogenicity of nitroso-2,6-dimethylmorpholine in rats. *Cancer Lett.*, **10**: 325–31.
Lijinsky, W., Saavedra, J. E., Reuber, M. D. & Singer, S. S. (1982e). Esophageal carcinogenesis in Fischer 344 rats by nitrosomethylethylamines substituted in the ethyl group. *J. Natl. Cancer Inst.*, **68**: 681–4.
Lijinsky, W. & Schmähl, D. (1978). Carcinogenicity of N-nitroso derivatives of N-methylcarbamate insecticides in rats. *Ecotoxicol. Envir. Safety*, **2**: 413–19.
Lijinsky, W. & Singer, G. M. (1974). Formation of nitrosamines from tertiary amines and nitrous acid. In: N-*nitroso Compounds in the Environment*, Lyon, *IARC Scientific Publication* No. 9, 111–14.
Lijinsky, W., Singer, G. M. & Kovatch, R. M. (1985e). Similar carcinogenic effects in rats of 1-ethyl-1-nitroso-3-hydroxyethylurea and 1-hydroxyethyl-1-nitroso-3-ethylurea. *Carcinogenesis*, **6**: 641–3.

Lijinsky, W., Singer, G. M. & Reuber, M. D. (1981d). The effect of 4-substitution on the carcinogenicity of nitrosopiperidine. *Carcinogenesis*, **2**: 1045–8.

Lijinsky, W., Singer, G. M., Saavedra, J. E. & Reuber, M. D. (1984e). Carcinogenesis in rats by asymmetric nitrosamines containing the allyl group. *Cancer Lett.*, **22**: 281–8.

Lijinsky, W. & Taylor, H. W. (1975a). Carcinogenicity of methylated nitrosopiperidines. *Intern. J. Cancer*, **16**: 318–22.

Lijinsky, W. & Taylor, H. W. (1975b). Carcinogenicity of N-nitroso-3,4-dichloro- and N-nitroso-3,4-dibromopiperidine in rats. *Cancer Res.*, **35**: 3209–11.

Lijinsky, W. & Taylor, H. W. (1975c). Increased carcinogenicity of 2,6-dimethylnitrosomorpholine compared with nitrosomorpholine in rats. *Cancer Res.*, **35**: 2123–5.

Lijinsky, W. & Taylor, H. W. (1975d). Induction of neurogenic tumors by nitrosotrialkylureas in rats. *Z. Krebsforsch.*, **93**: 315–21.

Lijinsky, W. & Taylor, H. W. (1975e). Induction of urinary bladder tumors in rats by administration of nitrosomethyldodecylamine. *Cancer Res.*, **35**: 958–61.

Lijinsky, W. & Taylor, H. W. (1975f). Tumorigenesis by oxygenated nitrosopiperidines. *J. Natl. Cancer Inst.*, **55**: 705–8.

Lijinsky, W. & Taylor, H. W. (1975g). Carcinogenicity of methylated dinitrosopiperazines in rats. *Cancer Res.*, **35**: 1270–3.

Lijinsky, W. & Taylor, H. W. (1976a). Carcinogenicity of two unsaturated derivatives of nitrosopiperidine in Sprague-Dawley rats. *J. Natl. Cancer Inst.*, **57**: 1315–17.

Lijinsky, W. & Taylor, H. W. (1976b). Carcinogenesis in Sprague-Dawley rats of N-nitroso-N-alkylcarbamate esters. *Cancer Lett.*, **1**: 275–9.

Lijinsky, W. & Taylor, H. W. (1976c). Carcinogenicity tests of N-nitroso derivatives of two drugs, phenmetrazine and methylphenidate. *Cancer Lett.*, **1**: 359–63.

Lijinsky, W. & Taylor, H. W. (1976d). The effect of substituents on the carcinogenicity of N-nitrosopyrrolidine in Sprague-Dawley rats. *Cancer Res.*, **36**: 1988–90.

Lijinsky, W. & Taylor, H. W. (1977a). Carcinogenesis tests of nitroso-N-methylpiperazine, 2,3,5,6-tetramethyldinitrosopiperazine, nitrosoisonipecotic acid and nitrosomethoxymethylamine in rats. *Z. Krebsforsch.*, **89**: 31–6.

Lijinsky, W. & Taylor, H. W. (1977b). Carcinogenicity of nitrosoazetidine and tetradeuteronitrosoazetidine in Sprague-Dawley rats. *Z. Krebsforsch.*, **89**: 215–19.

Lijinsky, W. & Taylor, H. W. (1977c). Nitrosamines and their precursors in food. *Cold Spring Harbor Symposium on the Origins of Human Cancer*, Book C., pp. 1579–90.

Lijinsky, W. & Taylor, H. W. (1977d). Transplacental chronic toxicity test of carbaryl with nitrite in rats. *Fd. Cosmet. Toxicol.*, **15**: 229–32.

Lijinsky, W. & Taylor, H. W. (1978a). Carcinogenicity of 4-chloronitrosopiperidine in Sprague-Dawley rats. *Z. Krebsforsch.*, **92**: 217–20.

Lijinsky, W. & Taylor, H. W. (1978b). Comparison of bladder carcinogenesis by nitrosomethyldodecylamine in Sprague-Dawley and Fischer rats carrying transplanted bladder tissue. *Cancer Lett.*, **5**: 215–18.

Lijinsky, W. Taylor, H. W. (1978c). Relative carcinogenic effectiveness of derivatives of diethylnitrosamine in rats. *Cancer Res.*, **38**: 2391–4.

Lijinsky, W. & Taylor, H. W. (1978d). The change in carcinogenic effectiveness of some cyclic nitrosamines at different doses. *Z. Krebsforsch.*, **92**: 221–5.

Lijinsky, W. & Taylor, H. W. (1978e). Carcinogenicity tests in rats of two nitrosamines of high molecular weight, nitrosododecamethyleneimine and nitrosodi-*n*-octylamine. *Ecotoxicol. Env. Safety*, **2**: 407–11.

Lijinsky, W. & Taylor, H. W. (1979a). Carcinogenicity of chlorinated nitrosotrialkylureas in rats. *J. Cancer Res. Clin. Oncol.*, **94**: 131–7.

Lijinsky, W. & Taylor, H. W. (1979b). Carcinogenicity of methylated derivatives of nitrosodiethylamine and related compounds in Sprague-Dawley rats. *J. Natl. Cancer Inst.* **62**: 407–10.

Lijinsky, W. & Taylor, H. W. & Keefer, L. K. (1976). Reduction of rat liver carcinogenicity of nitrosomorpholine by alpha deuterium substitution. *J. Natl. Cancer Inst.*, **57**: 1311–13.

Lijinsky, W. & Taylor, H. W., Snyder, C. & Nettesheim, P. (1973c). Malignant tumors of liver and lung in rats fed aminopyrine or heptamethyleneimine together with nitrite. *Nature*, **244**: 176–8.

Lijinsky, W., Thomas, B. J. & Kovatch, R. M. (1988g). Effects of feminization of male F344 rats on tumor induction and on nucleic acid alkylation by nitrosobis-(2-oxopropyl)-amine. *Chemico.-Biol. Interactions*, **66**: 111–19.

Lijinsky, W., Tomatis, L. & Wenyon, C. E. M. (1969). Lung tumors in rats treated with *N*-nitrosoheptamethyleneimine and *N*-nitrosooctamethyleneimine. *Proc. Soc. Exper. Biol. and Med.*, **130**: 945–9.

Lijinsky, W. & Winter, C. (1981). Skin tumors induced by painting nitrosoalkylureas on mouse skin. *J. Cancer Res. & Clinical ONcology*, **102**: 13–20.

Loeppky, R. N., McKinley, W. A., Hazlitt, L. G., Beedle, E. C., DeArman, S. K. & Gnewuch, C. T. (1980). The fragmentation and transformation of β-oxidised *N*-nitrosamines in relation to their analysis and occurrence. *IARC Scientific Publication* No. 31: 15–30.

Loeppky, R. N., McKinley, W. A., Hazlitt, L. G. & Outram, J. R. (1982). Base-induced fragmentation of β-hydroxynitrosamines. *J. Org. Chem.*, **47**: 4833–41.

Loeppky, R. N. & Outram, J. R. (1982). A biochemical retroaldol cleavage of β-hydroxy nitrosamines. *IARC Scientific Publication* No. 41: 459–72.

Loeppky, R. N., Tomasik, W. & Millard, T. G. (1984). Ester-mediated nitrosamine formation from nitrite and secondary or tertiary amines. *IARC Scientific Publication* No. 57: 353–63.

Love, L. A., Lijinsky, W., Keefer, L. K. & Garcia, H. (1977). Chronic oral administration of 1-nitrosopiperazine at high doses to MRC rats. *Z. Krebsforsch.*, **89**: 69–73.

Loveless, A. (1969). Possible relevance of O^6-alkylation of deoxyguanosine to the mutagenicity and carcinogenicity of nitrosamines and nitrosamides. *Nature*, **223**: 206–7.

Ludeke, B., Schubert, M., Yamada, Y., Lijinsky, W. & Kleihues, P. (1991a). DNA hydroxyethylation by hydroxyethylnitrosoureas in relation to their organ-specific carcinogenicity in rats. *Chem-Biol. Interactions*, in press.

Ludeke, B., Meier, T. & Kleihues, P. (1991b). Bioactivation of asymmetric *N*-dialkylnitrosamines in rat tissues derived from the ventral entoderm. *IARC Scientific Publication* No. 105, 286–93.

Lyle, R. E., Fribush, H. M., Singer, S., Saavedra, J. E., Lyle, G. G., Barton, R., Yoder, S. & Jacobson, M. K. (1979). Stereochemical effects on *N*-nitrosamine chemistry. *In*: N-*Nitrosamines, ACS Monograph* No. 101: 39–56.

Lyle, R. E., Gunn, V. E., Jacobson, M. K. & Lijinsky, W. (1983). The synthesis of α-acetoxybenzylbenzylnitrosamine. *Org. Prep. and Proc. Int.*, **15**: 57–62.

Lyle, R. E., Saavedra, J. E., Lyle, G. G., Fribush, M. M., Marshall, J. L.,

Lijinsky, W. & Singer, G. M. (1976). Conformational stereospecificity in electrophilic reactions with cyclic anions. *Tet. Letters*, 4431–4.
Maekawa, T., Ogiu, T., Onodera, H., Furuta, F., Matsuoka, C., Mochizuki, M., Anjo, T., Okada, M. & Odashima, S. (1982). Carcinogenicity of N-alkyl-N-(acetoxymethyl)nitrosamines after subcutaneous injections in F-344 rats. *J. Cancer Res. Clin. Oncol.*, **104**: 13–21.
Magee, P. N. & Barnes, J. M. (1956). The production of malignant primary hepatic tumours in the rat by feeding dimethylnitrosamine. *Br. J. Cancer*, **10**: 114–22.
Magee, P. N. & Barnes, J. M. (1967). Carcinogenic nitroso compounds. *Advances in Cancer Research*, **10**: 163–246.
Magee, P. N. & Farber, E. (1962). Toxic liver injury and carcinogenesis Methylation of rat liver nucleic acids by dimethylnitrosamine *in vitro*. *Biochem. J.*, **83**: 114–24.
Magee, P. N. & Hultin, T. (1962). Toxic liver injury and carcinogenesis Methylation of proteins of rat liver slices by dimethylnitrosamine *in vitro*. *Biochem. J.*, **83**: 106–13.
Magee, P. N. & Lee, K. Y. (1963). Experimental toxic liver injury by some nitrosamines. *Ann. N.Y. Acad. Sci.*, **104**: 916–25.
Malling, H. V. (1966). Mutagenicity of two potent carcinogens, dimethylnitrosamine and diethylnitrosamine in *Neurospora crassa*. *Mutat. Res.*, **3**: 537–40.
Malling, H. V. (1971). Dimethylnitrosamine: Formation of mutagenic compounds by interaction with mouse liver microsomes. *Mutat. Res.*, **13**: 425–9.
Mandell, J. & Greenberg, J. (1960). A new chemical mutagen for bacteria, 1-methyl-3-nitro-1-nitrosoguanidine. *Biochem. Biophys. Res. Comm.*, **3**: 575–77.
Mangino, M. M. & Scanlan, R. A. (1984). Rapid formation of N-nitrosodimethylamine from gramine, a naturally occurring precursor in barley malt. *IARC Scientific Publication* No. 57: 337–46.
Manson, D., Cox, P. J. & Jarman, M. (1978). Metabolism of N-nitrosomorpholine by the rat *in vivo* and by rat liver microsomes and its oxidation by the Fenton system. *Chem.-Biol. Interact.*, **20**: 341–54.
Marckwald, W. & Droste-Huelshorff, A. F. (1898). Die Darstellung sekundäre Amine aus Sulsamiden. *Ber. Dtsch. Chem. Ges.*, **31**: 3261–6.
Martino, P. E., Diaz-gomez, M. I., Tamayo, D., Lopez, A. J. & Castro, J. A. (1988). Studies on the mechanism of the acute and carcinogenic effects of N-nitrosodimethylamine on mink liver. *J. Toxicol. Env. Health*, **23**: 183–92.
McCann, J., Choi, E., Yamasaki, E. & Ames, B. N. (1975). Detection of carcinogens as mutagens in the *Salmonella*/microsome test assay of 300 chemicals. *Proc. Natl. Acad. Sci. U.S.A.*, **72**: 5135–9.
McGlashan, N. D., Walters, C. L. & McLean, A. E. M. (1968). Nitrosamines in African alcoholic spirits and esophageal cancer. *Lancet* ii, 1017.
McKay, A. F. (1948). A new method of preparation of diazomethane. *J. Amer. Chem. Soc.*, **70**: 1974–5.
McKay, A. F. & Wright, G. F. (1947). Preparation and properties of N-methyl-N-nitroso-N'-nitroguanidine. *J. Amer. Chem. Soc.*, **69**: 3028–30.
McLean, A. E. M. & Verschuuren, H. G. (1969). Effect of diet and microsomal enzyme induction on the toxicity of dimethylnitrosamine. *Brit. J. Exp. Path.*, **50**: 22–5.
Mehta, R., Labuc, G. E. & Archer, M. C. (1984). Tissues and species specificity of the microsomal metabolism of N-nitroso-methylbenzylamine. *IARC Scientific Publication* No. 57: 473–8.
Mehta, R. D. & Von Borstel, R. C. (1984). Genetic activity in yeast assays of

reputed nonmutagenic, carcinogenic N-nitroso compounds and methapyrilene hydrochloride. *IARC Scientific Publication* No. 57: 721–9.

Mennel, H. D. & Zülch, K. J. (1972). Zur Morphologie transplacentar erzeugter neurogener Tumoren bem Goldhamster. *Acta Neuropath.* (*Berlin*), **21**: 140–53.

Mergens, W. J., Kamm, J. J. & Newmark, H. L. (1978). Alpha-tocopherol: Uses in preventing nitrosamine formation. *IARC Scientific Publication* No. 19, 199–212.

Mergens, W. J., Vane, F. M., Tannenbaum, S. R., Green, L. & Skipper, P. L. (1979). In vitro nitrosation of methapyrilene. *J. Pharm. Sci.*, **68**: 827–32.

Michejda, C. J., Andrews, A. W. & Koepke, S. R. (1979). Derivatives of side-chain hydroxylated nitrosamines direct acting mutagens in *Salmonella typhimurium*. *Mutation Res.*, **67**: 301–8.

Michejda, C. J. & Koepke, S. R. (1978). Powerful anchimeric effect of the N-nitroso group. *J. Amer. Chem. Soc.*, **100**: 1959–60.

Michejda, C. J., Koepke, S. R., Droeger-Keopke, M. B. & Bosan, W. (1987). Recent findings on the metabolism of β-hydroxyalkylnitrosamines. *IARC Scientific Publication* No. 84: 77–82.

Michejda, C. J., Kroeger-Koepke, M. B., Koepke, S. R., Magee, P. N. & Chu, C. (1982). Nitrogen formation during in vivo and in vitro metabolism of N-nitrosamines. *Banbury Report* No. 12, Cold Spring Harbor, N.Y., 69–85.

Mico, B. A., Swagzdis, J. E., Hu, H. S., Keefer, L. K., Odfield, N. F. & Garland, W. A. (1985). Low dose in vivo pharmacokinetic and deuterium isotope effect studies of N-nitrosodimethylamine in rats. *Cancer Res.*, **45**: 6280–5.

Mirvish, S. S. (1975). Formation of N-nitroso compounds: Chemistry, kinetics, and in vivo occurrence. *Toxicol. Appl. Pharmacol.*, **31**: 325–51.

Mirvish, S. S. & Garcia, H. (1973). 1-Nitroso-5,6-dihydrouracil: Induction of liver cell carcinomas and kidney adenomas in the rat. *Z. Krebsforsch.*, **79**: 304–8.

Mirvish, S. S., Cardesa, A., Wallcave, L. & Shubik, P. (1975). Induction of mouse lung adenomas by amines or ureas plus nitrite, and by N-nitroso compounds: Effects of ascorbate, gallic acid, thiocyanate, and caffeine. *J. Natl. Cancer Inst.*, **55**: 633–6.

Mirvish, S. S., Issenberg, P. & Sornson, H. C. (1976). Air–water and ether–water distribution of N-nitroso compounds: implications for laboratory safety, analytic methodology, and carcinogenicity for the rat esophagus, nose and liver. *J. Natl. Cancer Inst.*, **56**: 1125–9.

Mirvish, S. S., Ji, C. & Rosinsky, S. (1988). Hydroxy metabolites of methyl-*n*-amylnitrosamine produced by rat esophagus, stomach, liver, and other tissues of the neonatal to adult rat and hamster. *Cancer Res.*, **48**: 5663–8.

Mirvish, S. S., Wallcave, L., Eagen, M. & Shubik, P. (1972). Ascorbate–nitrite reaction: Possible means of blocking the formation of carcinogenic N-nitroso compounds. *Science*, **177**: 65–8.

Mirvish, S. S., Wang, M.-Y., Smith, J. W., Deshpande, A. D., Makary, M. H. & Issenberg, P. (1985). β- to ω-Hydroxylation of the esophageal carcinogen methyl-*n*-amylnitrosamine by the rat esophagus and related tissues. *Cancer Res.*, **45**: 577–83.

Mochizuki, M., Anjo, T. & Okada, M. (1980). Isolation and characterization of N-alkyl-N(hydroxymethyl)nitrosamines by deoxygenation. *Tetrahedron Lett.*, **21**: 3693–6.

Mochizuki, M., Anjo, T., Takeda, K., Suzuki, E., Sekiguchi, N., Huang, G. F.

& Okada, M. (1982). Chemistry and mutagenicity of α-hydroxy nitrosamines. *IARC Scientific Publication* No. 41: 553–9.
Mochizuki, M., Ikarashi, A., Suzuki, A., Sekiguchi, N., Anjo, T. & Okada, M. (1987). Chemical and mutagenic properties of 2-phosphonooxynitrosamines. *IARC Scientific Publication* No. 84: 165–9.
Mohr, U. & Althoff, J. (1965). Die diaplacentare Wirkung des Cancerogens Diäthylnitrosamin bei der Maus. *Z. Krebsforsch.*, **67**: 152–5.
Mohr, U., Althoff, J. & Wrba, H. (1965). Diaplazentare Wirkung des Carcinogens Diäthylnitrosamin beim Goldhamster. *Z. Krebsforsch.*, **66**: 536–40.
Mohr, U., Reznik, G., Emminger, E. & Lijinsky, W. (1977). Induction of pancreatic duct carcinomas in the Syrian Golden Hamster with 2,6-dimethylnitrosomorpholine. *J. Natl. Cancer Inst.*, **58**: 429–32.
Montesano, R. & Magee, P. N. (1970). Metabolism of dimethylnitrosamine by human liver slices *in vitro*. *Nature*, **228**: 173–4.
Montgomery, J. A., James, R., McCaleb, G. S. & Johnston, T. P. (1967). The modes of decomposition of 1,3-bis-(2-chloroethyl)-1-nitrosourea and related compounds. *J. Med. Chem.*, **10**: 668–74.
Mori, Y., Takahashi, H., Yamazaki, H., Toyoshi, K., Makino T., Yokose, Y. & Konishi, Y. (1984). Distribution, metabolism and excretion of *N*-nitroso bis-(2-hydroxypropyl)amine in Wistar rats. *Carcinogenesis*, **5**: 1443–7.
Mori, Y., Yamazaki, H., Toyoshi, K., Maruyama, H. & Konishi, Y. (1986). Activation of carcinogenic *N*-nitrosopropylamines to mutagens by lung and pancreas S9 fractions from various animal species and man. *Mutation Res.*, **160**: 159–69.
Moss, R. A., Landon, J. J., Luchter, K. M. & Mamantov, A. (1972). A flexible and directed synthesis of azoxyalkanes. *J. Amer. Chem. Soc.*, **94**: 4392–4.
Nagel, D. L., Lewis, R., Fischer, M., Stansbury, K., Stepan, K. & Lawson, T. A. (1987). β-Oxidized *N*-nitrosoalkylcarbamates as models for DNA alkylation by *N*-nitrosobis(2-oxopropyl)amine in Syrian hamsters. *IARC Scientific Publication* No. 84: 71–4.
National Academy of Sciences (1981). The health effects of nitrate, nitrite, and *N*-nitroso compounds. Washington, DC.
Neurath, G. B., Dünger, M. & Pein, F. G. (1976). Nitrosation of nornicotine and nicotine in gaseous mixtures and aqueous solutions. *IARC Scientific Publication* No. 14: 227–36.
Newman, M. S. & Kutner, A. (1951). New reactions involving alkaline treatment of 3-nitro-2-oxazolidones. *J. Amer. Chem. Soc.*, **73**: 4199–204.
Nixon, J. E., Wales, J. H., Scanlan, R. A., Bills, D. D. & Sinnhuber, R. O. (1976). Null carcinogenic effect of large doses of nitrosoproline and nitrosohydroxyproline in Wistar rats. *Food Cosmet. Toxicol.*, **14**: 133–5.
Ogiu, T., Nakadate, M. & Odashima, S. (1975). Induction of leukemias and digestive tract tumors in Donryu rats by 1-propyl-1-nitrosourea. *J. Natl. Cancer Inst.*, **54**: 887–93.
Ohshima, H. & Bartsch, H. (1981). Quantitative estimation of endogenous nitrosation in humans by monitoring *N*-nitrosoproline excreted in the urine. *Cancer Res.*, **41**: 3658–62.
Okada, M. (1984). Comparative metabolism of *N*-nitrosamines in relation to their organ and species specificity. *IARC Scientific Publication* No. 57: 401–9.
Okada, M., Mochizuki, M., Anjo, T., Sone, T., Wakabayashi, Y. & Suzuki, E. (1980). Formation, deoxygenation and mutagenicity of α-hydroperoxydialkylnitrosamines. *IARC Scientific Publication* No. 31: 71–82.
Okada, M., Suzuki, E. & Hashimoto, Y. (1976a). Carcinogenicity of *N*-

nitrosamines related to N-butyl-N-(4-hydroxybutyl)nitrosamine and N,N-dibutylnitrosamine in ACI/N rats. *GANN*, **67**: 825–34.

Okada, M., Suzuki, E. & Mochizuki, M. (1976b). Possible important role of urinary N-methyl-N-(3-carboxypropyl)nitrosamine in the induction of bladder tumors in rats by N-methyl-N-dodecylnitrosamine. *GANN*, **67**: 771–2.

Okun, J. D. & Archer, M. C. (1977). Kinetics of nitrosamine formation in the presence of micelle-forming surfactants. *J. Natl. Cancer Inst.*, **58**: 409–11.

Ong, J. T. M. & Rutherford, B. S. (1980). Some factors affecting the rat of N-nitrosodiethanolamine formation from 2-bromo-2-nitropropane-1,3-diol and ethanolamines. *J. Soc. Cosmet. Chem.*, **31**: 153–9.

Oshiro, Y., Zielinski, W. L. & Keefer, L. K. (1975). The deuterium isotope effect on the metabolism of dimethylnitrosamine *in vivo*. *Proc. Amer. Assoc. Cancer Res.*, **16**: 8.

Osske, G., Warzok, R. & Schneider, J. (1972). Diaplazentare Tumorinduktion durch endogen gebildeten N-Äthyl-N-Nitrosoharnstoff bei Ratten. *Arch. Geschwulstforsch.*, **40**: 244–7.

Park, K. K. & Archer, M. C. (1978). Microsomal metabolism of N-nitrosodi-n-propylamine: Formation of products resulting from α- and β-oxidation. *Chem.-Biol. Interact.*, **22**: 83–90.

Pegg, A. E. (1983a). Alkylation and subsequent repair of DNA after exposure to dimethylnitrosamine and related carcinogens. *Rev. Biochem. Toxicol.*, **5**: 83–133.

Pegg, A. E. (1983b). Properties of the O^6-alkylguanine-DNA repair system of mammalian cells. *IARC Scientific Publication* No. 57: 575–80.

Pegg, A. E. & Lijinsky, W. (1984). Saturation of repair system for O^6-methylguanine in rat liver DNA by pretreatment with cyclic nitrosamines. *Chem.-Biol. Interactions*, **51**: 365–70.

Pegg, A. E., Schicchitano, D. & Dolan, M. E. (1984). Comparison of the rates of repair of O^6-alkylguanines in DNA by rat liver and bacterial O^6-alkylguanine-DNA alkyltransferase. *Cancer Res.*, **44**: 3806–11.

Peto, R., Gray, R., Brantom, P. & Grasso, P. (1984). Nitrosamine carcinogenesis in 5120 rodents: chronic administration of sixteen different concentrations of NDEA, NDMA, NPYR and NPIP in the water of 4440 inbred rats, with parallel studies on NDEA alone of the effect of age of starting (3, 6 or 20 weeks) and of species (rats, mice or hamsters). *IARC Scientific Publication* No. 57: 627–65.

Piegorsch, W. W. & Hoel, D. G. (1988). Exploring relationships between mutagenic and carcinogenic potencies. *Mutation Res.*, **196**: 161–75.

Pielsticker, K., Wieser, D., Mohr, U. & Wrba, H. (1967). Diaplazentar induzierte Nierentumoren bei der Ratte. *Z. Krebsforsch.*, **69**: 345–50.

Pienta, R. J. (1980). Transformation of Syrian Hamster embryo cells by diverse chemicals and correlation with their reported carcinogenic and mutagenic activities. In: *Chemical Mutagens*, vol. 6 (ed. F. J. deSerres & A. Hollaender), Plenum Publishing Corp., New York, pp. 175–202.

Poirier, S., Hubert, A., de-Thé, G., Ohshima, H., Bourgade, M.-C. & Bartsch, H. (1987). Occurrence of volatile nitrosamines in food samples collected in three high-risk areas for nasopharyngeal carcinoma. *IARC Scientific Publication* No. 84: 415–19.

Polo, J. & Chow, Y. L. (1976). Efficient photolytic degradation of nitrosamines. *J. Natl. Cancer Inst.*, **56**: 997–1001.

Polonski, T. & Prajer, K. (1976). Optical activity of nitrosamines. A new sector rule for the N-nitroso chromophore. *Tetrahedron*, **32**: 847–53.

Pour, P. (1978). A new and advantageous model for colorectal cancer. *Cancer Lett.*, **4**: 293–8.
Pour, P. (1983). Prostatic cancer induced in MRC rats by *N*-nitrosobis(2-oxopropyl)amine and *N*-nitrosobis(2-hydroxypropyl)amine. *Carcinogenesis*, **4**: 49–55.
Pour, P. (1986a). Induction of exocrine pancreatic, bile duc, and thyroid gland tumors in offspring of Syrian hamsters treated with *N*-nitrosobis(2-oxopropyl)amine during pregnancy. *Cancer Res.*, **46**: 3663–6.
Pour, P. (1986b). Transplacental induction of gonadal tumors in rats by a nitrosamine. *Cancer Res.*, **46**: 4135–8.
Pour, P., Althoff, J., Gingell, R., Kupper, R., Kruger, F. W. & Mohr, U. (1976). *N*-nitroso-bis(2-acetoxypropyl)amine as a further pancreatic carcinogen in Syrian Golden Hamsters. *Cancer Res.*, **36**: 2877–84.
Pour, P., Althoff, J., Krüger, F. W. & Mohr, U. (1977). A potent pancreatic carcinogen in Syrian Hamsters: *N*-Nitrosobis(2-oxopropyl)amine. *J. Natl. Cancer Inst.*, **58**: 1449–53.
Pour, P., Gingell, R., Langenbach, R., Nagel, D., Grandjean, C., Lawson, T. & Salmasi, S. (1980). Carcinogenicity of *N*-nitrosomethyl(2-oxopropyl)amine in Syrian Hamsters. *Cancer Res.*, **40**: 3585–90.
Pour, P. M., Grandjean, C. J. & Knepper, S. (1985). Selective induction of nasal cavity tumors in rats by diallylnitrosamine. *J. Cancer Res. Clin. Oncol.*, **109**: 5–8.
Pour, P., Krüger, F. W., Althoff, J., Cardesa, A. & Mohr, U. (1974). The effect of beta-oxidized nitrosamines on Syrian Golden Hamsters. 3. 2,2′-Dihydroxydi-*n*-propylnitrosamine. *J. Natl. Cancer Inst.*, **54**: 141–6.
Pour, P. & Raha, C. R. (1981). Pancreatic carcinogenic effect of *N*-nitrosobis-(2-oxobutyl)-amine and *N*-nitroso-(2-oxobutyl)(2-oxopropyl)amine in Syrian Hamster. *Cancer Lett.*, **12**: 223–9.
Pour, P. & Wallcave, L. (1981). The carcinogenicity of *N*-nitrosodiethanolamine, an environmental pollutant, in Syrian Hamsters. *Cancer Lett.*, **14**: 23–7.
Pour, P., Wallcave, L., Gingell, R., Nagel, D., Lawson, T., Salmasi, S. & Tines, S. (1979). Carcinogenic effect of *N*-nitroso(2-hydroxypropyl)-(2-oxopropyl)amine, a postulated proximate pancreatic carcinogen in Syrian Hamster. *Cancer Res.*, **39**: 3828–33.
Pour, P., Wallcave, L. & Nagel, D. (1981). The effect of *N*-nitroso-2-methoxy-2,6-dimethylmorpholine on endocrine and exocrine and exocrine pancreas of Syrian Hamsters. *Cancer Lett.*, **13**: 233–40.
Preussmann, R., Druckrey, H. & Bucheler, J. (1968). Carcinogene Wirkung von Phenyl-nitrosoharnstoff. *Z. Krebsforsch.*, **71**: 63–5.
Preussmann, R. & Eisenbrand, G. (1984). *N*-Nitroso carcinogens in the environment. In: *Chemical Carcinogens*, ed. C. E. Searle, *Amer. Chem. Soc. Monograph* No. 182, pp. 829–68.
Preussmann, R., Habs, M., Habs, H. & Schmähl, D. (1982). Carcinogenicity of *N*-nitrosodiethanolamine in rats at five different dose levels. *Cancer Res.*, **42**: 5167–71.
Preussmann, R., Habs, M., Pool, B., Stummeyer, D., Lijinsky, W. & Reuber, M. D. (1981). Fluoro-substituted *N*-nitrosamines. 1. Inactivity of *N*-nitroso-bis-(2,2,2-trifluoroethyl)amine in carcinogenicity and mutagenicity tests. *Carcinogenesis*, **2**: 753–6.
Preussmann, R., Schmähl, D. & Eisenbrand, G. (1977). Carcinogenicity of *N*-nitrosopyrrolidine: Dose–response study in rats. *Z. Krebsforsch.*, **90**: 161–6.
Preussmann, R. & Stewart, B. W. (1984). *N*-Nitroso carcinogens. In: *Chemical*

Carcinogens, ed. C. E. Searle, *American Chemical Society Monograph* No. 182, pp. 643–828.
Prival, M. J. & Mitchell, V. D. (1981). Influence of microsomal and cytosolic fractions from rat, mouse and hamster liver on the mutagenicity of dimethylnitrosamine in the *Salmonella* plate incorporation assay. *Cancer Res.*, **41**: 4361–2.
Prival, M. J. & Mitchell, V. D. (1984). Dimethylnitrosamine demethylase and the mutagenicity of dimethylnitrosamine: Effects of rodent liver fractions and dimethylsulfoxide. In: *Genotoxicology of* N-*Nitroso Compounds*, ed. T. K. Rao et al., Plenum Press, New York, pp. 149–65.
Prokopczyk, B., Rivenson, A., Bertinato, P., Brunnemann, K. D. & Hoffmann, D. (1987). A study of betel quid carcinogenesis.V. 3-(methylnitrosamino)-propionitrile: occurrence in saliva, carcinogenicity and DNA methylation in F344 rats. *Cancer Res.*, **47**: 467–71.
Quarles, J. M., Schenley, C. K. & Tennant, R. W. (1975). Use of a transplacental host-mediated culture system for assay of transformation by nitroso compounds. *Int. Res. Commun. System (Cancer)*, **3**: 22.
Quarles, J. M., Sega, M. W., Schenley, C. K. & Lijinsky, W. (1979). Transformation of hamster fetal cells by nitrosated pesticides in transplacental assay. *Cancer Res.*, **39**: 4525–33.
Qin, X., Nakatsuru, Y., Kohyama, K. & Ishikawa, T. (1990). DNA adduct formation and unscheduled DNA synthesis in rat esophagus *in vivo* after treatment with N-methyl-N-nitrosourea. *Carcinogenesis*, **11**: 235–8.
Raineri, R., Poiley, J. A., Andrews, A. W., Pienta, R. & Lijinsky, W. (1981). Greater effectiveness of hepatocyte and liver S9 preparations from hamsters than rat preparations in activating N-nitroso compounds to metabolites mutagenic to *Salmonella*. *J. Natl. Cancer Inst.*, **67**: 1117–22.
Rainey, W. T., Christie, W. H. & Lijinsky, W. (1978). Mass spectrometry of N-nitroso compounds. *Biomedical Mass Spectrometry.*, **5**: 395–408.
Rainey, W. T., Christie, W. H., Pritchard, C. A. & Lijinsky, W. (1976). Mass spectra of N-nitroso compounds. *Oak Ridge National Laboratory Technical Report* TM-5500.
Rao, M. S. & Pour, P. (1978). Development of biliary and hepatic neoplasms in guinea pigs treated with N-nitrosobis(2-oxopropyl)amine. *Cancer Lett.*, **5**: 31–4.
Rao, M. S. & Reddy, I. K. (1977). Induction of malignant vascular tumors of the liver in guinea pigs treated with 2,2'-dihydroxy-di-n-propylamine. *J. Natl. Cancer Inst.*, **58**: 387–92.
Rao, M. S., Scarpelli, D. G. & Lijinsky, W. (1981). Carcinogenesis in Syrian Hamsters by N-nitroso-2,6-dimethylmorpholine, its *cis* and *trans* isomers, and the effect of deuterium labeling. *Carcinogenesis*, **2**: 731–5.
Rao, T. K. (1984). Structural basis for mutagenic activity of N-nitrosamines in the *Salmonella* histidine reversion assay. In: *Genotoxicology of* N-*Nitroso Compounds*, ed. T. K. Rao, W. Lijinsky & J. L. Epler, Plenum Press, New York, pp. 45–58.
Rao, T. K., Epler, J. L. & Lijinsky, W. (1982). Structure–activity studies with N-nitrosamines using *Salmonella typhimurium* and *E. coli*. *IARC Scientific Publication* No. 41: 543–51.
Rao, T. K., Lijinsky, W. & Epler, J. L. (1984). *Genotoxicology of* N-*nitroso Compounds*, Plenum Press, New York.
Rao, T. K., Young, J. A., Lijinsky, W. & Epler, J. L. (1979). Mutagenicity of aliphatic nitrosamines in *Salmonella typhimurium*. *Mutation Res.*, **66**: 1–9.
Rayman, M. P., Challis, B. C., Cox, P. J. & Jarman, M. (1975). Oxidation of

N-nitrosopiperidine in the Udenfriend model system and its metabolism by rat liver microsomes. *Biochem. Pharmacol.*, **24**: 621–6.

Renger, B., Kalinowski, H. O. & Seebach, D. (1977). Regio- and stereochemical course of the reactions of substituted 2-lithio-N-nitropiperidines. *Chem. Ber.*, **110**: 1866–78.

Reynolds, C. A. & Thomson, C. (1985). Theoretical investigation of possible intermediates in chemical carcinogenesis by N-nitrosamines. In: *Molecular Basis of Cancer, Part A: Macromolecular Structure. Carcinogens and Oncogenes*, pp. 239–48, Alan R. Liss, New York.

Reznik, G., Lijinsky, W. & Mohr, U. (1978a). Carcinogenicity of subcutaneously injected N-nitrosoheptamethyleneimine in European Hamsters. *J. Natl. Cancer Inst.*, **61**: 239–43.

Reznik, G., Mohr, U. & Krüger, F. W. (1975). Carcinogenic effects of di-n-propylnitrosamine, β-hydroxypropyl-n-propylnitrosamine, and methyl-n-propylnitrosamine on Sprague-Dawley rats. *J. Natl. Cancer Inst.*, **54**: 937–43.

Reznik, G., Mohr, U. & Lijinsky, W. (1978b). Carcinogenic effect of N-nitroso-2,6-dimethylmorpholine in Syrian Golden Hamsters. *J. Natl. Cancer Inst.*, **60**: 371–8.

Reznik-Schüller, H. M. (1983). Cancer induced in the respiratory tract of rodents by N–nitroso compounds. In: *Comparative respiratory Tract Carcinogenesis*, vol. II, CRC Press, Boca Raton, Florida, pp. 109–34.

Reznik-Schüller, H. M. & Gregg, M. (1983). Ultrastructure of nitrosoheptamethyleneimine induced lung tumors in Fischer rats. *Anticancer Res.*, **6**: 381–4.

Reznik-Schüller, H. M. & Hague, B. F. (1981). Autoradiographic study of the distribution of bound radioactivity in the respiratory tract of Syrian Hamsters given N-[^{3}H]nitrosodiethylamine. *Cancer Res.*, **41**: 2147–50.

Reznik-Schüller, H. M. & Lijinsky, W. (1979). In vivo autoradiography and nitrosoheptamethyleneimine carcinogenesis in hamsters. *Cancer Res.*, **39**: 72–4.

Rhoades, J. W. & Johnson, D. E. (1972). N-Dimethylnitrosamine in tobacco smoke condensate. *Nature*, **236**: 307–8.

Rice, J. M., London, W. T., Palmer, A. E., Sly, D. L. & Williams, G. M. (1977). Direct and transplacental carcinogenesis by ethylnitrosourea in the patas monkey (*Erythrocebus patas*). *Proc. Am. Assoc. Cancer Res.*, **18**: 53.

Richter, E., Lorck, C. & Wiessler, M. (1988a). Intestinal first-pass metabolism of nitrosamines. 3. Identification of metabolites of N-nitrosodiamylamine. *Carcinogenesis*, **9**: 507–9.

Richter, E., Zwickenpflug, W. & Wiesler, M. (1988b). Intestinal first pass metabolism of nitrosamines. 2. Metabolism of N-nitrosodibutylamine in isolated perfused rat small intestine segments. *Carcinogenesis.*, **9**: 499–506.

Ridd, J. H. (1961). Nitrosation, diazotisation, and deamination. *Quart. Rev. Chem. Soc.*, **15**: 418–41.

Rivenson, A., Hoffmann, D., Prokopczyk, B., Amin, S. & Hecht, S. S. (1988). Induction of lung and exocrine pancreas tumors in F344 rats by tobacco-specific and areca-derived N-nitrosamines. *Cancer Res.*, **48**: 6912–17.

Röper, H., Röper. S. & Meyer, B. (1984). Amadori- and N-nitrosoamadori compounds and their pyrolysis products chemical, analytical and biological aspects. *IARC Scientific Publication* No. 57: 101–11.

Rosenkranz, H. S. & Klopman, G. (1987). Computer automated structure evaluation of the carcinogenicity of N-nitrosothiazolidine and N-nitrosothiazolidine-4-carboxylic acid. *Fd. Chem. Toxicol.*, **25**: 253–6.

Ross, A. E., Keefer, L. & Lijinsky, W. (1971) Alkylation of nucleic acids of rat liver and lung by deuterated N-nitrosodiethylamine *in vivo*. *J. Natl. Cancer Inst.*, **47**: 789–95.

Ross, A. E. & Mirvish, S. S. (1977). Metabolism of N-nitrosohexamethyleneimine to give 1,6-hexanediol bound to rat liver nucleic acids. *J. Natl. Cancer Inst.*, **58**: 651–5.
Russell, L. B. & Montgomery, C. S. (1982). Supermutagenicity of ethylnitrosourea in the mouse spot test. Comparison with methylnitrosourea and ethylnitrosourethane. *Mutat. Res.*, **92**: 193–204.
Russell, W. L., Kelly, E. M., Hunsucker, P. R., Bangham, J. W., Maddux, S. C. & Phipps, E. L. (1979). Specific-locus test shows ethylnitrosourea to be the most potent mutagen in the mouse. *Proc. Natl. Acad. Sci. U.S.A.*, **76**: 5818–19.
Rustia, M. (1974). Multiple carcinogenic effects of the ethylnitrosourea precursors ethylurea and sodium nitrite in hamsters. *Cancer Res.*, **34**: 3232–44.
Saavedra, J. E. (1978). Oxidative decarboxylation of nitrosamino acids: A synthetic approach to cyclic α-acetoxynitrosamines. *Tetrahedron Lett.*, 1923–6.
Saavedra, J. E. (1979). Oxidation of nitrosamines. I. Formation of N-nitrosoimminium ions through the oxidative decarboxylation of N-nitrosoproline, N-nitrosopipecolic acid and N-nitrososarcosine. *J. Org. Chem.*, **44**: 4511–15.
Saavedra, J. E. (1983). Lithiation of α-nitrosaminoalkyl ethers. Synthetic equivalents of α-primary amino carbanions. *J. Org. Chem.*, **48**: 2388–92.
Saavedra, J. E. (1987a). Recent synthetic application of N-nitrosamines and related compounds. *Org. Prep. & Proc. Int.*, **19**: 83–159.
Saavedra, J. E. (1987b). Heteroatom-substituted sp³ carbonionic synthons. *In*: *Umpoled Synthons: a Survey of Sources and Uses in Synthesis*, T. Hase, ed., pp. 101–43, John Wiley and Sons, New York.
Saavedra, J. E. (1990). Decomposition of 1-(nitrosoalkyl)-3-(2-hydroxyalkyl)ureas. *J. Org. Chem.*, **55**: 6373–4.
Saavedra, J. E., Farnsworth, D. & Pei, G. P. (1988). Oxidation of β-hydroxynitrosamines and β-oxo derivatives with CrO_3/Celite or CrO_3/Florisil in non-aqueous media. *Synth. Commun.*, **18**: 313–22.
Saavedra, J. E., Farnsworth, D. W. & Farrelly, J. G. (1989). β-ketonitrosamines. Synthetic equivalents of α-methylene alkylamino anions ($^-CH_2NHR$). *Synth. Commun.*, **19**: 1147–56.
Sakshaug, J., Sögnen, E., Hansen, M. A. & Koppang, N. (1965). Dimethylnitrosamine: Its hepatotoxic effect in sheep and its occurrence in toxic batches of herring meal. *Nature*, **206**: 1261–2.
Sander, J. (1970). Induktion maligner Tumoren bei Ratten durch orale Gabe von N,N-Dimethylharstoff und Nitrit. *Arzneimittel-Forsch.*, **20**: 418–19.
Sander, J. & Bürkle, G. (1969). Induktion maligner Tumoren bei Ratten durch gleichzeitige Verfutterung von Nitrit und sekundaren Aminen. *Z. Krebsforsch.*, **73**: 54–66.
Sander, J. & Bürkle, G. (1971). Induktion maligner Tumoren bei Ratten durch oraler Gabe von 2-Imidazolinon und Nitrit. *Z. Krebsforsch.*, **75**: 301–4.
Sander, J. & Schweinsberg, F. (1972). Wechselbeziehungen zwischen Nitrat, Nitrit und kanzerogenen N-Nitrosoverbindungen. 2 Mitteilung: Untersuchungen uber die Entstehung von Nitrosaminen und Nitrosamiden im Menschen, im Tier und in Nahrungsmitteln. *Zbl. Bakt. Hyg.*, I. Abt. Orig. B **156**, 321–40.
Sander, J. & Seif, F. (1969). Bakterielle Reduktion von Nitrat im Magen des Menschen als Ursache einer Nitrosaminbildung. *Arzneimittel.-Forsch.* **19**, 1091–3.
Scanlan, R. A. (1974). N-Nitrosamines in foods. *CRC Critical Reviews in Food Technology*, **5**: 357–402.
Scanlan, R. A., Farrelly, J. G., Hecker, L. I. & Lijinsky, W. (1980). Lack of

metabolism of 2,6-dimethyldinitrosopiperazine by microsomes and post-microsomal supernatant prepared from the rat esophagus and non-glandular stomach. *Cancer Lett.*, **10**: 293–9.
Schmähl, D. & Habs, M. (1980). Carcinogenicity of *N*-nitroso compounds. *Oncology*, **37**: 237–42.
Schmähl, D., Habs, M. & Ivankovic, S. (1978). Carcinogenesis of *N*-nitrosodiethylamine in chickens and domestic cats. *Internatl. J. Cancer*, **22**: 552–7.
Schmähl, D., Krüger, F. W., Habs, M. & Diehl, B. (1976). Influence of disulfiram on the organotropy of the carcinogenic effect of dimethylnitrosamine and diethylnitrosamine in rats. *Z. Krebsforsch.*, **85**: 271–6.
Schmidtpeter, A. (1963). Reaction of nitrosamines with electrophiles. I. Alkylation of nitrosamines. *Tetrahedron Lett.*, 1421–4.
Schoental, R. (1960). Carcinogenic action of diazomethane and of nitroso-*N*-methylurethane. *Nature*, **188**: 420.
Schoental, R. (1966). Carcinogenic activity of *N*-methyl-*N*-nitroso-*N'*-nitroguanidine. *Nature*, **209**: 726–7.
Schreiber, H., Nettesheim. P., Lijinsky, W., Richter, C. B. & Walburg, H. E. (1972). Induction of lung cancer in germ free, specific pathogen free and infected rats by *N*-nitrosoheptamethyleneimine: Enhancement by respiratory infection. *J. Natl. Cancer Inst.*, **49**: 1107–14.
Schueler, F. W. & Hanna, C. (1952). A synthesis of unsymmetrical dimethyl hydrazine using lithium aluminum hydride. *J. Amer. Chem. Soc.*, **73**: 4996.
Seebach, D. & Enders, D. (1975). Reversion of the poles of reactivity of amines. Nucleophilic secondary α-aminoalkylation via metalated nitrosamines. *Angew. Chem.*, **87**: 1–18.
Sen, N. P., Donaldson, B., Iyengar, J. R. & Panalaks, T. (1973). Nitrosopyrrolidine and dimethylnitrosamine in bacon. *Nature*, **241**: 473–4.
Sen, N. P., Donaldson, B., Seaman, S., Collins, B. & Iyengar, J. Y. (1977). Recent nitrosamine analyses in cooked bacon. *Cancer Inst. Food Sci. Technol. J.*, **10**: A13–15.
Sen, N. P., Seaman, S. & McPherson, M. (1980). Further studies on the occurence of volatile and non-volatile nitrosamines in foods. *IARC Scientific Publication* No. 31, 457–65.
Sen, N. P., Seaman, S. & Tessier, L. (1982). A rapid and sensitive method for the determination of non-volatile *N*-nitroso compounds in foods and human urine: Recent data concerning volatile *N*-nitrosamines in dried foods and malt-based beverages. *IARC Scientific Publication* No. 41: 185–97.
Shimada, H., Yakushi, K., Ikarashi, A., Mochizuki, M., Suzuki, E., Okada, M., Yokoyama, S., Miyazawa, T. & Hayatsu, H. (1987). Activation of *N*-nitrosodialkylamines by near-ultraviolet irradiation: Formation of directly-acting mutagens and DNA-damaging products. *IARC Scientific Publication* No. 84: 364–6.
Shimkin, M. B., Weisburger, J. H., Weisburger, E. K., Gubareff, N. & Suntzeff, V. (1966). Bioassay of 29 alkylating chemicals by the pulmonary-tumor response in strain A mice. *J. Natl. Cancer Inst.*, **36**: 915–35.
Shubik, P. & Sicé, J. (1956). Chemical carcinogenesis as a chronic toxicity test. *Cancer Res.*, **16**: 728–42.
Sieh, D. H., Andrews, A. W. & Michejda, C. J. (1980). Mutagenicity of trialkyltriazes: Mutagenic potency of alkydiazonium ions, the putative ultimate carcinogens from dialkylnitrosamines. *Mutation Res.*, **73**: 227–35.
Singer, B. (1975). The chemical effects of nucleic acid alkylation and their relation to mutagenesis and carcinogenesis. *Progress in Nucleic Acid Research and Mol. Biol.*, **15**: 219–84.

Singer, B. (1976). All oxygens in nucleic acids react with carcinogenic ethylating agents. *Nature*, **264**: 333–9.
Singer, G. M. (1980). The mechanism of nitrosation of tertiary amines. *IARC Scientific Publication* No. 31: 139–51.
Singer, G. M. & Lijinsky, W. (1979). Relative extents of hydrogen–deuterium exchange of nitrosamines: Relevance to biological isotope effect studies. *Cancer Lett.*, **8**: 29–34.
Singer, G. M., Lijinsky, W., Buettner, L. & McClusky, G. A. (1981). Relationship of rat urinary metabolites of N-nitrosomethyl-N-alkylamine to bladder carcinogenesis. *Cancer Res.*, **41**: 4942–6.
Singer, G. M. & MacIntosh, W. A. (1984). Urinary metabolits of some alicyclic nitrosamines. *IARC Scientific Publication* No. 57: 459–63.
Singer, G. M., Reuber, M. D., Mangino, M. M. & Lijinsky, W. (1984). The effect of 3-methyl substitution on carcinogenesis by nitroso-4-piperidone. *Carcinogenesis*, **5**: 1351–3.
Singer, G. M., Singer, S. S. & Schmidt, D. G. (1977). A nitrosamide specific detector for use with high pressure liquid chromatography. *J. Chromatog.*, **133**: 59–66.
Singer, G. M., Taylor, H. W. & Lijinsky, W. (1977). Liposolubility as an aspect of nitrosamine carcinogenicity: Quantitative correlations and qualitative observations. *Chem.-Biol. Interactions.*, **19**: 133–42.
Singer, S. S. (1980). Transnitrosation by nitrosamines and nitrosoureas. *IARC Scientific Publication* No. 31: 111–17.
Singer, S. S. (1982). Thermolysis of trialkylnitrosoureas: Formation of an unusual product. *J. Org. Chem.*, **47**: 3839–44.
Singer, S. S. (1984). Decomposition of N-nitrosohydroxyalkylureas and N-nitrosooxazolidones in aqueous buffer. *IARC Scientific Publication* No. 57: 371–5.
Singer, S. S. & Cole, B. B. (1981). Reactions of nitrosoureas and related compounds in dilute aqueous acid: Transnitrosation to piperidine and sulfamic acid. *J. Org. Chem.*, **46**: 3461–6.
Singer, S. S., Lijinsky, W. & Singer, G. M. (1978). Transnitrosation. An important aspect of the chemistry of aliphatic nitrosamines. In: *Environmental Aspects of N-Nitroso Compounds*, *IARC Scientific Publication* No. 19: 175–8.
Singer, S. S., Singer, G. M., Saavedra, J. E., Reuber, M. D. & Lijinsky, W. (1981). Carcinogenesis by derivatives of 1-nitroso-3,5-dimethyl-piperazine in rats. *Cancer Res.*, **41**: 1034–8.
Smith, P. A. S. & Loeppky, R. N. (1967). Nitrosative cleavage of tertiary amines. *J. Amer. Chem. Soc.*, **89**: 1147–57.
Smith, P. A. S. & Pars, M. G. (1959). Nitrosative cleavage of N',N'-dialkylhydrazides and tertiary amines. *J. Org. Chem.*, **24**: 1325–32.
Smith, R. H., Kovatch, R. M., Lijinsky, W. & Michejda, C. J. (1987). Tumors in F344 rats by oral administration of 1,3-diethyltriazene. *Cancer Lett.*, **35**: 129–132.
Snyder, C. M., Farrelly, J. G. & Lijinsky, W. (1977). Metabolism of three cyclic nitrosamines in sprague-Dawley rats. *Cancer Res.*, **37**: 3530–2.
Solleveld, H. A., Haseman, J. K. & McConnell, E. E. (1984). Natural history of body weight gain, survival, and neoplasia in the F344 rat. *J. Natl. Cancer Inst.*, **72**: 929–40.
Solonina, V. A. (1906). The decomposition of the nitrates of tertiary amines. *J. Russian Physico-Chem. Soc.*, **38**: 1286.
Somogyi, A., Conney, A. H., Kuntzman, R. & Solymoss, B. (1972). Protection

against dimethylnitrosamine toxicity by pregnenalone-16α-carbonitrile. *Nature New Biology*, **237**: 61–3.

Sparrow, A. (1973)., Hazards of chemical carcinogens and mutagens. *Science*, **181**: 700.

Spiegelhalder, B., Eisenbrand, G. & Preussmann, R. (1976). Influence of dietary nitrate on nitrite content of human saliva: Possible relevance to *in vivo* formation of N-nitroso compounds. *Food Cosmet. Toxicol.*, **14**: 545–8.

Spiegelhalder, B., Eisenbrand, G. & Preussmann, R. (1979). Contamination of beer with trace quantities of N-nitrosodimethylamine. *Food Cosmet. Toxicol.*, **17**: 29–31.

Spiegelhalder, B., Eisenbrand, G. & Preussmann, R. (1980). Occurrence of volatile nitrosamines in food: A survey of the West German Market. *IARC Scientific Publication* No. 31: 467–79.

Spiegelhalder, B., Eisenbrand, G. & Preussmann, R. (1980). Volatile nitrosamines in food. *Oncology*, **37**: 211–16.

Spiegelhalder, B. & Preussmann, R. (1982). Nitrosamines and rubber. *IARC Scientific Publication* No. 41: 231–43.

Spiegelhalder, B. & Preussmann, R. (1983). Occupational nitrosamine exposure. I. Rubber and tyre industry. *Carcinogenesis*, **4**: 1147–52.

Stephany, R. W., Freudenthal, J. & Schuller, P. L. (1978). N-Nitroso-5-methyl-1,3-oxazolidine identified as an impurity in a commercial cutting fluid. *Rev. Trav. Chim. Pays-Bas*, **97**: 177–8.

Sterzel, W. & Eisenbrand, G. (1986). N-Nitrosodiethanolamine is activated in the rat to an ultimate genotoxic metabolite by sulfotransferase. *J. Cancer Res. Clin. Oncol.*, **111**: 20–4.

Stewart, B. W., Swann, P. F., Holsman, J. W. & Magee, P. N. (1974). Cellular injury and carcinogenesis. Evidence for the alkylation of rat liver nucleic acids *in vivo* by N-nitrosomorpholine. *Z. Krebsforsch.*, **82**: 1–12.

Streeter, A. J., Nims, R. W., Hrabie, J. A., Heur, Y.-H. & Keefer, L. K. (1989). Sex differences in the single-dose toxicokinetics of N-nitrosomethyl (2-hydroxyethyl)amine in the rat. *Cancer Res.*, **49**: 1783–9.

Strickland, P. T., Lijinsky, W., Thomas, B. J. & Kovatch, R. M. (1988). Strain comparison of systemic N-nitrosohexamethyleneimine carcinogenesis in BALB/c, Sencar and CD-1 mice. *Cancer Lett.*, **41**: 139–46.

Sugimura, T. & Fujimura, S. (1967). Tumour production in glandular stomach of rat by N-methyl-N'-nitro-N-nitrosoguanidine. *Nature*, **216**: 943–4.

Sugimura, T., Fujimura, S., Nagao, M., Yokoshima, T. & Hasegawa, M. (1968). Reaction of N-methyl-N-nitrosoguanidine with protein. *Biochim. Biophys. Acta.*, **170**: 427–9.

Süssmuth, R., Haerlin, R. & Lingens, F. (1972). The mode of action of N-methyl-N'-nitro-N-nitrosoguanidine in mutagenesis. VII. The transfer of the methyl group of N-methyl-N'-nitrosoguanidine. *Biochim. Biophys. Acta.*, **269**: 276–81.

Suzuki, Y., Matsuyama, M. & Ogiu, T. (1984). Morphologic characteristics of thymic lymphomas induced by N-nitroso-N-propylurea in F344 rats. *J. Natl. Cancer Inst.*, **72**: 367–73.

Svoboda, D. & Higginson, J. (1968). A comparison of ultrastructural changes in rat liver due to chemical carcinogens. *Cancer Res.*, **28**: 1703–33.

Swann, P. F., Coe, A. M. & Mace, R. (1984). Ethanol and dimethylnitrosamine and diethylnitrosamine metabolism and distribution in the rat. Possible relevance to the influence of ethanol on human cancer incidence. *Carcinogenesis*, **5**: 1337–43.

Swann, P. F., Mace, R., Angeles, R. M. & Keefer, L. K. (1983). Deuterium

isotope effect on metabolism of N-nitrosodimethylamine *in vivo* in rat. *Carcinogenesis*, **44**: 821–6.
Swann, P. F. & Magee, P. N. (1968). Nitrosamine induced carcinogenesis. The alkylation of nucleic acids of the rat by N-methyl-N-nitrosourea, dimethylnitrosamine, dimethylsulphate and methylmethanesulphonate. *Biochem. J.*, **110**: 39–47.
Swann, P. F. & Magee, P. N. (1971). Nitrosamine induced carcinogenesis. The alkylation of N^7 of guanine of nucleic acids of the rat by diethylnitrosamine, N-ethyl-N-nitrosourea and ethyl methanesulphonate. *Biochem. J.*, **125**: 841–7.
Swann, P. F. & McLean, A. E. M. (1968). The effect of diet on the toxic and carcinogenic action of dimethylnitrosamine. *Biochem. J. (Proc.).*, **107**: 14–15.
Swenberg, J. M., Dyroff, M. C., Bedell, M. A., Popp, J. A., Huh, N., Kirstein, U. & Rajeweky, M. F. (1984). O^4-Ethyldeoxythymidine, but not O^6-ethyldeoxyguanosine, accumulates in hepatocyte DNA of rats exposed continously to diethylnitrosamine. *Proc. Natl. Acad. Sci. U.S.A.*, **81**: 1692–5.
Swenson, D. H., Fei, J. V. & Lawley, P. D. (1979). Synthesis of 1,(2-hydroxyethyl)-1-nitrosourea and comparison of its carcinogenicity with that of 1-ethyl-1-nitrosourea. *J. Natl. Cancer Inst.*, **63**: 1469–73.
Takahashi, M., Kurokawa, Y., Maekawa, A., Kokubo, T., Eurukawa, F., Mochizuki, M., Anjo, T. & Okada, M. (1982). Comparative carcinogenicities of model compounds of metabolically activated N,N-dibutylnitrosamine in rats. *GANN*, 73: 687–94.
Tanaka, T. (1973). Transplacental induction of tumors and malformations in rats treated with some chemical carcinogens. *IARC Scientific Publication* No. 4: 100–11.
Tannenbaum, S. R., Fett, D., Young, V. R., Land, P. D. & Bruce, W. R. (1978). Nitrite and nitrate are formed by endogenous synthesis in the human intestine. *Science.*, **200**: 1487–9.
Tannenbaum, S. R., Sinskey, A. I., Weisman, M. & Bishop, W. W. (1974). Nitrite in human saliva: Its possible relationship to nitrosamine formation. *J. Natl. Cancer Inst.*, **53**: 74–84.
Taylor, H. W. & Lijinsky, W. (1975*a*) Tumor induction in rats by feeding heptamethyleneimine and nitrite in water. *Cancer Res.*, **35**: 812–15.
Taylor, H. W. & Lijinsky, W. (1975*b*). Tumor induction in rats by feeding aminopyrine or oxytetracycline with nitrite. *Int. J. Cancer*, **16**: 211–15.
Taylor., H. W., Lijinsky, W., Nettesheim, P. & Snyder, C. M. (1974). Alteration of tumor response in rat liver by carbon tetrachloride. *Cancer Res.*, **34**: 3391–5.
Tennant, R. W., Margolin, B. H., Shelley, M. D., Zeiger, E., Haseman, J. K., Spalding, J., Caspary, W., Resnick, M., Stasiewicz, S., Anderson, B. & Minor, R. (1987). Prediction of chemical carcinogenicity in rodents from *in vitro* genetic toxicity assays. *Science*, **236**: 933–41.
Thomas, B. J. & Lijinsky, W. (1988). Alkylation of liver DNA by bis-(2-oxopropyl)-nitrosamine in rats of different ages. *GANN*, **79**: 1039–42.
Thomas, B. J., Lijinsky, W. & Kovatch, R. M. (1988). The induction of bladder tumors in F344 rats by intravesicular administration of some nitrosamines. *GANN*, **79**: 309–13.
Thomas, C. & Schmähl, D. (1963). Zur Morphologie der durch Diäthylnitrosamin erzeugen Lebertumoren bei der Maus und dem Meerschweinchen. *Z. Krebsforsch.*, **65**: 531–6.
Tjälve, H. & Castonguay, A. (1983). The *in vivo* tissue disposition and *in vitro* target tissue metabolism of the tobacco-specific carcinogen 4-

(methylnitrosamino)-1-(3-pyridyl)-1-butanone in Syrian Golden Hamsters. *Carcinogenesis.*, **4**: 1259–65.

Tjälve, H., Castonguay, A. & Rivenson, A. (1985). Microautoradiographic localization of bound metabolites in the nasal cavities of F344 rats treated with the tobacco-specific carcinogen 4-(methylnitrosamino)-1-(3-pyridyl)-1-butanone. *J. Natl. Cancer Inst.*, **74**: 185–9.

Tomatis, L. & Cefis, F. (1967). The effects of multiple and single administration of dimethylnitrosamine to hamsters. *Tumori.*, **53**: 447–52.

Toth, B. (1973). 1,1-Dimethylhydrazine (unsymmetrical) carcinogenesis in mice. Light microscopic and ultrastructural studies on neoplastic blood vessels. *J. Natl. Cancer Inst.*, **50**: 181–94.

Tsuda, M., Frank, N., Sato, S. & Sugimura, T. (1988). Marked increase in the urinary level of N-nitrosothioproline after ingestion of cod with vegetables. *Cancer Res.*, **48**: 4049–52.

Tsuda, M., Nagai, A., Suzuki, H., Hayashi, T., Ikeda, M., Kuratsune, M., Sato, S. & Sugimura, T. (1987). Effect of cigarette smoking and dietary factors on the amounts of N-nitrosothiazolidine 4-carboxylic acid and N-nitroso-2-methylthiazolidine 4-carboxylic acid in human urine. *IARC Scientific Publication* No. 84: 446–50.

Tu, Y. Y., Peng, R., Chang, Z. F. & Yang, C. S. (1983). Induction of a high affinity nitrosamine demethylase in rat liver microsomes by acetone and isopropanol. *Chem.-Biol. Interact.*, **44**: 247–60.

Tu, Y. Y., Sonnenberg, J., Lewis, K. F. & Yang, C. S. (1981). Pyrazole-induced cytochrome P-450 in rat liver microsomes: an isozyme with high affinity for dimethylnitrosamine. *Biochem. Biophys. Res. Commun.*, **103**: 905–12.

Underwood, B. & Lijinsky, W. (1982). Comparative metabolism of 2,6-dimethylnitrosomorpholine in rats, hamsters and guinea pigs. *Cancer Res.*, **42**: 54–8.

Underwood, B. & Lijinsky, W. (1984a). Comparative metabolism of the *cis* and *trans* isomers of N-nitroso-2,6-dimethylmorpholine in rats, hamsters and guinea pigs. *Chem.-Biol. Interactions.*, **50**: 175–88.

Underwood, B. N. & Lijinsky, W. (1984b). Metabolism of nitrosobis(2-oxopropyl)amine (BOP), nitrosobis-(2-hydroxypropyl)amine and nitrosobis(2-hydroxypropyl)(2-oxopropyl)amine in rats and hamsters. *Proc. Am. Assoc. Cancer Res.*, **25**: 115.

Urban, H. & Danz, M. (1976). Tumorigenicity of T.I.T. dissolved in DMSO *Arch. Geschwulstforsch.*, **46**: 657–62.

Van Romburgh, P. (1886). L'action de la chaleur sur les azetates des amines de la serie grasse. *Rec. Trav. Chim. Pays-Bas*, **5**: 246–51.

Von Brüning, G. (1888). Über Methylhydrazin. *Chem. Ber.*, **21**: 1809.

Von Hofe, E., Grahmann, F., Keefer, L. K., Lijinsky, W., Nelson, V. & Kleihues, P. (1986a). Methylation versus ethylation of DNA in target and non-target tissues of Fischer 344 rats treated with N-nitrosomethylethylamine. *Cancer Research*, **46**: 1038–42.

Von Hofe, E., Kleihues, P. & Keefer, L. K. (1986b). Extent of DNA 2-hydroxyethylation by N-nitrosomethylethylamine and N-nitrosodiethylamine. *Carcinogenesis*, **7**: 1335–7.

Von Hofe, E., Schmerold, J., Lijinsky, W. & Kleihues, P. (1987). DNA methylation in rat tissues by a series of homologous aliphatic nitrosamines ranging from N-nitrosodimethylamine to N-nitrosomethyldodecylamine. *Carcinogenesis*, **8**: 1337–41.

Von Pechmann, H. (1895). Über Diazomethan. *Chem. Ber.*, **28**: 855–61.

Walker, E. A., Castegnaro, M., Garren, L., Toussaint, G., Kowalski, B.

(1979). Intake of volatile nitrosamines from consumption of alcohols. *J. Natl. Cancer Inst.*, **63**: 947–51.

Walters, C. L. (1983). Group analysis of N-nitroso compounds. In: *Das Nitrosamin-Problem*, Verlag Chemie, Weinheim, pp. 93–6.

Watanabe, K., Reddy, S. Q., Weisburger, J. H. & Kritchevsky, D. (1979). Effect of dietary alfalfa, pectin, and wheat bran on azoxymethane or methylnitrosourea-induced colon carcinogenesis in F344 rats. *J. Natl. Cancer Inst.*, **63**: 141–45.

Waynforth, H. B. & Magee, P. N. (1975). The effect of various doses and schedules of administration of N-methyl-N-nitrosourea, with and without croton oil promotion, on skin papilloma production in Balb/c mice. *GANN Monogr. Cancer Res.*, **17**, 439–48.

Weil, C. S. (1952). Tables for convenient calculations of median-effective dose (LD_{50} or Ed_{50}) and instructions in their use. *Biometrics*, **8**: 249–63.

Weinstein, D., Katz, M. & Katzmer, S. (1987). Use of a rat/hamster S_9 mixture in the Ames mutagenicity assay. *Environ. Mutagen.*, **3**: 1–9.

Weinstein, I. B. (1988). The origins of human cancer: Molecular mechanisms of carcinogenesis and their implications for cancer prevention and treatment. Twenty-Seventh G.H.A. Clowes Memorial Award Lecture. *Cancer Res.*, **48**: 4135–43.

Wenke, G. & Hoffmann, D. (1983). A study of betel quid carcinogenesis on the in vitro N-Nitrosation of arecoline. *Carcinogenesis*, **4**: 169–72.

Wenke, G., Rivenson, A. & Hoffmann, D. (1984). A study of betel quid carcinogenesis. 3. 3'-(Methylnitrosamino)propionitrile, a powerful carcinogen in F344 rats. *Carcinogenesis.*, **5**: 1137–40.

Werner, E. A. (1919). The constitution of carbamides. IX. The interaction of nitrous acid and mono-substituted ureas. The preparation of diazomethane, diazoethane, diazo-n-butane and diazo-isopentane from the respective nitrosoureas. *J. Chem. Soc.*, **115**: 1093–102.

White, I. N. H., Smith, A. G. & Farmer, P. B. (1983). Formation of N-alkylated protoporphyrin IX in the livers of mice after diethylnitrosamine treatment. *Biochem. J.*, **212**: 599–608.

Wiessler, M. (1974). Synthesis of α-functional nitrosamines. *Angew. Chem.*, **13**: 743–4.

Wiessler, M. & Schmähl, D. (1973). Zur carcinogenen wirkung von N-Nitrosoverbindungen. 2. S(+) und R(−) Nitroso-2-methylpiperidin. *Z. Krebsforsch.*, **79**: 118–22.

Wiessler, M. & Schmähl, D. (1976a). Zur carcinogenen wirkung von N-Nitrosoverbindungen. V. Acetoxymethyl-methylnitrosamin. *Z. Krebsforsch.*, **85**: 47–9.

Wiessler, M. & Schmähl, D. (1976b). Zur carcinogenen Wirkung von N-Nitroso-Verbindungen. 6th Communication: Methoxymethyl-methylnitrosamin, 1-(Methoxy)-äthyl-äthylnitrosamin, Methoxymethyl-äthylnitrosamin, 1-(Methoxy)-äthyl-methylnitrosamin und N-nitrosooxazolidin. *Z. Krebforsch.*, **88**, 25–31.

Winn, D. M. (1984). Tobacco chewing and snuff dipping: An association with human cancer. *IARC Scientific Publication* No. 57: 837–49.

World Health Organization (1977). *Environmental Health Criteria 5: Nitrates, Nitrites and N-nitroso Compounds*, Geneva.

Wurdeman, R. L., Church, K. M. & Gold, B. (1989). DNA methylation by N-methyl-N-nitrosourea, N-methyl-N'-nitro-N-nitrosoguanidine, N-nitroso (1-acetoxyethyl)methylamine, and diazomethane: Mechanisms for the formation of N^7-methylguanine in sequence-characterized 5'-^{32}P-end-labeled DNA. *J. Amer. Chem. Soc.*, **111**: 6408–12.

Wykypiel, W. & Seebach, D. (1980). A nitrosamine route to macrostomine. *Tetrahedron Lett.*, **21**: 1927–30.

Yahagi, T., Nagao, M., Seino, Y., Matsushima, T., Sugimura, T. & Okada, M. (1977). Mutagenicities of *N*-nitrosamines on *Salmonella*. *Mutation Res.*, **48**: 121–30.

Yang, C. S., Tu, Y. Y., Hong, J. & Patten, C. (1984). Metabolism of nitrosamines by cytochrome P_{450} isozymes. *IARC Scientific Publication* No. 57: 423–8.

Yarita, T. & Nettesheim, P. (1978). Effects of carcinogen dose on characteristics of tracheal tumor response induced by *N*-nitroso-*N*-methylurea in hamsters. *Int. J. Cancer*, **22**: 298–305.

Zabeszhinskii, M. & Balanski, R. M. (1977). Induction of lung tumors in mice by diamylnitrosamine. *Bull. Exp. Biol. Med.*, **83**: 71–3.

Zarbl, H., Sukumar, S., Arthur, A. V., Martin-Zanca, D. & Barbacid, M. (1985). Direct mutagenesis of Ha-ras-1 oncogenes by *N*-nitroso-*N*-methylurea during initiation of mammary carcinogenesis in rats. *Nature.*, **315**: 382–5.

Zedeck, M. S., Frank, N. & Wiessler, M. (1979). Metabolism of the colon carcinogen methylazoxymethanol-acetate. *Front. Gastrointest. Res.*, **4**: 32–7.

Zeiger, E. (1987). Carcinogenicity of mutagens: Predictive capability of the *Salmonella* mutagenesis assay for rodent carcinogenicity. *Cancer Res.*, **47**: 1287–96.

Zeiger, E., Legator, M. S. & Lijinsky, W. (1972). Mutagenicity of *N*-nitrosopiperazines for *Salmonella typhimurium* in the host-mediated assay. *Cancer Res.*, **32**: 1598–9.

Zeller, W. J., Frühauf, S., Chen, G., Eisenbrand, G. & Lijinsky, W. (1989). Biological activity of hydroxylated chloroethylnitrosoureas. *Cancer Res.*, **49**: 3267–70.

Zingmark, P. A. & Rappe, C. (1976). On the formation of *N*-nitrosodiethanolamine from a grinding fluid under simulated gastric conditions. *AMBIO*, **5**: 80–1.

Zweig, G., Selim, S., Hummel, R., Mittelman, A., Wright, Jr., D. P., Law, Jr., C. & Regelman, E. (1980). Analytical survey of *N*-nitroso contaminants in pesticide products. *IARC Scientific Publication* No. 31: 555–64.

Index

For economy the *N*- is omitted. Asymmetric nitroso compounds are indexed according to the alkyl group of interest (e.g. hydroxyethylnitroso-), but sometimes in the text the name begins nitroso-, as is the case with symmetrical compounds and cyclic nitrosamines. Bold numbers refer to carcinogenesis data in Tables.

Accelerators of nitrosation 12
　of metabolism 124
Acetaldehyde 12, 75, 107, 130, 229
Acetone 75, 88
α-Acetoxybutylnitroso-3-
　carboxypropylamine 348
α-Acetoxydibenzylnitrosamine 235, **293**
Acetoxymethylnitrosoethylamine 346
α-Acetoxynitrosamines 79, 85, 113, 346, 385, 408
α-Acetoxynitrosodimethylamine 87, 145, 175, 187, 220, **282**, 319, 346, 385
α-Acetoxynitrosopyrrolidine 114
4-Acetyl-3,5-dimethylnitrosopiperazine **299**, 331
Acrolein 236
Activating enzymes for nitrosamines 401
Acute toxic effects 2, 166, 190
　dose 253
Acyclic nitrosamines, metabolism of 120
O-Acylation of nitrosamines 89
Adipic acid 118
Aflatoxins 3, 176
Agricultural chemicals 5, 35, 41, 46
Alcohol dehydrogenase 146, 236
Alcoholic beverages 124
Aldehydes 30, 50, 100, 130, 142, 146, 150, 381
Aldicarb 7, 72
　-nitroso 72, 248, **314**

Alkali, nitrosation in 3, 7
Alkaloid 5, 20, 31, 34, 42, 43, 85
Alkanediazotates 56
N-Alkylamides 26, 40
Alkylating agents 1, 2, 73, 84, 85, 93, 100, 106, 128, 130, 151, 156, 184, 192, 202, 203, 220, 250, 251, 260, 280, 368, 407
Alkylation of DNA 85, 86, 93, 101, 106, 107, 128, 138, 143, 151, 154, 156, 159–62, 167–72, 220, 240, 252, 255, 256, 260, 270, 272, 279, 280, 334, 342, 354, 356, 362, 366, 370, 379, 390, 407, 410
Alkylation of nitrosamines 79, 83, 89
　proteins 1, 94, 101
N-Alkylcarbamate esters 7
Alkyldiazohydroxide 151
Alkyldiazonium ion 75, 78, 89, 93, 97, 99, 110, 125, 129, 151, 161, 171, 187, 203, 204, 220, 226, 228, 229, 252, 266, 408
Alkylguanidines 224
O^6-Alkylguanine 160, 171
7-Alkylguanosine 152
Alkylhydrazides 14
Alkylnitrosocarbamates 14, 52, 57, 60, 62–3, 66, 69, 72–3, 97, 147, 151, 161, 167, 186, 192, 198, 203, 224, 244, 247, 265, 267, 355, 375, 385–6, 412

454   Index

Alkylnitrosoformamides 147
Alkylnitrosoguanidines 69, 151, 203, 224,
    247, 265–7
Alkylnitrosoureas 17, 26, 52, 57, 58–63,
    66, 69, 70, 73–5, 78, 89, 94–7, 103,
    147, 149, 151, 161, 163, 166–7, 174,
    186–7, 190, 192, 197, 203, 224, 230,
    240, 246–7, 256, 260, 265–6, 272,
    319, 334, 342, 346, 350, 352, 354–7,
    361, 363, 367, 369–70, 372, 375,
    381, 385–6, 402, 408, 412
  CNS tumours induced transplacentally
    by 352, 393–4
Alkylnitrosourethanes 190
Alkyltransferase 171
Allantoin 40
Allyl group, resistance to oxidation 374
Allylnitrosamines 227
Allylnitroso-2,3-dihydroxypropylamine
    291, 326, 383
Allylnitroso-2-hydroxypropylamine **291**,
    326, 337, 374
Allylnitroso-2-oxopropylamine **291**, 337,
    374
Allylnitrosourea 266, **305**, 351, 357, 369,
    372, 386
Alpha oxidation 87, 113, 118, 121, 130,
    132, 138, 145–6, 150, 184, 193, 217,
    220, 228, 234, 236, 270, 319, 340–1,
    344
Alveolar-bronchiolar tumours of lung
    343
Amadori reaction 24
Amines 3
  nitrosatable 5
  secondary 5, 9, 11, 20, 30, 33, 37–8,
    40, 42, 76, 405
  tertiary 8, 9, 11, 20–2, 31, 33, 41–2,
    48–9, 76
Aminoacetone 139
Aminopyrine 9, 41, 44, 47, 49
n-Amylnitrosourea **305**, 350–1, 357, 372,
    378, 386
α-Anions of nitrosamines 64
Antihistaminic 9, 34
Arecoline 43
Arginine 40, 69, 225
Aroclor, as enzyme inducer 122, 124
Aromatic amines 205, 405
  bladder tumours by 154
Ascorbic acid, as nitrosation inhibitor 12,
    36, 76–7
Asymmetric nitrosamines 230, 255
Athymic nude mice 246, 248
Atmospheric nitrosamines 50
Axial protons 116, 317
Azoyalkanes 56, 100, 106, 125, 219
Azoxymethane 107, 124, 164, 170, 172,
    188, 191, 220, 342, 363, 386

Azoxymethanol 107

Bacon 5, 23, 25, 36, 53, 76
Bacterial mutagenesis assay 229
Bacteriophage induction *see* prophage
    induction
Baeyer-Villiger oxidation 81–2, 143, 145
Barley 22, 43
Bathochromic shift in absorbance 60–3
Baygon 7, 72
  N-nitroso- 72, **314**
BCNU (bis-chloroethylnitrosourea) 73,
    200, 352
Beer, NDMA in 5, 21, 24–5
Benzaldehyde 177, 322
Benzo[a]pyrene 248
4-Benzoyl-3,5-dimethylnitrosopiperazine
    299, 331
Benzylnitrosourea 78, **305**
Betaine 20
Beta oxidation 110–1, 118, 130–3, 146,
    149, 274, 340–1, 359, 372
Beta-oxidised alkylnitrosamines 231, 368
Betel nut 43
Bile ducts 373
Bis-(2-hydroxypropyl)amine 132, 134,
    139
Bis-(2-oxopropyl)amine 132, 134, 139, 142
Bladder, adsorption of N-nitroso
    compounds from 34, 406
  infected, nitrite in 11, 34, 406
  mucosa 111–12, 124, 150, 198
  tumours 30, 112–13, 131, 150, 247,
    254, 256, 281, 335–6, 358–63, 410
Blood–brain barrier 354
Blood, measurement of nitrosamines in
    53
Brain 153, 163
  tumours of 152, 160, 171
  ethylation of DNA in 152, 160, 171
Bronopol 29
n-Butylnitroso-3-carboxypropylamine
    112, 359
4-*tert*-Butylnitrosopiperidine 115, 262,
    295, 328, 339
n-Butylnitrosourea 260, 267, **305**, 350–1,
    357, 372, 378, 386
iso-Butylnitrosourea 378
sec-Butylnitrosourea 78, 378
BUX-10 72
  N-nitroso- 72, **314**

Cabbage, pickled, nitrosamines in 25
Cancer therapeutic agent 200
Cancer therapy 73, 85, 352
Carbanion 77–8, 83, 85, 147, 187, 263,
    270
Carbaryl 7, 40, 72, 225, 355, 397
  N-nitroso- 7, 72, 226, 248–9, **314**, 375

Index

Carbocations 252
Carbofuran 7, 72
  N-nitroso- 72, 227, 248, **314**
Carbon tetrachloride 175, 205, 334
α-Carbon of nitrosamines 78–9, 264, 281
β-Carbon, oxidation of 273
Carbonates 75
Carbonium ions 252
Carbonyl compounds 129, 151, 189
  as catalysts of nitrosation 7, 12, 21, 50
Carboxyl group, reduction of nitrosamine carcinogenicity 319
Carcinogens not mutagenic 204
CCNU (cyclohexylchloroethylnitrosourea) 73, 200, 352
Central nervous system 185
Centrilobular necrosis of liver 173, 175, 191, 195
Cervix, tumours of in hamsters 371–2, 385
Cheese, nitrosamines in 24–5
Chloral 12
Chlordiazepoxide 40
Chlorinated trialkylnitrosoureas 200
Chloroethylnitrosodiethylurea **315**, 352
Chloroethylnitrosodimethylurea 200, **315**, 352
Chloroethylnitrosohydroxyethylurea (HECNU) 73, 199, 200, **312**, 352
Chloroethylnitroso-2-hydroxypropylurea 199, 200, **313**
Chloroethylnitrosoureas 192, 199, 200
Chloroform, for nitrosamine extraction 51
3-Chloronitrosopiperidine **295**, 330
4-Chloronitrosopiperidine 249, **295**, 330
Chlorozotocin 352
Chlorpheniramine 44, 47
Chlorpromazine 44
Cholangiocellular tumours in liver 340
Cholangiomas 396
Choline 20, 92
Cholinesterase inhibitors 7, 70, 225
Chromic oxide 81
Cigarette smoker, nitrosamine exposure 406
Cimetidine 39, 69
α-Cleavage
Clostridium botulinum 23
Colon 107
  tumours 106, 124, 357
Convulsions induced by cyclic nitrosamines 185, 189, 195
Corrosion inhibitor, nitrite as 7
Cosmetics, nitrosamines in 29, 218, 254
Cured meat 5, 10, 13, 24, 28, 51
Cutting oils, nitrosamines in 7, 20, 28–30, 42, 76, 218, 254, 405
Cyano group 236

Cyclic nitrosamines 15–16, 30, 57–8, 66, 86–7, 100, 113, 129, 131, 148, 152, 155, 159, 184–5, 189, 192, 195, 219, 224, 228, 256, 262, 281, 319, 338, 349, 356, 36–1, 368, 408
Cyclizine 44
4-Cyclohexylnitrosopiperidine 115, 239, 262, **295**, 328, 339
Cyclohexylnitrosourea 74, 78, 378
Cytochrome-P450 92, 100, 102–3, 122, 125
Cytosol 131, 206

Deamination 91
Demethylase 123
Denitrosation 121, 240
Detoxifying enzymes 121
Deuterium labelling, by exchange 66, 143, 145, 160, 187, 202, 263
  effects on carcinogenesis 133, 138, 150, 160, 270, 272–6
  effects on metabolism 103, 279, 320
  effects on toxicity 193
Dialkylarylureas 9, 47, 74
Dialkylcarbamates 75
Dialkyldisulphides 9
Dialkylglycine 75
Dialkylhydrazines 13–14, 83
Dialkylnitrosoureas 58, 73–5, 95, 357, 362, 367, 375, 380, 385
Dialkylthiocarbamates 9
Dialkyltriazenes 219
Diazoalkanes 1, 14, 19, 56–7, 69, 93, 99
Diazomethane 19, 56, 99, 143, 202
Diazotisation 15
3,4-Dibromonitrosopiperidine **295**, 330, 383
Di-*n*-butylamine 26
3,4-Dichloronitrosopiperidine 249, **295**, 330, 383
3,4-Dichloronitrosopyrrolidine **293**, 326
2,4-Dichlorophenoxyacetic acid 6, 28
2,6-Dicyanonitrosopiperidine 318
Diethanolamine 6, 7, 28
Diethylamine 6, 26, 37
Diethylnitrosourea 204, 240, 260, **308**, 351, 358, 362, 368, 371, 386
Diethylstilboestrol 364
Diethyltriazene 127
Differentiated cells 205
Dimethylamine 6, 11, 20–1, 26, 28, 49, 121, 176
4-Dimethylaminoazobenzene 8
Dimethylaminoethylnitrosoethylurea **311**
Dimethylaryltriazenes 276
2,5-Dimethyldinitrosopiperazine 263, **301**, 317, 332
2,6-Dimethyldinitrosopiperazine 117, 120, 129, 263, 273, **300**, 317, 332, 341, 349, 356, 392

Dimethyldodecylamine 30, 42
Dimethylformamide 188
Dimethylhydrazine 14, 56, 83, 188
2,6-Dimethylmorpholine 26
2,6-Dimethylmorpholine-3-one 132
2,6-Dimethylnitrosomorpholine 87, 113, 118, 120, 131–2, 134, 136, 138–9, 141, 196–7, 245, 263, **297**, 340, 349, 361, 365, 372
  cis 132–3, 136, 263, **298**
  trans 132–3, 136, 263, **298**, 332
  deuterium labelled 132, 273
3,5-Dimethylnitrosopiperazine **299**, 369, 382
2,6-Dimethylnitrosopiperidine 62, 78, 90, 281, **294**
3,5-Dimethylnitrosopiperidine 120, 263, **294**, 328
  cis & *trans* **294**, 328
2,5-Dimethylnitrosopyrrolidine 16, 61, 246, 281, **293**
Dimethylnitrosourea 58, 113, 186, 275, **307**, 351–2, 358, 362–4, 368, 370, 383, 386
Dimethyl sulphate 162
Dimethylsulphoxide 188
Dimethyltetradecylamine 30
Dimethylurea 40
Dinitrosohomopiperazine 229, **301**, 332, 387, 390
Dinitrosopiperazine 17, 35, 55, 62, 185, 195–6, 245–6, 249, 262–3, **300**, 316–17, 331, 341, 349
Diphenhydramine 44, 48
Diphenylamine 11, 15, 37
Disulphiram 9, 44, 46–7, 108, 124
DNA adducts 15, 113, 152–3, 196, 219, 272, 320
DNA repair *see* Repair of alkylated DNA
Dog, nervous system tumours in 352
  toxicity of NDMA 175, 191
Dose-response study 252, 386
Drosophila, mutagenesis 218
Drugs, nitrosation of 41
Duodenum, tumours in 266, 357

Electrophile 93, 154, 187, 196, 348
Enamine 43–4
Endogenous formation of nitrosamines 10, 34, 406
Endoplasmic reticulum 108, 122
Endothelial cells 385, 412
Enzyme inducers 228
Epithelial cells 148, 242
Erythorbate 36, 76
Esophagitis 194
Esophagus 155, 217, 238, 256, 259
  DNA alkylation 333, 356, 408
  human, metabolism in 153

  metabolism in 117, 150, 153
  tumours 16–17, 29, 31, 37, 48, 58, 79, 99, 112–16, 120, 127–9, 132, 138, 153–6, 187, 192–6, 217–18, 231, 252, 258, 263, 275, 278, 317, 320, 359–60, 387–8, 390
Esophagus cancer in China 10
Estradiol 361, 367
Ethanol 381
  reduces NDMA metabolism 108, 124
Ethylation 130, 151
  of DNA 80, 151, 156, 220, 276, 394
Ethyldiazonium ion 130, 159, 186, 203, 227, 375
Ethyleneimine 19
Ethyleneurea 40
7-Ethylguanine 152, 159
O^6-Ethylguanosine 152, 159
Ethylmethanesulphonate 152
Ethylnitrosoalkylamines 188
Ethylnitrosodimethylurea 204, **314**, 371
Ethylnitrosoethoxyamine 239
Ethylnitrosoformamide 175, 224
Ethylnitrosohydroxyethylurea **309–10**, 342, 352, 354, 358, 362, 368, 371, 386
Ethylnitrosomethylurea 204, **309**, 351, 354, 358, 362, 367–8, 371, 386
Ethylnitrosonitroguanidine 162, 225, 227, **315**, 355, 375
Ethylnitroso-2-oxopropylurea **310**, 352, 358, 362, 368, 371, 386
Ethylnitrosourea 58, 62, 74–5, 86, 93, 95, 151–3, 160–3, 166, 169, 171, 174, 197, 202–3, 227, 240, 243, 247, 260, **302–3**, 350–4, 357–8, 361, 370–1, 375, 383, 386, 393–4, 410
Ethylnitrosourethane 62–3, 70, 96, 162, 198, 203, 225–7, 266, **314**, 362, 365, 375
Ethylnitrosovinylamine 184
Ethylurea 40, 69
  feeding test with nitrite 397
European hamsters 344

Feminised male rats 360
Fenuron 47, 74
Fetal cell cultures 248
Fibroblasts, transformation of 174, 242, 246–7
  transfection 166
Fibrosarcomas 248
First-pass metabolism 103, 344
Fish, nitrosamines in 25
Fish meal, toxic nitrosamines in 3, 20
Fluorine substitution 231
2-Fluoroethylnitrosourea 198–9, **305**, 379
Food additives 5, 22, 41

Forestomach tumours 7, 59, 60, 72, 127, 162, 198, 200, 248, 267, 355–7
Formaldehyde 121, 126, 142–3, 147, 174, 229, 380–1
Frameshift mutagen 217

β-Galactosidase 240
Gas-phase nitrosation 5, 6, 50
Gastric juice 37
Gastric secretion of HCl 5
Gastrointestinal tract tumours 354
Gerbils, ethylation of DNA in 171
Germ cells of mice 202–3
Glandular stomach tumours 73, 162, 202, 267, 354–5
Glomerulosclerosis 199, 363
Glucuronides 81, 87, 103, 136, 139
Glutaric acid 118
Glutathione 12, 36
Glyceraldehyde 326
Glycerol 326
Glycylglycine 40
Gramine 5, 22, 43
Griess reagent 15
Guanidine 77
Guinea pig stomach 10, 132
  toxicity of NDMA in 175, 191
  resistance to tumourigens 347, 355, 359, 402

Haloalkylnitrosoureas 193, 363
Halogenated nitrosamines 68, 79, 85, 201
Hamster V79 cells 244
  embryo cells 247
Hemangiosarcomas (of liver) 251, 328, 334, 339–42
Hemiacetal 114–15
Hemoglobin, reaction with nitrite 10
Hemophilus influenzae 226, 249
Hepatitis 297
Hepatocytes, metabolism by 144, 243, 245
Hepatotoxicity 55, 102, 388
Heptamethyleneimine 39, 40
Herbicides 6, 9, 28
Herring meal, toxic NDMA in 176
1,6-Hexanediol 115
n-Hexylnitrosourea **306**, 350–1, 357, 372, 378, 386
HGPRT locus in CHO cells 246
Histones, reaction of alkylnitrosamides 94
Hordenine 5, 22, 43
Hormones 123, 407
Host-mediated activation 248
Human organ culture, nitrosamine metabolism 153
Humans, cancer risk estimates 30, 47, 49, 50, 154

Hydrazines 1, 8, 13, 77
Hydrazoic acid 77
Hydrochlorothiazide 40
Hydrophilicity 189
Hydroxyacetone 88
Hydroxyalkylamines 29, 30, 50
Hydroxyalkylnitrosamines 146, 220
Hydroxyalkylnitrosoureas 57, 75, 379
4-Hydroxybutylnitroso-*n*-butylamine 57, 112, 186, 337, 358, 398
1-Hydroxydi-*n*-butylamine 348
2-Hydroxyethoxyacetic acid 132
Hydroxyethylation 130, 167, 171, 235, 278, 370
Hydroxyethyldiazonium ion 131, 365
Hydroxyethylguanine 117, 171
O^6-Hydroxyethylguanosine 167
Hydroxyethylnitrosoallylamine **291**, 337
Hydroxyethylnitrosochloroethylurea 163, 167–8, **312**, 342, 365–6, 372
Hydroxyethylnitroso-2,3-dihydroxypropylamine **291**, 337
Hydroxyethylnitrosoethylamine 185, 191, 364
Hydroxyethylnitrosoethylurea 163, 167–8, **312**, 342, 352–4, 358, 361, 366–8, 371
Hydroxyethylnitroso-2-oxopropylamine **291**, 337, 374
Hydroxyethylnitrosourea 59, 75, 167–8, 226, **303**, 351, 357, 362, 365, 368–9, 372, 379, 385
α-Hydroxylation 103
β-Hydroxylation 115
2′-Hydroxy-1′-methyl-2-ethoxypropionic acid 132
Hydroxymethylnitrosomethylamine 82, 87–8, 121, 145, 174
α-Hydroxynitrosamine 81, 85, 87–9, 103, 114, 242, 319
β-Hydroxynitrosamines 91–2, 100
Hydroxynitrosonornicotine 118
3-Hydroxynitrosopyrrolidine 118, **293**, 338
Hydroxyproline 39, 64
2-Hydroxypropyldiazonium ion 143, 226, 374
2-Hydroxypropylnitrosoethanolamine 100, **290**, 337, 374
2-Hydroxypropylnitrosochloroethylurea **313**, 342, 362, 365, 369
2-Hydroxypropylnitroso-2,3-dihydroxypropylamine **292**, 324, 337, 356, 373, 384, 392
2-Hydroxypropylnitroso-2-oxopropylamine 80, 129–33, 136, 138–41, 197, 274, **291**, 324, 336, 347, 357, 361, 365, 367, 373
2-Hydroxypropylnitrosourea 75, 226, **306**, 351, 369, 374, 379

3-Hydroxypropylnitrosourea 149, **306**, 351, 355, 362, 367, 379

Ileum, tumours of 357
Iminium ion 49
Iminodiacetic acid 6
Immunological properties 17
Inhalation of nitrosamines 381
Inhibitos of nitrosation 12
Initiation of tumours 148, 153, 375
Insecticides 7, 9, 70, 198, 203, 225, 248
Intestine, tumours of 162, 267
Intravesicular treatment 34, 50, 188, 198, 259, 360–2, 367
Iodoacetic acid 205
Isooctane 62
Isopropanol 149
Isopropanolamine 132, 139

Jejunum, tumours of 357

Ketene 177
Ketones 30
β-Ketonitrosamine 91
Kidney 153, 157, 199, 238
  pelvis, transitional cell neoplasms 365–6
  tumours 29, 103, 106–7, 123–4, 188, 199, 343, 363–6, 395
    mesenchymal 106, 124, 155, 173, 363, 363–4, 395
Knoop oxidation 147

Lambda repressor 240
Landrin 72
  N-nitroso 72, 227, **314**
Lead acetate 79
Leather manufacture 6, 28
Leaving group 131
Lebanon bologna, nitrate in 23
Leucocytes 249
Liposolubility/lipophilicity 88, 190, 226, 230, 324, 345
Liquid preincubation 229
Liver 1, 152–3, 185, 194, 220
  toxicity 3, 20–1, 44, 121, 123, 173, 188, 190, 195, 197, 330, 390
  tumours 16, 29, 31, 37, 47–8, 91, 99, 103, 123–4, 127–32, 138, 150, 154–9, 163, 167, 172–3, 184, 188, 196–7, 206, 219, 229, 235, 251–2, 255, 272, 276, 280–1, 334–44, 387–8, 390
Lucanthone 44, 46, 48
Lung 31, 157, 185, 194, 238
  adenomas in mice 38, 118, 350
  congestion 193, 195
  tumours 29, 31, 39, 106, 113, 117, 124, 132, 150, 157, 174, 184, 195–6, 200, 330, 343–52, 373

Macostomine, synthesis of 85
Maillard reaction 51
Maleic hydrazide, diethanolamine salt 28
Malt 5, 21–2, 43
Mammalian cells *in vitro* 224
Mammalian enzymes 224
Mammary gland 94, 153, 410
  carcinomas 58, 162, 166, 174, 200, 240, 266, 370, 410
  fibroadenomas 370
Mass spectrometry 24, 51–2, 66, 86, 98–9, 163
Meat, smoked 25
Mechanisms of carcinogenesis 407
Median lethal dose (LD$_{50}$) 173, 177
Median time of death with tumours 390
Membranes 92, 205
Mesothelioma 162, 267, 383
Metalation 79, 83
Metal ions 20
Methapyrilene 9, 44, 46, 48, 219
Methine 78, 273, 275–6
Methomyl 7, 72
  N-nitroso 72, 226, **314**
2-Methoxy-2,6-dimethylnitrosomorpholine 341
Methoxyethyldiazonium ion 236
1-Methoxyethylnitrosoethylamine 346
1-Methoxyethylnitrosomethylamine 345
Methoxymethylnitrosoethylamine 345
Methoxymethylnitrosomethylamine 346
Methoxyethylnitrosourea 236, **304**, 351, 357, 362, 368–9, 372
Methylacetamide 40
Methylamine 121, 139
Methylating agent 1, 19, 113, 145, 159, 162, 164, 171, 174, 202–3, 275, 364–5, 369
Methylation of DNA 17, 80, 82, 107, 127, 136, 138, 146, 154–7, 160–3, 170, 174, 187, 202, 220, 252, 260, 272, 280, 322, 347–8, 356, 361–4, 407
  of protein 154, 252, 407
Methylbenzylamine 37, 40
N-Methylcarbamate esters 70. 198, 225, 248
S-Methylcysteine 101
Methyldiazonium ion 86, 107, 121, 127, 130, 144, 147, 155, 159, 174, 177, 186, 193, 195, 217, 226–7, 231, 364–5, 375
2-Methyldinitrosopiperazine 249, 263, **300**, 316, 332
Methylene chloride 51, 58, 60, 62
α-Methylene 62, 78, 81, 93, 126, 138, 145, 151, 231, 236–7, 270, 275, 278, 321–3
Methylguanidine 38, 40, 69, 225
7-Methylguanine 98–9, 101, 155, 157
O^6-Methylguanine 99, 130, 155, 157, 344
  transferase 159, 163

Methylhistidine 101
Methyl isocyanate 7
Methyl methanesulphonate 99, 162
Methylnitrosoalkylamines 17, 108, 110, 112, 127, 139, 146, 155–7, 161, 177, 188, 192, 194–6, 224, 230, 256, 258, 260, 275, 344, 356, 359–61, 366
Methylnitrosaminoacetonitrile 186, 318
Methylnitrosamino acids 147
   excretion in urine 359
Methylnitrosaminopropionaldehyde 43
Methylnitrosaminopropionitrile 43, 322
4-(Methylnitrosamino)-1-(3-pyridyl)-1-butanol *see* NNAl
4-(Methylnitrosamino)-1-(3-pyridyl)-1-butanone *see* NNK
Methylnitrosoacetamide 175, 224
Methylnitrosoallylamine 184, 322
Methylnitroso-*n*-amylamine 127, 150, 155, 157, 184, 193, 229, 258–9, **285**, 322, 334, 345
Methylnitrosobenzylamine 37, 128, 177, 184, 191, 194, 231, 240, 262, **287**, 321, 384–5
Methylnitrosobis-(2-chloroethyl)urea 200, **315**, 350, 352
Methylnitroso-*n*-butylamine 113, 125, 155, 157, 188, 192–6, 229, 240, 244, 251–2, 258, 270, 274–5, **284–5**, 323, 334, 345, 356, 360, 384
Methylnitroso-*tert*-butylamine 67, 78–9, 245, 252, 270, **285**
Methylnitrosocarbamate esters 355
Methylnitrosocyclohexylamine 78, 99, 129, 147, 184, 194, 217, 231, 240, 274, **287**, 322
Methylnitroso-*n*-decylamine **286**, 335
Methylnitrosodiethylurea 244, 275–6, **313**, 352
Methylnitroso-2,3-dihydroxypropylamine 231, **289**, 326, 337, 346
Methylnitroso-*N'*,*N'*-dimethylaminoethylamine **284**, 322
Methylnitroso-*n*-dodecylamine 8, 30, 42, 66, 113, 148, 190, 230, 274, **286**, 335, 345, 359–60
Methylnitrosoethylamine 16–17, 32–3, 86, 107, 124–6, 131, 146–9, 156–9, 164, 170, 172, 174, 184, 186, 188–93, 196, 217, 220, 224, 229, 240, 245, 258, 261, 276, **283**, 320–1, 335, 344, 364, 408
Methylnitrosoethylurea 58, 164, **307**, 351–2, 358, 364, 368–9, 383, 386
Methylnitrosoformamide 87, 175, 187, 220
Methylnitrosoguanidine 69, 225
Methylnitroso-*n*-heptylamine 129, 184, 193, 254, 261, **286**, 323, 335, 345, 356, 383

Methylnitroso-*n*-hexylamine 113, 129, 155, 193, 259, **285**, 323, 335, 345, 356, 360, 383–4, 392
Methylnitrosohydroxyethylamine 80, 92, 103, 130–1, 171, 188–9, 231, 279, **284**, 326, 337
Methylnitrosohydroxyethylurea 74
Methylnitroso-2-hydroxypropylamine 91, 112, 139, 195, 197, 231, **288**, 324, 337, 346, 359, 366, 374
Methylnitroso-3-hydroxypropylamine 80, 231, **288**, 346
Methylnitrosomethoxyamine 239, **289**
2-Methylnitrosomorpholine **297**, 332, 339, 349, 356, 384
Methylnitrosoneopentylamine 194, 231, **285**, 322
Methylnitroso-*N'*-nitroguanidine (MNNG) 14, 19, 56, 69, 94, 162, 167, 202, 225–7, 243, 247, **315**, 352, 355, 357, 375
Methylnitroso-*n*-nonylamine **286**, 335, 382, 408
Methylnitroso-*n*-octadecylamine 111, 230
Methylnitroso-*n*-octylamine 57, 112, 129, 239, 261, **286**, 323, 335, 345, 356, 408
2-Methylnitrosooxazolidine 240, **296**, 340
5-Methylnitrosooxazolidine 30, **296**, 340
5-Methylnitrosooxazolidone 72, 226, 266, 375, 379
Methylnitroso-2-oxopropylamine 81–2, 88, 111–12, 127, 131, 139–40, 144–5, 147, 161, 164, 195, 197, 216, 231, 275, **288**, 323, 346, 359, 361, 366, 374, 383
Methylnitrosophenylamine *see* nitrosomethylaniline
Methylnitroso-2-phenylethylamine 184, 194, 229, 231, 244, 262, 274–5, **287**, 321, 384, 392
1-Methyl-4-nitrosopiperazine 13, 28, 90, 185, 246, **299**, 369, 381–2
2-Methylnitrosopiperidine 78, 240, 281, **294**, 338
3-Methylnitrosopiperidine 249, 263, **294**
4-Methylnitrosopiperidine 263, **294**
3-Methylnitroso-4-piperidone **296**, 329, 338, 384
Methylnitroso-*iso*-propylamine 231
Methylnitroso-*n*-propylamine 17, 125–6, 129, 149, 155, 157–8, 188, 193, 229, 240, 244, 258, 278–9, **284**, 322, 334, 345, 356, 397
Methylnitroso-*n*-tetradecylamine 8, 30, 230, **287**, 335
Methylnitrosotoluenesulphonamide 19
Methylnitrosotrifluoroethylamine 231, 237, **287**, 321
Methylnitroso-*n*-undecylamine 67, 245, **286**, 335

Methylnitrosourea 5, 14, 19, 55–8, 73–5, 86, 94, 99, 152–3, 160–4, 166–7, 170, 172, 174, 197–8, 202–3, 226–7, 243, 249, 255, 260, 275, **302**, 351–3, 355, 357–8, 361–2, 369–71, 375, 384, 386, 394, 399, 410
Methylnitrosourethane 19, 55–7, 70, 96, 162, 170, 172, 186, 198, 225–7, 248, 266, **314**, 352–3, 362, 365, 375, 397
Methylnitrosovinylamine 92, 177, 323
N-Methyl oxidation 322
Methylphenidate 9, 39
Methylphenylamine 26
N-Methyl phenylcarbamate, N-nitroso 227, **314**
3-Methyl-2-phenylnitrosomorpholine *see* nitrosophenmetrazine
Methylurea 40, 121
Microsomal activation/oxidation 16, 122
Microsomal enzymes 205–6, 220, 229, 242, 335
Microsomes 122–3, 125, 144, 206
Milk, nitrosamines in 24–5
Mink 21
Mixed function oxides 108, 122, 129, 146, 205–6, 270
Molecular biology of carcinogenesis 2, 371
Monkeys, nervous system 352
Monoalkylnitrosamine 151, 228
Mononuclear cell leukemia in rats 368
Monuron 47, 74
Morpholine 6, 12–13, 26, 37, 40, 69, 90
Mortality rate from tumours 276
Mouse lymphoma cells 243, 250
Mouse skin carcinogenesis 147
Mustard gas 2, 98, 202
Mutagenic activity 202
Mutational fingerprint 86
Myoglobin 10, 22

Nasal mucosa 238, 280–2
 tumours 29, 31, 48, 103, 120, 124, 128–9, 219, 231, 234, 236, 254–5, 272, 317, 337–8, 380, 390
NDMA-demethylase 108, 122–3
Nephropathy 192, 198–9
Nervous system 74, 256
 tumours 58, 60, 96, 113, 162–3, 166, 174, 200, 266, 275, 352–4
Newborn animals 150
Nicotine 5, 31, 33, 42–3
Nitramine 13, 83
Nitrate, endogenous formation 10
 reduction to nitrite 4, 5, 10–11, 22–3, 31, 34, 50
Nitric oxide 49
Nitrilotriacetic acid 8, 39

Nitrite 3, 4, 7, 20, 22–3, 27–9, 31, 35–41, 48, 50, 69, 76, 91, 121, 176, 225, 406
Nitrite esters 52, 63
Nitrogen 121
Nitrogen mustards 202
Nitrogen oxides 5, 6, 20, 22, 27–8, 31, 33, 43–4, 76
Nitrogen-specific detector 51
Nitrogen trioxide 40
Nitrolic acids 52
Nitrosamine-activating enzymes 205, 381
Nitrosamino acids 58, 64, 68–9, 79, 85, 110
Nitrosaminoaldehyde 228
Nitrosaminoketone 110
Nitrosatable drugs 11
Nitrosative dealkylation 31, 42, 49
Nitrosiminodiacetaldehyde 236
Nitrosiminodiacetic acid 6, 8, 20, 33, 80, 186, **292**, 318, 322
Nitrosiminodiacetonitrile 186, 236, **292**, 318
Nitrosiminodipropionitrile 236, **292**
Nitrosolactam 86, 224
Nitrosopeptides 86
C-Nitroso compounds 52, 77
Nitroso-
 anabasine 32–3, **295**, 329
 anatabine 32–3, **295**
 atrazine 90
 azetidine 16, 19, 55, 57, 185, 189, 195, 245, 262, 273, **293**, 338, 349, 365
 azetidine-carboxylic acid 64
 bis-(2-chloroethyl)amine 186, 189, **292**
 bis-(diethoxyethyl)amine 80, 236, **290**, 336
 bis-(2-ethoxyethyl)amine 236, **290**
 bis-(2-hydroxypropyl)amine 8, 57–8, 68, 80–1, 91, 106, 132–3, 138–41, 146, 186, 189, 216–18, 235, 241, 245, 258, **292**, 324, 336, 347, 361, 367, 372–3, 398
  diacetate 347
 bis-(2-methoxyethyl)amine 236, **290**
 bis-(2-oxopropyl)amine 57, 68, 80, 131–2, 136, 138–41, 143, 146, 160, 164, 170–1, 186, 197, 206, 216, 235, 241–3, 274, **293**, 323, 336, 347, 357, 360–1, 365–7, 372–3, 396
 bis-(trifluoroethyl)amine 237, 258, **292**
 4-carbethoxypiperazine 341
 cimetidine 39, 69, 240, **315**
 dehydropiperidines 227
 diallylamine 184, 236, 239, 245, 253, 256, **283**, 326, 381
 di-n-amylamine (pentyl) 150, 184, 194, 234, 258, 323, 335, 344
 dibenzylamine 8, 184, 194, 234–5, 258

Index

di-*n*-butylamine 27, 57, 112, 148, 150, 184, 186, 194, 234, 245, 256, 260, 267, 281, **283**, 323, 335, 337, 344, 356, 358, 396, 398
di-*iso*-butylamine 234, 256, 281, **283**, 344, 381–2
di-*sec*-butylamine 234, 256, 280, **283**
dicyclohexylamine 184, 234, 258
di-*n*-decylamine 66
diethanolamine 7, 16, 27–9, 32, 42, 53, 57–8, 80, 92, 100, 103, 109–10, 117, 118, 130–2, 146, 167, 185, 189, 191, 218, 235–6, 254–5, **290**, 326, 328, 335, 364, 381, 387, 390, 405
  ethers of 335
  hemiacetal 109
diethylamine 2, 3, 6, 16–17, 25, 27, 32, 46, 55, 57, 62, 67, 79, 86, 113, 115, 117, 124–5, 130, 149–53, 156, 159–60, 171, 184–8, 191, 205, 229, 236, 240, 244–8, 252, 256, 260, 270, 277, 280, **282**, 328, 334, 341, 344, 356, 364, 386, 388, 393, 395, 399, 403, 408
  d_{10} 99
di-*n*-heptylamine 234
di-*n*-hexylamine 234
3,6-dihydrouracil 342, 365
dimethylamine 1, 3, 5, 9, 11–12, 17, 20–2, 25–33, 41–58, 79, 82–3, 86–7, 94, 98, 101–3, 106–8, 112–17, 120–8, 139, 145–8, 152–64, 170–7, 184, 186–92, 201, 205–6, 224, 229–30, 244–7, 251–2, 256, 258–60, 272, 277, 279–80, **282**, 334, 342–3, 356, 363, 367, 381, 386–7, 395, 397, 399, 407
  d_6 98
di-*n*-octylamine 234, 239, 258, **283**, 323
diphenylamine 6, 11–12, 27, 37, 78, 90, 184, 234, 245, 247, 254, **287**, 320, 360
di-*n*-propylamine 16, 27, 57, 61, 149, 184, 186, 244, 256, 260, 280, **283**, 322, 324, 335, 337, 344, 356, 397
di-*iso*-propylamine 61, 78, 184, 231, 256, 270, 280, **283**, 381
dithiazine 16, 62, **299**, 333, 341
dodecamethyleneimine 66, 185, 189, 235, 245, 262, **300**, 339, 355, 408
3,4-epoxypiperidine 196, **296**, 339
guvacoline 43, 239, 249, **296**
heptamethyleneimine 116, 118, 185, 195–6, 229, 245, 262, 272, **299**, 330, 339, 349, 356, 384, 390
hexamethyleneimine 16, 115–16, 118, 148, 185, 195, 244, 262, **299**, 330, 339, 349, 398
hydantoin 365

2-hydroxymorpholine 80, 118, 146, **298**
hydroxyproline 10, 13, 27–8, 90, **293**, 319, 384
indoline 341
isonipecotic acid **295**, 319
methylaniline 55, 78–9, 100, 126–8, 155, 195, 217–18, 227, 240, 262, 270, **287**, 320, 385
  4-fluoro- 218, **287**
  4-nitro- 218, 227, **287**
methylphenidate 10, 239, **295**, 317
morpholine 6, 12, 16–17, 21, 27, 30, 32, 37, 53, 57, 62, 113, 116–18, 129, 132, 138, 185, 189, 195, 245–6, 262, 272, **297**, 318, 332, 334, 339–40, 349, 365, 381, 384, 387–8, 405, 408
nipecotic acid 249
nornicotine 6, 27, 31–3, 42, 115, 276, **293**, 326, 349, 383
  N-oxide 349
octamethyleneimine 185, 195, 235, 262, **300**, 330, 339, 349, 408
oxazine 30
oxazolidine 30, 57, **296**, 340, 382, 408
oxazolidone 72, 75, 96, 167–8, 198, 226, 266, **315**, 350, 370, 375, 379
phenmetrazine 10, 38, 78, 239, 246, **298**, 318
phenylbenzylamine 13, 128, 217, **287**, 320, 322, 360, 384
pipecolic acid 249, **295**, 319
piperazine 28, 35, 90, 195, 245, 249, **299**, 341, 369
piperidine 17, 20, 25, 47, 55, 62, 67, 87, 115–18, 185, 195–6, 229, 244, 247, 249, 262, 281, **294**, 317, 328, 338, 349, 387, 389, 398
3-piperidinol 115, 117, 249, **296**, 329
4-piperidinol 62, 115, 117, **295**, 329
4-piperidone 62, 249, **295**, 329, 338
proline 10, 13, 27–8, 64, 67, 90, **293**, 318–19, 385
  measurement in urine 10, 36
pyrrolidine 5, 15–16, 23, 25, 27, 32–3, 55, 57, 62, 113, 115–17, 155, 185, 195, 220, 228, 245–6, 262, 270, 281, **293**, 319–20, 338, 349, 382, 387, 389, 408
pyrrolidone 86
3-pyrroline **294**, 338
sarcosine 147, 186, 318, 322
simazine 90
tetrahydrooxazine **297**, 332, 340
tetrahydrooxazone 72
1,2,3,4-tetrahydropyridine **296**, 328, 339
1,2,3,6-tetrahydropyridine **296**, 328, 339, 387, 390
thialdine 16, 62, 100, **299**, 333, 341

thiazolidine 16, **297**, 332, 340
thiomorpholine 16, 100, 195–6, **298**, 332
thioproline 36
tris-(2-chloroethyl)urea 200, **315**, 350, 352
Nitrous acid 5, 40, 42, 44, 249
Nitrous acidium ion 40
NMR spectrometry 54
NNAl 32–3, **289**, 347, 374
NNK 6, 27, 31–3, 42, 113, **289**, 337, 347, 383, 397
No-effect level of nitrosamines 387–8
Non-target organs, alkylation in 153
Non-volatile nitrosamines 24, 26
Nucleophiles 93, 100, 154, 197
Nucleophilic anions as catalysts 12, 20, 27, 42, 75
Nucleoprotein 205

Oedema 185
Omega oxidation 358–9
Oncogenes 166, 408, 410, 412
Oral cavity cancer 33
Organogenesis in fetus 352
Organ-specific carcinogenesis and chemical structure 412
Osteosarcomas 267
Ouabain-resistant cells 244
Oxadiazolium ion 80, 131, 188
Oxime 81
2-Oxopropyldiazonium ion 143, 374
2-Oxopropylnitrosamines 17, 92
2-Oxopropylnitroso-2,3-dihydroxypropylamine **291**, 324, 337, 356, 373, 384
2-Oxopropylnitrosourea 63, 143, **306**, 353, 364, 374
Oxytetracycline 9, 36, 44, 47

Pancreas 8, 17, 372–4
 ducts, tumours 132, 149, 153, 160, 197, 258, 273–4, 336, 372, 396
Pancreatitis 197
Partition coefficient 226
Peptides, nitrosation of 10
Pesticides 405
Pharmacokinetics 53, 374, 380, 407
Phenmetrazine 9, 10
 nitrosation *in vivo* 38
Phenobarbital as enzyme inducer 108, 122, 124
Phenols, as inhibitors of nitrosation 12, 36
Phenyldiazonium ion 79, 127–8, 155, 217, 242, 270, 320
Phenylethylnitrosourea 163, 266, 275, **305**, 342, 357, 378
Phenylhydrazone 81

Phenylnitrosobenzylamine *see* nitrosophenylbenzylamine
4-Phenylnitrosopiperidine 115, 240, 245, 262, **295**, 318, 328, 339
Phenylnitrosourea 74, 78, 217, 350, 378
Phosphate esters 103, 109, 187
 in DNA 93
Photochemical cleavage of *N*-nitroso compounds 50, 76
Pickling spice, nitrosamines in 20
Pimelic acid 118
Pipecolic acid 64
Piperazine 35, 40, 50
Piperidine 20, 26, 39, 40
Piperine 20, 47–8
Plasma membrane 58, 413
Plate test 205
Poisoning 1
Polarography 50, 98
Polynuclear hydrocarbons 2, 7, 31, 147, 205, 375, 405–6, 408
Porphyrins, hydroxyethylated 130, 149
Pregnenolone-16α-carbonitrile 124
Preincubation 205
Preneoplastic foci in liver 128
Proline 39, 64
 nitrosation of *in vivo* 36, 48
Prophage induction 239–41
β-Propiolactone 248
iso-Propyldiazonium ion 149
Propylnitrosamine, beta-oxidised 149, 336, 372–3
Propylnitroso-2-hydroxypropylamine 346, 398
Propylnitroso-2-oxopropylamine 347, 397
n-Propylnitrosourea 260, 369–70, 378
iso-Propylnitrosourea 74, 78, 378
Prostate gland 383
Protein starvation 189
Pseudoconhydrine, synthesis of 85
Pulsed doses (of nitrosamines) 124, 127
Pyelonephritis 198
Pyramidon *see* aminopyrine
Pyrazole 108, 125
Pyrrolidine 20, 26, 39
Pyruvic aldehyde 144, 187

Quinacrine 46

Rabbits 344, 347, 352, 355, 358
Rat liver microsomal fraction 205, 238
Rauch-Bier 21
Reactive metabolites 393
Receptors 16, 147, 162, 227, 260, 268, 328–9, 331, 348, 350, 362, 364, 368, 371–2, 380, 394, 401, 408–9
Repair of alkylated DNA 94, 130, 152, 154, 160, 163, 169, 171–2
Retroaldol cleavage 91

Retro-Claisen reaction 91
Rotamers of asymmetric nitrosamines 52, 63–4, 89, 127
Rubber 6, 28, 90, 254
  workers exposed to nitrosamines 387, 405

Saliva, mitrite in 5, 10–11, 23, 34, 406
Salmonella typhimurium assay 7, 45, 103, 204, 217, 224, 250
Saltpetre 22
Sarcomas 408
Sausages 5, 25
Sex hormones, kidney tumours in hamsters 364
Shampoo, nitrosamines in 8, 30
Sheep, toxic liver injury 3, 176
Sister chromatid exchange (SCE) 249
Skin 96, 169, 374–80
  tumours 96, 199, 265–6
  painting of mice, stomach tumours 350, 355, 375, 380
Skin absorption of nitrosamines 29, 30, 53, 175, 255
Smoked fish and nasal tumours 380
Smokeless tobacco 43, 76, 388, 405
Snuff 6, 27, 31–3, 404
Sodium nitrite, liver tumours 48
Solvent, NDMA as 56
Somatic mutation 251
Specific locus assay 203
Spleen 217, 384
  hemangiosarcomas in hamsters 17, 58–9, 343, 353, 374, 384–5, 412
Spontaneous tumours of rats 392
Squid, broiled, nitrosamines in 25
Steric hindrance 231
Stomach, nitrosamine formation 3, 9, 10, 23, 28, 35–7, 40, 44, 46–8, 53, 86
Stomach tumours *see* forestomach or glandular stomach
Streptomyces 56
Streptozotocin 5, 56, 342
Structure-activity relations 16
Subchronic toxicity 190
Subcutaneous injection, defects of 148
Sulphate esters 81, 87, 92, 103, 109, 131, 136, 139, 187
Sulphur-containing cyclic nitrosamines 332–3, 340–1
Sydnones, synthesis 85

Tannins 12, 36
Tetraalkylurea 75
Tetracyclines 44
2,3,5,6-Tetramethyldinitrosopiperazine 62, **301**, 316
2,2,6,6-Tetramethylnitrosopiperidine 62, 90, 281, **294**

Thermal energy analyzer (TEA) 15, 26, 52–3, 64, 68
Thin layer chromatography 51
Thiocyanate, accelerator of nitrosation 12, 27, 42
Thioguanine-resistant cells 244
Thiols 167, 266
Thioproline 36
Thiram (tetramethylthiuram disulphide) 9, 46–7
Thymus lymphoma 267, 331, 368, 370
Thyroid 367
  follicular cell tumours 132, 267, 336, 353, 367
Time-to-death-with-tumour 321, 387–8
Tobacco, amines in 5, 31, 42
  chewing 6, 31, 43, 404
  curing 5
  fermentation 31
  nitrosamines in 33, 254, 347
Tobacco smoke, nitrosamines in 5, 6, 31–3, 406
Tocopherols 12, 36
Toiletries, nitrosamines in 30
Tolazamide 45
Tolbutamide 40
Tongue, tumours 384
Trachea, tumours 338–9, 383–4
Transformed cell 16
Transforming activity *in vitro* 204, 246
Transnitrosation 12, 27, 35
Transplacental carcinogenesis 393–8
Transport forms of nitrosamines 394
Trialkylnitrosoureas 58, 68, 75–6, 95, 275, 357, 380, 385
  dealkylation of 74
Triazene 79, 127
Tribenzylamine 8, 41
Tridecylnitrosourea 378
Triethanolamine 7, 28–9, 42
Triethylamine 8
Triethylnitrosourea 68, 204, **313**, 371
Trifluoroethyldiazonium ion 237
Trimethylamine 8, 38, 176
  N-oxide 8, 21, 47, 176
3,4,5-Trimethylnitrosopiperazine 129, **299**, 349, 356, 369, 382, 392
Trimethylnitrosourea 58, 68, 186, **313**, 352
Tri-*iso*propanolamine 8
Tritium–carbon bond 279
Tubular cell neoplasms of kidney 364

Ultraviolet absorption 52, 60
Umpolung 83
Undecylnitrosourea 378
Urethane 248
Urethra 367
Urinary bladder *see* bladder

Uterus tumours 58, 162, 174, 266, 370-2
  endometrial stromal polyps 370

Vegetable oils 24
Vegetables, nitrate in 10
Vinyl chloride 405
Viruses 384
Vulcanisation of rubber 6, 26-7, 320

Whisky, NDMA in 5, 21

Woodworking, nasal cancer and 380

Xenobiotics, detoxication of 146
Xeroderma pigmentosum 169

Yeast, mutagenesis in 250

Zymbal gland tumours 162, 266, 385

For EU product safety concerns, contact us at Calle de José Abascal, 56–1°,
28003 Madrid, Spain or eugpsr@cambridge.org.

www.ingramcontent.com/pod-product-compliance
Lightning Source LLC
LaVergne TN
LVHW091527060526
838200LV00036B/509